Proceedings of the 2nd International Conference on the Health Effects of ω3 Polyunsaturated Fatty Acids in Seafoods, Washington, D.C., March 20–23, 1990

Health Effects of ω3 Polyunsaturated Fatty Acids in Seafoods

Volume Editors

Artemis P. Simopoulos

The Center for Genetics, Nutrition and Health, American Association for World Health, Washington, D.C., USA

Robert R. Kifer

National Marine Fisheries Service, National Oceanic and Atmospheric Administration, U.S. Department of Commerce, Charleston, S.C., USA

Roy E. Martin

National Fisheries Institute, Arlington, Va., USA

Stuart M. Barlow

International Association of Fish Meal Manufacturers, Hertfordshire, UK

104 figures and 79 tables, 1991

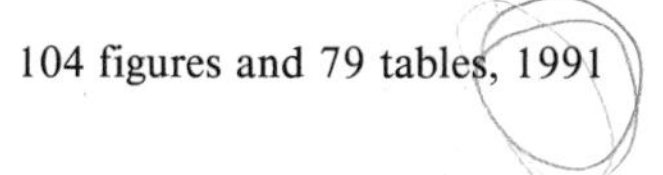

KARGER

Basel · München · Paris · London · New York · New Delhi · Bangkok · Singapore · Tokyo · Sydney

World Review of Nutrition and Dietetics

Last manuscripts released for publication in October 1990.

Library of Congress Cataloging-in-Publication Data
International Conference on the Health Effects of [omega] 3 Polyunsaturated Fatty Acids in Seafoods (2nd:1990: Washington, D.C.)
Health effects of [omega] 3 polyunsaturated fatty acids in seafoods: proceedings of the 2nd International Conference of the Health Effects on [omega] 3 Polyunsaturated Fatty Acids in Seafoods, Washington, D.C., March 20–23, 1990 / volume editors, Artemis P. Simopoulos ... [et al.].
(World review of nutrition and dietetics; vol. 66)
Includes bibliographical references and indexes.
1. Omega-3 fatty acids – Physiological effect – Congresses. 2. Omega-3 fatty acids – Therapeutic use – Congresses. 3. Fish oils – Health aspects – Congresses.
I. Simopoulos, Artemis P., 1933–. II. Title. III. Series.
[DNLM: 1. Coronary Diseases – prevention and control – congresses. 2. Fatty Acids, Omega-3 – metabolism – congresses. 3. Fish Oils – congresses. 4. Fishes – congresses]
ISBN 3–8055–5244–0

Bibliographic Indices
This publication is listed in bibliographic services, including Current Contents® and Index Medicus.

Drug Dosage
The authors and the publisher have exerted every effort to ensure that drug selection and dosage set forth in this text are in accord with current recommendations and practice at the time of publication. However, in view of ongoing research, changes in government regulations, and the constant flow of information relating to drug therapy and drug reactions, the reader is urged to check the package insert for each drug for any change in indications and dosage and for added warnings and precautions. This is particularly important when the recommended agent is a new and/or infrequently employed drug.

Printed in Switzerland by Thür AG Offsetdruck, Pratteln
ISBN 3–8055–5244–0

Contents

Biochemistry and Physiology

Advances in Mechanisms of ω3 Fatty Acids

Effects on Diseases

Cardiovascular I: Cell-Vessel Wall Interactions

Cardiovascular II: Heart

Cardiovascular III: Circulation – Blood Pressure

Rheumatoid Arthritis and Inflammatory Mediators

Diabetes

Psoriasis

Cancer

Appendix

Poster Session I (Abstracts)

Poster Session II (Abstracts)

Conference Organization

Co-chairmen

Artemis P. Simopoulos, MD
Stuart M. Barlow, PhD

Organizing Committee[1]

Stuart M. Barlow, PhD
International Association of Fish Meal Manufacturers

Robert R. Kifer, PhD
National Marine Fisheries Service – National Oceanic and Atmospheric Administration – U.S. Department of Commerce

Roy E. Martin, PhD
National Fisheries Institute

Artemis P. Simopoulos, MD
The Center for Genetics, Nutrition and Health
American Association for World Health

Under the Patronage of

Food and Agriculture Organization of the United Nations
World Health Organization

[1] In consultation with the members of the editorial board of the Omega 3 News.

Conference Sponsors

Sponsored by

National Marine Fisheries Service – National Oceanic and Atmospheric Administration – Department of Commerce, USA
The Center for Genetics, Nutrition and Health – American Association for World Health, USA
International Association of Fish Meal Manufacturers
National Fisheries Institute, USA
Food and Drug Administration, USA
National Cancer Institute, USA
National Institute on Alcohol Abuse and Alcoholism – ADAMHA, USA
National Institute of Allergy and Infectious Diseases, NIH, USA
National Institute of Arthritis and Musculoskeletal and Skin Diseases, NIH, USA
National Institute of Child Health and Human Development, NIH, USA
National Institute of Diabetes and Digestive and Kidney Diseases, NIH, USA
National Eye Institute, NIH, USA
U.S. Department of Agriculture, USA
DNA Plant Technology Corporation, USA
U.K. Sea Fish Industry Authority
Association of Fish Meal and Fish Oil Manufacturers in Denmark
Icelandic Association of Fish Meal Manufacturers
A/S Denofa OG Lilleborg Fabriker, Norway
Norsildmel, Norway

Additional Funds for the Publication were Contributed by

Canola Council of Canada
Council for Responsible Nutrition, USA
ENRECO, USA
Kraft General Foods
Laboratoires Lincoln, France

Preface

The Proceedings of the II International Conference on the Health Effects of ω3 Polyunsaturated Fatty Acids in Seafoods contain some exciting and provocative research findings. The conference was under the patronage of the Food and Agricultural Organization of the United Nations and the World Health Organization, and was cosponsored by government agencies, professional societies, international organizations and industry.

The conference goals were to review the latest research data on ω3 fatty acid metabolism and their relationship to ω6 and ω9 fatty acids in relation to: (1) essentiality of ω3 fatty acids in growth and development, membrane structure and function; and (2) cardiovascular disease, hypertension, diabetes, cancer, arthritis, psoriasis, and other inflammatory and autoimmune diseases.

We feel that the conference goals were fulfilled.

In the short intervening 5-year period from the first conference on the Health Effects of Polyunsaturated Fatty Acids in Seafoods, held from June 24 to 26, 1985 in Washington, D.C., research undertakings and publications increased almost exponentially. This strong indication of worldwide funding and support by medical research institutions has allowed remarkable progress in the exploration of the role of ω3 fatty acids in: (1) being essential for growth and in developmental processes, (2) membrane structure and function, and (3) control and/or prevention of disease states – cardiovascular disease, hypertension, diabetes, cancer, arthritis, psoriasis, and inflammatory and autoimmune disorders.

The role of ω3 fatty acids as being essential in growth and development has been expanded over the past 5 years. The scientific evidence is consistent with the fact that a recommended dietary allowance (RDA) for ω3 fatty acids for infants be made. A recommendation came from the conference that infant formula should contain 0.7–1.3% of its energy as docosahexaenoic acid (DHA). Speakers at the conference also called on the WHO and FAO to organize a further meeting on recommended daily allowances of fatty acids in order to make some recommendations on ω3 fatty acids in the light of the considerable knowledge now available.

The role of eicosapentaenoic acid (EPA) and docosahexaenoic acid (DHA) in atherosclerosis and hypertension continues to dominate the research support. Several of the factors involved in the development of cardiovascular disease such as prostaglandins, thromboxanes, leukotrienes, plasminogen activating factor, endothelium-derived relaxing factor, platelet-derived growth factor, interleukins and fibrinogen, are modulated by ω3 fatty acids. In practically all of them ω3 fatty acids have been shown to have a beneficial effect, which should not be unexpected since ω3 fatty acids replace ω6 fatty acids in practically all cell membranes and have also been shown to influence gene expression.

Nowhere else has the progress been so dramatic as in the understanding of the mechanisms of inflammation and autoimmune disorders, particularly in interleukin metabolism. It is therefore not surprising that there are a number of animal and human studies in which ω3 fatty acids have been shown to affect the course of rheumatoid arthritis, psoriasis, lupus erythematosus, ulcerative colitis, and others.

With respect to cancer, the original findings that ω3 fatty acids decrease the size and the number of tumors in animal models has been repeatedly confirmed, and human studies are in progress with patients to determine the effects of ω3 fatty acids in preventing the spread of metastases in patients with breast cancer.

A very exciting and new area of research is uncovering the synergistic effects of olive oil (18:1 ω9) and linseed oil (18:3 ω3) in increasing the incorporation of EPA into the cell, whereas ω6 fatty acids compete with ω3 fatty acids. This and other findings illustrate the importance of the need to study the various fatty acids and the appropriate proportions to each other in terms of the chain length, saturation, degree of unsaturation and location of the double bonds.

Another most interesting finding is in a new area of research in which ω3 fatty acids are used as adjuvants to drug therapy. In cardiovascular, inflammatory and autoimmune disorders supplementation with ω3 fatty acids has been shown to improve or enhance the effects of drug treatment.

In some studies the effect is synergistic. This effect is further strengthened by reduction in energy intake or exercise or both. In other studies, the addition of ω3 fatty acids appears to reduce the drug dose and thus decrease the toxic effects of drugs.

All these aspects are presented most eloquently in these conference proceedings of the II International Conference of the Health Effects of ω3 Polyunsaturated Fatty Acids in Seafoods.

These past 5 years of expanded research have not only enhanced our knowledge about ω3 fatty acids, but also about ω6, ω9, and saturated fatty acids. To deal with this expanded research area, it was decided to form an International Society for the Study of Fatty Acids and Lipids. The Society was established in March 1990 at the conclusion of the conference.

The conference consisted of the following panels:

Panel A: Growth and Development in Infants
Panel B: Requirements of Adults and the Elderly
Panel C: Advances in Mechanisms of ω3 Fatty Acids
Panel D: Cardiovascular I: Cell/Vessel Wall Interactions
Panel E: Cardiovascular II: Heart
Panel F: Cardiovascular III: Circulation – Blood Pressure
Panel G: Rheumatoid Arthritis and Inflammatory Mediators
Panel H: Diabetes
Panel I: Psoriasis
Panel J: Cancer

The conference brought together the pioneers in the field of ω3 fatty acid research. Drs. Hans Olaf Bang, Ralph Holman, Hugh Sinclair and Maurice Stansby made their presentations on the evening of March 22nd at the conference banquet. Their presentations are in response to the question posed to them, 'what is the most memorable event in your life relative to the ω3 fatty acids?' The proceedings begin with their presentations.

Without any doubt these four giants kept the interest in ω3 fatty acids research from being swept away by the storm that overtook western societies with the ω6 fatty acids, and the cholesterol phobia. The events as recounted by Drs. Bang, Holman and Stansby make it clear that:

> 'The search for truth is in one way hard and in another easy. For it is evident that no one can master it fully nor miss it wholly. But each adds a little to our knowledge of Nature, and from all the facts assembled there arises a certain grandeur.'

This, of course, is a quotation from Aristotle. It is this 'quest for/search for truth' that distinguishes science from other endeavors. Dr. Sinclair, an outstanding researcher and poet, elected to remit his experiences in a poem composed especially for the occasion, thus immortalizing the meeting and the establishment of the International Society for the Study of Fatty Acids and Lipids. The Proceedings of this conference are dedicated to his memory.

These presentations are followed by the Conference Statement, which in turn is followed by summaries by Jørn Dyerberg, chairman of the session summarizing the conference, and the panel chairmen: Norman Salem (Panel A: Growth and Development in Infants); Kristian Bjerve (Panel B: Requirements of Adults and Elderly); Alexander Leaf (Panel D: Cardiovascular I – Cell-Vessel Wall Interactions); Arne Nordoy (Panel E: Cardiovascular II – Heart); Howard Knapp (Panel F: Cardiovascular III – Circulation/Blood Pressure); Dwight Robinson (Panel G: Rheumatoid Arthritis and Inflammatory Mediators); Ritva Butrum (Panel J: Cancer).

Next is the keynote address on 'ω3 Fatty Acids: Research Advances and Support in the Field since June 1985 (Worldwide)' by Artemis P. Simopoulos, Robert Kifer and Arthur Wykes, followed by the panel presentations. The presentations by the panelists consist of a review paper in the field by one of the cochairmen, followed by papers on mechanisms and new investigations by the other panelists. Each panel cochairman was responsible for initiating discussions and summarizing the panel presentations. It was thought mandatory to provide ample time for discussions in order to discuss future research needs and the appropriate development of protocols, particularly in terms of mechanisms and dosage of ω3 fatty acids in both animal studies and clinical investigations. There were 86 papers given at the two poster sessions. The abstracts of these papers appear in the appendix.

These proceedings represent the most complete and up-to-date information on the role of ω3 fatty acids in health and disease. The review papers, research papers, and the summaries by the panel co-chairmen are outstanding and represent clear statements of accomplishments and future research needs. The proceedings should be of interest to all investigators interested in research on fatty acids and lipids, to physicians who take care of patients, to food scientists, dieticians, nutritionists, food producers and policy makers.

Artemis P. Simopoulos, MD
Robert R. Kifer, PhD
Roy E. Martin, PhD
Stuart M. Barlow, PhD

Dedication

Hugh MacDonald Sinclair 1910–1990
Advocate of the Essential Fatty Acids

Hugh MacDonald Sinclair, who has been identified with nutrition and the essential fatty acids throughout his scientific career, died at Oxford June 22, 1990 at the age of 80 years. He was born in Edinburgh February 4, 1910 into a family rich with history, which claimed as ancestors King Woldonius of Finland, the Vikings Nor and Gor, Thorfinn the Skull-Splitter, William Sinclair the Bold, Sir John Sinclair the epidemiologist, and one who discovered North America long before the time of Columbus. Hugh inherited regal bearing and the penchant for discovery.

While a medical student at Oriel College, Oxford, he read the text on vital statistics and learned that life expectancy of a 50 year old man had not changed since 1841. He concluded that medicine could do little more to keep a middle-aged man alive than it could in the middle ages, and that the probable cause was the composition of the dietary fat. This set the direction of his life. He went on to win three prizes while a medical student, including first place in physiology in 1932 when he became demonstrator in biochemistry. In 1937 he was chosen to teach physiology and biochemistry at Magdalen College. He worked with R.A. Peters on thiamine and took his DM in 1939. He remained at Magdalen throughout his life, serving as Fellow, Bursar and Vice President and in later years he became an Emeritus Fellow. Sinclair was a student of Sir Robert McCarrison, an early pioneer in the scientific approach to the study of nutrition and health, and he championed this approach to nutrition all his life, at a time when there were few who would listen. He was content to take quiet pleasure in knowing he was right, and he did not seek the limelight of the establishment.

Sinclair's early work on vitamin B was interrupted by work on chemical warfare, and in 1941 he organized the Oxford Nutrition Survey, the results of which prevented in Britain what had happened in Germany in World War I, and which assured that British wartime food policy was effective. During the war he visited The Royal Canadian Air Force and observed the absence of heart disease in Eskimos, leading him to conclusions about the need for dietary fats which were to give him notoriety and later, fame. With Sir Jack Drummond he led a survey team to Holland, and was in the first allied group to move into the famine-stricken areas of north and west Netherlands. Joining forces with a large group of Dutch scientists, they collected much nutritional information which was published in 1948. Sinclair also took a team to Düsseldorf to survey nutritional conditions in the British Zone. These experiences confirmed his interest in nutrition and gave informational background for his passion for the essential fatty acids. Sinclair was rewarded for his post-war nutritional surveys in Europe by an appointment as Chevalier of the Order of Orange-Nassau and his Medal of Freedom with Silver Palm from the United States.

Sinclair was a prolific reader, and he lived surrounded by books. His depth and breadth of knowledge became apparent at symposia and conferences around the world, when he was often called upon for information and opinion. He had realized the importance of the essential fatty acids discovered by Burr and Burr in 1929, and by the 40s he developed the concept that deficiency of essential fatty acids was the chief cause of heart disease in Western populations, and that they were necessary for normal health and circulatory function. In 1950 Groen, in the Netherlands, brought evidence to the First International Congress on Nutrition that saturated fats raised plasma cholesterol and that vegetable oils lowered it, concluding that the kind of fat or the fatty acid composition of dietary fat was related to heart disease. At that time he was attacked by an American leader in the field who believed that only the quantity of dietary fat was of importance. Sinclair defended Groen, and in 1956 wrote his classic letter to Lancet in which he showed that the essential fatty acids were needed to protect the arteries. In 1985 this paper was chosen a citation classic by *Current Contents.* He also gathered a variety of evidences which indicated that many diseases of man may be explained by low intakes of the essential polyunsaturated fatty acids. His very early recognition of the importance of these nutrients has been amply vindicated by a multitude of laboratory studies in recent decades. Much of the current enthusiasm for essential fatty acids stems from the insights of this man who was ahead of his time.

Sinclair was fearless in the pursuit of ideas, and he was his own guinea pig. In 1976 he conducted an experiment upon himself, in which he consumed most of a deep-frozen seal plus fish as his only food for 100 days. The seal, rich in ω3 fat, bestowed a distinctive odor on the kitchens at Magdalen College, and induced the longest bleeding times recorded for Western man, except in hemophilia. These hazardous studies upon himself led to his fame as 'the Eskimo Diet Doctor'.

Because of the controversial nature of his favorite subject, he found it difficult to find the necessary support for an adequate research laboratory in the academic world. He persuaded the university to form the Laboratory of Human Nutrition in 1947, and he held the readership in human nutrition from 1951 to 1958 when the university put the laboratory under other leadership. Sinclair stayed on at Magdalen in 'the wilderness', without laboratory, assistant, or adequate salary. Although he had been offered four chairs, he remained loyal to Oxford, and later built an independent International Institute of Human Nutrition at Sutton Courtenay 10 miles outside of the city. He never drew a salary and he contributed his large personal library and dedicated his property to this Institute. He served very successfully as Visiting Professor at the University of Reading for many years.

On one occasion Sinclair drove me at breakneck speed through the narrow country lanes of Oxon in his Aston Martin to his estate at Sutton Courtenay for a visit and a revelation of him as a country gentleman. He had made many of the concrete sculptures that graced the pools and paths of the garden, he kept peacocks and exotic birds and sponsored Tibetan students, and he had a life-long collection of rare and old books, including one about the ancestor who had discovered America.

In 1956 Sinclair organized the first international meeting on essential fatty acids, held at Oxford, which attracted all the leaders in this area of nutritional biochemistry. Conferences occurred with increasing frequency in the following decades, and Sinclair was often invited. At one of the early Deuel Conferences on Lipids he composed his first Hiawatha poem about lipids and lipid reasearchers, and they became popular satires on the scientific scene. They even inspired this writer to attempt them as a form of amusement. At the First International Congress on Essential Fatty Acids and Prostaglandins, held in Minneapolis in 1980, Hugh treated us to a parody on 'To be, or not to be'. At the Second International Conference he was the honored guest and we celebrated his 75th year. At the banquet held at Hatfield House he favored us with a humorous lecture on EFA and a somewhat modified text from King Henry V. Sinclair and that congress inspired the following lines:

Hatfield House! 'Lisbeth's House! What happened there?
We heard a long lecture by Dr. Sinclair.

What was his subject? What did he say?
History, inflation of words, EFA.

Miracles and wonders! What were they about?
Fish oils, the diet, the ague and gout.

What hope didst thou gain for our life and our pain?
Eat oils, make glandins, again and a gain.

At the II International Conference on the Health Effects of ω3 Polyunsaturated Fatty Acids in Seafoods in 1990, the proceedings of which are summarized in this book, Sinclair was also present, and although gravely ill, he entertained us with his poetry. Thus did this serious, imposing and tall Scotsman occasionally reveal his lighter side. Only his classic education and wide experience, together with his wide knowledge of the nutritional lipids and the people who worked with them, made it possible.

This past spring Sinclair reached his 80th birthday, and the McCarrison Society, of which he was president in 1983, produced a memoir in his honor. The Biochemical Society held a conference on fatty acids in his honor at Bath in April, at which he gave the introductory talk. Shortly afterward he became so ill that he had to be driven back to Oxford that morning for a corrective operation. His closest friend, Dr. Brian B. Lloyd, wrote of him to Dr. Simopoulos on June 11 that 'he has been an absolute brick in the last weeks, absurdly undemanding, never complaining, very grateful for help, full of beans and 100% on the ball intellectually. He is still trying to meet his last obligations, and is sorry to be sending his manuscript at the last minute'. The 'manuscript' consisting of prose, rhyme and poem appears in these proceedings.

Eleven days later Hugh Sinclair expired, ending a good and productive life. The common man on the street has already benefitted nutritionally from his insights, and Sinclair has grown from 'a voice crying in the wilderness' to a well-known Prophet. Sinclair must have felt gratified many times to know that what he predicted 40 years ago had become reality in his lifetime, that his scientific peers held him in respect and love, that his service has been a benefit to mankind, and that his contributions will be forever.

Ralph T. Holman

Special Presentations[1]

Simopoulos AP, Kifer RR, Martin RE, Barlow SM (eds): Health Effects of ω3 Polyunsaturated Fatty Acids in Seafoods. World Rev Nutr Diet. Basel, Karger, 1991, vol 66, pp 1–3

Speech at the Dinner

Hans Olaf Bang

Lyngby, Denmark

Dr. Simopoulos!
Dear Colleagues!
Ladies and Gentlemen!

Thank you, Dr. Simopoulos, for your kind words of introduction. As you heard, my name is BANG. This name may get you to think of the presently popular and much mentioned THE BIG BANG. But I must disappoint you: This is definitely not me! If I should propose a more proper nick-name, it might be ESKIMO-BANG, because I owe my research development and presence here to the Greenland Eskimos. And may I be immodest enough to say that at least several of you are present here primarily also thanks to the Greenland Eskimos and secondly – of course – to the Organizing Committee of this conference.

I was asked to focus in my speech on the most happy moments of my life. Well, let me try: I was married more than fifty years ago, and I am still married to the same sweet girl. I have two nice daughters and three lovely grandchildren, and this has given me many happy experiences in my life.

[1] Editor's note: These four special presentations were given at the II International Conference on the Health Effects of ω3 Polyunsaturated Fatty Acids in Seafoods banquet on March 22, 1990, in response to the question, 'What is the most memorable event in relation to ω3 fatty acids in your life (career)?'.

But this may not be interesting to you, so I shall turn over to the happiest moment in my life as a researcher. And it will of course be connected to the omega-3 domain.

But I may at first give you a short survey of the omega-3 story as I see it: In the late sixties Dr. Jørn Dyerberg and I decided to try and find the cause of the known rarity of thrombotic diseases, especially ischaemic heart disease in the Greenland Eskimos. We had four wonderful travels to the northern part of this beautiful huge island, on one of them accompanied by Professor Sinclair. We found a high level of EPA (and DHA) in the blood and food of the Eskimos as compared with Danish controls, and this was an original observation.

Just at that time, in the middle of the seventies, the Nobel laureate Dr. John Vane together with Dr. S. Moncada published their observation of prostacyclin. This prostanoid together with thromboxane has arachidonic acid as precursor. It was natural for us to ask whether the Eskimos could 'use' the fatty acid EPA, like arachidonic acid with twenty carbon atoms, but with five double bonds, as a precursor of platelet-active prostanoids, and if this be the case, if the balance between the pro-aggregatory thromboxane and the anti-aggregatory prostacyclin coming from EPA was shifted into an anti-aggregatory direction.

And now I come to the happiest moment of my life as a researcher: Jørn Dyerberg showed in his coagulation laboratory that EPA induces decreased platelet aggregability. I remember our enthusiasm that day as it was yesterday! We felt that we had not only found the explanation of the rarity of thrombotic diseases in the Eskimos, but we speculated whether our observation might lead to prevention of these diseases also in other parts of the world. This seems to be so as we have heard during these fine days of the conference. Nobody had hitherto realized that the two sorts of polyunsaturated fatty acids, belonging to the 3 and 6 family, have very different biological and nutritional abilities.

But we had not phantasy enough to imagine the tremendous interest in our observation which has resulted in so much scientific – and positive! – work in most parts of the world since then, and not limited to thrombotic diseases, but – as we have heard – also involved in rheumatic arthritis, psoriasis, inflammatory diseases and other ailments of mankind. And more – and hopefully positive! – scientific work will follow.

I want to express my gratitude, and not only mine, but that of all the participants of this conference, to the Organizing Committee that has given all of us good and new knowledge about the fatty acids and not least, has given us the opportunity to meet fellow researchers, and may I say friends.

May I express my thanks primarily to Dr. Simopoulos and Stuart Barlow for their big work in arranging the conference, and to all the sponsors of it. And may I congratulate with the formation today of the new society for the study of fatty acids.

I feel that I can add THIS MOMENT to the list of the happy events of my life.

Thank you!

Simopoulos AP, Kifer RR, Martin RE, Barlow SM (eds): Health Effects of ω3 Polyunsaturated Fatty Acids in Seafoods. World Rev Nutr Diet. Basel, Karger, 1991, vol 66, pp 4–8

Highpoints in an Affair with Polyunsaturated Fatty Acids

Ralph T. Holman

Hormel Institute, University of Minnesota, Austin, Minn., USA

Just as one should choose well his parents to achieve longevity, so it is necessary to choose wisely one's mentor. Fortunately the first choice was made well for me, and I did put much thought into the choice of my 'Doctor Father', George Burr. That choice placed me in the best lipid laboratory in the United States in 1941. At the end of the war, the surge in graduate students to Burr's laboratory was so great that he encouraged me to step in for him when he was busy with administrative matters. One of those students was given one of Burr's pet subjects, the effect of dietary fat upon fatty acid composition of lipids of tissue, and I became involved with the experiment and the writing of the report [1]. That study showed that when corn oil was fed and linoleic acid was the only dietary polyunsaturated fatty acid (PUFA), the lipids of many tissues contained high levels of a tetraene (arachidonic acid), indicating that arachidonic acid is synthesized from linoleic acid. When fat-deficient diet was fed, this tetraene was replaced by a trienoic acid. When cod liver oil (containing ω3 PUFA) was fed, the pentaene and hexaene acids were elevated in tissue lipids, and the content of these ω3 acids in the lipids of the heart approached the content of them in the cod liver oil fed.

In 1946 I took an NRC-NAS post-doctoral fellowship for study of soybean lipoxidase at the Medical Nobel Institute, where I succeeded in isolating, crystallizing and characterizing this enzyme which catalyzes the peroxidation of linoleic acid. When the preparation was complete, a study of the molecular yield of conjugated peroxide from non-conjugated linoleate was made. I had borrowed five Warburg apparatuses, set at 5 different temperatures from 0 to 37 °C, and Sune Bergstrom and I measured oxygen uptake and diene conjugation, running between the respirometers and the spectrophotometer. The result was the discovery that at 0 °C the molecular yield was the theoretical value [2]. Only that once in a lifetime could I produce one paper in one day!

At Texas A&M my first graduate student and I took up the issue of the conversion of linoleic to arachidonic acid using pure single dietary supplements. We were able to prove that arachidonic acid was synthesized from linoleic acid but not from linolenic acid, and that pentaenoic and hexaenoic acids arose from linolenic acid [3]. After several years of studies of essential fatty acid (EFA) deficiency, an experiment was done comparing 7 diets having different levels of linoleic acid. When the trienoic acid of heart lipid was plotted against dietary linoleic acid, the high value found in EFA deficiency dropped asymptotically to a very low level as dietary linoleic acid increased, and the low level of tetraenoic acid rose asymptotically. In an attempt to express these two phenomena in one index, the triene/tetraene ratio was introduced as a measure of EFA requirement, but the study also revealed that dose level governed the tissue response in a regular fashion [4], permitting the measurement of dietary requirement for EFA [5]. A long series of studies using single pure fatty acids and gas chromatographic analysis confirmed and extended the earlier studies using alkaline isomerization, and experiments with pairs of pure fatty acids led to the recognition of the metabolic competition which exists between the members of the linoleic and linolenic acid metabolic cascades. It also led to the invention of the omega system of abbreviated nomenclature necessary to communicate simply and clearly about the many isomers and homologs of PUFA which occur in the ω6, ω3 and ω9 families of PUFA [6]. Our laboratory also was privileged to measure relative rates of desaturation and chain elongation at each of the steps in the metabolism of linoleic acid to arachidonic acid and to identify the principal pathway [7].

Although most of our work centered on the ω6 family of PUFA, we were able to contribute to knowledge of the requirement for linolenic acid. In our studies of the quantified requirement for various essential fatty acids, we conducted a study to set the quantified requirement for linolenic acid in the female rat and estimated it to be 0.5% of calories. Some time later we were asked to measure the EFA status of a girl who had lost much of her intestine and who was maintained on intravenous alimentation with an emulsion not containing ω3 acids, and who developed some neurological symptoms. An assay of her plasma phospholipids revealed a relative deficiency of linolenic acid and ω3 products from it. When she was given a soybean oil emulsion with both linoleic and linolenic acids, her neuropathies disappeared. From the dose level she received, we estimated that the level of linolenic acid required to restore the ω3 acids to normal was 0.5% of calories! [8]. This discovery preceded and contributed to the current great interest in ω3 fatty acids.

Through all these studies it had been apparent that each new step of knowledge had waited for the prior development of a new method of measurement: alkaline isomerization preceded the triene/tetraene ratio; gas chromatography led to the dose-response curves, metabolic pathways awaited radio-labelling of fatty acids. Thus, our program had walked upon two legs – first a step by development of methodology, then one applying the method to biology. It was natural then to develop interest in mass spectrometry when it appeared, and we are proud that we could apply it to the determination of structure of lipids, leading to more than 25 publications and to inclusion of about 1,000 of our spectra of lipid derivatives in the EPA/NIH Mass Spectra Data Base.

In recent years our attention has turned toward the effects of disease in humans upon the profile of polyunsaturated and other acids of plasma lipids. This program has required the use of computers to handle approximately 80 individual fatty acids and combinations of them which have been used as indices of EFA status, indicators of metabolic reactions and measures of physical state. The modern computer has proven to be the most mind-expanding of the many techniques which have developed within the past 40 years. It has opened the way to a comprehensible graphic presentation of very complex fatty acid data in the form of the normalcy ratio fatty acid profile [9], which has proven useful in detecting and describing abnormalities in fatty acid metabolism in a variety of diseases.

We scientists are often admonished to take time along life's road to smell the roses. About 20 years ago I took this advice seriously, and undertook a long-term study of the composition of fragrances of orchids, which are at least as exotic as roses. If justification for such a whim be necessary, let me explain that fragrances are merely volatile lipids, the products of the metabolism of lipids by intracellular membranes, which themselves are rich in lipids containing ω3 and ω6 PUFA.

Professor Bang, our first speaker this evening, told us an Eskimo story, and let me conclude my talk with one. Some years ago, Professor

Dyerberg introduced me to the medical investigators who examined and described several Greenland Eskimo mummies who were placed in a cave and kept cold naturally for 500 years. They provided us with samples of gluteal tissue from four mummies, and we subjected them to thin layer chromatography and gas chromatographic analysis. The lipids were found to be largely decomposed, with little triglyceride and much mono- and diglyceride and free fatty acids. The fatty acids were largely saturated and monounsaturated, with little trace of polyunsaturated acids remaining (unpublished data), even though Eskimos are reputed to be very polyunsaturated. In recent years there has been much talk of preserving, now, cadavers at low temperatures, in the anticipation that the future expansion of scientific knowledge hopefully will discover a means to revive them. I must forewarn advocates and candidates for such revivalism that if candidates were to be revived 500 years in the future, it is highly likely that they would be resurrected to a profound essential fatty acid deficiency!

Breathe, O breathe thy loving spirit
Into every troubled heart!
Let us all in thee inherit
Let us find that second start.
Take away our bent to clotting
Alpha and Omega Three
End of faith, as its beginning
Set our hearts at liberty!

RTH, with apologies to Charles Wesley
May 2, 1990

References

1 Rieckehoff IG, Holman RT, Burr GO: Polyethenoid fatty acid metabolism. Effects of dietary fat upon polyethenoid fatty acids of rat tissues. Arch Biochem 1949;20: 331–340.
2 Bergstrom S, Holman RT: Total conjugation of linoleic acid in oxidation with lipoxidase. Nature 1948;161:55–56.
3 Widmer C, Holman RT: Polyethenoid fatty acid metabolism II. Deposition of polyunsaturated fatty acids in fat deficient rats upon single fatty acid supplementation. Arch Biochem 1950;25:1–12.
4 Holman RT: The ratio of trienoic:tetraenoic acids in tissue lipids as a measure of essential fatty acid requirement. J Nutr 1960;70:405–410.

5 Caster WO, Ahn P, Hill EG, Mohrhauer H, Holman RT: Determination of linoleate requirement of swine by a new method of estimating nutritional requirement. J Nutr 1962;78:147–154.
6 Holman RT, Mohrhauer H: A hypothesis involving competitive inhibitions in the metabolism of polyunsaturated fatty acids. Acta Chem Scand 1963;17:S84–S90.
7 Marcel YL, Christiansen K, Holman RT: The preferred metabolic pathway from linoleic acid to arachidonic acid in vitro. Biochim Biophys Acta 1968;164:25–34.
8 Holman RT, Johnson SB, Hatch TF: A case of linolenic acid deficiency involving neurological abnormalities. Am J Clin Nutr 1982;35:617–623.
9 Holman RT, Johnson SB, Kokmen E: Deficiencies of polyunsaturated fatty acids and replacement by non-essential fatty acids in plasma lipids in multiple sclerosis. Proc Natl Acad Sci USA 1989;86:4720–4724.

Simopoulos AP, Kifer RR, Martin RE, Barlow SM (eds): Health Effects of ω3 Polyunsaturated Fatty Acids in Seafoods. World Rev Nutr Diet. Basel, Karger, 1991, vol 66, pp 9–11

Highlights in the Study of ω3 Essential Fatty Acids

Hugh Sinclair

International Nutrition Foundation, Abingdon, Oxon, UK

I was interested in essential fatty acids (EFA) at about the same time as Ralph Holman. In 1937 I visited Dr. H.M. Evans, who with George Burr had just discovered them. Burr had moved to Minnesota, but Osmo Turpeinen was with Evans and showing the activity of arachidonic acid which Burr had denied. I wanted to work on EFA in view of the possibility that deficiency of them might account for the rise of Western diseases. I also met Hanson, McQuarrie and others working on them. It was a great privilege to become a close friend of Dr. Evans; he and Otto Warburg are the greatest biochemists of this century.

A second highlight was my visit to Eskimos in 1944 during which I noted the absence of any trace of arcus senilis and the prevalence of epistaxis.

A third highlight was in 1960 when I came in contact with Professor Notevarp of Trondheim since we were both interested in EPA and DHA. In 1966 the Scandinavian Council for Applied Research held a conference in Trondheim where I gave two lectures on EFA, one being on EPA and one on DHA. Professor Notevarp presented a paper, subsequently published, in which he showed that these two fatty acids were significantly lowered in the serum of atherosclerotic patients compared with healthy controls. He kindly gave me some of his pure EPA which I took back to Oxford to test for platelet adhesiveness on volunteers, using the Salzman glass bead technique; it was greatly decreased. I sent him the results, pointing out that they explained the absence of ischemic heart disease and arcus senilis in Eskimos and the prevalence of epistaxis.

A fourth highlight was in 1975 when Dr. Bang, who with Dr. Dyerberg had made two visits to Eskimos in northwest Greenland to study diet and lipids, kindly invited me to join them for their next one and asked for suggestions. On September 4, 1975 I sent some to Dr. Bang, adding 'I venture to suggest one observation that could be of crucial interest ... aggregation of platelets ... in view of the very high eicosapentaenoic acid you found in plasma I would expect that Eskimo platelets are relatively non-sticky'; this expectation was based on the earlier studies done with Notevarp. I took equipment for bleeding times but the weather broke and we had to leave. Although Dr. Bang generously invited me to return to measure them, I could not get the finances. Hence the last highlight: I did the Eskimo diet experiment on myself; though I should add that attending this excellent conference is a highlight too.

Dr. Simopoulos asked for a poem. Why not a nursery rhyme?

Artemis had a little fish,
Its fats were polyene,
And every time she ate a dish
It kept her vessels clean.

Dave Kritchevsky ate some bran,
He thought it good for you.
It cleans you out as nothing can
And stops you eating too.

Michael Crawford had a brain,
Rhinoceros it came from;
He thought it lacked the lipid chain
That intellect depends on.

Retinal rods ensure you can see;
Retinal rods have omega 3
Lipids; without you'd be blind as a bat
So ensure they are there in your daily fat.

Or following Humpty Dumpty:

Bang and Dyerberg had a dog-sledge,
Bang and Dyerberg at the world's edge.
Studying Eskimos in the deep snow
Found their cholesterol levels were low.

But I am the only person here who has entered second childhood, so nursery rhymes are inappropriate. We might consider how Longfellow would have handled this conference.

From a turret in her office,
Washington D.C. the city,
Artemis the organizer
Named Simopoulos, the huntress,
Made her bow and called the wise men;
Bid them join a mighty pow-wow
To discuss the fish oil acids.
So the fatty chemists gathered,
So the workers of all nations
Came to District of Columbia:
NIH grantees in Rollses
Central Europeans hitch-hiked.

Hiawatha journeyed thither,
Nutriture by Minnehaha,
With their little sons and daughters
And technician Wishi-Washi.
Opening the fish-oil conf'rence
Artemis reviewed their functions

Gracefully described how helpful
Fish oils were to stop diseases:
Modulating prostaglandins,
Curing clinical disorders,
Heart disease and hypertension,
Cancer, rheumatoid arthritis,
Skin disorders like psoriasis,
Multiple sclerosis also:
All are greatly helped by fish oils.

Then some papers showed quite clearly
Fish is good for brains and eyesight.
Hiawatha then concluded:
Politicians need more fishes,
Though perhaps a distant Russian
Gorbachev drinks fish oil daily;
Elderly Norwegian ladies
Had dry skin from lack of fish oil.
Then Icelandic stress and aging
Seized-up joints and diabetes
Scaly skin and other ailments
That afflicted western man but
Not the mighty fish-consumers,
Hiawatha's northern neighbors,
Who abhor the Indian naming:
Eskimo means raw-fish eater.
Inuit or 'man' is their name
Which is much to be prefer-red;
Inuit with seal and salmon
Do not suffer from diseases
That afflict the western nations.

Now experiment's important,
And some talked of cultured heart cells
'Mr. Fit', and aging studies:
Tried to stress their youthful subjects
Gazing at a pretty female.
Hiawatha liked these studies
Better than the biochemic
Details shown by Howard Sprecher,
Who with metabolic mast'ry
Summarized the paths of fish oil
And the piscine convolutions
Fatty acids can engage in;
As the caribou in summer
Bitten by the big mosquitoes
Hurry off in all directions
So the fish-oil fatty acids
Move through multifarious pathways.

Now to get these fatty acids
– *Omega* is bad for scansion,
O*mega*-3 is rather worse, but
These are dragged in willy-nilly,
Science conquers rules of verse. –
Just eat pilchards, trout and herring,
Salmon, sardines, catfish, dogfish;
Hiawatha favoured seal meat,
And a little walrus also;
Some took capsules for their fish oil
Took MaxEPA with their dinner.

Michael Crawford gave a lecture
Showing that the newborn baby
Needs some fish-oil fatty acids
For its brain's fast-growing structure.
Then Bill Lands in retrovisions
Took us back to schoolday lessons
Introduced Michaelis constants
Long repressed by Hiawatha
Though recalled along with Menten
By technician Wishi-Washi.

Having thus established clearly
Fish is good for healthy living
Then the conference decided
To promote this useful message
With a newly formed committee:
A society was needed
To promote the seafood message.

Back to snowland Hiawatha
Went on dogsledge with a bottle
Of MaxEPA in his stomach
Rattling as he journeyed northwards
Like the pebbles on the seashore;
And rejoices that he soon would
Catch a seal and eat the carcass
With his wife and little children
And technician Wishi-Washi
Living healthfully thereafter
From these useful fatty acids.

No surprise to Hiawatha
Were these findings at the congress.

Simopoulos AP, Kifer RR, Martin RE, Barlow SM (eds): Health Effects of ω3 Polyunsaturated Fatty Acids in Seafoods. World Rev Nutr Diet. Basel, Karger, 1991, vol 66, pp 12–14

Comments on Clinical Tests of Dr. Averly Nelson

Maurice Stansby

Northwest and Alaska Fisheries Research Institute, NMFS, Seattle, Wash., USA

Dr. Averly Nelson, a physician, conducted a 19-year clinical test on the effects of frequent consumption of fish in the diet upon heart attacks. This study, probably because it was published in relatively obscure journals, is not widely known. Because I was in close touch with Dr. Nelson, especially at the early stages of his work and because he died of cancer soon after the completion of the work in 1973, I have been trying to give his results more publicity.

Dr. Nelson began practice as a physician in the late 1940s in Seattle. Early in his career he was convinced that diet could have an important bearing upon diseases such as heart attacks. In order to prepare himself to better understand such effects, he temporarily abandoned his role as physician in 1946 and studied such subjects as nutrition and biochemistry at the University of California at Los Angeles (UCLA) where he received a degree of Master of Medical Sciences. While at UCLA he was greatly impressed by several of his professors' reports of findings that Norwegians, during the Second World War while their country was occupied by the Nazis, ate much more fish and less meat than previously and that during this period their incidence of heart attacks declined considerably. Nelson returned to Seattle and resumed his role as a practicing physician. He began to consider carrying out a long term study of the effects of diet upon heart disease.

Early in 1952 Nelson began a study comparing the number of fatal heart attacks occurring between his patients following a diet which included (1) a minimum of 3 meals per week in which fatty fish such as salmon, herring or rainbow trout and (2) his control group where little or no fish was consumed. This diet experiment continued to early 1971. During this period Dr. Nelson was able to keep his patients on their fish diet by means of several techniques not ordinarily employed by other physicians. He had frequent, regular meetings with his patients and members of their families stressing need for adhering to their diets. Once per year a special meeting was held with all patients in which specialists came in to discuss various aspects. For example, I attended such meetings and discussed how in our own research involving animals who consumed fish in their diet greatly held to a minimum the level of serum cholesterol. Dr. Nelson allowed dieted patients to go off their diet for 4 consecutive meals, once per week. Tests on serum cholesterol levels showed that very little difference in cholesterol levels resulted.

In 1966 my position changed to another line of work and I ceased having contact with Dr. Averly Nelson, and I knew nothing about how the results of his test were developing. On January 29, 1973, Dr. Nelson sent me a reprint of his final paper covering these studies which appeared in the December 1972 issue of 'Geriatrics' [1]. I was amazed to see the tremendous effects of the fish diet upon his patients. The control patients who did not consume fish in their diets had (on a percent basis) 4½ times as many fatal heart attacks over those using at least 3 meals per week of fatty fish! I feel that this was the day when I received the greatest thrill of my career with regard to effects of fish oil used for human consumption. Of course during the conduct of Dr. Nelson's clinical test it was believed that the benefits of fish oil related not to ω3 fatty acid effects, but only to the more general effects involving lowering of serum cholesterol levels. By 1973, however, when I first saw the results, there was sufficient knowledge on ω3 effects that it seemed most likely that the splendid findings were largely a result of the presence of ω3 fatty acids in the diet.

Very few people who might have been interested in Dr. Nelson's findings knew about his paper because 'Geriatrics' is not ordinarily read by those specializing in heart disease. With the death later in 1973 of Dr. Nelson, it soon became apparent to me that something needed to be done to get Nelson's results to those in the medical field who should be aware of them. Wherever possible I have cited his work in papers of my own. In 1981, I made a poster presentation at a scientific meeting, and I reviewed his work in the form of a short chapter in a book [2]. Nevertheless I still find many who are unaware of Nelson's results. I, therefore, welcome this opportunity to reach some of you, who probably had not previously been aware of this important work.

References

1 Nelson AM: Diet therapy in coronary disease – effect on mortality of high-protein, high-seafood, fat controlled diet. Geriatrics 1972;27:103–106.
2 Stansby M: A clinical study on the role of fish oil in alleviating human heart disease; in Barlow SM, Stansby ME (eds): Nutritional Evaluation of Long-Chain Fatty Acids in Fish Oil. London, Academic Press, 1982, pp 263–266.

Simopoulos AP, Kifer RR, Martin RE, Barlow SM (eds): Health Effects of ω3 Polyunsaturated Fatty Acids in Seafoods. World Rev Nutr Diet. Basel, Karger, 1991, vol 66, p 15

Conference Statement

The 2nd International Conference on the Health Effects of ω3 Polyunsaturated Fatty Acids in Seafoods included over 100 presentations by scientists from 25 countires during March 20–23. About 300 people discussed the new research results that led to several important recommendations.

(1) Based upon clear evidence of an essential role for ω3 fatty acids in human development and health, the scientists recommended that all infant formulae and diets for humans should include ω3 fatty acids, and they expressed concern that steps be taken to stop marketing enteral and parenteral formulas that fail to include any ω3 fatty acids.

(2) The researchers urged that the appropriate government agencies officially recognize the vitally important differences between ω3 and ω6 polyunsaturated fatty acids. Estimates of the average ω3 nutrition consumption in the USA presented by USDA scientists agreed with new nutrient measurements reported from a NHLBI study, with both studies indicating inadequate supplies of ω3 fatty acids in the typical American diet.

(3) New evidence with an extremely high level of statistical precision, from the NHLBI study, suggests that the daily dietary intake of 0.5–1.0 g of long-chain ω3 fatty acids per day reduces the risk of cardiovascular death in middle-aged American men by about 40%, and some new data suggests that ω3 fatty acids may also decrease cancer mortality.

(4) The research reports make it increasingly evident that eating ω3 fatty acids can have beneficial effects on chronic inflammatory and cardiovascular diseases.

(5) An international committee was established to help coordinate and enhance the rapid continued progress in this important area of biomedical research.

Conference and Panel Summaries

Simopoulos AP, Kifer RR, Martin RE, Barlow SM (eds): Health Effects of ω3 Polyunsaturated Fatty Acids in Seafoods. World Rev Nutr Diet. Basel, Karger, 1991, vol 66, pp 16–19

Conference Summary and Future Directions

Jørn Dyerberg

Medi-Lab a.s., Copenhagen, Denmark

As the last speaker and chairman of the final session I would like to begin by saying that the conference was anticipated by many with great expectations, expectations that I am glad to say were fulfilled. Dr. Michael Tillman of the National Oceanic and Atmospheric Administration (U.S. Department of Commerce) opened the conference with the following comments: '... the progress in this field has been made possible by 3 factors: (1) creative and dedicated scientists committed to the task of evaluating the role of ω3 fatty acids in human health; (2) a worldwide commitment of financial resources to this research effort; and (3) a cooperative spirit among the many, many players whether they be from the public, private, or industrial sectors. I encourage each of you to remain committed to the effort, to maintain the credibility of our science, and to wish each of you an enjoyable and successful conference.' Dr. Carlos Daza extended a greeting on behalf of the Director-General of the World Health Organization, Dr. Hiroshi Nakajima, and stated: 'There is no doubt that we need more data on ω3 polyunsaturated fatty acids from experimental, clinical, epidemiological and preventive studies ... We feel that in the future PAHO/WHO will be actively involved in some of the activities which will be discussed in this international conference and look forward to the success of the International Society for the Study of Fatty Acids and Lipids which is being organized.' Mr. Dawson, representing the Food and Agriculture Organization (FAO) of the United Nations, in his opening address said 'that the conference besides the scientific aspects should also focus on the dietary implications involved,' saying that 'FAO is looking forward to receiving your recommendations'.

The field of *ω3 fatty acids research* has had an exponential growth of publications since the mid 80's, especially from human studies as summarized by Dr. Simopoulos. This has paralleled an enormous growth of knowledge obtained especially during the last five years. One of the new and most promising items is the increase in the publications of cancer studies in the world literature.

Let me first highlight the first two sessions on essentiality. The first item *ω3 fatty acid deficiency* was covered by Drs. Connor, Bjerve, and Carlson. We now know that ω3 fatty acid deficiency exists and, consequently, that our diet *should* contain ω3 fatty acids. This leads us to the first conclusion of this conference namely that tube feeding products and mothers' milk substitutes devoid of ω3 fatty acids should be abandoned. From there is only one step to Recommended Daily Allowances (RDA) a touchy subject to deal with indeed. On this issue Dr. Pedersen stated that 'with the large amount of knowledge now available on the role of essential fatty acids and in particular of ω3 polyunsaturated fatty acids in human nutrition, a conference should be convened to propose RDA's in official dietary recommendations. Dr. Pedersen suggested the following RDA's of 5% of energy to be linoleic acid and 1% of energy to be alpha-linolenic acid and/or long-chain ω3 polyunsaturated fatty acids. It should be emphasized that for fish oils the borderlines among essentiality, nutrition, pharmacology and even toxicology are not established.

Basic biochemical problems were extensively covered by Howard Sprecher and William E. Lands. A general comment could be that the present clinical results surely justify an even enhanced intensity of basic research in ω3 fatty acids.

Cardiovascular. Peter C. Weber gave an extensive survey of ω3 fatty acids in cardiovascular disorders. One of his conclusions was that 'if the beneficial effects of ω3 fatty acids observed in epidemiological studies and in secondary intervention studies are confirmed, then the biological effector(s) must be sought among mostly unknown modulatory effects produced by *low* doses of ω3 polyunsaturated fatty acids'.

The startling results presented to us by Dr. Dolecek from the Multiple Risk Factor Intervention Trial (MRFIT) and Dr. Burr from the Diet and Reinfarction Trial (DART) strongly support this view.

One good candidate among these hitherto unknown factors is the enhancement of endothelial derived relaxing factor (EDRF) production –

or endothelium-derived hyperpolarizing factor (EDHF) to be more precise via the p-450 pathway – induced in endothelial cells exposed to eicosapentaenoic acid (EPA) as presented by Vanhoutte.

Other beneficial effects of ω3 fatty acids include reduced free radical production by macrophages and reduced platelet-derived growth factor (PDGF) production as presented by Fisher and Fox. John Charnock revealed to us yet another mechanism, namely the attenuated susceptibility to ventricular arrhythmias induced by fish oil feeding to primates. His results strongly support the proposed mechanisms for the DART results of Burr referred to previously.

The next panel in session III focused on *vascular reactivity.* The conclusion is that fish diets or fish oil supplementation do lower blood pressure albeit moderately in both normotensives and hypertensives, as most convincingly demonstrated in the Tromsø study presented by Dr. Bonaa. This should lead to a widespread use of fish oils in hypertension, both alone and in combination with anti-hypertensive drugs.

A reason for this – as stressed by Dr. Singer – is that one of the 'side effects' of fish oil is a normalization of the blood lipid profile, a problem of great concern in anti-hypertensive drug treatment.

The overall effect of fish oils on *plasma lipids* was dealt with by Tom Sanders, stressing the use of fish oil in type III hyperlipoproteinaemia and the increase of HDL-2 after fish oil supplementation.

An unsolved problem is the effect of fish oils on Lp(a). One reason for this may be that Lp(a) is not affected by fish oils in persons with low and normal Lp(a) levels but that in persons with high Lp(a) levels, long-term fish oil supplementation may reduce Lp(a), as presented by Dr. Schmidt from our group.

Inflammatory Diseases. The conclusion of the present status of ω3 fatty acids in inflammatory diseases as exemplified by rheumatoid arthritis and psoriasis is that ω3 fatty acids have a role to play in therapy *as have other dietary corrections and modulations.* One example is di-homo-gamma linolenic acid. With respect to mechanisms, new exciting data on ω3 fatty acids affecting cytokines were presented by Dr. Dinarello.

In trying to clarify the effects of such dietary modulations the composition of placebo-oils or -diets constitutes a major unsolved problem.

Diabetes is another disease of potential interest for ω3 fatty acid administration. It should be investigated with respect to fish oil supplementation by distinguishing whether the patients have non-insulin-dependent diabetes mellitus (NIDDM) or insulin-dependent diabetes mellitus

(IDDM). This distinction is essential because ω3 fatty acid supplementation of NIDDM is controversial, due to concern about worsening of glucose homeostasis, whereas IDDM might constitute a new field for ω3 fatty acid supplementation. The data presented by Dr. Jensen on reduced transcapillary albumin escape rate (TER) after fish oil supplementation in patients with IDDM with nephropathy are certainly promising.

One of the most fascinating fields in ω3 research is their role in *Neoplastic Disorders.* Epidemiological evidence, supported by evidence of mechanistic involvement of ω3 fatty acids in the attenuation of growth of neoplastic cells certainly offers the possibility of new and encouraging therapeutic means. This is reflected in the rapidly growing number of publications in this field. Great caution in the interpretation of these data should of course be exerted. Presently I think it is fair to conclude that ω3 fatty acid-enriched diets do *not* enhance cancer risk, as has been suggested by some, based on speculations on the effect of oxidation products of ω3 fatty acids. The MRFIT data and the data presented at the conference strongly suggest that the opposite could be the case.

Finally it should be mentioned that, in spite of a very restricted amount of time for presentation, the poster sessions were of an outstanding quality, and maybe were the foci for most of the new data on this conference.

In general, in relation to what has been achieved at the present conference and for objectives for future conferences, we should aim at unsolved problems by having groups discuss specific items, study objectives, and study design in order to develop protocol drafts, and maybe even consider how to obtain funds for performing these studies.

I would like to finish by addressing our friends in the commercial part of our new Fatty Acid society. It is one of your obligations to raise money for the necessary clinical studies, the results of which are of mutual interest and of which we both depend. In closing, it is my privilege to express my own, my co-chairmen and the conference attendants' respect and deep felt thanks to all and everyone that contributed to the success of this Second International Conference on the Health Effects of ω3 Polyunsaturated Fatty Acids in Seafood.

This of course addresses especially the organizing committee, Stuart Barlow, Robert Kifer, Roy Martin and Artemis Simopoulos, and among them most of all to the conference's two chairpersons, Artemis and Stuart. You have done a wonderful job and I ask everyone to join me in giving you a big hand.

Summary of Panel A

Simopoulos AP, Kifer RR, Martin RE, Barlow SM (eds): Health Effects of ω3 Polyunsaturated Fatty Acids in Seafoods. World Rev Nutr Diet. Basel, Karger, 1991, vol 66, pp 20–25

Growth and Development in Infants

Norman Salem Jr[a], *Susan E. Carlson*[b]

[a] Laboratory of Clinical Studies, NIAAA, Bethesda, Md.;
[b] Newborn Center, University of Tennessee, Memphis, Tenn., USA

The central question in this conference is, 'are ω3 fatty acids essential nutrients for mammals, and, in particular, for humans?'. There have been many new and important additions to our knowledge concerning this issue since the first international conference in this series in 1985. There are two separate lines of inquiry that bear upon this fundamental question. These approaches involve the study of the biological consequences of either ω3 deficiency or supplementation. In the former, the pathology associated with ω3 deprivation is described, whereas, in the latter, possible health benefits of ω3 supplementation of the diet is studied. It is also generally true that deficiency studies have focused upon the nervous system and retina and the functions of these organ systems whereas supplementation studies have focused upon the prevention or treatment of diseases relating to the periphery and, for example, involving the cardiovascular system, cancer, inflammation or metabolic disorders. Two separate arguments are implicit in these lines of inquiry. In the ω3 deficiency studies, it would be argued that the absence of these fats in the diet leads to poorer performance in tasks related to the functions of critical organs or to a diseased state. In contrast, ω3 supplementation studies are designed to discern whether there is a health benefit in such a diet and this may include the prevention of disease, treatment of a disease or a more optimal state of healthfulness. If either argument proves to be correct, then one may reasonably argue for the inclusion ω3 fatty acids as essential nutrients. This first panel has focused upon the deficiency approach; sessions to follow will deal with the supplementation studies.

Studies of ω3 deficiencies have focused upon organs wherein the levels of this family of fatty acids are some of the highest found in nature, i.e., the brain and retina [1]. In contrast to the situation in peripheral organs, 20:5 ω3 is at a trace level in the brain. Also, there is little or no 18:3 ω3 in the nervous system. The ω3 fatty acids in the nervous system (retinal tissue is embryologically brain gray matter) are almost entirely composed of 22:6 ω3, but there is also a small amount of 22:5 ω3. Therefore, one may narrow our question to, 'is 22:6 ω3 essential in the nervous system and retina?'.

In a review written for the 1985 meeting in this series, the evidence for considering 22:6 ω3 as an essential nutrient was examined [2]. The evidence available at that time suggested that it was essential since it was known that there were losses in discriminant learning ability and abnormalities in the electroretinograms of rats after ω3 depletion of their brains and retinas, respectively. Also, there were several neurological diseases or syndromes that were associated with losses in brain 22:6 ω3. In addition, there was some data linking human sperm function to the level of 22:6 ω3; this is another site of 22:6 ω3 concentration and the role of ω3 fatty acids in reproductive function is deserving of greater scrutiny than it is receiving in the present research environment.

At the 1990 meeting, it was clear that there had been much progress in describing the losses in functions associated with ω3 insufficiency and this work had been extended to both monkeys and humans. Connor et al. [2] had found that ω3 deprivation led to visual acuity losses in monkeys and restoration of the brain 22:6 ω3 during adolescence was ineffective in reversing this loss. Carlson reviewed a 1989 report from her laboratory indicating that premature infants fed formulas without long-chain ω3 fatty acids have poorer visual acuity in the first half of infancy in comparison to those fed formula supplemented with 0.2% 22:6 ω3. The small amount of fish oil included in this formula prevented a decline in erythrocyte phospholipid 22:6 ω3 which usually occurs beginning immediately after premature birth. New data were presented from follow-up of these infants into the last half of infancy. In infant groups fed either the fish oil supplemented or control (containing 18:3 ω3) formulas, there was a 3- to 4-fold range in the plasma phospholipid 20:4 ω6 concentration. The fish oil supplemented group had a reduction in their plasma and erythrocyte phospholipid 20:4 ω6. Since a positive correlation was found between phospholipid

20:4 ω6 and both growth and performance on cognitive tests, caution was expressed about the use of fish oil in formulas. It was pointed out, however, that ω3 supplementation accounted for no more than 15% of the variance in plasma phospholipid 20:4 ω6, and that the relationship between 20:4 ω6 and growth and performance reached significance only in the control group. The early results suggest, therefore, that 20:4 ω6 does not directly affect these outcomes but that it is a marker for some other variable that does.

Uauy and colleagues presented results that were complimentary to these in their studies of various visual system parameters of very low birth weight neonates and their relationship to the fatty acid composition of formulas. They found that visually evoked potentials (VEP) were better in the human milk/fish oil-fed group and that after 57 weeks (postconceptional age) both groups receiving long-chain ω3 fatty acids had significantly higher visual acuity (lower log MAR) than groups fed formulas with either 18:2 ω6 or 18:3 ω3. Cone function did not appear to be affected by these dietary differences. The authors interpreted these results as indicating that the 22:6 ω3 was the essential component in fish oil-supplemented formula leading to the visual acuity differences in early human development. It is also of great nutritional significance that the 18 carbon precursor to 22:6 ω3 was unable to support optimal visual system function. This suggests that premature infants are unable to desaturate and elongate 18:3 ω3 to 22:6 ω3 in sufficient quantities. This was supported by erythrocyte lipid compositional data that indicated that long-chain ω3 PUFA was *decreasing* between 36 and 57 weeks even though their formulas contained the shorter chain precursor to 22:6 ω3.

More severe deficiencies in ω3 fatty acids can be induced in experimental animals, and Bourre and coworkers have compared a variety of stimuli in soybean- and sunflower oil-fed rats. They reported enzymatic changes, electroretinogram abnormalities, altered response to neurotoxic agents and learning acquisition and extinction deficiencies in the low ω3 dietary group (sunflower oil). Another interesting facet of the research in this group was the report that membranes from the ω3-deficient group are less susceptible to fluidization by an ethanol challenge. Also, in this symposium, Martinez reported on a severe deficiency in brain, liver and kidney 22:6 ω3 in Zellweger's syndrome, a disorder that is thought to be a peroxisomal disorder. She suggested that the loss of brain 22:6 ω3 may be directly related to the profound neurological disturbances and early deaths of these infants. Another contribution of importance for the study of human lipid neurochemistry was the control data presented on the fatty acid concentrations in human infant brain from 26 weeks postconception to the 4–8 year old range. These data indicate a different rate of accretion of essential fatty acids during the prenatal and postnatal periods. The curves describing the ω3, ω6 and ω9 fatty acid accretion were quadratic during fetal development with the fastest rate of increase during the end of gestation. However, after birth, the accretion of these acids in the brain was increasing in a nearly linear fashion. These data reflect on the vital question as to whether there is a crucial period in human brain development wherein 22:6 ω3 is an essential nutrient and must be supplied in the proper quantities in order to provide for normal nervous system function. This syndrome as well as Pseudo-Zellweger are two more neurological disorders associated with 22:6 ω3 loss that may be added to those previously reviewed [3].

Finally, it is appropriate, although perhaps a bit premature, to ask, 'what are the critical cellular processes requiring 22:6 ω3 and what are the molecular mechanisms mediating these processes?'. These questions have been considered in recent review chapters [1, 3] and will only be briefly discussed in this summary. There are two approaches to the problem that are being pursued, and these are (1) the search for potent bioactive substances formed from 22:6 ω3 that are analogous to eicosanoids and (2) the delineation of an essential role for the fatty acid in phospholipids contained in cell membranes. Once these processes are understood, it would then be essential to understand how these functions are modulated in various dietary states. In the poster session associated with this meeting, data are presented by Kim et al. indicating that the brain is capable of producing hydroxylated 20:4 ω6 and 22:6 ω3 but that these compounds do *not* appear to be formed by a lipoxygenase. Furthermore, the overall level of activity is very low in the brain. Most of the 'lipoxygenase' activity present in brain slices appears to be due to the presence of cerebral microvessels and platelets present in unperfused brains. No prostanoids derived from 22:6 ω3 have as yet been found. Karanian et al. show that human platelet activity is comparable in magnitude to the rate of ω6 fatty acid hydroxylation and that the 22:6 ω3 metabolites are more potent antagonists of smooth muscle contractility than are the analogous eicosanoids. The case for the 22:6 ω3-derived bioactive substances is therefore being made in the peripheral circulation but there is still no convincing evidence that these compounds subserve any physiologically meaningful role in the brain or retina.

What we are left with in attempting an explanation of the ω3 deficiency effects noted above is a perturbation of biomembrane function. The reader is reminded that in a diet containing adequate ω6 but very low in ω3 fatty acids, the brain 22:6 ω3 declines and is replaced primarily with 22:5 ω6 and, to a lesser extent, with 20:3 ω9. This substitution is problematical for membrane biophysicists as most would predict little change in the overall physical state of the membrane in this case. Salem et al. [3, 4] have proposed a model of cell membrane structure encompassing a specialization in its structure with respect to polyunsaturated polar lipids. In this model, polyunsaturates are selectively localized in microenvironments on the interior leaflet of the plasma membrane and in some cases, in the proximity of membrane proteins where they serve to modulate both the physical environment and the activity of these macromolecular complexes. This concentration of polar PUFA species provides an amplifying mechanism that accentuates the physical and biological effects of these lipids. When these species are absent, such as in ω3 deficiency, there is a change in the function of many of the membrane complexes of proteins and lipids due to changes in their composition. Such a deficiency may be caused not only by dietary deprivation during early development but also by chronic ethanol abuse or by an inherited enzymatic defect. Conversely, an ω3 rich diet may alter the composition of membrane macromolecular complexes in the opposite direction and one might surmise that this would also alter the activity of many of these cellular processes. Hullin et al. presented a poster showing that when one compares rats on a corn oil- to those on a fish oil-based diet, that most of the excess ω3 fatty acids that accumulate in erythrocyte PE are localized in the plasmalogen pool on the interior leaflet of the plasma membrane. It is clear that there is a need for basic research regarding the functions of 22:6 ω6 especially with respect to its role in cell membranes and the mechanisms of its possible modulation of protein activities.

Another measure of the research progress in the area of ω3 essentiality is the extent to which the conference participants are convinced of this assertion. At the 1985 meeting, there was no general agreement regarding the question as to whether 22:6 ω3 should be included in milk formulas; it was regarded as a rather arbitrary addition. In contrast, a poll of conference participants after the 'Growth and Development in Infants' panel indicated that it was nearly unanimous that 22:6 ω3 should be added to milk formulas. According to this measure, there has certainly been a marked change in sentiment by experts in the field!

References

1 Salem N Jr: Omega-3 fatty acids: molecular and biochemical aspects; in Spiller G, Scala J (eds): New Protective Roles of Selected Nutrients in Human Nutrition. New York, Liss, 1989, pp 109–228.
2 Connor WE, Neuringer M, Reisbick S: Essentiality of ω3 fatty acids: Evidence from the primate model and implications for human nutrition; in Simopoulos AP, Kifer RR, Martin RE, Barlow SM (eds): Health Effects of ω3 Polyunsaturated Fatty Acids in Seafoods. World Rev Nutr Diet. Basel, Karger, 1991, vol 66, pp 118–132.
3 Salem N Jr, Kim HY, Yergey J: Decosahexaenoic acid: Membrane function and metabolism; in Simopoulos AP, Kifer RR, Martin RE (eds): The Health Effects of Polyunsaturated Fatty Acids in Seafoods. New York, Academic Press, 1986, pp 263–317.
4 Salem N Jr, Shingu T, Kim HY, Hullin F, Bougnoux P, Karanian JW: Specialization in membrane structure and metabolism with respect to polyunsaturated lipids; in Karnovsky M, Leaf A (eds): Aberrations in Membrane Structure und Function. New York, Liss, 1988, pp 319–333.

Simopoulos AP, Kifer RR, Martin RE, Barlow SM (eds): Health Effects of ω3 Polyunsaturated Fatty Acids in Seafoods. World Rev Nutr Diet. Basel, Karger, 1991, vol 66, pp 26–30

Requirements of Adults and Elderly

Kristian S. Bjerve

Department of Clinical Chemistry, University of Trondheim, Regional Hospital, Trondheim, Norway

Background and Summary

In 1978, WHO/FAO stated that 'Because of the possible specific role of this family of fatty acids in the specialized tissues, α-linolenic acid should be considered as an essential dietary constituent. However, more research is needed in clarifying the role of the α-linolenic acid family in human nutrition' [1].

Since then, a growing body of evidence has showed that ω3 fatty acids are one of the essential nutrients in man. The present conference reviewed data on both essentiality as well as function of ω3 fatty acids. The data show that ω3 fatty acids are essential for maintaining normal function in the brain, retina, immune competent cells, the myocardium, as well as in other tissues.

Although the essentiality of ω3 fatty acids in man has now been established, this is still not unequivocally and ubiquitously reflected in national and international recommended dietary allowances (RDAs). At present, two dietary recommendations specify the amount of ω3 fatty acids required. The Nordic countries have given recommended dietary allowances stating that linolenic acid and long-chain ω3 fatty acids should provide at least 0.5% of the energy [2]. The European Society of Paediatric Gastroenterology and Nutrition has published recommendations stating that linolenic acid should provide at least 0.5% of calories, or 55 mg per 100 kcal in formulas for the preterm infant [3]. The latest edition of the American RDAs published by The National Research Council includes α-linolenic acid as an essential fatty acid. However, they conclude that 'especially because essential fatty acid deficiency has been observed exclusively in patients with medical problems affecting fat intake or absorption, the subcommittee has not established an RDA for ω3 or ω6 fatty acids. ... The possibility of establishing RDAs for these fatty acids should be considered in the near future' [4].

During recent years several reports have appeared describing ω3 fatty acid deficiency in a total of 10 adults and children on long-term, parenteral or gastric tube formula feeding [for a review, see ref. 5]. It is reason to believe that all patients fed exclusively on artificial parenteral or gastric tube formulas containing none or too little ω3 fatty acids will be at risk of developing overt ω3 fatty acid deficiency. Since overt deficiency appears only after long-standing lack of dietary ω3 fatty acids, these patients have had a long period of ω3 fatty acid malnutrition before they develop symptoms of overt deficiency. During such periods of ω3 fatty acid malnutrition, the patients are probably less capable of handling stress situations like intercurrent infections, inflammations, and other diseases, and are probably more prone to develop pressure ulcers [6].

In 1989, only 1 out of 12 peroral nutrients listed in the Norwegian Drug Catalogue provided by the drug industry contained information on the content of ω3 fatty acids. This one preparation contained 0.02% of energy as linolenic acid, which is far below the dietary requirements in man. Two of the preparations contained soybean oil, but gave no information on the ratio between ω6 and ω3 fatty acids. Two of the formulas were low-fat nutrients with an unspecified source of fat, while 7 formulas contained lard, MCT, palm oil, corn oil, sunflower oil, or coconut oil, alone or in unspecified mixtures. Thus, several of the formulas marketed in Norway do not supply even the minimum amounts of ω3 fatty acids required in man. It is reason to believe that the same is true in most other countries [7]. Still, these formulas are marketed as mother's milk substitutes in newborn and preterm infants, or for the treatment of malnutrition and fat malabsorption, total enteral or parenteral feeding, as well as pre- and post-operative treatment of critically ill patients, all of which are situations where it is important to ensure optimal nutrition for the patient.

There is now consensus that ω3 fatty acids are essential nutrients in man, and that they must be present in formulas used for human nutrition. Inadequate supplies affect brain and visual development, impair the immune system and probably affect cell growth and tissue regeneration [5,

8, 9]. In this situation, the present lack of consensus on how much ω3 fatty acids should be included, and the question whether long-chain ω3 fatty acids also must be included, is one of the major obstacles in the process of ensuring optimal supplies of ω3 fatty acids for the critically ill patient as well as for newborn and preterm infants.

Observations from the 10 reported patients with ω3 fatty acid deficiency indicate that the optimal supply of linolenic acid is 900–1,200 mg/day, or 0.54–1.2% of energy, and the minimal dietary requirement of ω3 fatty acids was estimated to be 0.2–0.3% [5]. It should be noted, however, that the value 0.54% of energy was obtained in a patient receiving a relatively high energy diet. Human milk supplies on the average 0.35% of energy from linolenic acid and 0.35% of energy from long-chain ω3 fatty acids [1]. Functionally, EPA and DHA are the most important ω3 fatty acids. Linolenic acid is elongated and desaturated to eicosapentaenoic (EPA) and docosahexaenoic acid (DHA) rather slowly in man [7], so that EPA and DHA are 2–10 times more effective than linolenic acid [5, 10]. When this is taken into account, the data on linolenic acid requirements obtained from patients with ω3 fatty acid deficiency correspond rather well with the amounts of ω3 fatty acids supplied from human milk.

Conclusions and Future Perspectives

The Nordic [2] and European [3] RDAs state that linolenic acid should at least contain 0.5% of energy as ω3 fatty acids. However, the data available today indicate that this is not sufficient to maintain an optimal supply of ω3 fatty acids [5, 11], and may lead to impaired visual function [8]. It is now appropriate to conclude that nutrient formulas containing less than 0.5% of energy from ω3 fatty acids should not any longer be marketed, and the industry should be advised and encouraged to provide high-quality formulas containing at least 1% of energy from ω3 fatty acids, and with a ω6/ω3 ratio which is not excessive, probably in the range of 4–8. Some formulas on the market today already fulfill these requirements, and the medical profession and nutritionists should be advised to use these formulas in order to protect patients from getting an iatrogenic ω3 fatty acid malnutrition.

Further research is required to settle whether future RDAs also should include EPA and/or DHA from the ω3 family or arachidonic acid from the ω6 family, and to study whether the main guidelines for ω3 RDAs should be adjusted for special groups of patients such as preterm infants and patients with extensive burn injuries [11, 12]. The use of ω3 deficient primates as an experimental model should be explored further. So far, the primate model has been successfully used to demonstrate the essentiality of ω3 fatty acids for the normal function of vision and brain. In the future, the primate model should also be used to investigate the role of ω3 fatty acids in conditions of animal stress such as infections, inflammations, wound and bone healing. This could provide valuable information on the dietary requirements of ω3 fatty acids in the newborn and preterm infants as well as in the severely ill patient.

Acknowledgements

The present work has been supported by grants from The Norwegian Research Council for Science and the Humanities, The Nordic Insulin Fund, The Norwegian Diabetics Society, The Norwegian Cancer Society and The Norwegian Council for Heart and Cardiovascular Diseases.

References

1 FAO/WHO: Dietary fats and oils in human nutrition. Paper 3. Food and Agriculture Organization, World Health Organization, Rome, 1978.
2 Nordic Nutrition Recommendations, 2nd ed., 1988.
3 European Society of Paediatric Gastroenterology and Nutrition. Acta Paed Scand 1987; suppl 336: pp 1–14.
4 Recommended Dietary Allowances, 10th ed., Washington, National Academy Press, 1989, pp 44–51.
5 Bjerve KS: N–3 Fatty acid deficiency in man: Pathogenetic mechanisms and dietary requirements. N–3 News 1989;4:1–4.
6 Bjerve KS, Fischer S, Wammer F, Egeland T: Alpha-linolenic acid and long-chain n–3 fatty acid supplementation in three patients having n–3 fatty acid deficiency. Effect on lymphocyte function, plasma and red cell lipids, and prostanoid formation. Am J Clin Nutr 1989;49:290–300.
7 Heim T: How to meet the lipid requirements of the premature infant. Ped Clin North Am 1985;32:289–317.
8 Carlson SE, Cooke RJ, Peeples JM, Werkman SH, Tolley EA: Docosahexaenoate (DHA) and eicosapentaenoate (EPA) status of preterm infants: Relationship to visual acuity in n–3 supplemented and unsupplemented infants. Pediatr Res 1989; 25:285A.

9 Neuringer M, Connor WE: N–3 fatty acids in the brain and retina: Evidence for their essentiality. Nutr Review 1986;44:285–294.
10 Adam O, Wolfraum G, Zöllner N: Vergleich der Wirkung von Linolensäure und Eicosapentaensäure auf die Prostaglandinbiosynthese und Thrombozytenfunktion beim Menschen. Klin Wochenschr. 1986;64:274–280.
11 Carlson SE, Rhodes PG, Rao VS, Goldgar DE: Effect of fish oil supplementation on the n–3 fatty acid content of red blood cell membranes in preterm infants. Pediatr Res 1987;21:507–509.
12 Gottschlich MM, Alexander JW: Fat kinetics and recommended dietary intake in burns. J Parent Ent Nutr 1987;11:80–85.

Simopoulos AP, Kifer RR, Martin RE, Barlow SM (eds): Health Effects of ω3 Polyunsaturated Fatty Acids in Seafoods. World Rev Nutr Diet. Basel, Karger, 1991, vol 66, pp 31–33

Cardiovascular I: Cell-Vessel Wall Interactions

Alexander Leaf

Massachusetts General Hospital, Boston, Mass., USA

The aim of this section was to present some highlights of the new knowledge that has been gained regarding functional and biochemical effects of ω3 fatty acids since the first International Conference on the Health Effects of Polyunsaturated Fatty Acids in Seafoods in 1985.

In summary, the many additions to recent knowledge of cardiovascular actions, when membrane phospholipids are enriched with ω3 polyunsaturated fatty acids (PUFAs) include the following: (a) reduced vasospastic responses of blood vessel to several vasoconstrictor agonists; (b) increased vascular compliance; (c) reduced blood visocosity; (d) reduced albumen leakage in insulin dependent diabetes; (e) prevention of fatal ventricular arrhythmias in experimental animals following temporary or permanent coronary occlusion; (f) increased cardiac beta-receptor function; (g) increased post-ischemic coronary blood flow; (h) reduced vascular intimal hyperplasia and prevention of experimental atherosclerosis in animals.

Dr. Weber summarized the present status of ω3 PUFAs as having many actions within cells that modulate the intracellular signal transmission and, either directly or via interfering with the formation of arachidonic acid-derived products of cyclooxygenase, lipoxygenase, and possibly the C-P450 enzymes, induce changes in the profile of eicosanoids which are transmitters of many important messages within cells. Some of these involve alterations in i.e. Ca^{++}, inositolpolyphosphates, and gene expression.

Dr. Vanhoutte presented his studies on the endothelial derived relaxation factors (EDRF) and provided evidence that endothelial cells play a major role in controlling vascular actions. EDRF is now known to be nitric oxide, but another factor, endothelium-derived hyperpolarizing factor (EDHF), has not yet been defined structurally. It may be a C-P450-related arachidonic acid metabolite. EDRF and EDHF act together with prostacyclin to maintain a nonthrombotic endothelial surface. ω3 enrichment of endothelial cells potentiates the release of vascular relaxing substances by these cells (most probably of EDHF) and helps to explain the antithrombotic and antiatherosclerotic effects of these fatty acids.

Marc Fisher reviewed his evidence that human neutrophils and monocytes, when enriched with ω3 fatty acids by dietary fish oil supplements, show a decreased production of superoxide when stimulated by latex particles. The potential importance of this effect of ω3 fatty acids to reduce oxidative damage to cells within the developing atheroma and perhaps in reducing oxidative changes in the LDL particle was pointed out. However, several investigators reported increases in plasma malonyldialdehyde after fish oil feedings indicating more, rather than less, oxygen free radical damage to unsaturated fatty acids. Thus the generalizability of these reported changes in leucocyte production of superoxide by their specialized NADPH-oxidase complex to the many other sources of oxygen free radicals in the body cannot yet be made.

Dr. Hallaq presented evidence that responses of neonatal cardiac myocytes to the toxic effects of ouabain – a potent arrhythmogenic agent – could be prevented by enrichment of the myocytes with eicosapentaenoic acid. The resistance to accelerated heart rate, reduced amplitude of contraction and final lethal contracture from ouabain toxicity resulted from preventing toxic levels of cytosolic calcium. The reduced calcium levels in turn resulted from a reduced rate of influx of calcium into the cells when ω3 fatty acids enriched the sarcolemma phospholipids.

Dr. Fox reviewed his studies on the reduction in platelet-derived growth factor (PDGF) production by cultured bovine endothelial cells when the latter were enriched with ω3 fatty acids. The inhibition of PDGF production interestingly was present when the cells were grown in media containing fetal calf serum but the effect is lost if calf or adult human serum replaces the fetal serum in the incubation medium. An oxidative process is apparently necessary for the inhibitory action of ω3 fatty acids on PDGF production. Dr. Fox showed evidence that ceruloplasmin, which is an oxygen free radical scavenger, prevents this inhibitory action and since it is present only in small amounts in fetal serum, but abundant in adult serum, could explain the serum effects on PDGF production.

It is clear that much new and important understanding of potential antiatherosclerotic actions of ω3 fatty acids has been forthcoming and the pace of research in this area is accelerating but there is a great need for continued and additional research support. Some important problems needing study include:

(a) With the discovery of novel metabolic pathways of AA, EPA and DHA via the C-P450 pathway, new mechanisms of action of these fatty acid products in intracellular signalling need to be evaluated.

(b) Determination if other oxygen free radical producing systems besides the leucocyte NADPH-oxidase system are inhibited by ω3 fatty acid supplementation.

(c) Does DHA enrichment of cardiac myocytes also stabilize these cells against arrhythmogenic stress?

(d) The changes in membrane lipid composition by the loading dose of EPA used in Dr. Hallaq's study were so modest as to suggest some eicosanoid action [e.g. (a)] rather than a physical change in the cardiac cell membranes in preventing ouabain toxicity. What metabolite of EPA might this be?

(e) Are the in vitro effects of PDGF inhibition seen in bovine endothelial cells occurring as well in humans on diets supplemented with ω3 PUFAs?

(f) What arachidonic acid- or eicosapentaenoic acid-derived oxygenated metabolite (C-P450 product?) is the second endothelial derived relaxation factor, EDHF?

(g) There are contradictory reports on actions of ω3 PUFAs on the endogenous fibrinolytic system, specifically on tissue plasminogen activator, on tissue plasminogen activator inhibitor(s), and on fibrinogen. In studies with fish oil preparations high in DHA, fibrinogen seems to decrease. Since these are very important factors in atherogenesis the correct results (or the differential effects of either EPA or DHA) are important to determine.

(h) Although the majority of hypercholesterolemic patients (type IIA) show no significant effects of ω3 PUFAs on total LDL cholesterol, are there physical changes in the lipoproteins which may be significant for the development of atherosclerosis?

(i) Our understanding of how ω3 PUFAs can affect atherosclerosis is totally dependent upon our understanding of the atherogenic process itself. Therefore, all research that furthers our understanding of atherogenesis needs support and stimulation.

Simopoulos AP, Kifer RR, Martin RE, Barlow SM (eds): Health Effects of ω3 Polyunsaturated Fatty Acids in Seafoods. World Rev Nutr Diet. Basel, Karger, 1991, vol 66, pp 34–38

Cardiovascular II: Heart

A. Nordøy

Department of Medicine, Institute of Clinical Medicine, University of Tromsø, Tromsø, Norway

Epidemiology

The early studies among Greenland Eskimos [1] and Japanese fishermen [2] indicated a strong association between a high intake of eicosapentaenoic (EPA) and docosahexaenoic (DHA) acids of the ω3 family and a low incidence of coronary heart disease (CHD). These populations were characterized by diets extremely high in their content of these fatty acids combined with a low intake of saturated fatty acids and a low intake also of ω6 fatty acids compared with population groups in the Western world. Indirect evidence from observations in Norway during the 2nd World War, retrospectively, suggested that this association also could be obtained in Western population groups [for review see 3]. More recently, the study of Kromhout et al. [4] from Holland reported an inverse relationship between the amount of fish eaten each week and the mortality from CHD. The consumption of as little as 35 g of fish per day was associated with a 50% lower mortality than with no fish consumed. This observation was confirmed in two other studies [5, 6] and also in the recent study by Dolecek [7] on data accumulated from the multiple risk factor intervention trial (MRFIT) presented at this meeting. Common for all these later studies has been the rather substantial effect of a very low intake of ω3 polyunsaturated fatty acids (PUFA) compared with no such intake. Surprisingly, the mortality of CHD seems not to have been positively affected by a much higher consumption of fish oils in other western population groups [8–11]. In some of these studies it has been suggested that a combined high intake of saturated fats may have abolished the beneficial effects of the fish oils

[10]. At present, the mechanisms explaining the variable associations between mortality from CHD and the intake of low, medium or high amounts of dietary fish oils is not clear.

Intervention Studies

The present study of Burr et al. [12] is the only intervention study of significance so far. Eating fatty fish in moderate amounts (about 300 g weekly) reduced the mortality from all causes by about 30%. The cardiac events were not reduced possibly suggesting that the increased survival actually reflected a reduced number of patients with sudden deaths. Unfortunately, the numbers of patients dying from sudden death was not registered in this study. The effects on total mortality reported in this study were not associated with any significant changes in blood cholesterol and HDL-cholesterol levels.

A series of studies have reported effects on plasma lipid levels and hemostatic parameters by dietary supplementation of fish oils [for review see 13]. These changes may, over prolonged periods of time, have significant effects on the risk profile for developing CHD. A recent study carried out in a metabolic ward seems to confirm that when 40% of the calorie-intake came from fats and mainly from saturated fatty acids, the effects of supplementary fish oils on the coronary risk profile were mostly abolished compared with those observed when similar amounts of fish oils were given in a low-fat diet (25% energy) [14].

Other Risk-Factors for CHD

Significant effects on blood pressure in patients with a stable, mild hypertension, documented in a randomised 10-week dietary supplementation trial [15], indicate that dietary supplementation of 6 g per day of 85% EPA/DHA acid may have potent effects on this well documented risk-factor. It is interesting to observe that this supplementation of fish oils did not change mean blood pressure in the subjects who consumed, by habit, three or more fish meals per week. Recent studies also seem to indicate that dietary supplementation with 4 g of ω3 PUFA for 9 months significantly reduced the concentration of lipoprotein Lp(a) in healthy subjects with high levels of this lipoprotein [16]. This effect on a risk factor for developing CHD is of special interest relating the effects of ω3 PUFA both to LDL (atherosclerosis) and plasminogen (thrombosis), both factors directly relating to the development of coronary vascular disease.

Experimental Studies

A series of animal studies using a variety of models have confirmed that in many species, the dietary supplementation of fish oils, with subsequent changes in the fatty acid composition of the myocardium, the endothelial cells and the blood cells, induces protection of the ischemic myocardium by many possible mechanisms as reviewed by Nestel [17]. The prevention of serious ventricular arrhythmias in rats and marmoset monkeys subjected to ligation of a major coronary artery or to electrical stimulation, by dietary fish oils for lengthy periods may be of significance [for review see 18]. This experimental evidence supports epidemiological and intervention studies in man. However, such serious arrhythmias have not been prevented in other species (pig, dog) using other models and also using shorter feeding periods. The studies of β_1-receptors of cardiac sarcolemma in rats indicating decreased affinity in animals fed cod liver oil have been associated with a low ratio of AA/DHA in the cardiac phospholipids [19]. A *high* such ratio has frequently been observed in sudden cardiac death in man contrary to that observed in people of the same age who died in accidents [19]. Others have observed modifications induced by fish oil supplement also in the alpha-adrenoreceptor function in rat heart. Finally, the potentiation of endothelium-dependent relaxations in the blood vessels of animals fed fish oils [20] adds to the potential of EPA and DHA as cardio-protective agents.

Conclusions

Increasing evidence strongly indicates that EPA and DHA given as dietary supplements have preventive effects on the development of coronary heart disease both relating to the coronary circulation and the myocardium. Recent studies stress the importance of evaluating in detail the effects of these fatty acids in relation to the effects of the other dietary, saturated and unsaturated (mono, ω6) fatty acids.

Perspectives

The goals for future research in this specific area should include:

(1) Additional secondary prevention trials, particularly relating to sudden death and serious cardiac arrhythmias.

(2) Prevention trials relating to the occurrence of complications to coronary surgery and percutaneous transluminal coronary angioplasty (PTCA).

(3) Angiographic and perfusion studies relating to the possible regression of vascular lesions.

(4) Further studies on the cellular level relating to the mechanisms involved in contraction and relaxation of the myocardium and the coronary vasculature.

(5) Which mechanisms explain the epidemiological observations indicating cardio-protective effects of low and very high intakes of ω3 fatty acids and the lack of effect by medium dosages?

References

1 Dyerberg J, Bang HO, Stoffersen HO, Moncada S, Vane J: Eicosapentaenoic acid and prevention of thrombosis and atherosclerosis. Lancet 1978;ii:117–119.

2 Hirai A, Terano T, Salto H, Tamura Y, Toshida S: Clinical and epidemiological studies of eicosapentaenoic acid in Japan; in Lands WEM (ed): Proc. AOCS short course on polyunsaturated fatty acids and eicosaenoids. Champaign, AOCS, 1987, pp 9–24.

3 Nordøy A, Goodnight S: Dietary lipids and thrombosis. Arteriosclerosis 1990;10: 149–163.

4 Kromhout D, Bosschieter EB, de Lezenne Coulander C: The inverse relation between fish consumption and 20-year mortality from coronary heart disease. N Engl J Med 1985;312:1205–1209.

5 Shekelle RB, Misell LU, Paul O, Shryock AM, Stamler J: Fish consumption and mortality from coronary heart disease. N Engl J Med 1985;313:820.

6 Norell SE, Ahlbom A, Feychting M, Pedersen NL: Fish consumption and mortality from coronary heart disease. Br Med J 1986;293:246.

7 Dolecek TA, Grandits G: Dietary polyunsaturated fatty acids and mortality in the multiple risk-factor Intervention Trial; in Simopoulos AP, Kifer RR, Martin RE, Barlow SM (eds): Health Effects of ω3 Polyunsaturated Fatty Acids in Seafoods. World Rev Nutr Diet. Basel, Karger, 1991, vol 66, pp 205–216.

8 Vollset SE, Hench I, Bjelke E: Fish consumption and mortality from coronary heart disease. N Engl J Med 1985;313:820–821.

9 Curb JD, Reed M: Fish consumption and mortality from coronary heart disease. N Engl J Med 1985;313:821.

10 Simonsen T, Vårtun A, Lyngmo V, Nordøy A: Coronary heart disease, serum lipids, platelets and dietary fish in two communities in northern Norway. Acta Med Scand 1987;222:237–245.

11 Hunter DJ, Kazda I, Chockallngam A, Fodor JG: Fish consumption and cardiovascular mortality in Canada: An interregional comparison. Am J Prev Med 1988;4: 5–6.

12 Burr ML, Fehily AM, Gilbert JF, et al: Effect of changes in fat, fish, and fibre intakes on death and myocardial reinfarction: diet and reinfarction trial (DART). Lancet 1989;ii:757–761.

13 Harris WS: Fish oils and plasma lipid and lipoprotein metabolism in humans: A critical review. Lipid Res 1989;30:785–807.
14 Nordøy A, Goodnight S, Connor WE: Do dietary saturated fatty acids counteract the antithrombotic effects of n–3 fatty acids. X Int Symp Drugs Affecting Lipid Metabolism (abstract). Houston, 1989;p 38.
15 Bønaa KB, Bjerve KS, Gran IT, Straume B, Thelle D: Effect of eicosapentaenoic and docosahexaenoic acids on blood pressure in hypertension. N Engl J Med 1990;322: 795–801.
16 Schmidt EB, Klausen IC, Kristensen SD, Lervang HH, Færgeman O, Dyerberg J: Effect of ω3 fatty acids on Lipoprotein (a); in Simopoulos AP, Kifer RR, Martin RE, Barlow SM (eds): Health Effects of ω3 Polyunsaturated Fatty Acids in Seafoods (abstract). World Rev Nutr Diet. Basel, Karger, 1991, vol 66, p 529.
17 Nestel PJ: Review: Fish oil and cardiac function; in Simopoulos AP, Kifer RR, Martin RE, Barlow SM (eds): Health Effects of ω3 Polyunsaturated Fatty Acids in Seafoods. World Rev Nutr Diet. Basel, Karger, 1991, vol 66, pp 268–277.
18 Charnock JS: The antiarrhythmic effects of fish oils; in Simopoulos AP, Kifer RR, Martin RE, Barlow SM (eds): Health Effects of ω3 Polyunsaturated Fatty Acids in Seafoods. World Rev Nutr Diet. Basel, Karger, 1991, vol 66, pp 278–291.
19 Gudbjarnason S, Benediksdóttir VE, Gudmundsdóttir E: Balance between ω3 and ω6 fatty acids in heart muscle in relation to diet, stress and ageing; in Simopoulos AP, Kifer RR, Martin RE, Barlow SM (eds): Health Effects of ω3 Polyunsaturated Fatty Acids in Seafoods. World Rev Nutr Diet. Basel, Karger, 1991, vol 66, pp 292–305.
20 Vanhoutte PM, Shimokawa H, Boulanger Ch: Fish oil and the platetet-blood vessel wall interaction; in Simopoulos AP, Kifer RR, Martin RE, Barlow SM (eds): Health Effects of ω3 Polyunsaturated Fatty Acids in Seafoods. World Rev Nutr Diet. Basel, Karger, 1991, vol 66, pp 233–244.

Summary of Panel F

Simopoulos AP, Kifer RR, Martin RE, Barlow SM (eds): Health Effects of ω3 Polyunsaturated Fatty Acids in Seafoods. World Rev Nutr Diet. Basel, Karger, 1991, vol 66, pp 39–43

Cardiovascular III: Circulation – Blood Pressure

Howard Knapp

Vanderbilt University, Nashville, Tenn., USA

With the theme of interrelated benefits of ω3 PUFA on numerous physiological processes in mind, Dr. Peter Singer presented the first paper of the panel, summarizing the clinical studies on blood pressure effects which have largely been published since the first conference in 1985. His overall impression was that studies in normotensive subjects have yielded highly variable results, usually with negligible or small reductions in blood pressure. As is often seen in studies of hypertension, it seems that patients with higher baseline blood pressure values have larger, more easily detected reductions in response to fish or fish oil supplementation. There is also some suggestion from the literature that hemodialysis patients may be more sensitive to the blood pressure-lowering effects of ω3 PUFA than other patients. Although some studies have suggested beneficial effects of ω3 PUFA on renal electrolyte handling in hypertensive subjects, the effects in dialysis patients must relate instead to alteration of vascular reflexes, and will require further investigation.

Dr. Singer also presented data suggesting that a fish-supplemented diet produced a greater decline in blood pressure than an equivalent amount of ω3 PUFA provided as encapsulated fish oil. During the question session, however, he responded to an inquiry about the statistical significance of this by noting that the number of subjects was small and that there were other differences between the supplements such as saturated fat content and other non-fish components (i.e., from the tomato pulp included in the canned mackerel). Therefore, further work on the optimal dose and formulation for delivering ω3 PUFA to hypertensive subjects is needed. An additional important topic just starting to be investigated is the interaction between ω3 PUFA and antihypertensive drugs, and recent data on the additive effects of fish supplements and propranolol was presented. There was a potentiation of blood pressure lowering by the combination of these two. In addition, the theoretically unfavorable increases in plasma triglycerides often seen during antihypertensive therapy were not evident. Thus, there is potential for beneficial modification of several cardiovascular risk factors at the same time in patients by adding ω3 supplements as adjuvants to the antihypertensive regimen.

Dr. Singer concluded his presentation by noting that although ω3 PUFA supplements should not yet be prescribed as sole therapy for patients with established, significant hypertension, there is still no evidence that standard pharmacotherapy of subjects with borderline hypertension confers a cardiac benefit, and such non-pharmacological therapy could be considered in this group. He recommended that future studies be more carefully designed in regard to blood pressure measurement, and better knowledge of the changes taking place in dietary components other than lipids, as well as how these alterations may relate to the ω3 PUFA effects. During the following discussion, Dr. Nestel amplified these comments with reference to his own study (presented in Poster Session I) showing the potentiation of ω3 hypotensive effects in the setting of sodium restriction. Other important areas for further work were concluded to be those of dose- and time-responsiveness of the ω3 effect, the interactions of ω3 PUFA with other hypotensive agents, and mechanistic studies both in regard to autacoids (prostaglandins/eicosanoids and other endogenous vasoactive substances) and hemodynamic parameters (cardiac effects, resistance in specific vascular beds).

To amplify Dr. Singer's comments on effects of ω3 on vascular control mechanisms, the next presentation, by Dr. David Mills, centered upon the speaker's studies in this area in both animal and man. Since excessive blood pressure response to stressful stimuli is believed to reflect a predisposition towards the development of hypertension, a primary prevention approach in altering dietary PUFA might usefully focus on modification of such variables. Much of the work carried out to date has been with the borderline hypertensive rat model, which develops hypertension in response to stress. In a variety of experimental conditions, both eicosapentaenoic and γ-linolenic acids exerted dampening effects on the response to stress. In studies on normotensive human subjects, the effects of ω3 and ω6 PUFA were different, with borage oil (23% 18:3 ω6), but not fish oil or olive oil, causing a reduction in the blood pressure and heart rate responses to stress. Interestingly, borage oil increased the forearm vasoconstrictor response to lower-body-negative-pressure maneuvers, while fish oil increased resting peripheral blood flow. These observations were consistent with a sensitization of the baroreceptor reflexes by 18:3 ω6, and a peripheral vasodilatory action of 20:5 ω3. During the following discussion, Dr. Geza Bruckner pointed out some agreement with his studies on peripheral capillary RBC velocity, as far as the fish oil effects were concerned. There was also some commentary by Dr. Michael Crawford as to the possible mechanisms of such effects, and a general agreement that more work was needed both on the mechanisms and in the studies of hypertensive humans.

Continuing on the relationship between hypertension and plasma lipids as risk factors modifiable by ω3 PUFA, Dr. Tom Sanders presented a summary of the biochemical and pharmacological effects of ω3 supplements on plasma lipids in man. Both the strong dose-relationship of the various ω3 effects, as well as a contrast with the effects of ω6 PUFA supplements, were illustrated. Controversial data on the changes taking place in plasma LDL and HDL fractions were summarized and rationalized as being due to examining effects in different patient groups, and an overall picture of the metabolic results of ω3 PUFA supplementation was presented into which the disparate observations fit in a sensible way. The therapeutic usefulness of ω3 PUFA preparations in patients with hypertriglyceridemia was reviewed, as was the lack of effects of such preparations in the treatment of hypercholesterolemia. A cautionary note was raised on both the high vitamin E requirement induced by giving highly enriched DHA (22:6 ω3) supplements as well as the alterations of glucose homeostasis caused by fish oil in patients with non-insulin-dependent diabetes. It was theorized that plasma EPA and/or DHA may possibly alter the rate or form of LDL oxidation in vivo, and thereby cause a reduced atherogenic potential not reflected in an actual lowering (or even despite an increase) of LDL. An additional theoretical benefit might relate to the tendency of EPA and DHA to replace ω6 eicosanoid precursor PUFA from plasma and blood cell phospholipid pools, and somehow this could be involved in the overall beneficial changes noted in platelet-vascular interactions.

The final presentation was made by Dr. Knapp, in which he discussed the problems in making mechanistic projections from epidemiologic data, particularly in regard to ω3 intake and population blood pressures. Consistent with the work of Dr. Nestel referred to earlier in normal subjects, cultural groups traditionally consuming large quantities of fish often had high sodium intake as well which obscured any possible hypotensive effect of ω3 PUFA. Possible mechanisms of ω3 PUFA effects were reviewed, starting with effects on the eicosanoid system. Although fish oil supplements increase prostacyclin synthesis in hypertensive subjects, the increase is not maintained as blood pressure falls, i.e., there does not need to be a continued high output of prostacyclin in order for blood pressure to be reduced. Also, the blunted response to high, but not moderate, infusion rates of phenylephrine was associated with a lower (instead of higher) synthesis of prostacyclin. These observations suggested that changes in prostacyclin do not fully explain the vascular effects of fish oil.

The effects of ω3 supplements on PGE synthesis were then reviewed, with the conclusion that there may be reductions in vivo which could have an adverse effect on patients with prior impairment of renal function. Other studies on ω3 supplements in a number of renal disease models were reviewed, with the demonstration that fish oil exerts adverse effects in rats with reduced renal mass, and that studies on this point in man are needed. The picture is less clear in models of immune nephritis, but there have been two negative small studies in human lupus thus far. Several animal studies and two small uncontrolled human studies suggest that fish oil may prevent/ameliorate the nephrotoxic/hypertensive effects of cyclosporin therapy. Although no studies have been designed to assess potential adverse effects of ω3 fatty acids in human renal disease, one small trial in renal transplant patients and one in a heterogeneous group of patients with chronic renal failure did not report any adverse effects associated with the fish oil supplements, which were at a relatively low dose. This important question is an obvious area for well-designed future studies.

Some final comments about the eicosanoid system indicated that reductions in thromboxane synthesis were unlikely to be of importance in ω3 hypotensive effects, except possibly in toxemia of pregnancy. No reports on ω3 therapy in this syndrome have been published to date, but such patients have excessive thromboxane synthesis, and several clinical trials have shown that low, platelet-selective doses of aspirin are very effective in lowering their blood pressure. Lastly, a number of speakers in several of the Panels have suggested that cytochrome P-450 products of arachidonic acid may be involved in the vascular effects of fish oil. Dr. Knapp presented the first data documenting the appearance in urine of such epoxides made from EPA during fish oil ingestion by humans, and indicated that this new area of eicosanoid research may hold the answer to many of the questions regarding fish oil's vascular effects.

The session was concluded with two slides summarizing the reports of ω3 fatty acid effects on plasma viscosity/fibrinogen concentrations and RBC/whole blood viscosity. If blood viscosity were reduced by fish oil, then this might provide the basis for many of the observed effects such as prolonged bleeding time, increased capillary blood flow, and lowered blood pressure. Unfortunately, there is considerable controversy regarding both of these possibilities, although the majority of reports find no change in plasma viscosity/fibrinogen levels.

Overall recommendations for ω3 PUFA as a primary therapy of established hypertension cannot be made at this time, but increased fish consumption would seem a reasonable addition to the non-pharmacologic means of attenuating the blood pressure rise in subjects with mild/borderline hypertension. It is hoped that the research discussed/proposed in today's session will define the optimal dose and form to be used in various types of hypertensive patients in the near future.

Summary of Panel G

Simopoulos AP, Kifer RR, Martin RE, Barlow SM (eds): Health Effects of ω3 Polyunsaturated Fatty Acids in Seafoods. World Rev Nutr Diet. Basel, Karger, 1991, vol 66, pp 44–47

Rheumatoid Arthritis and Inflammatory Mediators

Dwight R. Robinson[a], *Joel M. Kremer*[b]

[a] Harvard Medical School and the Arthritis Unit at the Massachusetts General Hospital, Boston, Mass., and
[b] Division of Rheumatology, Albany Medical College, Albany, N.Y., USA

It is now well-established that dietary ω3 fatty acids may modify inflammation in humans and in experimental models, but the mechanisms of these effects are unclear. Suppression of the levels of arachidonic acid by either essential fatty acid deficiency or by incorporation of alternative fatty acids, such as eicosapentaenoic acid (EPA), reduces the ability of tissues to produce eicosanoids, but other anti-inflammatory mechanisms may also be important when ω3 fatty acids are incorporated into tissues. It is also well established from work with experimental systems that not all inflammatory states are influenced in the same way by ω3 fatty acids. For example, murine autoimmune glomerulonephritis is markedly suppressed by dietary ω3 fatty acids, but the severity of Type II collagen arthritis is unaffected [1, 2]. The complexity of inflammation and the incomplete state of understanding of the mechanisms of inflammation dictate that the efficiency of any intervention, including ω3 fatty acids, must be tested in each disease state before generalizations may be drawn.

Several well-controlled studies of ω3 fatty acid supplements in patients with rheumatoid arthritis reviewed in this conference have documented anti-inflammatory effects on the basis of clinical parameters reflecting the severity of inflammation in diseased joints. However, at the level of these dietary supplements ca. 6 g of ω3 fatty acids daily, the clinical effects are only modest, and it remains to be seen whether a higher daily dose of ω3 fatty acids would yield more significant therapeutic effects. Recent studies with the inbred strain of mice, (NZB × NZW)F_1, have shown a dose-dependent alleviation of the spontaneous autoimmune glomerulonephritis in these animals by purified preparations of both eicosapentaenoic acid and docosahexaenoic acid (DHA), given separately as their ethyl esters. When these two ω3 fatty acids are administered together, a synergistic effect is seen. Thus, optimal suppression of autoimmune disease in this model requires two ω3 fatty acids, which are provided in fish oil preparations. The basis for the synergistic effects of these fatty acids in terms of alterations in inflammatory cell function or eicosanoid synthesis has not been investigated.

At least some of the anti-inflammatory effects of dietary marine lipids may be attributed to alterations in leukocyte function. Dr. Sperling has reviewed several studies of leukocyte function based on ex vivo studies of human peripheral blood neutrophils and monocytes from patients ingesting dietary supplements of various lipids. In normal volunteers, marine lipids suppress 5-lipoxygenase pathway products from both neutrophils and monocytes, and they suppress production of PAF from monocytes. In addition, the chemotactic response of neutrophils to transmembrane agonists is suppressed by dietary fish oils.

The effects of dietary marine lipids on leukocyte functions in normal volunteers differ from effects seen with leukocytes from patients with rheumatoid arthritis. In the latter, inhibition of 5-lipoxygenase products is limited to LTB_4, suggesting inhibition of the epoxide hydrolase step. In rheumatoid patients, monocyte PAF production is also inhibited but curiously, neutrophil chemotaxis is augmented by the fish oil dose, rather than suppressed as in normal volunteers. These differences could be related to specific alterations of the disease, or to medications taken by patients. Purified dietary ω3 fatty acid ethyl esters differ in their ability to alter leukocyte function. In human volunteers, EPA suppressed neutrophil LTB_4 synthesis and response to chemotactic agents, whereas little or no changes in these functions were seen with DHA [3]. Nonetheless, it is interesting to note that DHA ethyl ester is more effective than EPA ethyl ester in alleviating murine autoimmune glomerulonephritis (4).

Alteration of cell function and inflammatory reactions by marine lipids is generally assumed to be related to incorporation of ω3 fatty acids into cell phospholipids. Dr. Snyder points out that the ether-linked phospholipid subclasses are enriched in polyunsaturated fatty acids both in fish species as well as in human cells. Incorporation of ω3 fatty acids into

P388D cells leads to greater enrichment of their ethanolamine plasmalogens than other phospholipid subclasses, primarily associated with reductions of oleic acid rather than other polyunsaturated fatty acids, a pattern differing from other systems, where incorporation of ω3 fatty acids often reduces the contents of the ω6 species, pointing to the variability in patterns of fatty acid composition after exposure to ω3 fatty acids in different systems. Dr. Snyder's laboratory has also documented that ω3 fatty acid enrichment is associated with augmentation in the rate of plasmalogen formation from alkylysoglycerophosphoethanolamine. These observations may account, at least in part, for the facile incorporation of ω3 fatty acids into ether-linked phospholipids.

It is generally acknowledged that the protein mediators collectively referred to as cytokines are important for the development of inflammation, immune reactions, and other biological processes. Dr. Dinarello and coworkers have shown that dietary marine lipid supplements in human volunteers reduce the ability of peripheral blood mononuclear cells to produce IL-1α, IL-1β, and TNF following stimulation by lipopolysaccharide. Inhibition of cytokine production was observed after 6 weeks of the fish oil diet but it was remarkable that this inhibition persisted for ten weeks after discontinuing the fish oil, at a time when mononuclear cell content of ω3 fatty acids had essentially returned to the levels observed prior to fish oil ingestion. At twenty weeks after discontinuing the fish oil, cytokine production returned to pre-treatment oil levels. The mechanism of suppression of cytokine formation by dietary fish oil is unclear, but the observations have important implications for the effects of ω3 fatty acids in human health and disease. Among other implications, these findings provide a further caveat against the use of crossover designs in clinical trials.

In conclusion, many experimental studies have provided evidence that incorporation of alternative fatty acids into tissues may modify inflammatory and immune reactions, and ω3 fatty acids in particular are potential therapeutic agents for inflammatory diseases.

References

1 Prickett JD, Robinson DR, Steinberg AD: Effects of dietary enrichment with eicosapentaenoic acid upon autoimmune nephritis in female NZB × NZW/F_1 mice. Arthritis Rheumatol 1983;26:133–139.

2 Prickett JD, Trentham DE, Robinson DW: Dietary fish oil augments the induction of arthritis in rats immunized with Type II collagen. J Immunol 1984;132:725–729.

3 Terano T, Seya A, Hirai A, Saito H, Tamura Y, Yoshida S: Effect of oral administration of highly purified eicosapentaenoic acid and docosahexaenoic acid on eicosanoid formation and neutrophil function in healthy subject; in Lands WEM (ed): Proceedings of the AOCS Short Course on Polyunsaturated Fatty Acids and Eicosanoids. Champaign, American Oil Chemist's Society, 1987, pp 133–147.

4 Robinson DR, Tateno S, Knoell C, Olesiak W, Xu L, Hirai A, Guo M, Colvin RB: Dietary marine lipids suppress murine autoimmune disease. J Int Med 1989; 225(suppl 1):211–216.

Simopoulos AP, Kifer RR, Martin RE, Barlow SM (eds): Health Effects of ω3 Polyunsaturated Fatty Acids in Seafoods. World Rev Nutr Diet. Basel, Karger, 1991, vol 66, pp 48–50

Cancer

Ritva R. Butrum, Mark J. Messina

National Cancer Institute, Bethesda, Md., USA

The focus of this session was to review findings related to the potential of ω3 fatty acid consumption to reduce cancer incidence and mortality. Since the first International Conference on the Health Effects of Polyunsaturated Fatty Acids in Seafoods was held in 1985, an impressive amount of work has been conducted in this area. ω3 fatty acids have now been shown either directly or indirectly through the administration of fish oils to inhibit experimental cancer in several animal models and may even have a role in controlling tumor metastasis. Much insight into the probable mechanism(s) by which ω3 fatty acids exert their anti-cancer effects has been gained. Presented below is a brief summary highlighting the main points discussed by each of the speakers in this session. Findings are expressed on the basis of comparison to high ω6 fatty acid diets.

Dr. Galli, the first speaker in this session, presented a paper entitled 'ω3 fatty acids and cancer – an overview'. Dr. Galli reviewed the evidence for and against each of the suggested mechanisms through which ω3 fatty acids may influence cancer. Overall, the evidence appears to be strongest for two mechanisms, enhanced immunity and modification of eicosanoid synthesis. Other potential mechanisms cited by Dr. Galli, but for which the evidence is much weaker, include lipid peroxidation, membrane fluidity, intercellular communication, hormone secretion, hormone responsiveness and serum lipid levels.

Dr. Cave, who followed Dr. Galli, in his paper 'ω3 fatty acid diet effects tumorigenesis in experimental animals' thoroughly reviewed the effects of high fish oil diets or diets supplemented with ω3 fatty acids on experimental cancer in rodents. Among the findings cited by Dr. Cave are the following: ω3 fatty acids increase mammary tumor latency, decrease tumor incidence and/or number and slow the development of transplanted tumors. With respect to dosage, however, Dr. Cave noted that small amounts of fish oil are relatively ineffective when added to diets containing more than 15% corn oil. Similar observations have been made in the case of experimental colon cancer; ω3 fatty acids decrease tumor incidence and/or number and slow the development/spread of transplanted colon tumors. ω3 fatty acids also depress colonic mucosa PGE_2 levels. Although less work has been conducted in comparison to mammary and colon cancer, ω3 fatty acids also appear to inhibit experimental pancreatic and prostatic cancer. Pancreatic atypical acinar cell nodule formation was found to decrease as the ratio of ω3 to ω6 fatty acids increased even when this ratio was altered several months after tumor induction. Serum levels of TBX_2, PGE_2 and 6-keto-F_{1a} levels in response to ω3 fatty acids decrease. Finally, in immune-suppressed rodents, fish oil suppresses transplanted prostatic tumor development.

The third speaker, Dr. Blackburn, shifted somewhat from cancer prevention to an examination of the role of ω3 fatty acids in controlling tumor metastasis. Dr. Blackburn, in his paper 'ω3 fatty acids and cancer metastasis' suggested that ω3 fatty acids have the ability to control breast tumor growth by reducing protein synthesis and by modifying tumor protein degradation. Dr. Blackburn also hypothesized that ω3 fatty acids, by limiting platelet aggregation and by acting as potent vasodilators, could control breast tumor metastasis. Dr. Blackburn stressed the potential use of ω3 fatty acids as an adjuvant therapy in controlling tumor metastasis.

Dr. Fernandes followed Dr. Blackburn with a discussion of his in vitro work using MCF-7 cells. Dr. Fernandes, in his paper 'ω3 fatty acids and MCF-7 cells' reported the following findings in mice fed fish oil and injected with MCF-7 cells: (1) tumor development, rate of growth and/or volume is decreased; (2) serum estrogen levels are decreased; (3) spleen and tissue PGE_2 is decreased; (4) tumor levels of linoleic and arachidonic acid are decreased while levels of EPA and DHA are increased; and (5) C-myc RNA expression in tumors is decreased.

Interest in the potential anti-cancer effects of ω3 fatty acids has increased dramatically during the past 5 years. Findings presented by these authors indicate the wide ranging effects of consuming ω3 fatty acids in comparison to diets composed primarily of ω6 fatty acids. ω3 fatty acids inhibit experimental cancer; precisely how this occurs is still unclear but most attention has focused on the pattern of eicosanoid synthesis that results from the mutual competition between ω6 and ω3 fatty acids for elongation and desaturation. Among the many research needs raised by these authors include the following: (1) to differentiate between the direct effects of ω3 fatty acids on initiated cells and/or tumors from general processes potentially related to cancer such as immune function; (2) to better characterize membrane composition changes in response to ω3 fatty acid consumption; (3) to more precisely determine the extent to which ω3 fatty acids inhibit platelet aggregation; and (4) to establish dose response relationships between dietary lipid consumption and estrogen receptor negative and positive tumor cell growth. As investigation into these issues and others proceeds, our understanding the role of ω3 fatty acids in preventing and/or treating cancer will become more clear.

Simopoulos AP, Kifer RR, Martin RE, Barlow SM (eds): Health Effects of ω3 Polyunsaturated Fatty Acids in Seafoods. World Rev Nutr Diet. Basel, Karger, 1991, vol 66, pp 51–71

ω3 Fatty Acids: Research Advances and Support in the Field Since June 1985 (Worldwide)

A.P. Simopoulos[a], *R.R. Kifer*[b], *A.A. Wykes*[c]

[a] The Center for Genetics, Nutrition, and Health-AAWH, Washington, D.C.;
[b] National Marine Fisheries Service, U.S. Department of Commerce, Charleston, S.C.;
[c] National Library of Medicine, National Institutes of Health, Bethesda, Md., USA

Introduction

Research on the role of ω3 fatty acids in growth and development and in health and disease has expanded extensively over the past 5½ years. The 1985 conference on the Health Effects of Polyunsaturated Fatty Acids in Seafoods, held June 24–26, 1985 in Washington, D.C. established the fact that ω3 fatty acids of marine origin, eicosapentaenoic acid (EPA) and docosahexaenoic acid (DHA), play important roles in prostaglandin metabolism, thrombosis and atherosclerosis, immunology and inflammation, and membrane function [1]. The conference participants recommended

- the support of research on the role of ω3 fatty acids in growth and development and in health and disease, and the mechanisms involved; and
- the establishment of a 'test materials program' to specifically define nutritional requirements throughout the life cycle, dose, and type of ω3 fatty acid in intervention studies and in clinical trials.

In December 1985 the U.S. National Institutes of Health (NIH) and the US Alcohol, Drug Abuse, and Mental Health Administration (ADAMHA) developed the first program announcement on 'Biological Mechanisms of Omega-3 Fatty Acids in Health and Disease States'. Additional program announcements followed that provided the impetus and the funding for support of research in the field (table 1). A major factor in the expansion of this research was the establishment of the 'test materials laboratory' within

Table 1. RFAs and PAs by NIH, ADAMHA, December 6, 1985 to April 17, 1987

Date	Title	Type	Institute
Dec. 6, 1985	Biological Mechanisms of ω3 Fatty Acids in Health and Disease States	PA	NCC (NIADDK, NINCDS, NIAID, NICHD, NIGMS, NEI, NIEHS, NIA, NIAAA, NIMH)
June 1986	Studies of ω3 Polyunsaturated Fatty Acids in Thrombosis and Cardiovascular Disease	RFA	NHLBI
Aug. 22, 1986	The Role of ω3 Polyunsaturated Fatty Acids in Cancer Prevention	PA	NCI
April 17, 1987	The Role of ω3 Polyunsaturated Fatty Acids in Cancer Prevention (re-issued)	PA	NCI
Oct. 22, 1987	Fatty Acid Derived Mediators of Inflammation	RFA	NIAID

the U.S. Department of Commerce, Division of Fisheries. The Biomedical Test Materials (BTM) program has been designed to produce a long-term, consistent supply of quality-assured/quality-controlled test materials to researchers, in order to facilitate the evaluation of the role that ω3 fatty acids play in health and disease. The BTM program was established in December 1986 through the cooperation of the National Institutes of Health (NIH), the Alcohol, Drug Abuse, and Mental Health Administration (ADAMHA), and the National Oceanic and Atmospheric Administration/Department of Commerce (NOAA/DOC). The Fish Oil Test Materials Program is administered by the Division of Nutrition Research Coordination in the Office of Disease Prevention, NIH.

In order to gain a complete understanding of the expansion of research on ω3 fatty acids, and to develop an overview of the conduct of research worldwide, it was necessary to have information on trends and analyses of research support and performance. This was accomplished by:

(1) examination of NIH and ADAMHA research support by review and analysis of data from the Computer Retrieval and Information on Scientific Projects (CRISP);

(2) retrieval of publications from 1984 through 1989 using MEDLINE® at the National Library of Medicine (NLM) NIH; and

(3) review of conferences and workshop proceedings held around the world.

It is hoped that this approach holds up a mirror to the scientific literature and reflects what the scientific community itself is signalling as noteworthy.

This paper consists of six sections. The first section titled 'Highlights of Research Results' is a review of the research accomplishments over the past 5–6 years in terms of the role of ω3 fatty acids in four major areas:

(a) Essentiality (which refers to the role of ω3 fatty acids in growth and development throughout the life cycle);

(b) Functions and Mechanisms;

(c) ω3 Fatty Acids as an Adjuvant to Drug Therapy;

(d) Factors to be Considered in Protocol Development in Clinical Investigation.

The second section describes the Biological Test Materials Program. The third section reviews the support of research by NIH and ADAMHA from 1984 to 1989. The fourth section on conferences and workshops highlights their contributions in setting research agendas that led to the expansion of research. The fifth section reviews the publications from 1984 to 1989. The publications are classified into four major categories: Human, Animal, In vitro, and Other. They are further classified into (1) Growth and Development; (2) Cardiovascular and Associated Diseases; (3) Inflammatory/Autoimmune Diseases; (4) Cancer; (5) Skin Diseases; and (6) Miscellaneous Diseases/Reviews/Other Studies and Reports. The section on conclusions completes the paper.

Highlights of Research Results

Prior to 1985 research focused on the modification of the eicosanoid system by dietary ω3 fatty acids [1]. Today it is clear that some of the resulting cellular functional changes might be related more directly to the modification of membrane phospholipid fatty acid composition and their effects on those proteins embedded in the lipid bilayer. This new understanding of the additional functions of ω3 fatty acids has expanded the investigations to include research on malaria, AIDS, ulcerative colitis, the control of cancer and the control of cancer metastasis [2–5]. Animal and human studies indicate that ω3 fatty acids are essential for visual functional development of the retina and of the visual cortex in early infancy, particularly for the premature infant [6].

Essentiality

Results from studies on rodents, nonhuman primates, premature newborns and in patients receiving prolonged treatment with total parenteral nutrition (TPN) appear to converge towards acceptance that ω3 fatty acids are essential for growth and development and health throughout the life cycle [6–11].

Further evidence that man evolved on a diet that was rich in ω3 fatty acids is provided from the composition of chicken egg yolk from free ranging chickens that has a ratio of ω6 to ω3 of 1.3, whereas the USDA standard egg has a ratio of 19.4 [12]. Wild vegetables are also richer in ω3 fatty acids than cultivated ones [13]. Modern agriculture and aquaculture have changed considerably the ω3 fatty acid content of both animals and plants, leading to a Western diet that is deficient in ω3 fatty acids relative to the one human beings originally relied upon for daily sustenance when our genetic patterns were established [14, 15].

Functions and Mechanisms

Recent studies show that ω3 fatty acids play important roles in influencing gene expression [16]; are involved in the control of a number of metabolic diseases [17]; are essential to human health [17]; and have synergistic effects with drugs in the treatment of a number of conditions and disease states [18–24]. Restriction in dietary intake and exercise, such as swimming, potentiate the effects of fish oil [23, 24]. Furthermore, the presence of olive oil or linseed oil in the diet increases the incorporation of EPA and DHA into leukocytes in vitro and further facilitates the beneficial effects of fish oil [25].

Studies on atherosclerosis, inflammation and autoimmune diseases indicate that ω3 fatty acids appear to decrease or inhibit risk/precipitating factors in the development of these diseases (table 2). A number of new and exciting findings have emerged since 1985. ω3 fatty acids decrease such factors as platelet-activating factor (PAF), platelet-derived growth factor (PDGF), interleukin I, and lipoprotein (a) [Lp(a)], and increase endothelial-derived relaxing factor (EDRF) and interleukin II. The fact that ω3 fatty acids lower interleukin I in patients with arthritis and increase interleukin II is one of the most exciting new findings [42].

Table 2. Effects of dietary ω3 fatty acids on factors and mechanisms involved in the development of inflammation, atherosclerosis and immune diseases

Reduce or inhibit risk/precipitating factors	*Increase beneficial/protective*
Arachidonic acid [26]	Prostacyclin formation ($PGI_2 + PGI_3$) [41]
Platelet aggregation [27]	LTB_5 [29]
Thromboxane formation [28]	Interleukin II [42]
Monocyte/macrophage function [29]	Endothelial-derived relaxing factor (EDRF) [43]
Leukotriene formation (LTB_4) [29]	Fibrinolytic activity [44]
Formation of platelet activating factor (PAF) [30]	Red cell deformability [45]
Toxic oxygen metabolites [31]	HDL [39]
Interleukin-1 formation [32]	
Formation of tumor necrosis factor (TNF) [32]	
Platelet-derived growth factor-like protein (PDGF) [33]	
Intimal hyperplasia [34]	
Blood pressure/BP response [35]	
VLDL, LDL [36]	
Triglycerides [37]	
Lipoprotein (a) [Lp(a)] [24, 38]	
Fibrinogen [39]	
Blood viscosity [40]	

The new findings in relation to interleukin metabolism and gene expression indicate that in addition to their major effects on prostaglandin metabolism, ω3 fatty acids have other far reaching effects on intracellular cell communication. These results further indicate that it is very important to know and eventually understand the numerous inter- and intracellular factors that are influenced by ω3 fatty acids as well as the specific mechanisms involved. It is this type of information that will enable us to design appropriate clinical trials and precisely define the dose of ω3 fatty acids to be utilized, the type of fatty acid, and length of intervention required for effective usage(s) regarding their beneficial effects, minus any possible adverse reactions.

There is evidence that the effects of ω3 fatty acids on interleukin I persist for at least 18 weeks after cessation of fish oils [46]. Therefore, crossover studies may not be appropriate because of these prolonged effects. The same is probably true for studies involving hypertension. More is needed to be known about the mechanisms involved in increasing interleukin II since injections of interleukin II in genetically hypertensive rats lowers blood pressure [47].

The finding that ω3 fatty acids lower serum Lp(a) is of great importance because Lp(a) is a most atherogenic protein and is genetically determined [24, 38, 48–50]. Patients heterozygous for familial hypercholesterolemia with elevated serum Lp(a) are at higher risk for myocardial infarction (MI) at the same or even lower cholesterol concentrations than those with lower serum Lp(a) levels (below 30 mg/dl) [49–51]. It is therefore important to evaluate all hypocholesterolemic drugs for any effects on serum Lp(a). Certainly one should not use drugs that while lowering serum cholesterol raise serum Lp(a) levels. Of interest is the fact that the Lp(a) lowering effect of ω3 fatty acids is seen only in individuals with elevated serum Lp(a) levels, and not in subjects with non-elevated serum Lp(a) values [38]. Here then is another piece of evidence for either an increased requirement for ω3 fatty acids in certain individuals similar to vitamin dependency syndromes i.e., vitamin B_6 – a single gene defect; or multiple interactions involving genes and nutrients where the presence of two or more genes (increased Lp(a) levels and familial hypercholesterolemia) increase the requirement of one nutrient (ω3 fatty acids) for normal physiologic function(s) to correct the pathologic state(s). It is also quite possible that 'the deficiency state' relative to dietary ω3 fatty acids may in some manner influence the expression of the Lp(a) gene.

As indicated above Lp(a) is genetically determined and has been found in all racial groups. The death rate for cardiovascular disease varies substantially around the world. It is very low in Japan but high in Finland with the U.S. somewhere in between. In these populations genetic differences have been shown to exist in terms of apolipoprotein E (ApoE). The Finns with the highest death rate from cardiovascular disease have a high frequency of $ApoE_4$ that is associated with high serum cholesterol levels. The Japanese on the other hand have a high frequency of $ApoE_2$ that is associated with low serum cholesterol levels and in Minnesota (U.S.) the most frequent ApoE is $ApoE_3$ associated with moderate cholesterol levels [52]. Therefore, it would be of great value to know the frequency of elevated serum Lp(a) in Finland, the U.S. and Japan, since these populations have divergent dietary habits along with various combinations of ω3, ω6,

Table 3. Conditions in which ω3 fatty acids have been shown to have synergistic effects with drugs

Human studies	*Animal studies*
Hypertension [18]	Autoimmune disorders [23]
Arthritis [42, 46]	
Psoriasis [21]	
Ulcerative colitis [4]	
Restenosis [22]	

ω9 and saturated fatty acid intakes. It would be of further interest to investigate if there are any differences in the levels of serum Lp(a) between 'fishing' and 'farming' villagers in Japan.

ω3 Fatty Acids as an Adjuvant to Drug Therapy

ω3 fatty acids in combination with drugs for the treatment of diseases is an area of immense interest because it opens a new field in nutrition research, 'Nutrients in the Control of Metabolic and Autoimmune Disorders', that includes cardiovascular disease, arthritis, lupus erythematosus, psoriasis, ulcerative colitis and cancer [18–24, 42, 46]. It is possible that the ingestion or administration of ω3 fatty acids may lead to a decrease both in the dose of ω3's and in the therapeutic drug dose for patients, as a result of replacing the fatty acids of phospholipids in the cell membranes which modify enzymes, drug/hormone receptors, and certain other important functional proteins (table 3). The latter could very well promote a decrease in the side effects of a given drug therapy [18–24].

Factors to be Considered in Protocol Development in Clinical Investigations

A number of randomized double-blind controlled studies have been carried out in disease states such as following angioplasty; rheumatoid arthritis; psoriasis; atopic dermatitis; Raynaud's phenomenon; ulcerative colitis; and bronchial asthma. Inconsistencies in the results of some of the above studies appear to be due to variations in the design and duration of the studies; selection of proper controls; failure to determine ω3 and ω6 fatty acid status along with the status of other dietary factors, e.g. vitamin E; and also failure to consider the genetic variations of multifactorial diseases and consequently the need to study subgroups for the various hypercholesterolemias, as well as hypertension, diabetes (whether NIDDM or IDDM), and arthritis; and failure to control for the amount of exercise and for body mass index.

In many studies involving animals, differences in animal species were not taken into consideration in the design of the study. For example, the rate of oxidation of ω3 fatty acid in the spontaneous hypertensive rat (SHR) is faster than in the normotensive Wistar Kyoto rat (WKR). As a result, less of ω3 fatty acids is found in the tissues of the SHR.

In undertaking studies it is therefore essential to know the initial phospholipid content of cells and to know about related factors that pertain to the disease under study. Also, genetic variations must always be considered in designing experimental protocols. Patients and controls must always be matched for age and sex since there is evidence that women respond to ω3 fatty acids differently than men. To illustrate, the same dose of ω3 fatty acids lowers thromboxane 2 (TXA_2) more in women than in men [53].

Many of the inconsistent findings on the effects of ω3 fatty acids on LDL cholesterol in human subjects appear to be associated with inadequate dietary control of saturated and polyunsaturated fatty acids during the control and intervention period, as well as differences in life-long dietary intake of fatty acids, differences in disease state, differences in types of hyperlipidemia, and differences in the type of diabetes (NIDDM or IDDM) studied, amount of fish oil used, length of time administered, etc.

Furthermore, it is essential to measure tissue levels of various metabolites, and other factors influenced by ω3 fatty acids, since plasma levels change faster after cessation of treatment than tissue levels.

The Biomedical Test Materials Program

The objectives of the BTM Program are to develop and provide test materials necessary to attain a thorough understanding of the mechanisms and interactions of ω3 fatty acids; and to stimulate the conduct of well-designed clinical studies in order to assist in the interpretation of the action of ω3 fatty acids.

There are over 100 researchers throughout the world who use the BTM in animal and human studies as well as for in vitro studies. Table 4 shows the types of test materials available. For the past 2 years, the BTM

Table 4. Test materials currently available

Test materials
ω3 ethyl ester concentrate, prepared from menhaden oil, bulk packed or soft-gel encapsulated (80% ω3 fatty acids including EPA and DHA)
Ethyl esters of corn, olive, or safflower oil, bulk packed or soft-gel encapsulated
Deodorized menhaden oil, bulk packed or soft-gel encapsulated
Commercial preparations of corn, olive, or safflower oil, soft-gel encapsulated only
EPA ethyl ester (> 95% ethyl esters), prepared from menhaden oil, packaged in 1–5 g aliquots
DHA ethyl ester (> 95% ethyl esters), prepared from menhaden oil, packaged in 1–5 g aliquots

have been used in a number of studies directed towards a wide range of biomedical research areas worldwide. These test materials are being utilized by U.S. and foreign investigators for research principally on arthritis, cardiovascular diseases, diabetes, blood coagulation, hyperlipidemias, autoimmune disorders, kidney disorders, lipid metabolism, malaria, cancer, and skin disorders.

The most recent announcement of the availability of > 95% pure EPA and DHA, as two additional test materials, was made in March 1990. Production rate of ω3 concentrate was increased recently by an additional 50% and is expected to be increased another 50% during the next few months in order to maintain the supply of test materials needed for NIH/ADAMHA approved research studies. The 1989 production of test materials was: 10,000 kg of refined menhaden oil, 1,200 kg ω3 ethyl ester concentrate, and 500 g each of EPA and DHA. The 1990 production is expected to be 20,000–25,000 kg of refined oil, 3,600 kg of ω3 ethyl ester concentrate, and 4 kg of EPA and 2.5 kg of DHA. The program is currently utilized by 106 basic and clinical researchers. In addition to NIH/ADAMHA approved researchers, another 17 requests for test materials were granted to researchers in industry and academic institutions.

Research Support

The Computer Retrieval of Information on Scientific Projects (CRISP) is a large database maintained and operated by the NIH. CRISP contains comprehensive scientific and selected administrative data on research carried out by the U.S. Public Health Service (PHS) or supported by PHS

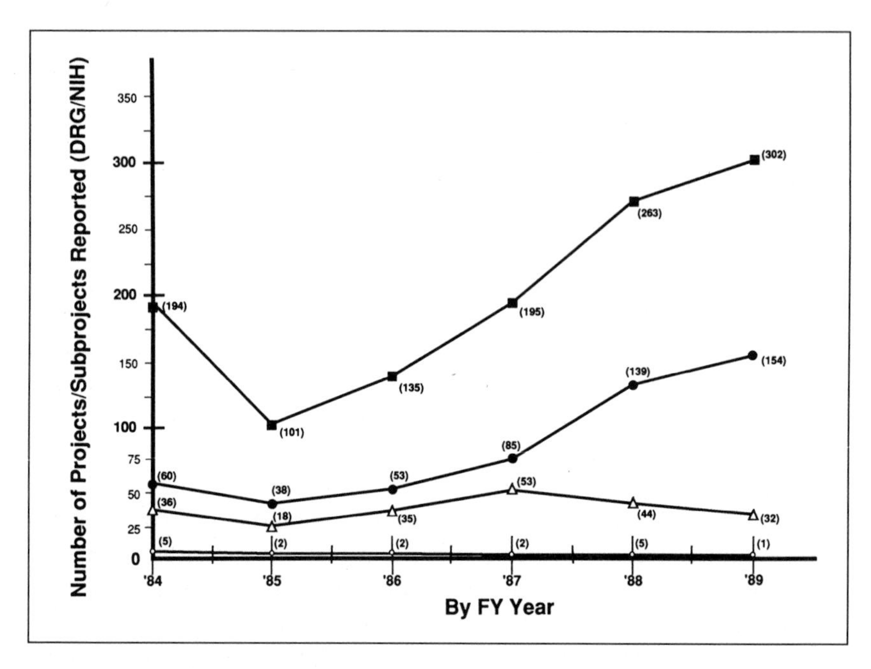

Fig. 1. Marine oil/fish oil and ω3 fatty acid studies supported by the National Institutes of Health (NIH/ADAMHA) from FY1984 to FY1989. ○ = Intramural (NIH) new projects; △ = Extramural (NIH) new projects; ● = extramural and intramural (NIH) total old and new projects; ■ = extramural and intramural (NIH) all studies using marine oils/fish oils or ω3 fatty acids for any purpose.

grants and contracts. Developed originally to meet the needs of NIH, it is an excellent, largely untapped resource for health information professionals at large, revealing new trends, methods, and techniques often before they appear in the published literature. CRISP uses its own controlled vocabulary, developed to permit indexing of new and active research areas. Queries can combine subject headings with a great variety of administrative data elements (e.g., research category or principal investigator's name, etc.). Output is available in a variety of formats and media. While information professionals cannot directly access the CRISP system, abridged CRISP records are merged into the FEDRIP (Federal Research in Progress) database, and FEDRIP is publicly accessible through DIALOG.

Computerized CRISP files go back to 1972. The capability of searching 'historically' is very useful in identifying research trends and areas receiving support. Reports of biomedical research in progress is a major information resource. While the published literature will continue to be foremost for the typical researcher, it is not the only information resource. In the area of biomedicine, the CRISP system can serve as a significant and unique repository of timely and critical information that deserves to be more widely utilized.

As can be seen from figure 1, the total number of projects and subprojects rose from 101 in FY1985 to 302 in FY1989. These numbers refer to all the projects funded by the National Institutes of Health (NIH) and ADAMHA, including those that are carried out in the intramural program of the NIH and ADAMHA using fish oils or ω3 fatty acids for any purpose. In other words, all these projects could include studies whose purpose to do research with fish oils could be primary, secondary, or tertiary. This rather startling increase most likely represents the response of the scientific community to the RFAs and PAs published by NIH and ADAMHA in 1985 and 1986 (table 1), and the availability of the Test Materials Program (table 4). The increase in NIH and ADAMHA research support is even better reflected in the line (curve) in figure 1 that indicates the number of *new* extramural projects rose from 18 in FY1985 to 53 in FY1987. It also shows that the peak was reached in 1987 and that in the absence of programmatic interest, there was a continuous steady decrease in the funding of new projects in FY1988 and FY1989 (fig. 1). No new RFA's or PA's have been published since 1987 (table 1).

Conferences and Workshops

Another evidence of the research interest in ω3 fatty acids and their relationship to ω6 and ω9 fatty acids is clearly seen by the large number of conferences and workshops that have been held since the memorable conference of June 1985 [1]. The growing number of conferences and workshops has resulted in bringing into focus important research findings, as well as setting research agendas for expanded research [17, 54]. Some of the conferences were held explicitly on ω3 fatty acids, while at the same time numerous workshops were included as part of national meetings such as the American Oil Chemists Society (AOCS), the Federation of American Societies of Experimental Biology (FASEB), or as part of an international meeting like the International Congress on Prostaglandins. Table 5 represents only a partial listing of just such meetings held in 1989.

Table 5. 1989 Conferences and workshops on ω3 polyunsaturated fatty acids (partial listing)

Title of meeting	Sponsor	Place	Date
Second Toronto Workshop on Essential Fatty Acids in Human Nutrition	University of Toronto	Toronto, Ontario	May 9–11, 1989
Effects of Dietary Fatty Acids on Serum Lipoproteins and Hemostasis	American Heart Association	Washington DC, USA	May 23–24, 1989
Advanced Research Workshop on Essential Fatty Acids and Infant Nutrition	NATO	Athens, Greece	May 26–27, 1989
Nutrition, Lipid Metabolism and Immunomodulations	University of Milan National Research Council	Castel Ivano, Italy	Sept. 22–23, 1989
Fats, Proteins and Micronutrients: Relevance of Animal Sources	University of Milan National Research Council	Castel Ivano, Italy	Oct. 4–5, 1989
Fish in Human Nutrition Research	University of Milan National Research Council	Castel Ivano, Italy	Oct, 17–18, 1989
New Perspectives on the Metabolic Role of ω3 Fatty Acids (Workshop)	Division of Research Grants NIH	Washington DC, USA	Oct. 18, 1989

Publications

The U.S. National Library of Medicine is the world's largest research library in a single scientific and professional field.

Its computerized biomedical literature file, MEDLINE® (MEDlars® onLINE) contains approximately 704,164 (February 1990) citations (many with abstracts) to biomedical reports and articles published in over 3,500 international biomedical journals during the current and two preceding years. Older material (back to 1966) is contained in the online MEDLINE® backfiles. The total file contains about 6.3 million records as of April 1990. Approximately 30,000 new citations are added each month.

The SDILINE® (Selective Dissemination of Information Online) file contains the current month's input of citations to the MEDLINE® database, as well as citations for the next printed edition of the monthly *Index Medicus®*.

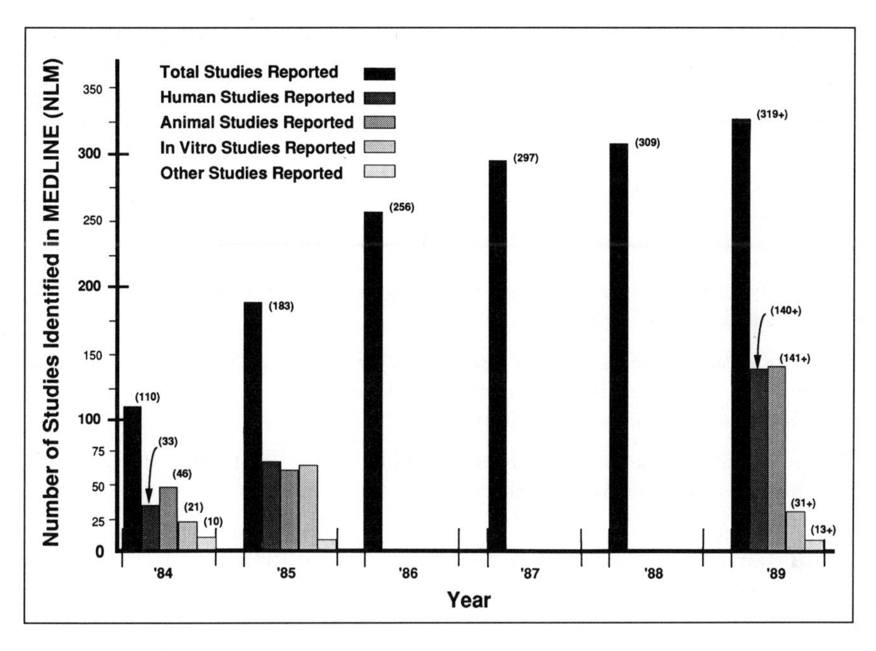

Fig. 2. Publications of marine oil/fish oil and ω3 fatty acid studies retrieved from MEDLINE® (NLM/NIH) from 1984 to 1989 (data as of January 10, 1990 from MEDLINE®. By June the total number of publications were 386).

MEDLARS II®, NLM's national interactive retrieval service containing the MEDLINE® and SDILINE® files, among others, is publicly available to users at minimum cost in the U.S. and overseas. Figure 2 illustrates the total number of publications per year (1984–1989) pertaining to fish oils and ω3 fatty acids that have been organized into 4 major categories: human studies; animal studies; in vitro studies; and other studies. In reviewing the citations and abstracts paper by paper, each publication was classified only once by the primary purpose of the study. In the event of mixed studies, the preference for classification was in the order of human studies then animal studies, followed by in vitro, then miscellaneous reports or studies. In 1984 there were only 110 publications, whereas by 1989 the number of publications in the field had almost tripled to 319 based on a January 10, 1990 MEDLINE® search output. By June 13, 1990 MEDLINE® search output, the publication numbers increased to 386. A total of 1,541 papers were published from 1984 to 1989 as determined via MED-

Table 6. Marine oils/fish oils and ω3 fatty acid research reports published in the biomedical literature (MEDLINE®, NLM)

Human studies (by selected research areas)		Number of MEDLINE® citations by year		
		1984	1985	1989[2]
1	Growth and development (GD) (early retinal and nerve develop., birth weights, premature births, etc.)	1	1	3+
2	Cardiovascular and associated diseases (CV)[1] (cardiac conditions, B.P., atherosclerosis, blood lipids, blood flow, platelet aggregation, etc.)	27	37	86+
3	Inflammatory/auto-immune diseases (INFL) (arthritis, immune diseases, other)	1	7	10+
4	Cancer (C) (breast, colon, pancreatic, liver, other)	0	1	4+
5	Skin diseases (SK) (psoriasis, eczema, other)	1	2	10+
6	Miscellaneous diseases/reviews/other studies and reports (MSC)[1] (malaria, membrane and metabolism studies, assays, etc.)	3	15	27+
Totals		33	63	140+

[1] Research areas with the highest number of studies reported.
[2] MEDLINE® search for 1989 was carried out on January 10, 1989, therefore the data for 1989 are incomplete since the input of citations to MEDLINE® searches lag by varying intervals of time.

LINE® searches completed by June 13, 1990. Another major difference, over the 6-year period, is the larger increase in the number of publications involving human studies, from 33 in 1984 to 140+ in 1989, so that by 1989 the number of publications involving primarily human or primarily animal studies was about the same, 140+ and 141+, respectively (fig. 2).

The publications for 1984, 1985 and 1989 were then subdivided into six subcategories or groupings: Growth and Development; Cardiovascular Diseases; Inflammation/Autoimmune Diseases; Cancer; Skin Diseases; and Miscellaneous Diseases. Within the publications involving human

Table 7. Marine oils/fish oils and ω3 fatty acid research reports published in the biomedical literature (MEDLINE®, NLM)

	Animal studies (by selected research areas)	Number of MEDLINE® citations by year		
		1984	1985	1989[2]
1	Growth and development (GD)	6	4	8+
2	Cardiovascular and associated diseases (CV)[1]	17	15	61+
3	Inflammatory/auto-immune diseases (INFL)	2	4	8+
4	Cancer (C)	2	2	15+
5	Skin diseases (SK)	0	1	3+
6	Miscellaneous diseases/reviews/other studies and reports (MSC)[1]	19	27	45+
Totals		46	53	141+

[1] Research areas with the highest number of studies reported.
[2] MEDLINE® search for 1989 was carried out on January 10, 1989, therefore the data for 1989 are incomplete since the input of citations to MEDLINE® searches lag by varying intervals of time.

studies (table 6), the cardiovascular disease subcategory included the largest number of studies in both 1984 and 1989, followed by publications in the miscellaneous diseases/reviews/other studies and reports group and the skin and inflammatory/autoimmune diseases grouping.

Under the animal studies category the cardiovascular disease grouping predominated in numbers, followed by studies in the miscellaneous group. Of interest is the fact that the subcategory cancer studies was in third position and gaining in numbers with time. Growth and development, and inflammation and autoimmune diseases groupings were in fourth position (table 7).

A major trend was noted in the numbers of publications in the in vitro studies category. There was a decrease from 60 in 1985 to 31+ in 1989. Within the in vitro publications category, miscellaneous studies led the way in frequency of occurrence, followed by cardiovascular and cancer studies (table 8).

It should be noted here that in September, 1989, the U.S. National Library of Medicine published a selective bibliography on the 'Health Benefits of Fish Oils' that included publications from January 1985 to May

Table 8. Marine oils/fish oils and ω3 fatty acid research reports published in the biomedical literature (MEDLINE®, NLM)

	In vitro studies (by selected research areas)	Number of MEDLINE® citations by year		
		1984	1985	1989[2]
1	Growth and development (GD)	0	3	0
2	Cardiovascular and associated diseases (CV)[1]	3	12	8+
3	Inflammatory/auto-immune diseases (INFL)	0	0	0
4	Cancer (C)	1	4	7+
5	Skin diseases (SK)	0	0	0
6	Miscellaneous diseases/reviews/other studies and reports (MSC)[1]	17	41	16+
Totals		21	60	31+

[1] Research areas with the highest number of studies reported.
[2] MEDLINE® search for 1989 was carried out on January 10, 1989, therefore the data for 1989 are incomplete since the input of citations to MEDLINE® searches lag by varying intervals of time.

1989 but was intentionally limited to human studies. Cited were 576 articles published in 155 journals from around the world. The bibliography is indicative of the explosive and expanding interest in the health benefits of fish oils and ω3 fatty acids to humans [55]. A yearly update of the bibliography with an expanded scope to include animal, in vitro, as well as human studies is planned for distribution via the National Technical Information Services (NTIS), Springfield, Va., U.S.A., and the National Library of Medicine, NIH, U.S.A.

Conclusions

The ω3 fatty acid sphere of influence is enormous since they appear to influence the fatty acid composition of membrane phospholipids of practically all cells in the body, and to influence gene expression. As can be seen from the topics of the papers and the authors of these Proceedings, research on ω3 fatty acids in health and disease (states) has expanded significantly around the world. Important new data indicate that ω3 fatty acids are essential for human health, and in the control of a number of clinical states.

Furthermore, research to date on ω3 fatty acids has assisted in focusing attention on the relationship of ω3 to ω6 and to ω9 fatty acids, and to saturated fats, at the molecular level, thus further enhancing the scientific basis of lipid and fatty acid nutrition and nutrition in general. Just as the studies with amino acids after World War II unravelled a number of aminoacidurias and genetic diseases due to enzyme deficiencies and inborn errors of metabolism, current and future research on fatty acids and their metabolic pathways will lead to the identification of genetic disorders as well as specific treatment regimens for them. New genetic diseases and syndromes will be assigned to either enzyme deficiencies or dependencies, and other abnormalities in ω3 and ω6 fatty acid metabolism.

The availability of the Biomedical Test Materials Program, the NIH Requests for Applications (RFA's) and Program Announcements (PA's) developed by NIH and ADAMHA and other government agencies worldwide, and the international conferences and workshops supported by NATO, industry and governments have contributed enormously to our current progress and accomplishments in the field of ω3 fatty acids.

References

1 Simopoulos AP, Kifer RR, Martin RE (eds): Health Effects of Polyunsaturated Fatty Acids in Seafoods. Proceedings from the Conference, June 1985. Orlando, Academic Press, 1986.

2 Levander OA, Ager AL, Morris VC, et al: Antimalarial activity of a marine ω3 free fatty acid concentrate (FFAC) in mice fed graded dietary levels of vitamin E (VE) (abstract); in Simopoulos AP, Kifer RR, Martin RE, Barlow SM (eds): Health Effects of ω3 Polyunsaturated Fatty Acids in Seafoods. World Rev Nutr Diet. Basel, Karger, 1991, vol. 66, p 535.

3 Sears B, Kahl P: Reduction of AZT-induced fatigue in AIDS by dietary supplementation with combinations of ω3 and ω6 fatty acids (abstract); in Simopoulos AP, Kifer RR, Martin RE, Barlow SM (eds): Health Effects of ω3 Polyunsaturated Fatty Acids in Seafoods. World Rev Nutr Diet. Basel, Karger, 1991, vol 66, p 534.

4 Stenson WF, Cort D, Beeken W, et al: A trial of fish oil supplemented diet in ulcerative colitis (abstract); in Simopoulos AP, Kifer RR, Martin RE, Barlow SM (eds): Health Effects of ω3 Polyunsaturated Fatty Acids in Seafoods. World Rev Nutr Diet. Basel, Karger, 1991, vol 66, p 533.

5 Man-Fan Wan J, Kanders B, Kowalchuk M, et al: ω3 fatty acids and cancer metastasis in humans; in Simopoulos AP, Kifer RR, Martin RE, Barlow SM (eds): Health

Table 1. Effect of various factors on red blood cell (wt %) phospholipid DHA in infants[1]

	DHA
Formula vs. human milk	↓
Preterm vs. term delivery	↓
High vs. low intakes of linoleic acid	↓ or –
High ratios of linoleic/linolenic acid	↓
Formula with 20–22 carbon ω3 fatty acids	↑

[1] All comparisons made between infants at the same postconceptional age.

rapidly in neural and retinal tissue. Furthermore, alternatives to mother's milk are available for feeding following birth. Unlike human milk, these currently contain neither AA nor DHA. Thus preterm infants may survive and grow without benefit of the normal maternal contribution of preformed very long chain polyunaturated fatty acids.

A variety of causes may lead to inadequate accumulation of DHA in the premature infant including (1) poor polyunsaturated fatty acid stores at birth, (2) formulas with high ratios of linoleic-to-linolenic acid, (3) formulas with very high concentrations of linoleic acid, and (4) inefficient conversion of linolenic to docosahexaenoic acid (table 1). At the same time that these risk factors for poor DHA status have been recognized, evidence has been accumulating that DHA plays an important role in retinal physiology, retinal function, and learning.

As noted earlier there is ample evidence that standard infant feeding practices result in biochemically significant reductions in red blood cell, neural, and retinal DHA following premature delivery. In attempting to determine if these biochemical losses represent a functionally significant effect in infants, several approaches have been taken: (1) to contrast the biochemistry seen in ω3 deficient primates with that observed in infants fed standard infant formula; (2) to feed preformed DHA and successfully maintain red blood cell and plasma phospholipid DHA in the range seen in cord blood/human milk-fed infants; and (3) to determine if functional improvements in visual acuity and retinal physiology occur with DHA supplementation.

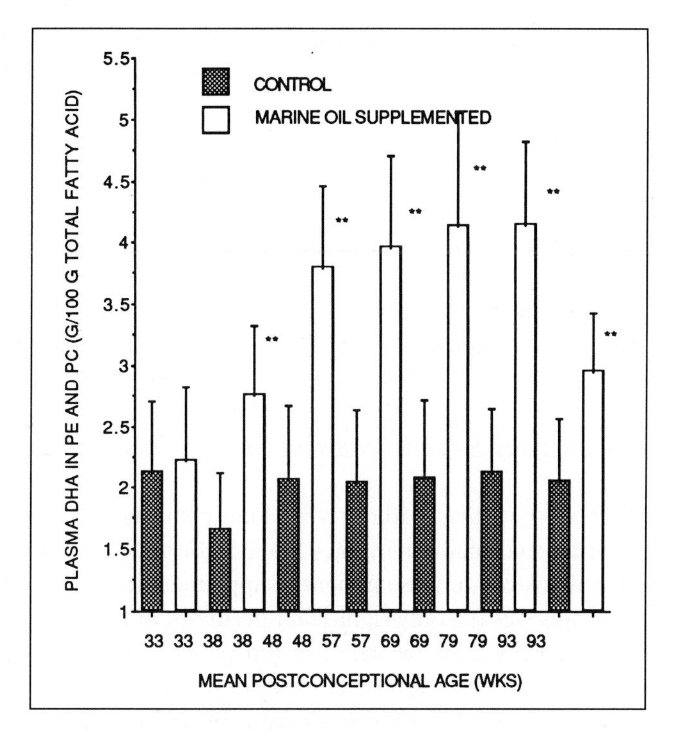

Fig. 2. Plasma PC and PE docosahexaenoic acid (DHA) (g/100 g total fatty acid) in infants fed formula with linolenic acid (control) or the same formula supplemented with very long chain ω3 fatty acids (marine oil-supplemented) at enrollment ($\overline{X}$ 33 weeks postconception, $\overline{X}$ postnatal age 3 weeks) and throughout infancy (31). Control and marine oil-supplemented formulas were fed through 79 weeks postconception (most infants continued to receive their assigned formula for another 2–3 weeks). Thereafter, infants received whole cows' milk as part of a mixed diet. Values are shown as mean ± SD. Marine oil-supplemented vs. control formula: ** $p < 0.001$.

ω3 Deficient Rhesus Monkeys versus Preterm Infants

Premature infants are more likely to be ω3 deficient than rhesus monkeys at birth because they are likely to receive less preformed DHA during gestation. On the other hand, most infant formulas contain adequate amounts of linolenic acid in contrast to experimental diets fed to monkeys to produce ω3 deficiency.

Four weeks after birth, the concentration of DHA in plasma phospholipids of preterm infants [30] was closer to that of ω3 deficient monkeys (1.8 vs. 0.8%) than to their controls (6.5%). While DHA decreased further with time in ω3 deficient monkeys, this did not occur in infants who received 2.5% of energy from linolenic acid. Nevertheless, DHA in plasma phospholipids did not increase during infancy unless the formula was supplemented with DHA (0.1% of energy) and eicosapentaenoic acid (EPA, 20:5 ω3, 0.2% of energy) (fig. 2).

Interventions to Prevent Declines in Red Blood Cell Phospholipid DHA
Declines in red blood cell and plasma DHA following birth can be prevented by adding small amounts of fish oil containing DHA and EPA to the diet as has been shown repeatedly [31–35].

Is DHA a Conditionally Essential Nutrient for Premature Infants?

Early data suggest that diets without very long chain ω3 fatty acids (DHA and EPA) limit visual acuity in infancy [34, 35]. Despite the complex neonatal and perinatal history of these patients, supplementation with very long chain ω3 fatty acids was the only significant predictor of red blood cell and plasma DHA and EPA, and the relationship between ω3 fatty acid status and acuity was found even in infants who did not receive DHA and EPA supplementation [35]. Other preliminary data suggest that electroretinogram responses in preterm infants are less than optimal as the result of ω3 deficient diets [36]. Uauy et al. [37] reported that fish oil supplementation results in visual evoked potential acuity like that of infants fed human milk by 57 weeks postconception. Thus, data accumulated so far suggest that limited ω3 accumulation limits visual acuity and the physiologic responses of the retina in some preterm infants. These studies, as yet published only in abstract form, provide the first suggestion that preterm infants, like other primates, experience alterations in visual acuity and electroretinogram responses when ω3 accumulation is limited.

Research Needs Related to Appropriateness of ω3 Supplementation for Preterm Infants

Preliminary studies with preterm infants demonstrate that there are clear differences in visual function (acuity, visual evoked potentials) in infants receiving very long chain ω3 fatty acids compared to controls fed reasonable levels of linolenic acid [34, 35, 37]. By these criteria there is an apparent need for preformed dietary very long chain ω3 fatty acids. Nevertheless, the appropriateness of fish oil supplementation as a means of supplying very long chain ω3 fatty acids has not been fully evaluated with regard to safety. Concerns fall primarily into two categories: (1) undesirable components of fish oil which may be transmitted to the infant along with DHA; e.g., oxidative metabolites, impurities of various sorts, minor organic components other than fatty acids, and other fatty acid components such as 22:1, and (2) the possible oversupply of DHA and EPA. The first category has been reviewed only recently by Salem [38]. The second will be discussed briefly here.

The adverse responses to fish oil feeding noted in adults have included increased bleeding times [39, 40] and reduced polymorphonuclear leukocyte and monocyte chemotaxis and reduced ability to generate oxidative metabolites involved in the killing of bacteria [41, 42]. Ultimately, the possibility of adverse effects of fish oil feeding in infants must be assessed in infants since adults have generally consumed much higher intakes (g/kg) of fish oil although often for shorter periods of time. Bleeding times are currently being determined by Uauy et al. and by Carlson et al. It does not appear that the low levels of supplementation used have introduced any clinically relevant changes in clotting times [Uauy, personal communication]. Furthermore, polymorphonuclear (PMN) leukocyte superoxide production and rate of generation have not been altered by the low level of ω3 supplementation fed in our study compared to controls in the first year of life [unpublished data].

Even low levels of fish oil reduce the arachidonic acid (AA) content (g/100 g of total fatty acid) of plasma and membrane phospholipids [unpublished data] and the absolute quantity of plasma phospholipid AA (μg/ml) (fig. 3). Too great a reduction could have adverse consequences in developing infants since AA is also found in high concentrations in all membranes and is the source of metabolically important prostaglandins and leukotrienes. The preterm infant may be at risk for inadequate AA accumulation for the same reasons as given for inadequate DHA accumulation (table 1) with the exception that 20–22 carbon ω3 PUFA decreases rather than increases AA. Although some decline in membrane phospholipid AA in the last trimester appears to be physiologically normal, preterm infants have lower red blood cell and plasma phospholipid AA than term infants when they are compared at the same postconceptional ages. The declines in plasma AA (μg/ml) occur in infants fed the usual preterm formula (our control) or that containing additional very long chain ω3 fatty

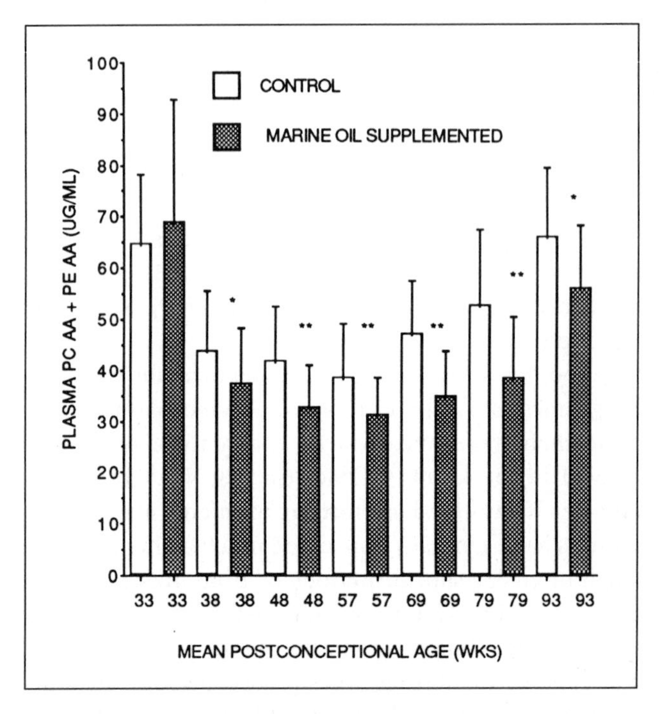

Fig. 3. Plasma PC and PC arachidonic acid (AA) (μg/ml) in infants fed regular formula with linolenic acid (control) or the same formula supplemented with very long chain ω3 fatty acids (marine oil-supplemented) beginning at enrollment (33 weeks postconception): Plasma PC AA (μg/ml) + plasma PE AA (μg/ml) = plasma PC + PE AA (μg/ml). Values are shown as mean ± SD. Marine oil-supplemented vs. control: * $p < 0.01$; ** $p < 0.001$.

acids (ω3 supplemented) (fig. 3). Prior administration of Intralipid (fig. 4) rather than postnatal age was a major predictor of plasma AA at enrollment. The effect of all these factors appears to be additive with premature infants having as little as half as much AA in red blood cell phospholipids as term, human milk-fed infants of the same postconceptional age.

Hrboticky et al. [11] have shown that AA in plasma and red blood cell phospholipids poorly reflect hepatic and neural changes in formula-fed and sow milk-fed piglets. However, major declines in AA of plasma and red blood cell phospholipids may be indicative of physiologically significant phenomena. Our early data suggest that red blood cell and plasma phospholipid AA positively predict growth and cognitive development

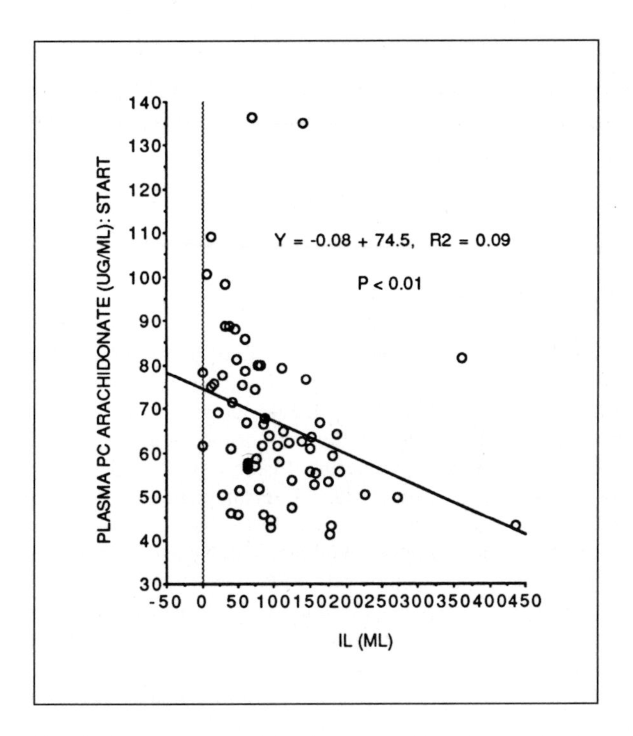

Fig. 4. Plasma phosphatidylcholine arachidonic acid (AA) at enrollment. Stepwise regression included Intralipid (IL, ml) received prior to enrollment, postnatal age (days) at enrollment, gestational age (weeks), red blood cells administered before enrollment, log of hours of mechanical ventilation, and gravida. Only intralipid administration significantly predicted plasma AA (μg/ml) at enrollment with higher levels of Intralipid associated with lower plasma AA.

(Bayley MDI, Fagan Infantest) in the first year of life [43]. Poor or no catch-up growth of preterm infants in the first year of life and differences in body composition have been noted previously. Georgieff et al. [44] reported that mean weight and length percentiles consistently fell from birth to 57 weeks and from 57 to 92 weeks postconception in a group of 30 infants, none of whom received very long chain ω3 fatty acids. Our data suggest that catch-up growth occurs from birth to 57 weeks postconception with a subsequent fall-off in length and weight similar to that reported by Georgieff et al. [44] during the remainder of infancy [unpublished data]. AA status or some related or unrelated nutritional factor may account for poor catch-up growth in the last half of infancy.

Currently, both AA and DHA status are being studied in the preterm infant. While DHA supplementation has been shown to improve visual function of these infants, it is necessary to know if it is desirable to simultaneously improve AA status as well. Here the answers are not so simple since reasonably priced sources of AA are not readily available.

References

1 O'Brien JS, Fillerup DL, Mean JF: Quantification of fatty acid and fatty aldehyde composition of ethanolamine choline and serine phosphoglycerides in human cerebral gray and white matter. J Lipid Res 1964;5:329–330.
2 Anderson RE, Maude MB, Zimmerman W: Lipids of ocular tissues. X. Lipid composition of subcellular fractions of bovine retina. Vision Res 1975;15:1087–1090.
3 Sinclair AJ, Crawford MA: The accumulation of arachidonate and docosahexaenoate in the developing rat brain. J Neurochem 1972;19:1753–1758.
4 Sinclair AJ, Crawford MA: The incorporation of linolenic acid and docosahexaenoic acid in liver and brain lipids of developing rats. FEBS Lett 1972;26:127–129.
5 Clandinin MT, Chappell JE, Leong S, et al: Intrauterine fatty acid accretion rates in human brain: Implications for fatty acid requirements. Early Hum Dev 1980;4: 121–129.
6 Martinez M, Ballabriga A: Effects of parenteral nutrition with high doses of linoleate on the developing human liver and brain. Lipids 1987;22:133–138.
7 Martinez M, Ballabriga A, Gil-Gibernou JJ: Lipids of the developing human retina: I. Total fatty acids, plasmalogens, and fatty acid composition of ethanolamine and choline phosphoglycerides. J Neurosci Res 1988;20:484–490.
8 Carlson SE, Rhodes PG, Ferguson MG: Docosahexaenoic acid status of preterm infants at birth and following feeding with human milk or formula. Am J Clin Nutr 1986;44:798–804.
9 Putnam JC, Carlson SE, DeVoe PW, et al: The effect of variations in dietary fatty acids on the fatty acid composition of erythrocyte phosphatidylcholine and phosphatidylethanolamine in human infants. Am J Clin Nutr 1982;36:106–114.
10 Peeples J, Carlson S, Cooke R, et al: Long-term supplementation of preterm infants with docosahexaenoate (DHA) and eicosapentaenoate (EPA): Effect on red blood cell phosphatidylethanolamine (PE) fatty acid composition. Fed Proc 1989;3: A1056.
11 Hrboticky N, MacKinnon MJ, Innis SM: Effect of a vegetable oil formula rich in linoleic acid on tissue fatty acid accretion in the brain, liver, plasma and erythrocytes of infant piglets. Am J Clin Nutr 1990;51:173–182.
12 Forrest GL, Futterman S: Age-related changes in the retinal capillaries and the fatty acid composition of retinal tissue of normal and essential fatty acid deficient rats. Invest Ophthalmol 1972;11:760–764.
13 Mohrhauer H, Holman RT: Alteration of the fatty acid composition of brain lipids by varying levels of dietary essential fatty acids. J Neurochem 1963;10:523–530.
14 Sanders TAB, Mishy M, Naismith DJ: The influence of a maternal diet rich in

linoleic acid on brain and retinal docosahexaenoic acid in the rat. Br J Nutr 1984;51: 57–66.
15 Galli C, Trzeciak HI, Paoletti R: Effect of dietary fatty acids on the fatty acid composition of brain ethanolamine phosphoglyceride: Reciprocal replacement of n–6 and n–3 polyunsaturated fatty acids. Biochim Biophys Acta 1971;248:449–454.
16 Lamptey MS, Walker BL: A possible essential role for dietary linolenic acid in the development of the young rat. J Nutr 1976;106:86–93.
17 Yamamoto N, Saitoh M, Moriuchi A, et al: Effect of dietary alpha linolenate/linoleate balance on brain lipid compositions and learning ability in rats. J Lipid Res 1987;28:144–151.
18 Enslen M, Nouvelot A, Milon H: Effects of n–3 polyunsaturated fatty acid deficiency during development in the rat: Functional effects; in Lands WEM (ed): Proceedings of the AOCS Short Course on Polyunsaturated Fatty Acids and Eicosanoids. Champaign, American Oil Chemists Society, 1987, pp 495–497.
19 Benolken RM, Anderson RE, Wheeler TG: Membrane fatty acids associated with the electrical response in visual excitation. Science 1973;182:1253–1254.
20 Wheeler TG, Benolken RM, Anderson RE: Visual membranes: Specificity of fatty acid precursors for the electrical to illumination. Science 1975;188:1312–1314.
21 Okuyama H, Saitoh M, Naito Y, et al: Re-evaluation of the essentiality of alpha-linolenic acid in rats; in Lands WEM (ed): Proceedings of the AOCS Short Course on Polyunsaturated Fatty Acids and Eicosanoids. Champaign, American Oil Chemists Society, 1987, pp 296–300.
22 Cho Es, Kolder HE, Wertz PE, et al: Electroretinography and retinal phosphatidylcholine fatty acids in omega-3 fatty acid deficient rats. Brighton, XIII International Congress of Nutrition, 1985, p 104.
23 Leat WMF, Curtis R, Millichamp NJ, et al: Retinal function of rats and guinea pigs reared on diets low in essential fatty acids and supplemented with linoleic and linolenic acids. Ann Nutr Metabol 1986;30:166–174.
24 Neuringer M, Connor WE, Luck SL: Omega-3 fatty acid deficiency in rhesus monkeys: Depletion of retinal docosahexaenoic acid and abnormal electroretinograms. Am J Clin Nutr 1985;43:706.
25 Connor WE, Neuringer M, Lin D: The incorporation of docosahexaenoic acid into the brain of monkeys deficient in omega-3 essential fatty acids. Clin Res 1985;33: 598A.
26 Neuringer M, Connor WE, Luck SJ: Suppression of ERG-amplitude by repetitive stimulation in rhesus monkeys deficient in retinal docosahexaenoic acid. Invest Ophthalmol Vis Sci 1985;26(suppl 3):31.
27 Neuringer M, Connor WE, Daigle D, et al: Electroretinogram abnormalities in young rhesus monkeys deprived of omega-3 fatty acids during gestation and postnatal development or only postnatally. Invest Ophthalmol Vis Sci 1988;29(suppl 3): 145.
28 Neuringer M, Conner WE, Van Petten C, et al: Dietary omega-3 fatty acid deficiency and visual loss in infant rhesus monkeys. J Clin Invest 1984;73:272–276.
29 Salem N Jr, Kim H-Y, Yergey JA: Docosahexaenoic acid, membrane function and metabolism; in Simopoulos AP, Kifer RR, Martin RE (eds): Health Effects of Polyunsaturated Fatty Acids in Seafoods. New York, Academic Press, 1986, pp 263–317.

30 Carlson SE, Cooke RJ, Peeples J, et al: Docosahexaenoic acid (DHA) supplementation of preterm infants. Pediatr Res 1988;23:481A.
31 Carlson SE, Cooke RJ, Peeples JM, et al: High linolenic acid feeding accompanied by variations in docosahexaenoic and eicosapentaenoic acids in very low birth weight infants. Am J Clin Nutr (submitted).
32 Liu C-CF, Carlson SE, Rhodes PG, et al: Increase in plasma phospholipid docosahexaenoic and eicosapentaenoic acids as a reflection of their intake and mode of administration. Pediatr Res 1987;22:292–296.
33 Koletzko B, Schmidt E, Bremer HJ, et al: Dietary long chain polyunsaturates for premature infants. J Pediatr Gastroenterol Nutr 1987;6:997–998.
34 Carlson S, Cooke R, Werkman S, et al: Docosahexaenoate (DHA) and eicosapentaenoate (EPA) supplementation of preterm infants: Effects on phospholipid DHA and visual acuity. Fed Proc 1989;3:A1056.
35 Carlson SE, Cooke RJ, Peeples JM, et al: Docosahexaenoate (DHA) and eicosapentaenoate (EPA) status of preterm infants: Relationship to visual acuity in n–3 supplemented and unsupplemented infants. Pediatr Res 1989;25:285A.
36 Uauy R, Birch D, Birch E, et al: Effect of omega-3 fatty acid (FA) on retinal function of very low birth weight neonates. Fed Proc 1989;3:A1247.
37 Uauy R, Birch D, Birch E, et al: Effect of dietary omega-3 fatty acids (FA) on eye and brain development in very low birth weight neonates (VLBWN). Proceedings of the II International Conference on the Health Effects of ω3 Polyunsaturated Fatty Acids in Seafoods, Washington, D.C. 1990.
38 Salem N Jr: Omega-3 fatty acids: Molecular and biochemical aspects, in Spillar GA, Scala J (eds): New Protective Roles for Selected Nutrients. Current Topics in Nutrition and Disease. New York, Alan R. Liss, 1989, vol 22, pp 188–190.
39 Dyerberg J, Bang HO: Haemostatic function and platelet polyunsaturated fatty acids in Eskimos. Lancet 1979;2:433–435.
40 Hirai A, Terano T, Saito H, et al: Clinical and epidemiological studies of eicosapentaenoic acid in Japan; in Lands WEM (ed): Proceedings of the AOCS Short Course on Polyunsaturated Fatty Acids and Eicosanoids. Champaign, American Oil Chemists' Society, 1987, pp 9–24.
41 Lee TH, Hoover RL, Williams JD, et al: Effects of dietary enrichment with eicosapentaenoic and docosahexaenoic acids on in vitro neutrophil and monocyte leukotriene generation and neutrophil function. New Engl J Med 1985;312:1217–1224.
42 Strasser T, Fischer S, Weber PC: Leukotriene B5 is formed in human neutrophils after dietary supplementation with eicosapentaenoic acid. Proc Natl Acad Sci USA 1985;82:1540–1543.
43 Carlson SE, Werkman SH, Peeples JM, et al: Phospholipid arachidonic acid is a positive predictor of growth and cognitive performance in infancy following very low birth weight delivery (abstract). Proceedings of the II International Conference on the Health Effects of Omega-3 Polyunsaturated Fatty Acids in Seafoods, Washington, D.C., 1990.
44 Georgieff MK, Mills MM, Zempel CE, et al: Catch-up growth, muscle and fat accretion and body proportionality of infants one year after newborn intensive care. J Pediatr 1989;114:288–292.

S.E. Carlson, MD, Newborn Center, Departments of Pediatrics and Obstetrics and Gynecology, The University of Tennessee, Memphis, TN 38163 (USA)

Simopoulos AP, Kifer RR, Martin RE, Barlow SM (eds): Health Effects of ω3 Polyunsaturated Fatty Acids in Seafoods. World Rev Nutr Diet. Basel, Karger, 1991, vol 66, pp 87–102

Developmental Profiles of Polyunsaturated Fatty Acids in the Brain of Normal Infants and Patients with Peroxisomal Diseases: Severe Deficiency of Docosahexaenoic Acid in Zellweger's and Pseudo-Zellweger's Syndromes

Manuela Martinez

Autonomous University of Barcelona, Hospital Infantil Vall d'Hebron, Barcelona, Spain

Introduction

Neuronal membranes are very rich in ω3 and ω6 polyunsaturated fatty acids (PUFA), especially 22:6 ω3, 20:4 ω6 and, to a lesser extent, 22:4 ω6 and 22:5 ω6. Nerve endings and photoreceptor cells have a high proportion of 22:6 ω3 (docosahexaenoic acid), suggesting an important role of this fatty acid in neuronal transmission [1] and the visual process [2]. The myelin sheet, on the other hand, is particularly abundant in very long chain (VLC) saturated and monounsaturated fatty acids (24:0, 24:1, 26:0, 26:1). During development, the proportion of docosahexaenoic acid increases markedly in the ethanolamine phosphoglycerides of the human cerebrum [3] and retina [4], whereas that of arachidonic acid (20:4 ω6) decreases. However, the brain is a very complex mixture of different structures and lipids and, from the nutritional point of view, it is the total amount of ω3 and ω6 fatty acids that matters.

A quantitative study of the different brain PUFA throughout development is, therefore, the best way to assess the nutritional requirements of the infant brain. Such quantitative data are also invaluable as controls for pathological conditions. If performed carefully, a total fatty acid quantification gives enough information without the need for time-consuming fractionation techniques, which inevitably lead to substantial lipid losses.

Table 1. Total fatty acids in the human developing cerebrum

	26–32 weeks (n = 5)	33–37 weeks (n = 6)	38–42 weeks (n = 10)	0–3 months (n = 5)	4–9 months (n = 5)	1–2 years (n = 3)	4–8 years (n = 3)
14:0	1,262 ± 32	1,252 ± 83	1,220 ± 54	890 ± 114	652 ± 25	640 ± 31	693 ± 57
16:0	15,944 ± 330	20,519 ± 1,227	23,689 ± 1,061	22,122 ± 839	22,265 ± 749	24,044 ± 1,072	24,682 ± 298
18:0	11,195 ± 308	16,302 ± 1,104	20,346 ± 875	22,125 ± 1,282	26,797 ± 644	32,031 ± 2,167	33,514 ± 528
18:1 ω9	6,006 ± 134	7,765 ± 559	9,301 ± 338	10,213 ± 548	13,813 ± 623	20,542 ± 1,738	26,616 ± 892
18:1 ω7	1,917 ± 65	2,317 ± 170	2,838 ± 157	3,194 ± 213	3,911 ± 158	4,974 ± 581	6,382 ± 344
18:2 ω6	120 ± 7	193 ± 37	205 ± 16	311 ± 71	524 ± 114	697 ± 139	861 ± 131
20:1 ω9	195 ± 12	203 ± 19	222 ± 13	299 ± 43	442 ± 43	868 ± 123	1,682 ± 82
20:3 ω9	252 ± 24	237 ± 25	293 ± 37	616 ± 108	457 ± 91	316 ± 27	338 ± 58
20:3 ω6	261 ± 21	507 ± 41	770 ± 191	974 ± 82	1,407 ± 98	1,660 ± 115	1,544 ± 134
20:4 ω6	5,083 ± 137	6,137 ± 251	7,396 ± 326	7,604 ± 420	10,017 ± 687	10,907 ± 748	10,003 ± 388
22:4 ω6	2,451 ± 83	3,604 ± 138	4,428 ± 171	4,447 ± 118	5,514 ± 331	7,578 ± 756	7,148 ± 157
22:5 ω6	1,485 ± 118	1,790 ± 129	1,919 ± 143	1,955 ± 242	2,326 ± 153	1,689 ± 258	1,488 ± 293
22:5 ω3	59 ± 6	79 ± 7	138 ± 20	184 ± 35	183 ± 21	280 ± 25	221 ± 45
22:6 ω3	3,314 ± 99	4,315 ± 212	5,854 ± 273	6,082 ± 208	6,988 ± 224	9,582 ± 275	9,692 ± 111
22:0	17 ± 4	36 ± 8	57 ± 8	135 ± 15	302 ± 42	486 ± 77	456 ± 27
24:0	12 ± 3	38 ± 6	85 ± 10	183 ± 34	841 ± 144	1,948 ± 366	2,137 ± 300
24:1 ω9	43 ± 3	68 ± 9	125 ± 11	321 ± 48	1,045 ± 177	3,268 ± 570	5,760 ± 366
24:1 ω7	ND	6 ± 2	8 ± 4	49 ± 9	206 ± 42	604 ± 89	687 ± 92
26:0	ND	ND	4 ± 2	25 ± 5	110 ± 14	258 ± 50	246 ± 63
26:1 ω9	ND	ND	ND	21 ± 6	126 ± 23	505 ± 120	799 ± 48
26:1 ω7	ND	ND	3 ± 3	44 ± 8	226 ± 44	688 ± 134	715 ± 106
TFA	51,895 ± 558	68,210 ± 3,554	82,195 ± 2,910	84,884 ± 2,901	101,053 ± 2,763	128,581 ± 9,637	142,805 ± 1,416
TP	2,748 ± 86	4,136 ± 232	5,472 ± 203	6,727 ± 453	9,766 ± 553	14,634 ± 1,118	16,567 ± 256
16DMA/16	0.092 ± 0.001	0.094 ± 0.003	0.098 ± 0.003	0.106 ± 0.003	0.154 ± 0.01	0.229 ± 0.013	0.212 ± 0.014
18DMA/18	0.089 ± 0.003	0.115 ± 0.005	0.132 ± 0.003	0.159 ± 0.007	0.177 ± 0.005	0.173 ± 0.006	0.156 ± 0.005

All the values (except for the ratios) are nmol/g of brain wet tissue.

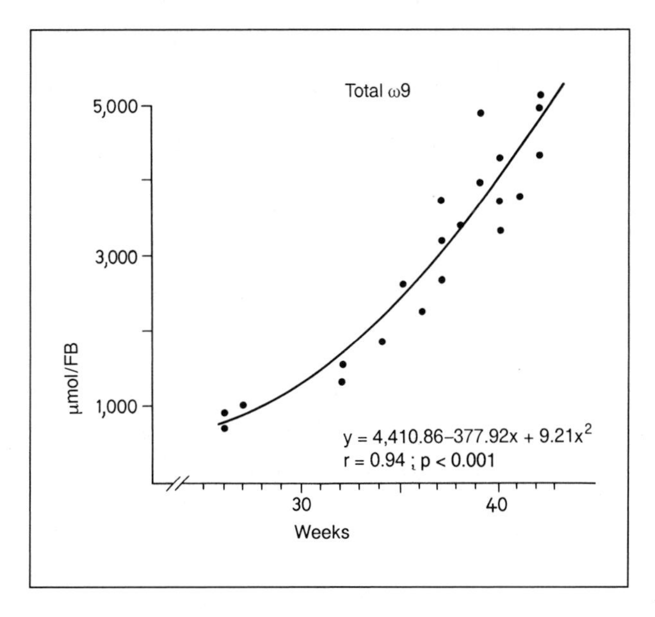

Fig. 3. Total amount of ω9 fatty acids per whole cerebrum during prenatal life (x, y and n as in previous figures).

Prenatal Development. Figures 1–3 show the total accretion of fatty acids of the three families, ω3, ω6 and ω9, per whole cerebrum, during the prenatal period. The prenatal population was constituted by 21 newborn infants with gestational ages ranging from 26 to 42 weeks, who died during the first hours of life, before taking any food. The fatty acids were quantified in a homogenate of whole cerebrum, the total amounts being calculated from the actual weight of each individual forebrain. This gave curves with little dispersion and a very high statistical significance. It can be seen from the figures that the total accretion of ω3, ω6 and ω9 fatty acid is of a quadratic type during prenatal life, the increase being most rapid towards the end of gestation.

Postnatal Development. The postnatal accretion of total ω3, ω6 and ω9 unsaturated fatty acids is depicted in figures 4–6, respectively. The equations in the figures (computed exclusively with the postnatal values) show a much nearer linear increase of ω3 and ω6 fatty acids than during prenatal life, which is totally so in the case of ω9 fatty acids. However, at

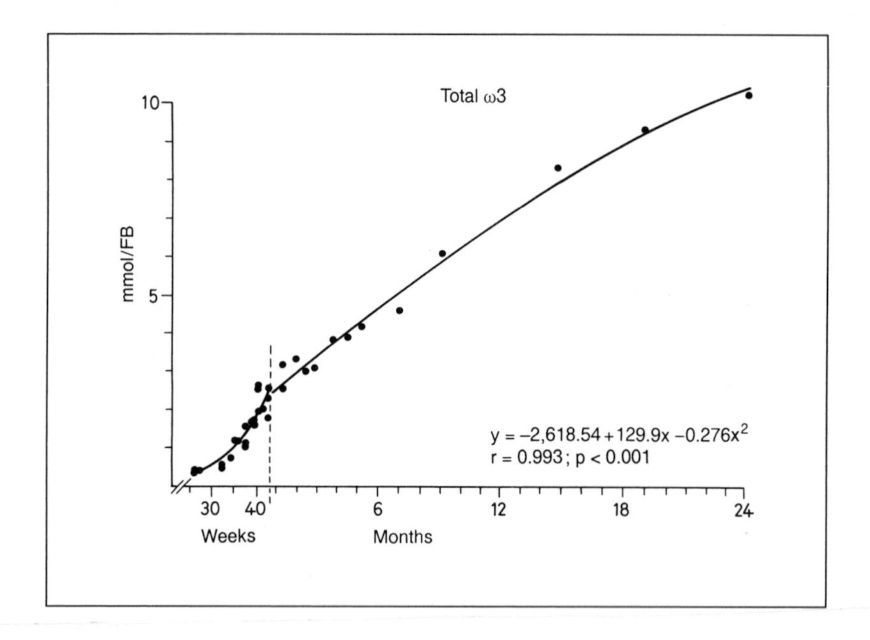

Fig. 4. Postnatal accretion of total ω3 acids per whole cerebrum. The equation shown has been computed exclusively with the postnatal cases (n = 13). As in prenatal life, irrespectively of the larger scale units, the results of the equation (y) are given in μmol per whole cerebrum; x = postconceptional age, in weeks.

no time can the tendency to decrease or level off found by Clandinin et al. [8, 9] be discerned in the present series of cases. On the contrary, the increase is constant and very significant until 2 years of age.

Of the individual fatty acids, 22:6 ω3 has virtually the same profile as total ω3 fatty acids, as it should be, since more than 95% of ω3 fatty acids is docosahexaenoate. Due to its metabolic importance, the accretion of arachidonate is shown individually in figure 7. As for myelin fatty acids, the total accretion of nervonic (24:1 ω9) and cerotic (26:0) acids is shown in figures 8 and 9, respectively. It is interesting to point out that, in contrast to most fatty acids, 24:1 ω9 shows a more rapid postnatal than prenatal accretion, the slope of the curve increasing with postnatal age. This is even more pronounced for 26:0, a fatty acid that only appears towards the end of gestation, and then increases very fast as myelination progresses. The accretion curves of 24:1 ω9 and 26:0 show the value of these two fatty acids as myelin markers.

5

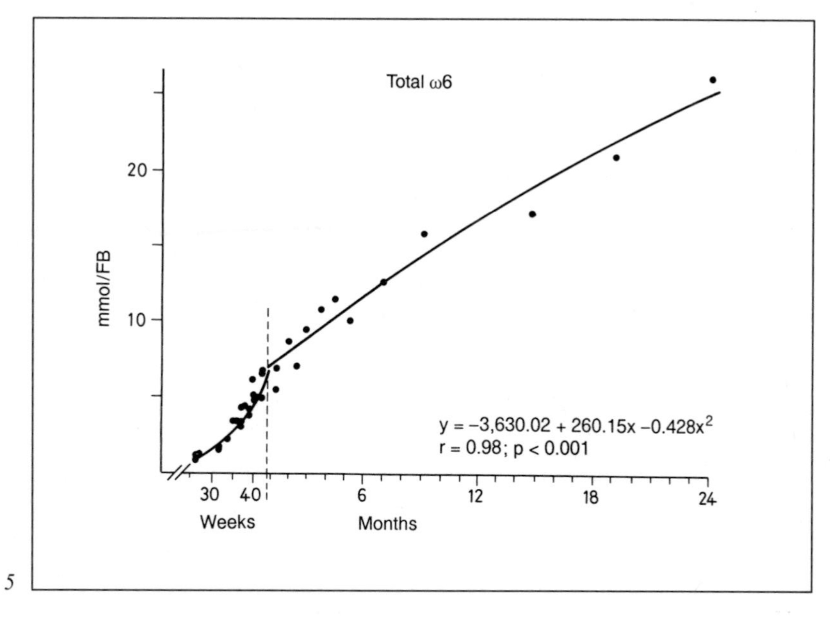

6

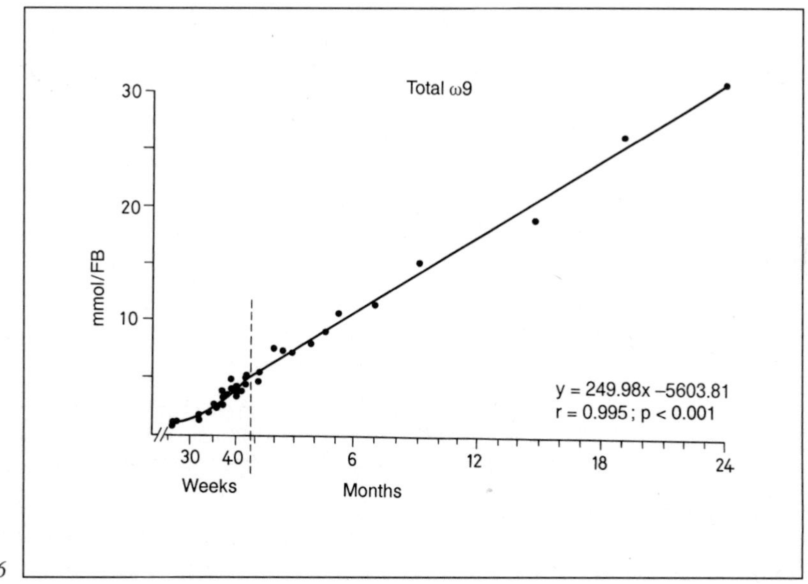

Fig. 5. Postnatal accretion of total ω6 acids per whole forebrain (see figure 4 for additional explanations).

Fig. 6. Postnatal accretion of total ω9 fatty acids. Notice that, for the oleic family, the postnatal profile is a straight line (x and y, like in previous figures).

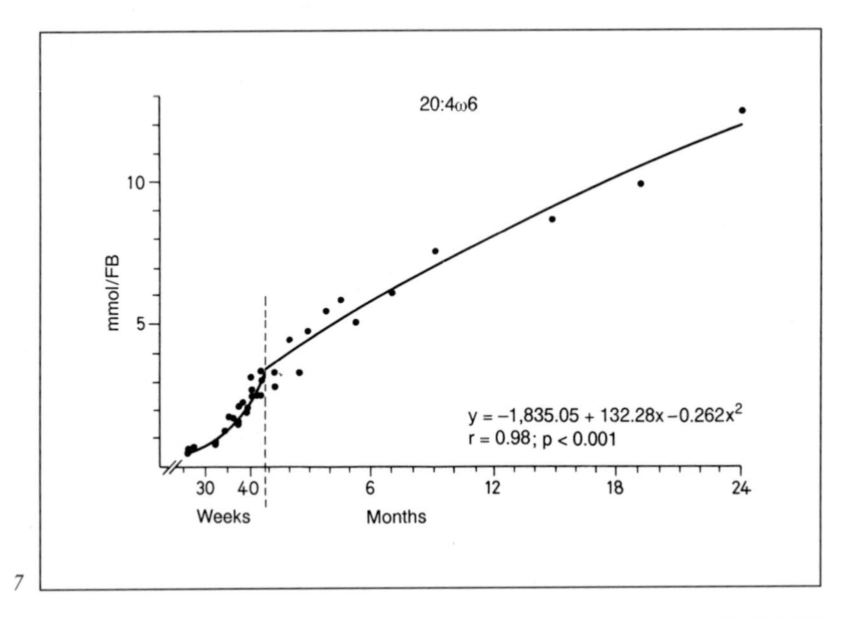

7

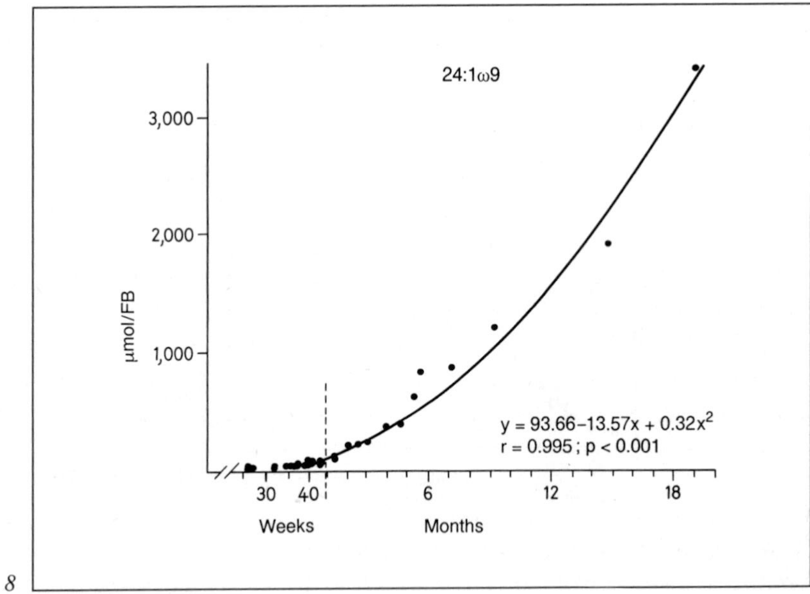

8

Fig. 7. Total profile of arachidonic acid per whole forebrain. The equation shown refers to postnatal life.

Fig. 8. Total profile of nervonic acid per whole cerebrum. Notice the increasing slope after term.

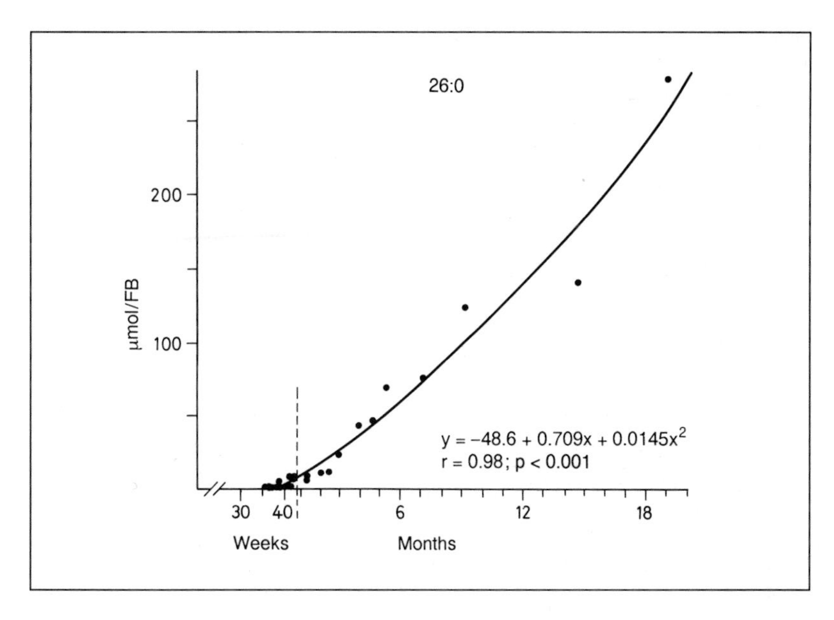

Fig. 9. Total amount of cerotic acid per whole forebrain. There is practically no 26:0 during prenatal life, the postnatal increase being very fast, and similar to that of nervonic acid.

Polyunsaturated Fatty Acid Profiles in Peroxisomal Disorders

Zellweger's syndrome, the prototype of generalized peroxisomal disorders, is a fatal genetic disease, causing severe neurological symptoms from birth, such as hypotonia, seizures, profound psychomotor retardation, and visual impairment. Death occurs usually before 6 months of age, and post-mortem examination reveals dysmyelination, gliosis and neuronal heterotopias. In Zellweger's syndrome, liver peroxisomes are virtually absent [10], and some lipid enzymes, known to be located in these organelles, are deficient [11, 12]. As a consequence of these deficiencies there is tissue accumulation of very long chain fatty acids [13] and a decrease in the plasmalogen levels [14]. In pseudo-Zellweger's syndrome, on the other hand, peroxisomes are present [15] although the clinical picture is very similar to that of classical Zellweger's syndrome. At the biochemical level, pseudo-Zellweger's syndrome has been described as an isolated deficiency of peroxisomal 3-oxoacyl-CoA thiolase [16]: there is accumulation of 26:0 and 26:1 but the plasmalogen levels are normal.

Table 2. Main polyunsaturated fatty acids in the brain of some peroxisomal disorders and neurologically normal controls

	Controls 1–15 months (n = 10)	ZS			psZS
		3 months	4 months	5 months	11 months
18:2 ω6	446 ± 72	1,028	1,796	2,172	1,091
20:3 ω6	1,214 ± 88	2,379	3,639	3,651	2,604
20:4 ω6	8,715 ± 410	7,314	11,791	11,374	7,926
22:4 ω6	5,088 ± 244	2,020	3,237	3,568	5,792
22:5 ω6	2,053 ± 161	181	513	145	1,184
22:5 ω3	196 ± 71	102	191	292	241
22:6 ω3	6,763 ± 296	1,478	1,639	1,391	3,143
22:6 ω3/22:5 ω3	38.59 ± 4.29	14.44	8.56	4.77	13.05
22:5 ω6/22:4 ω6	0.41 ± 0.03	0.09	0.16	0.04	0.20
22:6 ω3/22:4 ω6	1.33 ± 0.04	0.73	0.51	0.39	0.54

The fatty acid values are nmol/g of wet tissue.

Table 2 shows the most interesting polyunsaturated fatty acids and indices in the brains of four peroxisomal patients: three with classical Zellweger's syndrome, and one with pseudo-Zellweger's syndrome. It can be seen that the most important abnormality was a significant decrease in the brain levels of 22:6 ω3 in all four peroxisomal patients compared with the normal values for a wide range of ages. Docosapentaenoic acid (22:5 ω6) was also decreased in all cases, especially in the Zellweger patients. The fact that both PUFA with 22 carbon atoms, 22:6 ω3 and 22:5 ω6, were markedly decreased indicates a deficient Δ^4 desaturation system (22:5 ω3 → 22:6 ω3 and 22:4 ω6 → 22:5 ω6), as shown by the reduced indices 22:6 ω3/22:5 ω3 and 22:5 ω6/22:4 ω6. The combined ratio 22:6 ω3/22:4 ω6 was most significantly decreased in all patients. Table 2 shows that some precursor fatty acids, especially 18:2 ω6 and 20:3 ω6, were significantly increased, probably as a consequence of the blockade at the Δ^4-desaturation level.

Figures 10 and 11 show the levels of docosahexaenoic acid in the liver and kidney of all the patients, compared with control values. It can be seen how important the deficiency of this crucial fatty acid was in the peroxisomal patients, especially in the liver, where the levels of 22:6 ω3 were virtually negligible in the most affected cases. It is interesting to point out that the 22:6 ω3 deficiency was almost as severe in pseudo-Zellweger's as

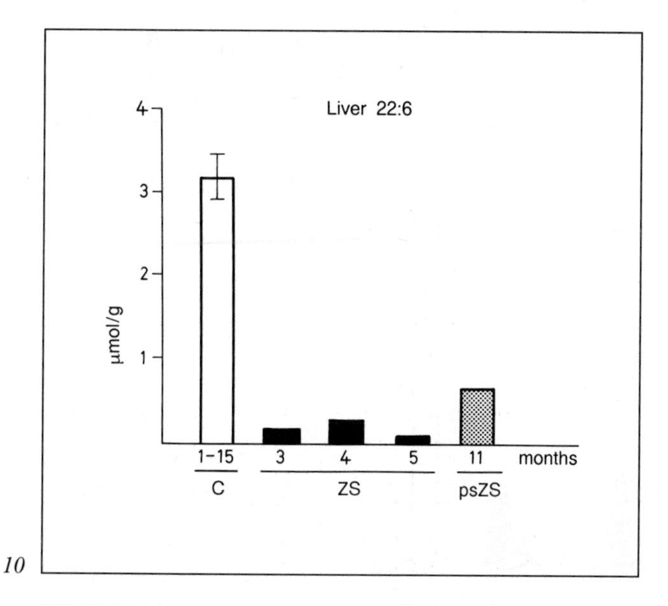

10

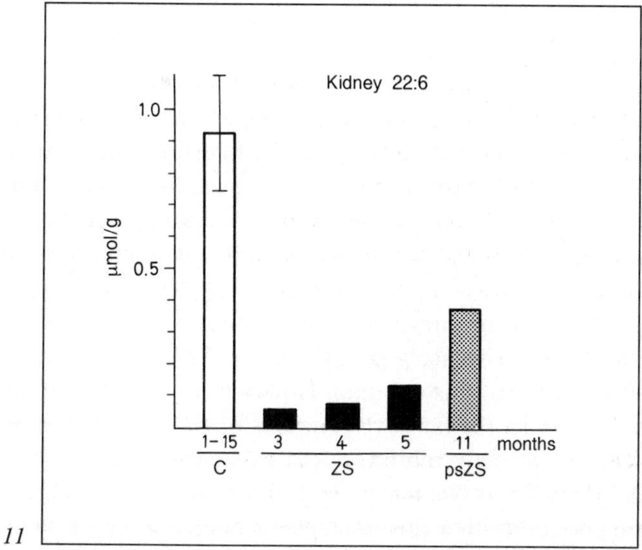

11

Fig. 10. Liver concentration of docosahexaenoic acid, in µmol per gram of wet tissue, in normal controls with ages ranging between 1 and 15 postnatal months and 4 peroxisomal patients. C = Normal controls; ZS = Zellweger's syndrome; psZS = pseudo-Zellweger's syndrome.

Fig. 11. Concentration of docosahexaenoic acid in the renal tissue of normal infants and 4 peroxisomal patients (for further details, see figure 10).

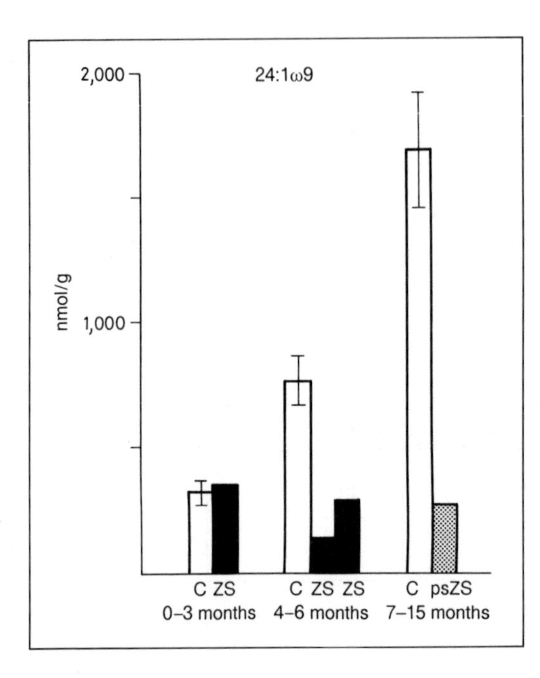

Fig. 12. Nervonic acid concentration in the brain of 4 peroxisomal patients, in nmol per gram of wet tissue, compared with age-matched controls (same notations as in figure 10).

in classical Zellweger's syndrome, especially if one considers that the pseudo-Zellweger patient was older than the others. The question why a patient with a normal number of peroxisomes should have such important reduction in the tissue levels of docosahexaenoic acid is intriguing, but it could contribute to explain the clinical similarity of Zellweger's and pseudo-Zellweger's syndromes.

Δ^4-Desaturase is one of the enzymes which has not been fully characterized to date and whose existence is merely deduced by its products of desaturation (see Sprecher H., this volume). However, the fact that in the peroxisomal patients studied so far both products of Δ^4-desaturation are simultaneously reduced strongly reinforces the existence of this enzyme system. PUFA desaturation is known to take place in the endoplasmic reticulum, and the possibility that this organelle is altered in peroxisomal disorders cannot be ruled out [10]. Alternatively, Δ^4-desaturase could be one of the many peroxisomal enzymes and be deficient even in apparently normal peroxisomes, as has been the case for 3-oxoacyl-CoA thiolase in pseudo-Zellweger's syndrome [16].

Figures 12–14 show some myelin fatty acids in the brain of the peroxisomal patients, compared with age-matched controls. It is interesting to

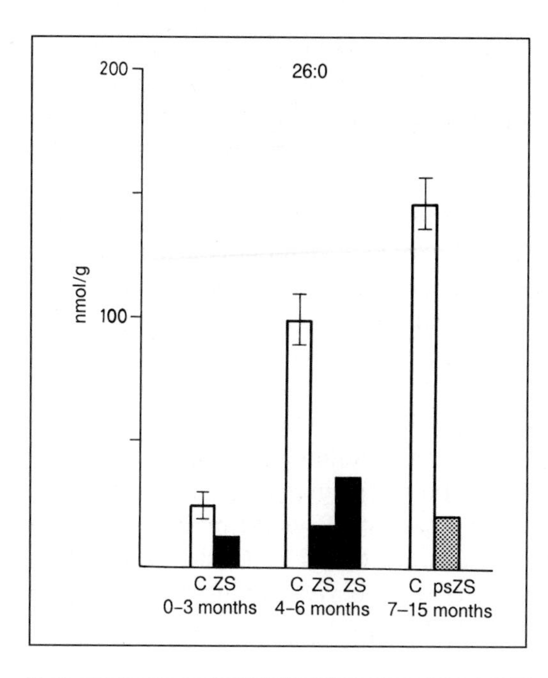

Fig. 13. Cerotic acid concentration in the brain of peroxisomal patients and neurologically normal controls. The concentration of 26:0 goes down as the disease progresses.

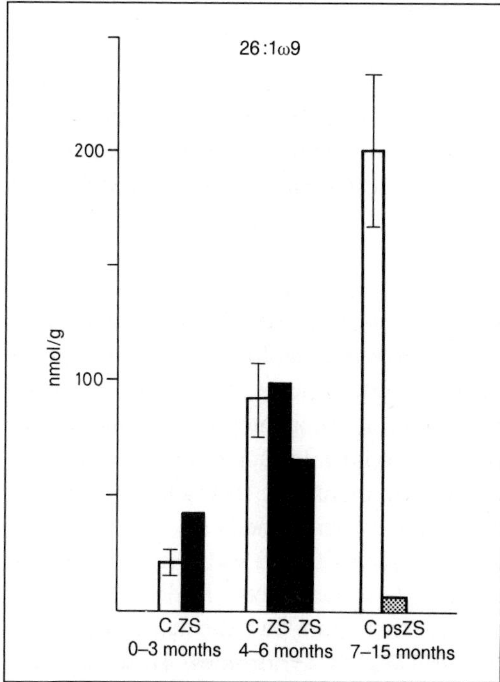

Fig. 14. Brain concentration of 26:1 ω9 in peroxisomal patients, compared with normal values for the age. Notice that the level of 26:1 ω9 is higher than that of 26:0 in the youngest patients but (like 26:0) it decreases as demyelination progresses and it has virtually disappeared in the oldest, pseudo-Zellweger patient.

point out that, in contrast to other tissues, in the brain of the Zellweger patients 26:0 was not at all increased. On the contrary, this fatty acid was significantly decreased in a direct proportion to the age of the patients, exactly paralleling the changes of the typical myelin constituent nervonic acid (fig. 12). Similarly, 26:1 ω9 was very significantly decreased in the oldest patients. This shows that, in peroxisomal disorders, demyelination predominates over accumulation of very long chain fatty acids and, as it should be, the decrease of these fatty acids is proportional to age.

These data suggest that accumulation of very long chain fatty acids can hardly explain any damage in the peroxisomal brain, since 26:0 and 26:1 are only increased in other tissues and they are either normal or significantly decreased in the peroxisomal brain. On the other hand, given the importance of 22:6 ω3 in neuronal and photoreceptor membranes, the deficiency of docosahexaenoic acid could explain much of the neurological and visual symptoms in peroxisomal patients. Even the myelination problems could, at least partially, be explained by a 22:6 ω3 deficiency, since the docosahexaenoate content of oligodendrocytes and myelin seems to be particularly affected in ω3 deprivation [17].

Conclusions

It can be concluded that the total concentration of polyunsaturated fatty acids in brain tissue keeps quite constant within each age range and, unless there are subtle changes in isolated structures, any alteration can be detected by simply comparing the total fatty acid composition of the tissue against age-matched controls.

When considering the total accretion of the main families of unsaturated fatty acids, ω3, ω6 and ω9, per whole cerebrum, very significant quadratic curves are obtained during prenatal life, followed by more linear increases during the postnatal period until, at least, 2 years of age. At no time does the accretion of PUFA show any tendency to diminish before that age. Therefore, the supply of essential fatty acids must be carefully controlled so that the nutritional requirements are met, and an adequate balance between ω3 and ω6 fatty acids is maintained.

Of the many biochemical abnormalities found so far in peroxisomal disorders, the dramatic decrease in the tissue levels of docosahexaenoic acid is, probably, the alteration that can explain best the neurological picture in Zellweger's and pseudo-Zellweger's syndromes. It seems very improbable that normal constituents of myelin sphingophospholipids, such as 26:0 and 26:1, can cause any neurological trouble, especially since these fatty acids are *not* increased in the peroxisomal brain. On the other hand, such a severe deficiency of 22:6 ω3 as shown in this chapter, far beyond what can be found in ω3 deprivation or imbalance, could indeed cause serious membrane alterations, especially in the central nervous system and retina, where docosahexaenoate seems to play a very important role. The existence of such profound alterations in the PUFA composition of tissues opens new avenues for research in the field of peroxisomal disorders. While waiting for some light to clarify this fascinating issue, it seems worthwhile testing the possible beneficial effects of a diet rich in docosahexaenoic acid in peroxisomal patients.

References

1 Breckenridge WC, Gombos G, Morgan IG: The lipid composition of adult rat brain synaptosomal plasma membranes. Biochim Biophys Acta 1972;266:695–707.
2 Anderson RE, Benolken RM, Dudley PA, Landis DJ, Wheeler TG: Polyunsaturated fatty acids of photoreceptor membranes. Exp Eye Res 1974;18:205–213.
3 Martinez M, Conde C, Ballabriga A: Some chemical aspects of human brain development. II. Phosphoglyceride fatty acids. Pediat Res 1974;8:93–102.
4 Martinez M, Gil-Bibernau JJ, Ballabriga A: Lipids of the developing human retina. I. Total fatty acids, plasmalogens, and fatty acid composition of ethanolamine and choline phosphoglycerides. J Neurosci Res 1988;20:484–490.
5 Lepage G, Roy CC: Direct transesterification of all classes of lipids in a one-step reaction. J Lipid Res 1986;27:114–120.
6 Martinez M: Polyunsaturated fatty acid changes suggesting a new enzymatic defect in Zellweger syndrome. Lipids 1989;24:261–265.
7 Martinez M: Severe deficiency of docosahexaenoic acid in peroxisomal disorders: A defect of Δ4 desaturation? Neurology 1990;40:1292–1298.
8 Clandinin MT, Chappell JE, Leong S, Heim T, Swyer PR, Chance JW: Intrauterine fatty acid accretion rates in human brain: implications for fatty acid requirements. Early Hum Dev 1980;4:121–129.
9 Clandinin MT, Chappell JE, Leong S, Heim T, Swyer PR, Chance GW: Extrauterine fatty acid accretion in infant brain: Implications for fatty acid requirements. Early Hum, Dev 1980;4:131–138.
10 Goldfischer S, Moore CL, Johnson AB, et al: Peroxisomal and mitochondrial defects in the cerebro-hepato-renal syndrome. Science 1973;182:62–64.
11 Singh I, Moser AB, Goldfischer S, Moser HW: Lignoceric acid is oxidized in the peroxisome: Implications for the Zellweger cerebro-hepato-renal syndrome and adrenoleukodystrophy. Proc Natl Acad Sci USA 1984;81:4203–4207.
12 Datta NS, Wilson GN, Hajra AK: Deficiency of enzymes catalyzing the biosynthesis of glycerol-ether lipids in Zellweger syndrome. N Engl J Med 1984;311:1080–1083.

13 Brown FR, McAdams AJ, Cumins JW, et al: Cerebro-hepato-renal (Zellweger) syndrome and neonatal adrenoleukodystrophy: Similarities in phenotype and accumulation of very long chain fatty acids. Johns Hopkins Med J 1982;151:344–361.
14 Heymans HSA, Schutgens RBH, Tan R, van den Bosch H, Borst P: Severe plasmalogen deficiency in tissues of infants without peroxisomes (Zellweger syndrome). Nature 1983;306:69–70.
15 Goldfischer S, Collins J, Rapin I, et al: Pseudo-Zellweger syndrome: Deficiencies in several peroxisomal oxidative activities. J Pediatr 1986;108:25–32.
16 Schram AW, Goldfischer S, van Roermund CWT, et al: Human peroxisomal 3-oxoacyl-coenzyme A thiolase deficiency. Proc Natl Acad Sci USA 1987;84:2494–2496.
17 Bourre JM, Pascal G, Durand G, Masson M, Sumont O, Piciotti M: Alterations in the fatty acid composition of rat brain cells (neurons, astrocytes, and oligodendrocytes) and of subcellular fractions (myelin and synaptosomes) induced by a diet devoid of n–3 fatty acids. J Neurochem 1984;43:342–348.

Manuela Martinez, MD, Autonomous University of Barcelona,
Hospital Infantil Vall d'Hebron, E–08035 Barcelona (Spain)

Simopoulos AP, Kifer RR, Martin RE, Barlow SM (eds): Health Effects of ω3 Polyunsaturated Fatty Acids in Seafoods. World Rev Nutr Diet. Basel, Karger, 1991, vol 66, pp 103–117

Essentiality of ω3 Fatty Acids for Brain Structure and Function[1]

Jean-Marie Bourre[a], *Odile Dumont*[a], *Michèle Piciotti*[a], *Michel Clément*[a], *Jean Chaudière*[a], *Michelle Bonneil*[b], *Gilles Nalbone*[b], *Huguette Lafont*[b], *Gérard Pascal*[c], *Georges Durand*[c]

[a]INSERM Unité 26, Hôpital Fernand Widal, Paris;
[b]INSERM Unité 130, Marseille; and [c]INRA, CNRZ Jouy-en-Josas, France

Introduction

The brain is the tissue that contains the highest concentration of lipids, immediately after adipose tissue. These lipids are practically all structural (and not energetic). They participate directly in the structure and hence the function of cerebral membranes. Cerebral development is genetically programmed, the renewal of neurones and oligodendrocytes is nil, and that of nervous membranes is often very slow. Therefore, during differentiation and multiplication, cells require adequate supplies of nutrients, especially lipids and particularly polyunsaturated fatty acids (PUFA). A lipid abnormality leads to an alteration in the function of membranes.

Saturated and monounsaturated fatty acids are mainly synthesized by nerve tissue itself, via complex mechanisms that differ according to cell type and to organelle [1]. In the nervous system, on average, one fatty acid out of three is polyunsaturated. In fact, the polyunsaturated fatty acids present in the membranes are not the dietary precursors (linoleic and alpha-linolenic acids) but longer and more desaturated chains (mainly 20:4 ω6 and 22:6 ω3). These control the composition of membranes and hence their fluidity, and, as a result, their enzymatic activity, the binding

[1] This work was supported by INSERM, INRA, ONIDOL and CETIOM.

Table 1. Composition of polyunsaturated fatty acids of cerebral cells and organelles

	Total PUFAs	20:4 ω6	22:6 ω3
Neurones	32	15	8
Synaptosomes	33	18	12
Astrocytes	29	10	11
Oligodendrocytes	20	9	5
Myelin	15	9	5
Capillaries	35	16	10
Mitochondria	30	16	12
Microsomes	29	11	12
Retina	45	5	35
Photoreceptor membrane	65	4	56
Peripheral nerve	10	7	2
Schwann cells	22	11	5

Results are expressed as percentages of total fatty acids (mg%). Animals were fed standard labchow diet, containing both ω3 and ω6 fatty acids.

between molecules and their receptors, cellular interactions, and the transport of nutrients. As far as the nervous system is concerned, these fatty acids can also influence certain electrophysiological parameters as well as learning functions. Dietary polyunsaturated fatty acids to a great extent determine membrane levels of these fatty acids [2, 3] and are particularly important for ensuring harmonious cerebral development [4]. There are many reports on the influence of polyunsaturated fatty acids on the structure and function of the nervous system [5–22]. However, polyunsaturated fatty acids of the ω3 series play a very special role in membranes, especially in the nervous system: all cerebral cells and organelles are extremely rich in these fatty acids (table 1). It is therefore extremely important to know precisely what quantity should be supplied by the diet.

It is clear that consumption of fish oils containing ω3 PUFA may have beneficial effects on ischemic heart disease and thrombosis [23–25]. However, as the ingestion of large amounts of ω3 PUFA in experimental animals gives rise to adverse effects, it is eventually possible that a diet abundant in fish oil may be harmful in man. Not much is known about human susceptibility to ω3 PUFA with respect to disturbances in vitamin E metabolism. Interestingly, during development and aging, in the rat peripheral nerve ω3 fatty acid content and vitamin E content are not correlated [26]. The PUFA composition of the diet regulates the fatty acid composition of the liver endoplasmic reticulum [27], and this in turn is an important factor controlling the rate and extent of lipid peroxidation in vitro and possibly in vivo [28, 29]. The replacement of cell membrane ω6 fatty acids by dietary ω3 fatty acids and subsequent alterations of membrane composition remain to be elucidated. Maintenance of membrane fluidity within narrow limits is presumably a prerequisite for proper functioning of the cell. Lipids play a key role in determining membrane fluidity, and changes in lipid and fatty acid composition have been reported to alter important cellular functions [8, 30]. Therefore, dietary modification of membrane phospholipids by fish oil feeding may have significant effects.

Brain contains high amounts of ω3 PUFA, and it is well known that fish oil alters the PUFA composition of various organs, especially liver and heart [31, 32].

Animal Experiments

Diet deficient in ω3: Two groups of wistar rats were fed for several generations with a semi-synthetic diet containing either sunflower oil or peanut oil. Diet containing ω3: Two other groups of rats were fed diets containing either soybean oil or rapeseed oil. Rapeseed oil-fed animals were compared with peanut-fed animals, soybean animals with sunflower animals.

Results

Alpha-Linolenic Acid Controls the Composition of Nerve Membranes [33, 34]

In animals given the ω3-deficient diet, the cells and organelles showed a very marked deficiency in 22:6 ω3 which was generally compensated for by an excess of 22:5 ω6 (table 2). The ω3/ω6 ratio was 16 times lower in oligodendrocytes, 12 times lower in myelin, 2 times lower in neurones, 6 times lower in synaptosomes, 3 times lower in astrocytes, 7 times lower in mitochondria, and 5 times lower in microsomes [10].

The importance of fatty acids of the ω3 series has likewise been demonstrated by specifically studying certain phospholipids such as phosphatidylethanolamine in animals fed diets containing peanut or rapeseed oil [12].

Table 2. Quantities of 22:6 ω3 and 22:5 ω6 in the nervous system of ω3-deficient-diet animals, expressed as percentages of the non-ω3-deficient animals

	22:6 ω3	22:5 ω6
Neurones	28	214
Synaptosomes	27	1,088
Oligodendrocytes	10	240
Myelin	14	1,200
Astrocytes	47	344
Mitochondria	25	917
Microsomes	28	592
Retina	36	1,280
Sciatic nerve	28	1,000

Data from ref. 10.

The Rate of Recovery of Anomalies Is Extremely Slow

After switching from the ω3-deficient to the ω3-containing diet [35, 36], several months were needed before brain cells and organelles recovered normal levels of 22:6 ω3 and lost the excess 22:5 ω6. This slow recovery was the same whatever the cell or organelle. It could be expected that recuperation would not be rapid in myelin, which has a slow turnover. But it is very surprising that nerve terminals also have a very slow recovery although turnover of their membrane molecules is supposed to be rapid. It can be suggested that regulation of recuperation occurs either at the level of synthesis of chain ends in the liver (22:6 ω3 and 20:4 ω6) or transport across the blood-brain barrier, or the enzymatic activities of desaturation and elongation which are known to be very weak in the liver after birth [11, 22]. It is interesting to note that cerebral microvessel and capillaries also have a very slow rate of recuperation, even though they are in contact with plasma lipoproteins of normal composition, since the liver recuperates very rapidly (2 weeks).

Effects of Alpha-Linolenic Acid Deficiency on Enzymatic, Electrophysiological, Behavioral, and Toxicological Parameters

Enzymatic Activities. The activity of 5′-nucleotidase is decreased by 30% in whole brain, but not in myelin or in nerve terminals, signifying that its activity is probably altered considerably in cell membranes (table 3). These results are in agreement with those of Bernsohn et al. [37] who have

Table 3. Deficiency in alpha-linolenic acid and brain membrane enzyme activities

	Brain	Myelin	Nerve endings
5′ nucleotidase	0.70	0.74	1.20
Na^+K^+ATPase	0.95	1.10	0.55
CNPase	0.95	0.78	0.00

Values represent enzyme activities obtained with animals fed a ω3-deficient diet divided by those obtained with non-ω3-deficient animals. Data from reference 34.
5′-nucleotidase: EC 3.1.3.5; Na^+K^+ATPase: EC 3.6.1.3; CNPase (2′-3′-cyclic nucleotide-3′-phosphodiesterase): EC 3.1.4.37.

shown that a decrease in the activity of this enzyme produced by simultaneous deficiency in linoleic and alpha-linolenic acids is only corrected by the addition of alpha-linolenic acid to the diet.

Na-K-ATPase is reduced nearly by half in the nerve terminals of animals fed an ω3-deficient diet compared with those fed the ω3-containing diet. On the other hand, simultaneous deficiency in linoleic and alpha-linolenic acid leads to an increase in Na-K-ATPase activity [19]. This enzyme controls ion transport produced by nerve transmission. It consumes half the energy used by the brain.

It is interesting to note that CNPase (2′-3′-cyclic nucleotid 3′-phosphodiesterase), which is specific for myelin, decreases as a result of alpha-linolenic acid deficiency, even though the myelin membrane is considered to be very rigid and not very metabolically active. The activity of another enzyme, acetylcholine-esterase, is also modulated by dietary lipids [38].

Electroretinogram. In the retina, 22:6 ω3 level is high [6]. Prolonged deficiency in PUFAs induces changes in the distribution of membrane fatty acids in the retina which are associated with changes in the electroretinogram [17]. In 4-week-old animals, the threshold of detection (10 μV) of wave A required a light stimulation 10 times stronger than that of the non-ω3-deficient animals. In 6-week-old animals, electroretinogram changes were less marked; in adult animals, only the A wave remained abnormal [33, 34].

Learning Tests. A simultaneous deficiency in linoleic and alpha-linolenic acids affects the learning capacities of animals [15, 18], as does a

Table 4. Fatty acid composition[a] of brain lipids of Menhaden and olive oil rat groups (n = 15 each)

Fatty acid	Menhaden oil			Olive oil control		
	TL	PE	PC	TL	PE	PC
18:2 ω6	0.8 ± 0.8	0.6 ± 0.4	0.2 ± 0.1*	0.9 ± 1.2	0.9 ± 0.3	0.8 ± 0.3
20:4 ω6	9.1 ± 0.4	20.9 ± 1.3	6.3 ± 0.9	9.1 ± 0.6	20.8 ± 2.0	6.7 ± 0.9
22:4 ω6	0.7 ± 0.4**	5.5 ± 0.8	1.0 ± 0.4	6.4 ± 3.7	6.8 ± 0.9	0.7 ± 0.2
22:5 ω6	2.4 ± 0.6	4.7 ± 1.4	1.4 ± 0.2	2.3 ± 0.6	4.6 ± 1.0	0.6 ± 0.1
20:5 ω3	2.5 ± 0.5**	1.8 ± 0.6**	0.4 ± 0.15	0.4 ± 0.1	0.2 ± 0.06	0.2 ± 0.1
22:5 ω3	0.2	0.2	0.2	0.2	0.2	0.2
22:6 ω3	12.1 ± 0.9	14.9 ± 0.8	1.3 ± 0.2	10.4 ± 2.6	12.2 ± 2.4	1.4 ± 0.3
Saturated	45.0	34.6	59.6	43.7	37.9	60.8
Monounsaturated	26.2	16.5	29.5	26.2	15.8	28.6
Polyunsaturated	28.9	48.9	11.1	30.5	46.2	10.7

Thirty male rats of the Sprague-Dawley strain (Iffa-Credo, l'Arbresle, France), weighing 200 g (approximately 2 months) at the beginning of the experiment, were divided into two groups of 15 animals. All the animals were fed a rat chow diet ad libitum (containing 5% lipids). In addition, rats of the first group received a daily amount of 100 µl of Menhaden oil (MAX-EPA, Scherer, Strasbourg, France) per 100 g body weight, for 60 days through gastric intubation, while rats of the second group received the same amount of olive oil of current commercial quality (Puget, Marseille, France).
TL = Total lipids; PE = Phosphatidylethanolamine; PC = Phosphatidylcholine.
[a] Results are expressed as percentages of total fatty acids (mean ± SD).
* $p < 0.05$; ** $p < 0.01$.
Data from ref. 45.

selective deficiency in alpha-linolenic acid [39]. Though motor activity and open field tests were practically normal in animals fed the ω3-deficient diet, their learning capacities were severely perturbed, as shown by the shuttle box test. In the first session, animals fed the ω3-containing diet made a more rapid association between the light stimulus and the electric shock, since they avoided on average 7 shocks out of 30, whereas ω3-deficient diet animals only avoided 2. These differences diminished with further conditioning and disappeared at the fourth session [33, 34]. Very interestingly the extinction of learning capacities was significantly longer in animals deficient in dietary alpha-linolenic acid [40, 41].

Mortality in Animals Tested with the Neurotoxic Agent Triethyltin. The LD_{50} of animals fed the soybean oil or sunflower oil diets did not differ significantly (6.18 vs. 6.02 μl/kg, respectively). But the animals fed the sunflower oil diet died more rapidly than did those fed the soybean oil diet.

The fluidizing effect of ethanol shown with diphenylhexatriene (DPH) is also decreased significantly in animals fed sunflower oil [42]. Concomitantly, rats fed sunflower oil are more sensitive to ethanol-induced hypothermia, illustrating the importance of diet to membrane sensitivity and animal response to ethanol, regardless of the exact mechanisms. In nerve-ending membranes, fluidity is affected by the diet, depending on the membrane region. Feeding the sunflower oil diet compared to the soybean oil diet results in less fluidity in the surface polar part of the membranes probed by TMA- or PROP-DPH but greater fluidity in the apolar part of the membranes (probed by DPH) [42].

Minimum Requirement of Alpha-Linolenic Acid Needed in Cerebral Membranes

When diets with intermediate levels of alpha-linolenic acid were given, increasing the amount of 18:3 ω3 led to an overall increase in 22:6 ω3, and inversely a decrease in 22:5 ω6. In fact, in brain, levels of 22:6 ω3 increased linearly at an intake of 18:3 ω3 that varied from 0 to 200 mg/100 g diet and then reached a plateau (the opposite was observed for 22:5 ω6). (In liver, kidney, and muscle the same threshold was found but the plateau was less clear.)

These precursors, linoleic and alpha-linolenic acids, have to be elongated and desaturated by the liver into longer chains, which are in fact the essential fatty acids for the brain, as cell cultures seem to have demonstrated [43]. Nerve cells in culture differentiate, multiply, and capture and liberate neurotransmitters only if the medium contains 20:4 ω6 and 22:6 ω3, but not in the presence of 18:2 ω6 and 18:3 ω3 [43, 44].

Effects of Fish Oil

Effect of Pharmacological Doses [45]

Feeding rats a diet enriched in ω3 PUFAs (200 μl/day Menhaden oil) increased the content of eicosapentaenoic acid (20:5 ω3) in brain phospholipids (table). Conversely 22:4 ω6 was reduced. These changes were not associated with alterations in either vitamin E concentration or glutathione peroxidase and catalase activities in cerebrum and cerebellum. No increase in peroxidative damage was found. Interestingly, the major very-long-chain fatty acids (22:6 ω3 and 22:5 ω3) were not affected.

Significant differences were observed in total fatty acids where the percentage of 22:4 ω6 was 9 times lower in the Menhaden group, while the percentage of 20:5 ω3 was 6 times higher. In the phosphatidylethanolamine fraction which contains 45–50% of PUFAs, the percentage of 20:5 ω3 was 9 times higher in the Menhaden group, while the percentage of 22:6 ω3 was not significant in different groups. In the phosphatidylcholine fraction which only contains 11% of PUFAs, the percentage of 20:5 ω3 and 22:6 ω3 was higher in the Menhaden group, but these differences were minor when compared to those observed in the phosphatidylethanolamine fraction.

As 22:4 ω6 was decreased in total lipid from Menhaden oil-fed animals, but no significant alterations were found in neither phosphatidylethanolamine nor phosphatidylcholine, large alterations are speculated to be found in phosphatidylserine.

The main differences observed between groups in both total lipids and the phosphatidylethanolamine fraction did not change the percentage of total PUFAs which remained close to 29% in the total lipids of each group.

The vitamin E content as well as the activity of glutathione peroxidase and catalase in the brain and cerebellum were unchanged in both the Menhaden and olive oil groups [14].

Effect of Large Excess [46]

Table 5 shows that cod liver oil as well as salmon oil (15 and 12.5% in the diet, respectively) supplemented with alpha-tocopherol (100 mg/100 g) induce similar alterations in forebrain PUFA levels (but saturated and monounsaturated fatty acids are little affected if at all). Though the brain is considered to be well protected, an 8-week fish oil diet in 60-day-old animals increased the ω3 series and decreased the ω6 series. When cod liver oil- and salmon oil-fed animals are compared with those fed corn oil, the brain contents of 20:4 ω6, 22:4 ω6 and 22:5 ω6 were decreased by 16–19, 37–40, 64–79%, respectively, whereas brain 20:5 ω3, 22:5 ω3 and 22:6 ω3 were increased by 35–75, 14.1–22, 20–29%, respectively. In the cod liver-fed animals all liver fatty acids were significantly affected.

Effects of Increasing Amount of Fish Oil [47]

For brain, table 6 shows that 20:4 ω6 decreased proportionately with increasing dietary fish oil content (and decreasing corn oil). We have previously shown that 20:4 ω6 content is independent of dietary linoleic acid

Table 5. Fatty acid composition of the forebrain from animals fed the 4 oil diets. Comparison with liver

Fatty acid	Brain				Liver		
	control	corn	cod liver	salmon	control	corn	cod liver
C18:2 ω6	0.62	1.3*	0.82	0.85	14.0	30.6**	7.8***
C20:4 ω6	8.8	9.1	7.2**	7.7	20.36	18.34	3.2***
C22:4 ω6	3.15	3.5	2.1**	2.2**	0.52	1.16*	–
C22:5 ω6	0.75	0.78	0.17***	0.2***	1.0	1.06	–
C20:5 ω3	0.2	0.2	0.35	0.27	0.13	0.1	1.0***
C22:5 ω3	0.22	0.27	0.98***	0.65***	0.5	0.32	0.3
C22:6 ω3	10.7	10.8	13.0***	13.9***	3.72	2.9*	1.6*

Four groups of male Wistar rats (IFFA-Credo, l'Arbresle, France) weighing 190–210 g were housed two per cage. Group 1 (10 rats) was fed a diet containing 4.4% (w/w) fat consisting of a lard (2.2%) and corn oil (2.2%) mixture. Experimental animals were fed a 17% lipid diet: group 2, which served as control (10 rats) was fed 2% corn oil with 15% cod liver oil; group 3 (12 rats) receiced the same diet as group 2 but supplemented; group 4 (12 rats) received a salmon oil-enriched diet (12.5% w/w) supplemented with 4.5% of corn oil. Corn oil, as supplied, contained 45 mg of alpha-tocopherol per 100 g oil. Salmon oil was supplemented with 100 mg alpha-tocopherol/100 g oil. Therefore, the amount of vitamin E supplied by the salmon oil-enriched diet was 295 mg/kg of diet.
* $p < 0.05$; ** $p < 0.01$; *** $p < 0.001$.
Data from ref. 46.

content in excess of minimal level (0.3% of the calories) [48]; this is largely in all diets tested. Thus the decrease of 20:4 ω6 in brain membranes is only due to the increase of fish oil in the diet. 22:4 ω6 was less affected than arachidonic acid, 22:5 ω6 was reduced by about 60%. The high amount of dietary corn oil could increase 22:5 ω6, since its level in membranes parallels dietary excess of linoleic acid [48] as well as the deficiency in alpha-linolenic acid [16, 36, 37].

EPA was nearly undetectable in diet containing low amount of fish oil, and increased in diet containing 4% and more fish oil, but the content was still extremely low, even in the diet containing a very high amount of fish oil. 22:5 ω3 was increased 7-fold in 10% fish oil diet, but the brain content was always below 1% of the total fatty acids. 22:6 ω3 was increased by about 30%.

Table 6. Total lipid fatty acid composition (weight percent) of rat brain after various diets

Composition of the diet					
Salmon oil, g/100 g	–	1.5	4.0	7.0	10.0
Corn oil, g/100 g	10	8.5	6.0	3.0	–
ω6/ω3	75	10.0	3.0	1.0	0.1
Brain fatty acids, weight percent					
18:2 ω6	0.9[a]	1[a]	0.9[a]	0.7[a]	0.3[b]
20:4 ω6	10.2[a]	9.1[a]	8.5[a]	7.7[b]	7.8[b]
22:4 ω6	3.5[a]	2.7[b]	2.7[b]	3.0[b]	2.7[b]
22:5 ω6	0.8[a]	0.4[b]	0.3[b]	0.3[b]	0.3[b]
20:5 ω3	–	–	0.1[a]	0.2[a]	0.3[b]
22:5 ω3	0.1[a]	0.2[a]	0.4[b]	0.6[c]	0.7[c]
22:6 ω3	12.1[a]	12.3[a]	15.4[b]	14.6[b]	15.6[b]

Five groups of 12 male wistar rats (IFFA-Credo, l'Arbresle, France) weighing 190–200 g were housed two per cage. All groups received for 8 weeks the same semi-synthetic diet having the same total amount of lipids, but varying in fish oil (increasing salmon oil was compensated for by decreasing corn oil).
Values (means; n = 6) not bearing the same superscript letter are significantly different at $p < 0.05$. If no similar superscript appears values are not different. SD did not exceed 10% of the mean values.
Data from ref. 47.

Table 7 shows that in the liver, all polyunsaturated fatty acids were affected in diet containing high amount of fish oil; alterations were less important or minimal with diet containing 1.5% fish oil (except for EPA).

Discussion

Essentiality of Alpha-Linolenic Acid

A diet deficient in alpha-linolenic acid caused marked alterations in the fatty acid composition of all cellular and subcellular fractions examined. The total content (number of moles) of PUFAs was not altered, the marked decrease in 22:6 ω3 being compensated for by an increase in 22:5 ω6. This compensation is quantitative, but total unsaturation remains in deficit. It is evident that PUFAs control the fluidity of biological membranes, hence many of their activities. A specific deficiency in ω3 fatty acids perturbs the

Table 7. Total lipid fatty acid composition (weight percent) of liver of rats fed various diets

Composition of the diet					
Salmon oil, g/100 g	–	1.5	4.0	7.0	10.0
Corn oil, g/100 g	10	8.5	6.0	3.0	–
ω6/ω3	75	10.0	3.0	1.0	0.1
Liver fatty acids, weight percent					
18:2 ω6	18^{a}	19.4^{a}	15.4^{b}	10.4^{c}	2.2^{d}
20:4 ω6	14.1^{a}	13.2^{a}	7^{b}	6.4^{b}	5.9^{c}
22:4 ω6	0.8^{a}	0.1^{b}	0.1^{b}	0.1^{b}	0.2^{b}
22:5 ω6	0.5	–	–	0.2	0.2
20:5 ω3	0.2^{a}	1.9^{b}	2.9^{c}	5.8^{d}	9.5^{e}
22:5 ω3	–	0.85^{a}	1.3^{b}	2.2^{c}	2.7^{c}
22:6 ω3	5.5^{a}	5.7^{a}	5.4^{a}	8.3^{b}	9.7^{c}

For experimental procedure see table 6.
Values (means; n = 6) not bearing the same superscript letter are significantly different at $p < 0.05$. If no similar superscript appears values are not different. SD did not exceed 10% of the mean values.
Data from ref. 47.

activities of membrane enzymes, alters some electrophysiological activities as shown by the electroretinogram, and disturbs learning abilities. After switching from a deficient to a normal diet, the rate of recovery is remarkably slow: it is several months before brain cells and organelles recover normal levels of cervonic acid. This rate is the same for all organelles. It is therefore crucial to supply the fatty acids necessary for cerebral structures at the developmental stage. A deficiency is difficult to correct.

Bearing in mind that (1) the mean cerebral fatty acid composition of man is little different from that of the rat; (2) that in man a greater amount of brain is formed per day over a longer period; and (3) that the brain weight/total body weight ratio is greater in man, even taking body surface into consideration, it is evident that minimum levels in the rat are also, a fortiori, those in man. At any rate, for evident ethical reasons, it will always be impossible to determine the effects of increasing doses of dietary fatty acids on the composition of cerebral membranes in man. The minimum levels recommended are therefore: linoleic acid, 1,200 mg/100 g food intake (2.4% of calories); alpha-linolenic acid, 200 mg/100 g food intake (0.4% of calories).

Advantages (and Eventual Adverse Effects) of Very-Long-Chain Precursors Derived from Fish Oil

Since cerebral structures contain very-long-chain PUFAs, it might seem wise to provide these acids directly in the diet, especially since the ability of the organism to transform linoleic and alpha-linolenic precursors diminishes rapidly during development [11, 22]. If the diet of rats is supplemented with Menhaden oil (1% by weight added to the normal diet), our results indicate that the profile of cerebral fatty acids is little altered, peroxidized derivatives do not appear, and there is no change in the activity of enzymes that protect against peroxidation: cytosolic superoxide dismutase containing Cu and Zn and mitochondrial containing Mn, glutathione peroxidase, glutathione reductase, and catalase [45].

On the other hand, large quantities (up to 12%) of dietary fish oil, even supplemented with vitamin E, perturb the fatty acid profile of the liver as well as that of the brain. In brain, there is a deficiency of arachidonic acid and a marked decrease in 22:4 ω6 and 22:5 ω6, associated with excess 22:6 ω3 and 22:5 ω3.

As subtle changes in brain membrane PUFAs determined by dietary alterations in alpha-linolenic acid provoke alterations in brain membrane PUFAs, membrane fluidity, enzymatic activities, electrophysiological parameters, learning tests and resistance to poisons [33, 34], the question can be raised of whether increased fish oil intake leads to functional alterations in the nervous system [14].

Although blood parameters were normal, except triacylglycerol, in all animals, alterations in the membrane PUFAs could alter some liver or brain function, as yet of an undetermined nature.

References

1 Bourre JM: Origin of aliphatic chains in brain; in Baumann N (ed): Neurological Mutations Affecting Myelination. INSERM Symposium No 14. Amsterdam, Elsevier/North Holland Biomedical Press, 1980, pp 187–206.

2 Mead JF: The non-eicosanoid functions of the essential fatty acids. J Lipid Res 1984;25:1517–1521.

3 Holman RT: Control of polyunsaturated acids in tissue lipids. J Am Coll Nutr 1986; 5:183–211.

4 Menon NK, Dhopeshwarkar GA: Essential fatty acid deficiency and brain development. Prog Lipid Res 1982;21:309–326.

5 Alling C, Bruce A, Karlsson I, et al: Effect of maternal essential fatty acid supply on

fatty acid composition of brain, liver, muscle and serum in 21-day-old rats. J Nutr 1971;102:773–782.
6 Bazan N, Di Fazio De Escalante S, Careaga M, et al: High content of 22:6 (docosahexaenoate) and active (2-^{3}H) glycerol metabolism of phosphatidic acid from photoreceptor membranes. Biochim Biophys Acta 1982;712:702–706.
7 Bjerne KS, Mostad IL, Thoresen L: Alpha-linolenic acid deficiency in patients on long term gastric tube feeding: estimation of linolenic acid and long-chain unsaturated n-3 fatty acid requirement in man. Scand J Clin Nutr 1987;45:66–77.
8 Brenner RR: Effect of unsaturated acids on membrane structure and enzyme kinetics. Prog Lipid Res 1984;23:69–96.
9 Clandinin MT, Chappell JE, Leong S, et al: Intrauterine fatty acid accretion rates in human brain: implications for fatty acid requirements. Early Human Development 1980;4:121–129.
10 Bourre JM, Pascal G, Durand G, et al: Alterations in the fatty acid composition of rat brain cells (neurons, astrocytes and oligodendrocytes) and of subcellular fractions (myelin and synaptosomes) induced by a diet devoid of n-3 fatty acids. J Neurochem 1984;43:342–348.
11 Cook HW: In vitro formation of polyunsaturated fatty acids by desaturation in rat brain: some properties of the enzyme in developing brain and comparison with liver. J Neurochem 1978;30:1327–1334.
12 Kaare RN, Drevon CA: Dietary n-3 fatty acids and cardiovascular diseases. Arteriosclerosis 1986;6:352–355.
13 Crawford MA, Hassam AG, Stevens PA: Essential fatty acid requirements in pregnancy and lactation with special reference to brain development. Prog Lipid Res 1981;20:31–40.
14 Philbrick DJ, Mahadevappa VG, Ackman G, Holub J: Ingestion of fish oil or a derived n-3 fatty acid concentrate containing eicosapentaenoic acid (EPA) affects fatty acid compositions of individual phospholipids of rat brain, sciatic nerve and retina. J Nutr 1987;117:1663–1670.
15 Lamptey MS, Walker BL: Learning behaviour and brain lipid composition in rats subjected to essential fatty acid deficiency during gestation. Lactation and growth. J Nutr 1978;108:358–367.
16 Holman RT, Johnson SB, Hatch TF: A case of human linolenic acid deficiency involving neurological abnormalities. Am J Clin Nutr 1982;35:617–623.
17 Neuringer M, Connor WE: n-3 fatty acids in the brain and retina: evidence for their essentiality. Nutr Rev 1986;44:289–296.
18 Paoletti R, Galli C: Effect of essential fatty acid deficiency on the central nervous system in the growing rat; in Lipids malnutrition and the developing brain. Ciba Foundation Symposium. Amsterdam, Elsevier/North Holland, 1972, p 121–135.
19 Sun GY, Sun AY: Synaptosomal plasma membranes: acyl group composition of phosphoglycerides and (Na^{+} + K^{+})-ATPase activity during fatty acid deficiency. J Neurochem 1974;22:15–18.
20 Sprecher H: Biosynthetic pathways of polyunsaturated fatty acids.
21 Svennerholm L, Alling C, Bruce A, Karlsson I, Sapia O: Effects on offspring of maternal malnutrition in the rat; in Lipids, Malnutrition and the Developing brain. Ciba Foundation Symposium. Amsterdam, Elsevier/North Holland, 1972, pp 141–157.

22 Strouve-Vallet X, Pascaud M: Désaturation de l'acide linoléique par les microsomes du foie et du cerveau du rat en développement. Biochimie 1971;53:699–703.
23 Kinsella JE: Food components with potential therapeutic benefits: the n-3 polyunsaturated fatty acids of fish oil. Food Technol 1986;146:89–97.
24 Simopoulos AP: Summary of the conference on the health effects of polyunsaturated fatty acids in seafoods. J Nutr 1986;116:2350–2354.
25 Lands WE: Renewed questions about polyunsaturated fatty acids. Nutr Rev 1986; 44:189–195.
26 Clément M, Bourre JM: Alteration of alpha-tocopherol content in the developing and aging peripheral nervous system: persistence of high correlation with total and specific n-6 polyunsaturated fatty acids. J Neurochem 1990;54:2110–2111.
27 Tahin QS, Blum M, Carafoli E: The fatty acid composition of subcellular membrane of rat liver, heart and brain: diet induced modifications.Eur J Biochem 1981;121: 5–13.
28 Mounié J, Faye B, Magdalou J, Goudonnet H, Truchot R, Siest G: Modulation of UDPG-lucuronosyltransferase activity in rats by dietary lipids. J Nutr 1986;116: 2034–2043.
29 Hammer CT, Wills ED: The role of lipid components of other diet in the regulation of the fatty acid composition of the rat liver endoplasmic reticulum and lipid peroxidation. Biochem J 1978;174:585–593.
30 Stubbs CD, Smith AD: The modification of mammalian membrane polyunsaturated fatty acid composition in relation to membrane fluidity and function. Biochim Biophys Acta 1984;779:89–137.
31 Huang YS, Nassar BA, Horrobin DF: Changes of plasma lipids and long-chain n-3 and n-6 fatty acids in plasma, liver, heart and kidney phospholipids of rats fed variable levels of fish oil with or without cholesterol supplementation. Biochim Biophys Acta 1986;879:22–27.
32 Swanson JE, Kinsella JE: Dietary n-3 polyunsaturated fatty acids: modification of rat cardiac lipids and fatty acid composition. J Nutr 1986;116:514–523.
33 Bourre JM, Dumont O, Piciotti M, Pascal G, Durand G: Composition of nerve biomembranes and nutritional fatty acids. Nutrition 1989;5:266–270.
34 Bourre JM, François M, Weidner C, et al: The importance of dietary linolenic acid in the composition of nervous membranes. Control of enzymatic activity, amplitude of electrophysiological parameters, resistance to poisons, and performance of learning tasks. Linolenic acid requirement for developing brain and various organs. J Nutr 1989;119:1880–1892.
35 Homayoun P, Durand G, Pascal G, Bourre JM: Alteration in fatty acid composition of adult rat brain capillaries and choroid plexus induced by a diet deficient in (n-3) fatty acids. Slow recovery by substitution with a non deficient diet. J Neurochem 1988;51:45–48.
36 Bourre JM, Durand G, Pascal G, Youyou A: Brain cell and tissue recovery in rats made deficient in n-3 fatty acids by alteration of dietary fat. J Nutr 1989;119:15–22.
37 Bernshohn J, Spitz FJ: Linoleic and linolenic acid dependency of some brain membrane-bound enzymes after lipid deprivation in rats. Biochem Biophys Res Com 1974;57:293–298.

38 Foot M, Cruz TF, Clandinin MT: Influence of dietary fat on the lipid composition of rat brain synaptosomal and microsomal membranes. Biochem J 1982;208:631–640.
39 Lamptey M, Walker BK: A possible essential role for dietary linolenic acid in the development of the young rat. J Nutr 1976;106:86–93.
40 Yamamoto N, Sarton M, Moriuchi A, Nomura M, Okuyama H: Effect of dietary alpha-linolenate/linoleate balance on brain lipid compositions and learning ability of rats. J Lipid Res 1987;28:144–151.
41 Yamamoto N, Hashimoto A, Takemoto Y, Okuyama H, Nomura M, Kitailma R, Togashi T, Tamai Y: Effect of the dietary alpha-linolenate/linoleate balance on lipid compositions and learning ability of rats. II. Discrimination process, extinction process, and glycolipid compositions. J Lipid Res 1988;29:1013–1021.
42 Beaugé F, Zerouga M, Niel E, et al: Effects of dietary linolenate/linoleate balance on the neuronal membrane sensitivity to ethanol; in Kuriyama K, Takada A, Ishich Y (eds): Biomedical and social aspects of alcohol and alcoholism. Amsterdam, Elsevier/North Holland, 1988, p 291–294.
43 Bourre JM, Faivre A, Dumont O, et al: Effect of polyunsaturated fatty acids on fetal mouse brain cells in culture in a chemically defined medium. J Neurochem 1983;41:1234–1242.
44 Loudes C, Faivre A, Barret A, et al: Release of immunoreactive TRH in serum free culture of mouse hypothalamic cells. Dev Brain Res 1983;9:231–234.
45 Chaudière J, Clément M, Driss F, Bourre JM: Unaltered brain membranes after prolonged intake of highly oxidizable long-chain fatty acids of the (n-3) series. Neurosci Lett 1987;82:233–239.
46 Bourre JM, Bonneil M, Dumont O, et al: High dietary fish oil alters the brain polyunsaturated fatty acid composition. Biochim Biophys Acta 1988;960:458–461.
47 Bourre JM, Bonneil M, Dumont O, et al: Effect of increasing amounts of dietary fish oil on brain and liver. Biochim Biophys Acta 1990;1043:149–152.
48 Bourre JM, Piciotti M, Dumont O, et al: Dietary linoleic acid and polyunsaturated fatty acids in rat brain and other organs; minimal requirements of linoleic acid. Lipids 1990;25:465–472.

Jean-Marie Bourre, PhD, INSERM Unité 26, Hôpital Fernand-Widal,
200, rue du Fbg-St-Denis, F–75010 Paris (France)

Simopoulos AP, Kifer RR, Martin RE, Barlow SM (eds): Health Effects of ω3 Polyunsaturated Fatty Acids in Seafoods. World Rev Nutr Diet. Basel, Karger, 1991, vol 66, pp 118–132

Essentiality of ω3 Fatty Acids: Evidence from the Primate Model and Implications for Human Nutrition[1]

William E. Connor, Martha Neuringer, Sydney Reisbick

Oregon Health Sciences University, Department of Medicine, L-465, Portland, Oreg., USA

Introduction

A diet deficient in ω3 fatty acids leads to a triad of signs in the rhesus monkey: visual impairment, abnormalities of the electroretinogram, and polydipsia [1–4]. Profound biochemical changes in the fatty acid composition of the membranes of the retina, brain and other organs accompany these other disturbances [4, 5]. Low concentrations of ω3 fatty acids occur at birth in the plasma, red blood cells, and neural tissues of infants born from mothers fed an ω3 deficient diet [1]. Docosahexaenoic acid (DHA, 22:6 ω3), an ω3 fatty acid which is uniquely rich in neural membranes, is found in very low concentrations in these infant monkeys. These concentrations even become lower as the deficient diet is continued postnatally. By 4 weeks of age, visual impairment can be demonstrated and shortly after that abnormalities of the electroretinogram occur [1, 4]. Polydipsia develops later in life [3].

Docosahexaenoic acid (DHA), is the most prominent fatty acid of the retina, brain, and spermatozoa. It generally occupies the sn2 position of the phospholipid membranes of these organs [5]. DHA is especially concentrated in the outer segment membranes of the retinal rod and cone

[1] Supported by National Institute of Diabetes and Digestive and Kidney Disease research grant DK29930, General Clinical Research Center grant RR00334, Oregon Regional Primate Research Center core grant RR00063 and the Clinical Nutrition Research Unit DK40566.

Table 1. The differing characteristics of ω3 and ω6 essential fatty acid deficiencies

	ω3	ω6
Clinical features	normal skin, growth and reproduction reduced learning abnormal electroretinogram impaired vision polydipsia	growth retardation skin lesions reproductive failure fatty liver polydipsia
Biochemical markers	decreased 18:3 ω3 and 22:6 ω3 increased 22:4 ω6 and 22:5 ω6 increased 20:3 ω9 (only if ω6 also low)	decreased 18:2 ω6 and 20:4 ω6 increased 20:3 ω9 (only if ω3 also low)

photoreceptors and in the synaptosomes of the cerebral cortex. Other functions of the ω3 fatty acids are currently being described, such as serving as precursors of prostaglandins and eicosanoids and perhaps also as inhibitors of the ω6 series of prostaglandins.

In contrast to the more obvious clinical stigmata of ω6 fatty acid deficiency, the findings of ω3 deficiency are far more subtle as might be expected for a fatty acid class required for neural membranes. ω3 fatty acid deficient rhesus monkeys appear grossly normal even after long-term depletion. The fur and the skin do not show the conspicuous dermatitis found in ω6 fatty acid deficiency. Fatty liver does not occur either. Table 1 illustrates the characteristic differences, clinically and biochemically, of the two essential fatty acid series, ω6 and ω3. Historically, of course, ω6 fatty acid deficiency has been well known and described in animals and humans [7, 8]. Until recently ω3 fatty acid deficiency, because it is less conspicuous in its symptomatology, has been less well categorized and has at times been combined with ω6 fatty acid deficiency [9–12].

Methods

Rhesus monkeys were studied after a combination of maternal and postnatal deprivation of dietary ω3 fatty acids. Throughout pregnancy, adult females were given a semipurified diet with safflower oil as the only fat source [1]. Their infants were fed the same liquid diet from birth onwards. Safflower oil is particularly low in ω3 fatty acids with less than 0.3% of the total fatty acids as α-linolenic acid (18:3 ω3). Safflower oil is especially high in ω6 fatty acids and the ratio of ω6 to ω3 fatty acids was about 250:1. Mothers and

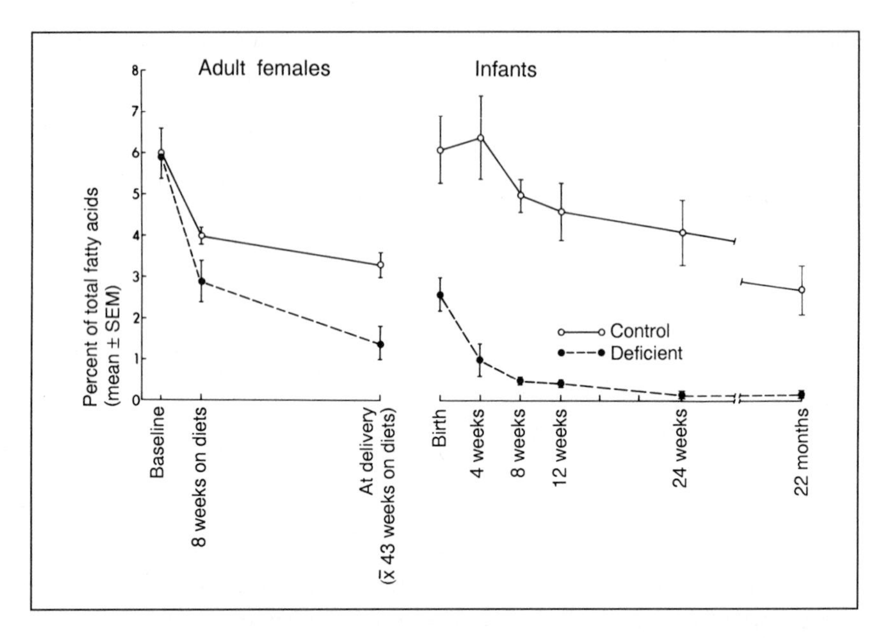

Fig. 1. Total ω3 fatty acids (weight percent of total fatty acids) in plasma phospholipids of adult female rhesus monkeys and their offspring fed control (soybean oil) and ω3 acid deficient (safflower oil) diets.

infants in the control group received diets with the same composition except that the fat source was soybean oil which provides 7.7% of total fatty acids as α-linolenic acid and which has an ω6 to ω3 ratio of 7:1. The more highly polyunsaturated fatty acids of the ω3 series such as eicosapentaenoic and docosahexaenoic were not present in these diets.

To test the reversibility of the deficiency, five of the deficient offspring were repleted with very long chain and highly polyunsaturated ω3 fatty acids from fish oil beginning at 10–24 months of age [13]. In the repletion diet, fish oil replaced 80% of the safflower oil, the remaining safflower oil providing ample ω6 linoleic acid (4.5% of calories) and the fish oil supplying large amounts of ω3 fatty acids, DHA and eicosapentaenoic acid (EPA, 20:5).

Plasma, erythrocyte and tissue samples, including whole retina and cerebral cortex, were anaylzed for fatty acid composition by capillary column gas-liquid chromatography after separation of total phospholipids or individual phospholipid classes by thin-layer chromatography [13]. In the repletion study, the fatty acids of plasma and erythrocytes were determined. In addition, in order to follow the time course of the biochemical changes, serial biopsies of the frontal cortical gray matter were obtained for fatty acid analysis after craniotomy.

Visual acuity was measured by the preferential-looking method as described previously [1]. The specific physiological effects of ω3 fatty acid deficiency on the function of the retina were examined by the electroretinogram [6].

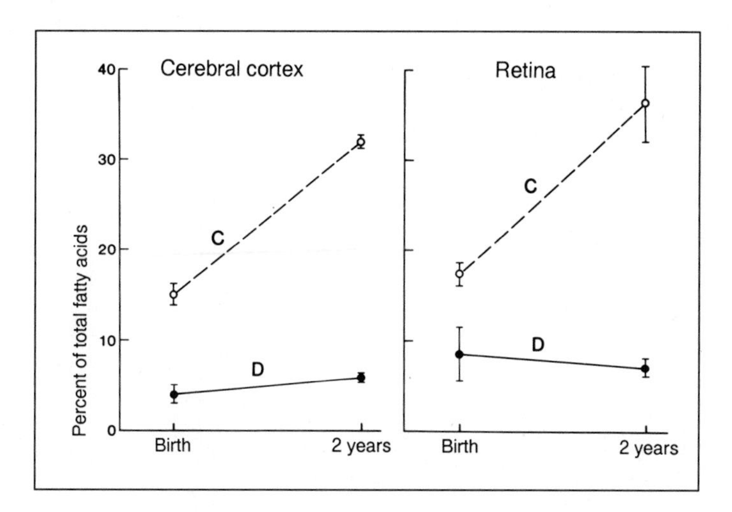

Fig. 2. The concentrations of docosahexaenoic acid in the cerebral cortex and retina of control animals (C) and deficient animals (D) as provided by birth and at 2 years. Note that at birth both retina and cerebral cortex had much higher levels of docosahexaenoic acid than deficient animals but these disparities increase greatly after 2 years of development.

Results

Fatty Acid Composition of Plasma and Tissues

Dietary deprivation of ω3 fatty acids resulted in low plasma and red blood cell concentrations of all ω3 fatty acids (fig. 1).

To be especially noted was the low concentration of ω3 fatty acids even at birth in the deficient monkeys. Their mothers also had ω3 fatty acid concentrations in plasma less than 50% of controls. In the deficient animals, the ω3 fatty acids were barely detectable by 24 weeks of age in the phospholipids of plasma and erythrocytes, in contrast to appreciable concentrations in the control animals.

The brain and retina had also very low levels of ω3 fatty acids and, in particular, docosahexaenoic acid. In deficient animals at or near birth, DHA concentrations in phosphatidylethanolamine were reduced by 50% in the retina and by 75% in the cerebral cortex compared to the control values (fig. 2). The proportion of DHA in both tissues doubled between birth and 22 months of age in control monkeys but failed to increase in the deficient group, so that by 22 months DHA concentrations in deficient monkeys were reduced to 15–20% of control values. It is notable that ω6 fatty acids,

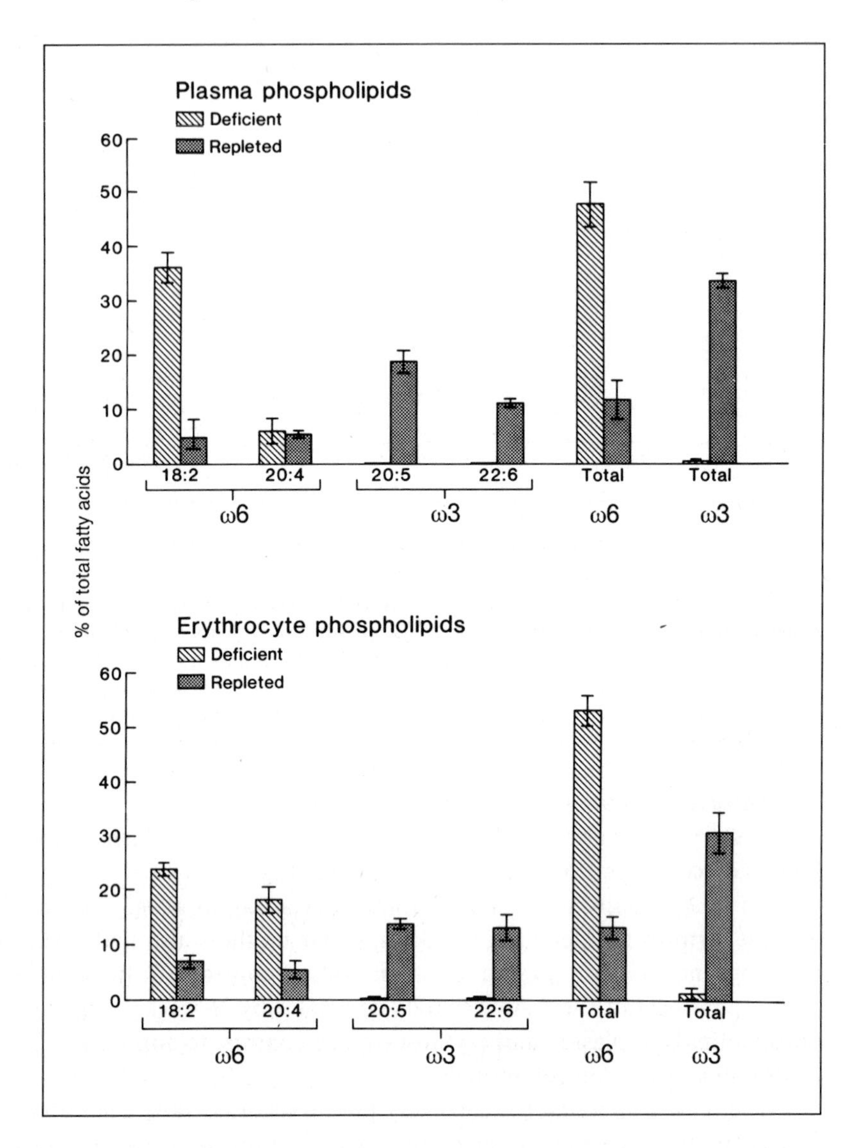

Fig. 3. The fatty acids of the plasma phospholipids and erythrocyte phospholipids in ω3 fatty acid deficient monkeys as compared with monkeys repleted with fish oil. Note the reciprocal relationships between the ω3 and ω6 fatty acids with ω3 fatty acids increasing after the fish oil repletion diet and the ω6 fatty acids diminishing greatly from the deficient state.

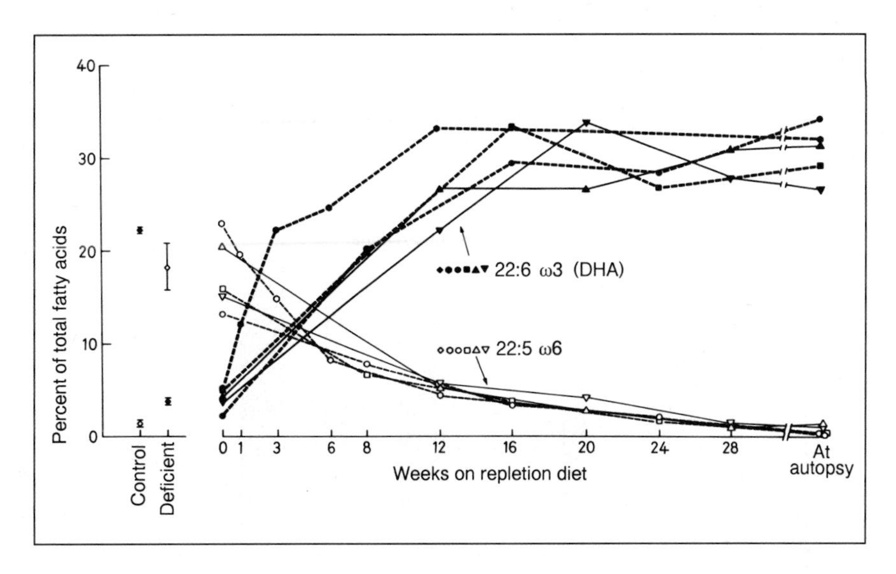

Fig. 4. The time course of fatty acid changes in phosphatidylethanolamine of the cerebral cortex of 5 juvenile monkeys fed fish oil for 43–129 weeks. As DHA increased, 22:5 ω6 decreased reciprocally. Levels of DHA and 22:5 ω6 in phosphatidylethanolamine of the frontal cortex of monkeys fed control (soybean oil) and deficient diets from previous study [4] are given for comparison. DHA and 22:5 ω6 in control monkeys were 22.3 ± 0.3 and 1.4 ± 0.3% of total fatty acids, respectively. DHA and 22:5 ω6 in the deficient monkeys were 3.8 ± 0.4 and 18.3 ± 2.5%, respectively.

22:5 ω6 particularly, compensatorily increased in response to the reduced DHA, so that polyunsaturation of the phospholipid membranes was preserved as much as possible. This ω6 fatty acid comprises less than 1% of total fatty acids in the tissue phospholipids of normal animals but in 22-month-old deficient animals rose to approximately 20% in the phosphatidylethanolamine of the cerebral cortex and nearly 30% in the retina.

After dietary repletion with fish oil, these changes in fatty acid composition were rapidly reversed as indicated by both changes in plasma phospholipids, red blood cells, and biopsy specimens of frontal cortex obtained by biopsy [13] (fig. 3). The changes in the fatty acid composition of the frontal cortex also occurred rapidly, as early as 1 week after fish oil supplementation. By 24–28 weeks the DHA in phosphatidylethanolamine increased from 4.2 to 29.3% of total fatty acids (fig. 4) compared to the 22.3% in soybean oil fed control animals. Of interest with regard to membrane function was the fact that EPA and 22:5 ω3 both increased from 0 to

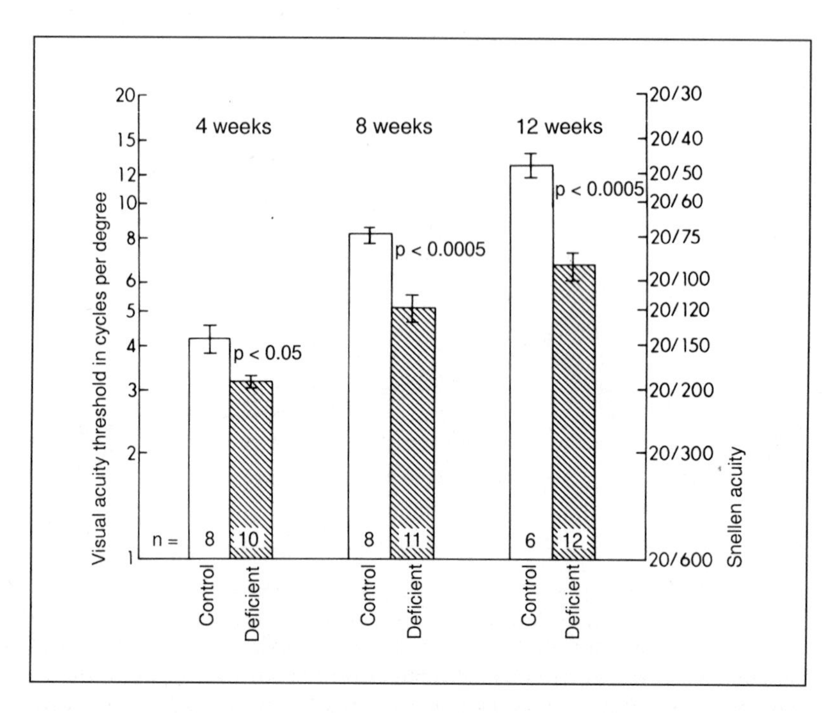

Fig. 5. Visual acuity thresholds (mean ± SEM) as determined by the preferential looking method for control and ω3 fatty acid-deficient infant monkeys. Thresholds are expressed in cycles per degree of visual angle and in the equivalent Snellen values. The p values for statistical significance were determined by Student's t test.

approximately 3% in the cerebral cortex. At the same time, the concentrations of 22:5 ω6 and other longer chain ω6 fatty acids decreased.

Visual Function and Electroretinograms

The visual acuity of ω3 fatty acid deficient infants was reduced by half at 8 and 12 weeks of age (fig. 5), as previously reported [1]. Furthermore, deficient monkeys developed a number of abnormalities in the electroretinogram (ERG) [6]. The timing of the ERG response was altered, with significant delays in the peak latency (time to the B-wave peak) of both cone and rod responses (fig. 6). In contrast to previous studies in ω3 fatty acid deficient rats, differences in response amplitudes were not detected at 7–24 months of age. However, more recent recordings of younger infants at 3–4 months of age have demonstrated clear differences in the A-wave

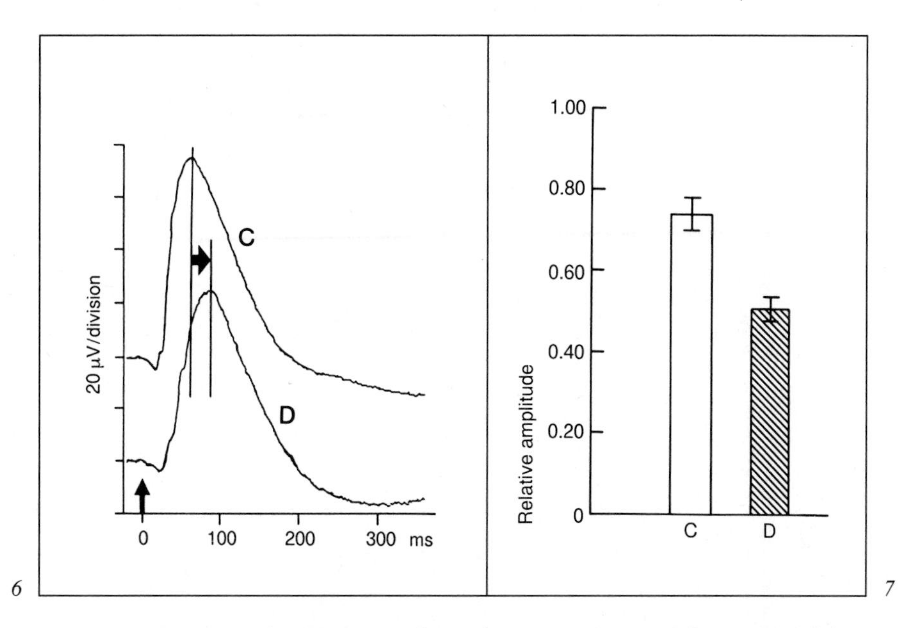

Fig. 6. Representative ERG wave forms from control and ω3 fatty acid deficient monkeys under conditions selectively stimulating the rod system. The vertical arrow indicates time of flash. B-wave peak latencies were delayed in the deficient group; vertical lines have been drawn through the peaks to aid comparison.

Fig. 7. Relative B-wave amplitude (mean ± SEM) of the ERG elicited by flashes at 3.2-second intervals, as a percent of the maximal amplitude produced at intervals of 20 s or more. Relative amplitude is significantly reduced in the deficient group ($p < 0.01$).

amplitudes of both rod and cone responses. The reason for the transient nature of this effect is unknown. Deficient animals also showed a specific abnormality in the rate of recovery of the ERG response after an initial bright flash. This effect was present at 3 months but increased in magnitude with age. With an interval of 3.2 s between flashes, response amplitude in the deficient animals was reduced nearly twice as much as in controls, relative to the maximal amplitude seen to the first flash or to flashes presented at long (≥ 20 s) intervals (fig. 7). Thus, recovery of the capacity to generate a full ERG response was significantly slowed.

In deficient monkeys repleted with fish oil, ERGs were recorded at 3, 6, and 9 months after the beginning of the repletion phase. Despite the increase in ω3 fatty acid levels in tissues, no improvement was seen in either peak latencies or in the ERG recovery function [13].

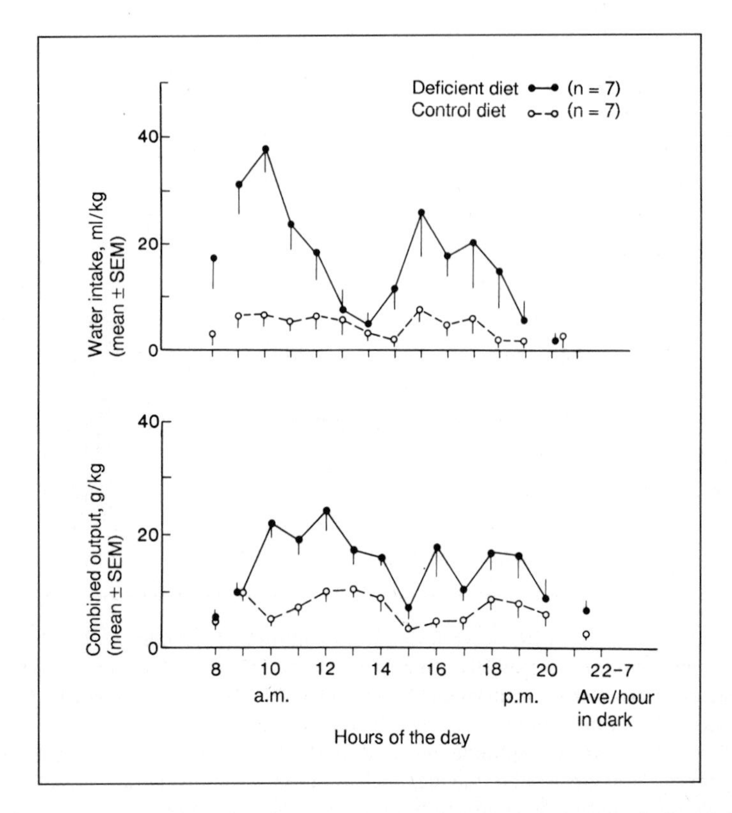

Fig. 8. The mean hourly water intake (ml/kg) over a 24-hour period and the hourly mean output of combined urine and feces (g/kg) over 24 h. The 'Ave/hour in dark' refers to the measurement taken at the end of the 10-hour dark period divided by 10.

Polydipsia in Deficient Monkeys

Cage behavior in ω3 fatty acid deficient and control monkeys was monitored by video taping. It was noticed that the deficient monkeys visited the water spouts of their cages much more often than did control monkeys, suggesting the possibility of greater water intake. A study of 24-hour water intakes was then carried out to confirm and quantify this difference [3]. Figure 8 depicts the mean water intake in milliliters per kilograms of body weight over 24 h. The intake was more than double for deficient over control monkeys (264 vs. 70 g/kg). The excretion or output was necessarily a combined output of feces and water because there was mixing of the soft stools and urine. The output was also more than double in deficient monkeys (268 vs. 121 g/kg) (fig. 8). These input and output studies were repeated on several different occasions with similar results in monkeys of varying ages from 1.8 months of age to young adults.

The mechanism of the polydipsia is not yet understood but several possible explanations can be eliminated. The effect is probably not caused by an osmotic imbalance. Dietary electrolytes and serum electrolytes were similar in both control and deficient animals. Fasting glucose levels were normal, indicating that polydipsia did not result from fluid losses as a result of diabetes. The blood urine nitrogen and creatinine levels were also equal and normal with no indication of renal insufficiency in either group.

Polydipsia does occur in ω6 dietary deficiency because of increased skin permeability and the resulting loss of water by evaporation. However, the skin and fur of the deficient animals appeared visually and histologically normal up to 2 years of age. The amount of ω6 fatty acids in the diet was actually 50% higher in deficient than in control monkeys. Increased fluid consumption activated by a diet deficient in calories was not present because the deficient and control monkeys received the same amount of food. Investigations to describe possible hormonal and prostaglandin mechanisms of the polydipsia are now in progress.

Discussion

This study has shown that dietary ω3 fatty acid deficiency leads to severe and progressive depletion of ω3 fatty acids from the plasma and from all tissues analyzed including red blood cells, liver, skin, fat, cerebral cortex and retina. In particular, the very long chain ω3 fatty acid, docosahexaenoic (DHA, 22:6), was selectively depleted from neural and retinal phospholipids and was replaced by ω6 fatty acids, in particular 22:5 ω6. Associated with these biochemical changes was a significant impairment in the development of visual acuity and abnormalities in the electroretinogram.

In the initial experiments, linolenic acid (18:3) was the only dietary source of ω3 fatty acids. In the control animals, DHA was synthesized by successive and multiple steps of desaturation and elongation from linolenic acid. DHA then selectively accumulated in brain and retinal phospholipids, particularly in the phosphatidylethanolamine and phosphatidylserine fractions [2]. The conversion of linolenic acid to DHA may occur in the liver, in the brain or retina, or in the placenta during fetal development. In the deficient monkeys, the dietary supply of linolenic acid was insufficient to support the synthesis of adequate levels of DHA, and high levels of dietary linoleic acid may also have suppressed the synthesis of DHA from the available linolenic acid [14]. However, during intrauterine life our experimental monkeys probably received some DHA directly from their mothers via the placenta. Although deficient mothers had been receiving the ω3 fatty acid-deficient diet for at least 2 months before conception, their stores of DHA probably were depleted very slowly. The plasmas of newborn deficient infants and their mothers revealed the presence of some DHA, the levels being consistently higher in the infant plasmas than in the mothers (selective biomagnification). However, plasma concentrations of DHA were much lower in the deficient infants than in control infants. Once the infants were removed from any maternal sources of ω3 fatty acids, their plasma levels fell rapidly. They were permanently separated from their mothers at birth, and therefore, did not have access to maternal milk, a natural source of DHA.

DHA is the predominant fatty acid in phospholipids of the retina, particularly in the outer segment membranes of the photoreceptor cells. This fatty acid is thought to be responsible for the special biophysical properties of the outer segment membranes, which contain the visual pigment [6]. Depletion of DHA from these membranes would be expected to alter their physical properties, and, therefore, the efficiency of the visual process. It is also possible that DHA has a more specific biochemical function in the retina. In addition, DHA is a major fatty acid of brain gray matter and especially of synaptic membranes which, like photoreceptor membranes, are excitable and highly fluid. Thus, the change in membrane phospholipid composition produced by ω3 fatty acid deficiency might alter the transmission of information through the brain's visual pathways as well as affecting the photoreceptive process in the retina.

A few cases of ω3 fatty acid deficiency have been described in humans. Holman et al. [9] recently reported a clinical case of peripheral neuropathy and blurred vision in a child receiving total parenteral nutrition. The symptoms were attributed to ω3 fatty acid deficiency because safflower oil, as in our study, was the sole fat source. Replacement of the safflower oil emulsion with linolenic acid-rich soybean oil was associated with recovery. However, it is difficult to be certain that the symptoms were due to ω3 fatty acid deficiency rather than to some other metabolic disturbance induced by long-term parenteral nutrition.

Most recently Bjerve et al. [10, 11] described 10 cases of 'linolenic acid deficieny', occurring in a number of nursing-home patients in Norway, some semicomatose, who were fed by gastric tube over several years. The diet of these patients was based on a commercially available powder supplement containing small amounts of corn oil (1.3 g/100 g) and mixed with skim milk. The authors reported that the patients developed very low plasma levels of ω3 fatty acids and a scaly dermatitis. Corn oil has an especially low content of the ω3 linolenic acid (0.3%) but, if supplied in quantity would have ample ω6 fatty acid (i.e. linoleic 18:2) to meet the needs for ω6 essential fatty acids. However, the diet furnished only 0.5% of energy as 18:2 ω6, plasma 18:2 ω6 was low (9–15% of total plasma fatty acids vs. 35% in healthy control subjects [12]) and levels of eicosatrienoic acid (20:3 ω9) were substantially elevated. These observations suggest that the patients were also deficient in ω6 fatty acids and that perhaps the skin lesions occurred from the ω6 deficiency since dermatitis has not been found in animals having pure ω3 fatty acid deficiency (see table 1). However, these 10 patients certainly did have ω3 fatty acid deficiency in addition to the ω6 deficiency, i.e. a combined essential fatty acid deficiency.

The findings of our study provide the first experimental evidence for a dietary requirement for ω3 fatty acids in primates and add to the extensive studies in rats emphasizing the essentiality of the ω3 fatty acid, linolenic acid [14]. ω3 fatty acids are essential nutrients for retinal and brain function, especially during fetal and postnatal development. Both biochemical and functional stigmata are characteristic of the ω3 fatty acid deficient state. A recent review details the experimental background of ω3 fatty acid deficiency in different species of animals [15].

Summary and Implications

In summary, ω3 fatty acid deficiency is characterized in rhesus monkeys by a triad of problems: impaired vision, abnormal electroretinograms, and polydipsia. Concomitant with the visual and ERG defects is the disturbed biochemistry of the retinal outer segment membranes: a great decrease in docosahexaenoic acid (22:6 ω3) in the retinal phospholipids and a reciprocal increase of the ω6 fatty acids, particularly 22:5 ω6. In the absence of ω3 fatty acids, the body produces the most similar polyunsaturated fatty acids possible from precursors, so that the amount of polyunsaturation of the tissue membranes is largely maintained. The cause of the polydipsia may also be related to the same altered biochemistry of the brain. Polydipsia does not seem mediated by abnormalities of the posterior pituitary hormones, renal disease or osmotic regulation. However, the precise mechanism of this phenomenon remains to be elucidated.

The diagnosis of ω3 fatty acid deficiency can be made at birth and later in life by fatty acid determinations of plasma and red cells of cord and peripheral venous blood. There will be low levels of the ω3 fatty acids 18:3 and 22:6 and high levels of the ω6 22:5, normally very low in amount in both plasma and tissues of primates. The deficiency state at any age can be prevented by the provision of adequate amounts of ω3 fatty acids in the diet. It is not yet known whether there should be several ω3 fatty acids in the diet for optimal development; i.e., linolenic acid and DHA, both being found in human milk. To be stressed is the point that once the deficiency state is well developed, by 10 months of age for example, biochemical correction in the tissues may not necessarily restore all the functional defects, especially, in our studies the abnormal electroretinogram.

The most critical periods of life for providing adequate ω3 fatty acids would be during pregnancy, with placental transfer of these fatty acids to the fetus; in infancy when there continues to be considerable accumulation of ω3 fatty acids in the brain and retina; during lactation to supply these fatty acids postnatally. The diet of nursing mothers will affect the ω3 content of their milk also [16]. If formula feeding is provided, there is the necessity of having adequate and balanced sources of both ω3 and ω6 fatty acids in the fats and oils of the formula. In all probability, there are continuing requirements for ω3 fatty acids during childhood and even in adult life. Elderly and other patients may be at risk for ω3 fatty acid deficiency. We suggest that the dietary requirements for ω3 fatty acids are from 0.5 to 1% of total calories and ideally might include both 18:3 and 22:6 as does human milk. The ratio of ω6 to ω3 fatty acids is important and should range from about 4 to 12. Too high a ratio, as is present in the ω3 fatty acid deficient safflower oil, might further accentuate the deficient state. To be noted is the fact that some infant formulas, particularly the dry powdered formulas containing corn and coconut oil, are marginal in their supply of ω3 fatty acids which are as low as 0.1–0.2% of total calories. The ratio of ω6 to ω3 fatty acids is also high, at 75 [Connor WE, Van Winkle S: The fatty acid composition of human infant formulas; to be submitted for publication].

These studies support the conclusion that there should be adequate amounts of both ω3 and ω6 fatty acids in the diet throughout life and that their ratio is of great importance. This is particularly important for infants fed artificial formulas whose fat sources include coconut oil, corn oil, or safflower oil, and which may not meet the criteria of having adequate ω3 fatty acids and not an excess of ω6 fatty acids. Ideally, the fat content and fatty acid composition of infant formulas should resemble human milk. This objective seems reasonable and technologically feasible.

References

1 Neuringer M, Connor WE, Van Petten C, Barstad L: Dietary omega-3 fatty acid deficiency and visual loss in infant rhesus monkeys. J Clin Invest 1984;73:272–276.
2 Connor WE, Neuringer M, Barstad L, Lin DS: Dietary deprivation of linolenic acid in rhesus monkeys: Effects on plasma and tissue fatty acid composition and visual function. Trans Assoc. Am Phys, Vol. 97, 1984.
3 Reisbick S, Neuringer M, Hasnain R, Connor WE: Polydipsia in rhesus monkeys deficient in omega-3 fatty acids. Physiol Behav 1990;47:315–323.
4 Neuringer MD, Connor WE, Lin DS, Barstad L, Luck S: Biochemical and functional effects of prenatal and postnatal omega-3 fatty acid deficiency on retina and brain in rhesus monkeys. Proc Natl Acad Sci 1986;83:4021–4025.
5 Lin DS, Connor WE, Anderson GJ, Neuringer M: The effects of dietary ω3 fatty acids upon the phospholipid molecular species of monkey brain. J. Neurochem. 190; (in press).
6 Neuringer M, Connor WE: Omega-3 fatty acids in the retina; in Galli C, Simopoulos AP (eds): Dietary ω3 and ω6 Fatty Acids: Biological Effects and Nutritional Essentiality. New York, Plenum Publishing, 1989, pp 177–190.
7 Brown WR, Hansen AE, Burr GD, McQuarrie I: Effects of prolonged use of extremely low-fat diet on an adult human subject. J Nutr 1938;16:511–524.
8 Holman RT: Essential fatty acid deficiency. Prog Chem Fats Other Lipids 1970;9: 275–339.
9 Holman RT, Johnson SB, Hatch TF: A case of human linolenic acid deficiency involving neurological abnormalities. Am J Clin Nutr 1982;35:617–623.
10 Bjerve KS, Mostad IL, Thoresen L: Alpha-linolenic acid deficiency in patients on long-term gastric tube feeding: Estimation of linolenic acid and long-chain unsaturated ω3 fatty acid requirement in man. Am J Clin Nutr 1987;45:66–77.
11 Bjerve KS, Fischer S, Alme K: Alpha-linolenic acid deficiency in man: Effect of ethyl linolenate on plasma and erythrocyte fatty acid composition and biosynthesis of prostanoids. Am J Clin Nutr 1987;46:570–576.
12 Wene JD, Connor WE, DenBesten L: The development of essential fatty acid deficiency in healthy men fed fat-free diets intravenously and orally. J Clin Invest 1975; 56:127–134.

13 Connor WE, Neuringer M, Lin DS: Dietary effects upon brain fatty acid composition: The reversibility of ω3 fatty acid deficiency and turnover of docosahexaenoic acid in the brain, erythrocytes and plasma of rhesus monkeys. J Lipid Res 1990;31:237–248.
14 Tinoco J: Dietary requirements and functions of α-linolenic acid in animals. Prog Lipid Res 1982;21:1–45.
15 Neuringer M, Anderson GJ, Connor WE: The essentiality of ω3 fatty acids for the development and function of the retina and brain. Ann Rev Nutr 1988;8:517–541.
16 Harris WS, Connor WE, Lindsey S: Will dietary omega-3 fatty acids change in composition of human milk? Am J Clin Nutr 1984;40:780–785.

William E. Connor, MD, Oregon Health Sciences University,
Department of Medicine, L-465, 3181 SW Sam Jackson Park Road,
Portland, OR 97201 (USA)

Simopoulos AP, Kifer RR, Martin RE, Barlow SM (eds): Health Effects of ω3 Polyunsaturated Fatty Acids in Seafoods. World Rev Nutr Diet. Basel, Karger, 1991, vol 66, pp 133–142

ω3 Fatty Acid Deficiency in Man: Implications for the Requirement of Alpha-Linolenic Acid and Long-Chain ω3 Fatty Acids

Kristian S. Bjerve

Department of Clinical Chemistry, University of Trondheim, Regional Hospital, Trondheim, Norway

Introduction

The first report on the essentiality of fats in children was published as early as 1919 by von Grøer [1]. However, it was not until the work of Burr and Burr [2, 3] that it was recognized that it was specifically a lack of dietary ω6 fatty acids that caused the development of essential fatty acid deficiency. The clinical symptoms and the risk of ω6 deficiency in infants were described extensively through the work of Hansen and coworkers [4]. From their data, as well as from those presented by Holman and coworkers [5–7], recommendations for dietary allowances could be given, and criteria for the diagnosis of ω6 deficiency were made. In 1977, FAO/WHO published internationally recommended dietary allowances for ω6 fatty acids [8]. These, together with similar national recommended dietary allowances, have made an important contribution in securing an optimal supply of essential fatty acids in patients that need prolonged parenteral and/or peroral feeding.

Essentiality of ω3 Fatty Acids

Unlike the ω6 fatty acids, the nutritional essentiality of ω3 fatty acids in man and mammals has until recently been uncertain. The ω3 acids can not be synthesized by man or mammals, but must be supplied through the diet. Dietary alpha-linolenic acid (18:3 ω3) can be further elongated and desaturated by man and mammals, and therefore also serves as a precursor for the long-chain ω3 fatty acids eicosapentaenoic acid (20:5 ω3) and docosahexaenoic acid (22:6 ω3), which are found in the cell membrane lipids of all mammals examined, and most abundantly in the retina and the brain [9, 10]. Eicosapentaenoic and docosahexaenoic acid are both dietary sources of essential ω3 fatty acids, as well as direct precursors of a wide range of biologically important lipid mediators which seem to have important roles in cellular signalling [11, 12], protection against cardiovascular disease [13–15], lowering blood pressure [16, 17] and modifying inflammatory response [18, 19].

In spite of this, it has been difficult to produce overt deficiency symptoms even after prolonged deprivation of ω3 fatty acids [10, 20]. However, it has been reported that ω3 fatty acid deficiency produces skin symptoms in the capuchin monkey [21], visual dysfunction in primates [22, 23], and learning impairment [24, 25] as well as perinatal mortality in the rat [26].

ω3 Fatty Acid Deficiency in Man

The first published case of ω3 fatty acid deficiency was a 6-year-old girl presenting with neurological and visual disturbances [27]. She had received parenteral nutrition with a high linoleic acid and low linolenic acid content as her only source of fat for 5 months, and the clinical symptoms disappeared after changing to a different preparation containing both linoleic and linolenic acid. In 1985, it was found that a powdered preparation used for gastric tube feeding in Norway supplied very low amounts of ω3 fatty acids when it was mixed with water, or with water and skimmed milk. A systematic search throughout Norway revealed a total of 8 patients who had received this low-fat, low-linolenic-acid formula for 2.5–12 years [28–30]. During this search, one 7-year-old girl who had received gastric tube feeding for several years using a different, low-linolenic-acid liquid formula was also found [31]. Together with the cases described by Holman et al. [27] and Stein et al. [32], a total of 11 cases with ω3 fatty acid deficiency have been reported. The patient described by Stein and coworkers received 7 g of linolenic acid per week, although it was later reported that the amount probably was lower [33]. This patient is therefore not included in the following discussion on dietary requirement of ω3 acids.

The intake of linolenic acid in the ω3 deficient patients ranged from 28 to 130 mg per day, while the intake of long-chain ω3 acids (20:5 ω3, 22:5 ω3 plus 22:6 ω3) ranged from 0 to 34 mg (table 1). Although most of

Table 1. Daily intake of ω3 fatty acids in patients developing ω3 fatty acid deficiency

Case	No.	Linolenic acid		Long-chain ω3 acids	
		mg/d	E%[1]	mg/d	E%
1	[27][2]	130	0.07	–	–
2	[28]	36	0.02	34	0.02
3	[28]	29	0.02	10	0.01
4	[28]	28	0.02	19	0.01
5	[28]	36	0.02	24	0.01
6	[29]	35	0.02	16	0.01
7	[30]	46	0.05	4	0.01
8	[30]	115	0.09	6	0.01
9	[30]	76	0.09	4	0.01
10	[31]	37	0.07	ND[3]	ND

[1] E% = Percent of energy intake.
[2] Data are collected from the references given in parenthesis.
[3] Not detected.

the patients had a very low linolenic acid intake for more than 2.5 years, case 1 and 7 received 130 and 46 mg linolenic acid, and 0 and 4 mg of long-chain ω3 acids, respectively, for 5 months only. This does, however, not indicate that this is sufficient to produce clinical symptoms of ω3 deficiency. Shorter or longer periods of caloric restriction resulting in a negative energy balance contributed in several of the reported patients with overt ω3 fatty acid deficiency.

Diagnosis of ω3 Fatty Acid Deficiency

The clinical symptoms reported in ω3 fatty acid deficiency are quite diverse, and all are nonspecific (table 2). Neurological and visual dysfunction are both generally acknowledged as symptoms of ω3 fatty acid deficiency. It has been questioned whether dermatological symptoms are part of a ω3 deficiency [20], and growth retardation has so far been reported only in one single patient [31]. The question whether skin symptoms can specifically be produced by ω3 fatty deficiency is intriguing. Several investigations and reviewers have stated that ω3 deficiency does not give skin symptoms. Nevertheless, Holman et al. [27] reported extensive abdominal

Table 2. Symptoms reported in ω3 fatty acid deficiency

Clinical	Biochemical
Impaired visual acuity [22, 23]	Decreased 20:5 ω3, 22:5 ω3 and 22:6 ω3 [22, 23, 27–31]
Abnormal retinograms [23]	Increased 22:4 ω6 and 22:5 ω6 [22, 23, 27–31]
Learning deficiency [24, 25]	Hyperreactivity of lymphocytes [30]
Growth retardation [31]	Impaired metabolism of ω6 fatty acids [29, 30, 35–37]
Neurological dysfunction [27]	
Hemorrhagic folliculitis [30]	
Dermatitis [27–30]	

References are given in parentheses.

and perineal dermatitis in his case, and all 8 adults reported by Bjerve and coworkers [28–30] showed some form of skin symptoms. Further, eicosapentaenoic acid is converted in the skin to 15-lipoxygenase products that inhibit leukotriene B_4 generation [34], and some of the ω3 deficient patients had skin symptoms that could be explained by a hyperreactivity of immune-competent cells. Since it is not possible to diagnose ω3 fatty acid deficiency through clinical symptoms, one must rely on a combination of biochemical and nutritional information (tables 2 and 3).

Pathogenesis of Clinical Symptoms

The clinical symptoms reported in ω3 deficiency are neurological and visual disturbances [22, 23, 27], dermatitis and folliculitis [28–30], and growth retardation [31]. The skin symptoms are different from those in ω6 deficiency. In case 7 (table 1) who had a combined ω3 and ω6 deficiency, supplementing with ω3 acids normalized both the hemorrhagic dermatitis and folliculitis, while a coarse, scaly dermatitis localized mainly to the legs was unaffected, and a high concentration of 20:3 ω9 in plasma and red cell lipids remained unchanged. These latter symptoms only normalized after supplementing with ω6 acids, strongly indicating that the skin symptoms in ω3 fatty acid deficiency are different from ω6 fatty acid deficiency, both clinically as well as biochemically. In fact, the skin might be only a secondary target organ in ω3 deficiency, mainly being affected through a hyperreactivity of immune-competent cells [30].

Table 3. Criteria for the diagnosis of ω3 fatty acid deficiency in man

(1)	Long-term feeding on a completely defined diet supplying alpha-linolenic acid at less than approximately 50–100 mg/day, and long-chain ω3 acids less than 20–30 mg/day.
(2)	Low intake of ω3 fatty acids should be confirmed by low concentration of 20:5 ω3 and 22:6 ω3 in plasma and/or cellular lipids.
(3)	Disappearance of clinical symptoms after supplementing with pure ω3 fatty acids to an otherwise unchanged diet.
(4)	Verify a concomitant increase in ω3 fatty acid concentration in plasma and/or cellular lipids.

All three patients investigated showed a hyperreactivity of isolated lymphocytes and a decreased level of T cells, which was normalized after supplementing with alpha-linolenic acid [30]. This could explain why the skin symptoms in some of the reported cases were characterized by inflammatory changes of the skin and hair follicles. One of the pathogenetic mechanisms of skin changes in ω3 fatty acid deficiency could therefore be an impaired function of immune-competent cells, leading to a state of immune system hyperreactivity.

One 7-year-old girl presented with growth retardation [31], and supplementing with ω3 fatty acids induced rapid growth and weight gain [31]. One of the adults presented with a treatment-resistant pressure ulcer [30] which began to heal 10 days after supplementing with pure alpha-linolenic acid. This suggests that ω3 acids in some way are essential for normal cell growth.

Finally, in several ω3 deficient patients, the ω3 fatty acid supplement caused an accumulation not only of ω3 acids, but also of ω6 acids [29, 30]. This suggests that ω3 deficiency impairs, or restricts, the accumulation of ω6 fatty acids, and that they are required for the normal metabolism and incorporation of ω6 fatty acids into membrane lipids. This is supported by the findings in chick nutritional encephalomalacia [35], where alpha-linolenic acid completely protects chicks against an otherwise lethal effect of ω6 acids [36, 37]. It must be emphasized, however, that the paucity of information on human ω3 fatty acid deficiency restrains any conclusions to what are the mechanisms behind the observed clinical symptoms. It is, however, very likely that at least some of the symptoms could be attributed to the well-documented effects of ω3 fatty acids in skin and on immune-competent cells [18, 19, 34].

Dietary Requirements of ω3 Fatty Acids

When addressing the question of recommended dietary allowances of ω3 fatty acids, three different topics should be considered. Firstly, what is the minimal dietary level necessary to avoid overt ω3 fatty acid deficiency? Secondly, what is the optimal dietary intake that will ensure health, normal cell function, and give optimal protection against disease? Thirdly, what is the optimal intake required to obtain maximal therapeutic effects in diseases beneficially affected by ω3 fatty acid intake? The first two points will be discussed below.

Although accurate estimates of minimal and optimal dietary requirements of ω3 fatty acids cannot be obtained from 10 heterogeneous cases of ω3 fatty acid deficiency, the paucity of available data from humans demands that the available data are studied and used in the best possible way. If the optimal dietary intake is defined as the intake giving a mid-normal concentration of ω3 fatty acids in plasma and red cell lipids, observations from 6 of the reported patients can be used to estimate optimal dietary intakes [27, 29–31], while all 10 reported patients can be used to estimate minimal requirements.

Using these data, the minimal dietary requirement of alpha-linolenic acid in the absence or near absence of long-chain ω3 fatty acids was estimated to 290–390 mg/day (table 4). Similarly, the minimal dietary requirement of long-chain ω3 fatty acids in the presence of a low intake of

Table 4. Estimates of dietary requirements of ω3 fatty acids in man

		Optimal requirement	Minimal requirement
Linolenic acid[1]	mg/day	860–990	290–390
	mg/day	1,220	–
	energy %	1.0–1.2	0.2–0.3
	energy %	0.54	–
Long-chain ω3[2]	mg/day	350–400	100–200
	energy %	0.4	0.1–0.2

The estimates are based on nutritional information and fatty acid changes observed in 10 patients with ω3 fatty acid deficiency after supplementing with ω3 fatty acids.

[1] In the absence of dietary long-chain ω3 fatty acids.

[2] Dietary intake of linolenic acid below 100 mg/day.

alpha-linolenic acid was estimated to 100–200 mg/day. Optimal dietary requirement of alpha-linolenic acid in the absence of long-chain ω3 acids was estimated to be 860–1,220 mg/day (table 4), while the optimal dietary requirement of long-chain ω3 fatty acids was estimated to be 350–400 mg/day. Long-chain ω3 fatty acids thus seem to be 2–3 times more effective than alpha-linolenic acid in curing and probably preventing clinical symptoms of ω3 fatty acid deficiency.

Implementing ω3 Dietary Recommendations

All patients reported so far have had an iatrogenic ω3 fatty acid deficiency caused by the use of commercially available nutrients used for enteral and parenteral feedings. These nutrients are used to treat patients in a wide variety of clinical diseases and conditions, many of them found in the very young or the very old patients. Further, they are partly used to treat life-lasting diseases, or used to treat patients that have previously been exposed to severe trauma or long-lasting malnutrition (table 5). If one now looks at what kind of declaration of essential fatty acid content is given by the manufacturers of these nutrients, it can be seen that of the 12 different brands listed in the Drug Catalogue of Norway (which is distributed to all Norwegian physicians by the drug industry), only one states its content of ω3 fatty acids (table 6). It is further also clear that this content is far below that of any future dietary recommendation of ω3 fatty acids. One brand only states that ω3 and ω6 fatty acids are present. The other 10 nutrients do not specify their content of ω3 fatty acids.

When one looks at the sources of fat used (table 6), it becomes clear that several of these nutrients can not supply sufficient amounts of essen-

Table 5. Indications for the use of artificial peroral nutrients according to information published by producers of preparations available in Norway 1989

Intolerance against	Mother's milk	Galactosemia
	Cow milk	Malabsorption of fat
	Lactose	Malnutrition
	Gluten	Mother's milk substitute
	Saccharose	Post-operative treatment
	Soya protein	Pre-operative treatment
		Total gastric tube feeding

Table 6. Declaration of essential fatty acid concentration in peroral nutrients in Norway

Brand	Fat, E%	ω6, E%	ω3, E%	Fat source
A	3.3–6.4	1.2	0.02	not given
B	35	present	present	MCT, soybean oil, evening primrose oil
C	33	6.6	not given	palm and sunflower oil
D	48	13.9	not given	corn oil, coconut oil
E	35	12.4	not given	MCT, corn oil
F	30	15.9	not given	MCT, sunflower oil
G	36	23.4	not given	soybean oil
H	47	16	not given	lard, palm oil, sunflower oil
I	35	not given	not given	corn oil
J	36	not given	not given	MCT, sunflower oil
K	2–8	not given	not given	not given
L	0.5	not given	not given	not given

Included are nutrients registered in Norway in 1989. The table is based on information given in the Norwegian Drug Catalogue, which is provided by the drug industry. Most manufacturers can give more detailed information on ω6 and ω3 fatty acid content upon request. E% = Percent of energy intake.

tial ω3 fatty acids. This is a serious situation when one considers the type of patients that will receive these nutrients (table 5). To secure critically ill patients a sufficient supply of essential ω3 fatty acids, especially those requiring long-term peroral and/or parenteral feeding, it is necessary that officially recommended dietary allowances are rewritten to include defined amounts of ω3 fatty acids. It is not sufficient just to state that ω3 fatty acids are essential nutrients. Presently, this has been done only in the Nordic Nutrition Recommendations [38]. Finally, such future dietary allowances for ω3 fatty acids must be implemented and used to secure that nutrients used to treat critically ill patients with long-standing disease will cover the needs of essential ω3 fatty acids.

References

1 von Grøer F: Zur Frage der praktischen Bedeutung des Nährwertbegriffes nebst einigen Bemerkungen über das Fettminimum des menschlichen Säuglings. Biochem Z 1919;93:311–329.

2 Burr GO, Burr MM: A new disease produced by the rigid exclusion of fat from the diet. J Biol Chem 1929;82:345–367.

3 Burr GO, Burr MM: On the nature and role of the fatty acids essential in nutrition. J Biol Chem 1930;86:587–619.
4 Hansen AE, Wiese HF, Boelsche AN, et al: Role of linoleic acid in infant nutrition. Pediatrics 1963;31(suppl):171–192.
5 Holman RT, Caster WO, Wiese HF: The essential fatty acid requirement of infants and the assessment of their dietary intake of linoleate by serum fatty acid analysis. Am J Clin Nutr 1964;14:70–75.
6 Paulsrud JR, Pensler L, Whitten CF, et al: Essential fatty acid deficiency in infants induced by fat-free intravenous feeding. Am J Clin Nutr 1972;25:897–904.
7 Holman RT: Essential fatty acids in human nutrition. Adv Exp Med Biol 1977;83: 515–534.
8 FAO/WHO: Dietary fats and oils in human nutrition. Paper 3. Geneva, Food and Agriculture Organization, World Health Organization, 1977.
9 Crawford MA, Casperd NM, Sinclair AJ: The long-chain metabolites of linoleic and linolenic acids in liver and brain in herbivores and carnivores. Comp Biochem Physiol 1976;54B:395–401.
10 Tinoco J: Dietary requirements and functions of alpha-linolenic acid in animals. Prog Lipid Res 1981;21:1–45.
11 Strasser T, Fischer S, Weber PC: Leukotriene B_5 is formed in human neutrophils after dietary supplementation with icosapentaenoic acid. Proc Natl Acid Sci, USA 1985;82:1540–1543.
12 Knapp HR, Reilly IAG, Alessandrini P, et al: In vivo indexes of platelet and vascular function during fish-oil administration in patients with atherosclerosis. N Engl J Med 1986;314:937–942.
13 Leaf A, Weber PC: Cardiovascular effects of n–3 fatty acids. N Engl J Med 1988; 318:549–557.
14 Dehmer GJ, Popma JJ, van den Berg EK, et al: Reduction in the rate of early restenosis after coronary angioplasty by a diet supplemented with n–3 fatty acids. N Engl J Med 1988;319:733–740.
15 Burr ML, Fehily AM, Gilbert JF, et al: Effect of changes in fat, fish, and fibre intakes on death and myocardial reinfarction: Diet and reinfarction trial (DART). Lancet 1989;ii:757–761.
16 Knapp HR, FitzGerald GA: The antihypertensive effects of fish oil. A controlled study of polyunsaturated fatty acid supplements in essential hypertension. N Engl J Med 1989;320:1037–1043.
17 Bønaa KH, Bjerve KS, Straume B, et al: Effect of eicosapentaenoic and docosahexaenoic acids on blood pressure in hypertension. A population based intervention trial from the Tromsø study. N Engl J Med 1990;322:795–801.
18 Kremer JM, Bigauette J, Michalek AV, et al: Effects of manipulation of dietary fatty acids on clinical manifestations of rheumatoid arthritis. Lancet 1985;i:185–187.
19 Lee TH, Hoover RL, Williams JD, et al: Effect of dietary enrichment with eicosapentaenoic and docosahexaenoic acids on in vitro neutrophil and monocyte leukotriene generation and neutrophil function. N Engl J Med 1985;312:1217–1224.
20 Anderson GJ, Connor WE: On the demonstration of omega–3 essential-fatty-acid deficiency in humans. Am J Clin Nutr 1989;49:585–587.
21 Fiennes RNT-W, Sinclair RN, Crawford MA: Essential fatty acid studies in primates. Linolenic acid requirements of Capuchins. J Med Primate 1973;2:155–169.

22 Neuringer M, Connor WE, van Petten C, et al: Dietary omega-3 fatty acid deficiency and visual loss in infant Rhesus monkeys. J Clin Invest 1984;73:272–276.
23 Neuringer M, Connor WE, Lin DS, et al: Biochemical and functional effects of prenatal and postnatal omega-3 fatty acid deficiency on retina and brain in rhesus monkeys. Proc Natl Acad Sci, USA 1986;83:4021–4025.
24 Yamamoto N, Hashimoto A, Takemoto Y, et al: Effect of the dietary alpha-linolenate/linoleate balance on lipid compositions and learning ability of rats. II. Discrimination process, extinction process, and glycolipid compositions. J Lipid Res 1988; 29:1013–1021.
25 Lamptey MS, Walker BL: A possible essential role for dietary linolenic acid in the development of the young rat. J Nutr 1976;106:86–93.
26 Guesnet P, Pascal G, Durand G: Dietary alpha-linolenic acid deficiency in the rat. Effects on reproduction and postnatal growth. Reprod Nutr Dev 1986;26:969–985.
27 Holman RT, Johnson SB, Hatch T: A case of human linolenic acid deficiency involving neurological abnormalities. Am J Clin Nutr 1982;35:617–623.
28 Bjerve KS, Løvold Mostad I, Thoresen L: Alpha-linolenic acid deficiency in patients on long term gastric tube feeding: Estimation of linolenic acid and long-chain unsaturated n–3 fatty acid requirement in man. Am J Clin Nutr 1987;45:66–77.
29 Bjerve KS, Fischer S, Alme K: Alpha-linolenic acid deficiency in man. Effect of ethyl linolenate on fatty acid composition and biosynthesis of eicosanoids. Am J Clin Nutr 1987;46:570–576.
30 Bjerve KS, Fischer S, Wammer F, et al: Alpha-linolenic acid and long-chain n–3 fatty acid supplementation in three patients having n–3 fatty acid deficiency. Effect on lymphocyte function, plasma and red cell lipids, and prostanoid formation. Am J Clin Nutr 1989;49:290–300.
31 Bjerve KS, Thoresen L, Børsting S: Linseed and cod liver oil induce rapid growth in a 7-year old girl with n–3 fatty acid deficiency. J Parent Enteral Nutr 1988;12: 521–525.
32 Stein TP, Marino PL, Harner RN, et al: Case report: Linoleate and possibly linolenate deficiency in a patient on long-term intravenous nutrition at home. J Am Coll Nutr 1983;2:241–247.
33 The Editor: Clinical nutrition cases: Combined EFA Deficiency in a patient on long-term TPN. Nutr Reviews 1986;44:301–305.
34 Ziboh VA: Implications of dietary oils and polyunsaturated fatty acids in the management of cutaneous disorders. Arch Dermatol 1989;125:241–245.
35 Dam H, Nielsen GK, Prange I, et al: Influence of linoleic and linolenic acid on symptoms of vitamin E deficiency in chicks. Nature 1958;182:802–803.
36 Budowski P, Crawford MA: Effect of dietary linoleic and alpha-linolenic acids on the fatty acid composition of brain lipids in the young chick. Prog Lipid Res 1986; 25:615–618.
37 Budowski P, Hawkey CM, Crawford MA: L'effet protecteur de l'acide alpha-linolenique sur l'encephalomalacie chez le poulet. Ann Nutr Alim 1980;34:389–400.
38 Nordic Nutrition Recommendations, 2nd Edition, 1988.

Prof. Kristian S. Bjerve, MD, PhD, Department of Clinical Chemistry,
University of Trondheim, Regional Hospital, N–7006 Trondheim (Norway)

Simopoulos AP, Kifer RR, Martin RE, Barlow SM (eds): Health Effects of ω3 Polyunsaturated Fatty Acids in Seafoods. World Rev Nutr Diet. Basel, Karger, 1991, vol 66, pp 143–160

Polyunsaturated Fatty Acids and the Health of the Elderly[1]

Daniel Rudman, Mary E. Cohan

Division of Geriatrics, Department of Medicine, Medical College of Wisconsin and the Clement J. Zablocki Veterans Affairs Medical Center, Milwaukee, Wisc., USA

History

The last four decades have seen much progress in the understanding and use of dietary fatty acids. In the 1950s, attention was focused mainly on how the *quantity* of fat in the diet influenced health and disease. In the 1960s, it was shown that switching from saturated and monounsaturated animal fat to vegetable polyunsaturated fat caused a lowering of the serum cholesterol level. During the 1970s, the bioactive eicosanoids were discovered. Early in the 1980s, attention was called to the ω3 fatty acids which are derived from sea foods. Also during the last decade the differential effects on serum lipid levels of saturated versus monounsaturated fatty acids were described. Now in 1990, a diet intervention must consider not only the quantity of fat, but also the proportions of saturated, monounsaturated, total ω6 and total ω3 fatty acids, and within the ω6 or ω3 families the ratio of C18, C20 and C22 components. Manipulation of the mix of dietary fatty acids changes the fatty acid composition of the phospholipids in cellular membrane systems throughout the body and alters the quantities and types of eicosanoids produced in many tissues [1–5]. Shifting the profile of eicosanoids impacts on hemostasis, inflammation, immunity, and the contractile state of blood vessels and bronchioles [6]. The incidence or severity of lipemia, hypertension, stroke, ischemic heart disease, rheumatoid arthritis, psoriasis and cancer in relation to short- or long-term polyunsaturated fatty acid intake have been favorite topics of investigation [7–14].

[1] Supported by PHS Grant ID31PE95008-01.

Table 1. Size of the elderly subgroups in the United States in 1985

Age years	Number of people	Elderly people %	Total population of 242,000,000, %
65–75	18,000,000	62	7.4
75–85	8,400,000	29	3.5
> 85+	2,700,000	9	1.1
Total	29,100,000	100	12.0

Most of the clinical studies of dietary polyunsaturated fatty acids (PUFA) have so far been conducted with subjects in the first two thirds of the human life span. The authors' assignment in this conference, on the other hand, is to discuss the potential applications of the presently available information about PUFA to the health of the elderly.

Demography and Pathophysiology of Old Age

Let us begin by considering the life curves in figure 1. In pre-civilization, the life curve of homo sapiens was that of 'random death'; less than 2% of each human generation lived past age 40. The effect of civilization, providing a protective environment, has been to shift the life curve progressively to the right. Theoretically, if environmental improvements continue, the life curve will approach a rectangular shape whereby over 90% of each generation will live to the genetically determined life span of 90–100 years, with nearly all deaths occurring in that decade [15].

The effect of shifting the life curve to the right is to alter the 'age mix' of the population, as shown in table 1. In 1900 only 5% of Americans were in the 'elderly' segment of 65+. In 1975, this proportion was 12%. In 2030, it will be 18%. Moreover, the 'oldest old' or 85+ contingent is now the most rapidly growing segment in our population [16].

The prolongation of the average length of life, however, has not been matched by an extension of the active or independent life span. For those who reach age 65, life expectancy is now 15–20 years, but only one half of these years on the average are free of serious disability [17, 18]. Figure 2 portrays the epidemiology of disability and shows the progressive increase, year by year, of physical impairment in the post-retirement population.

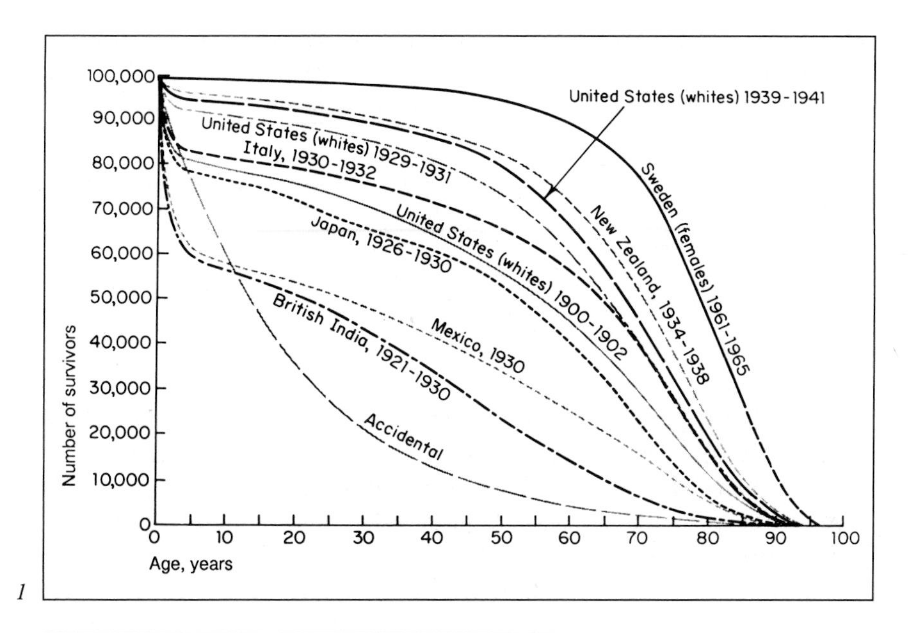

1

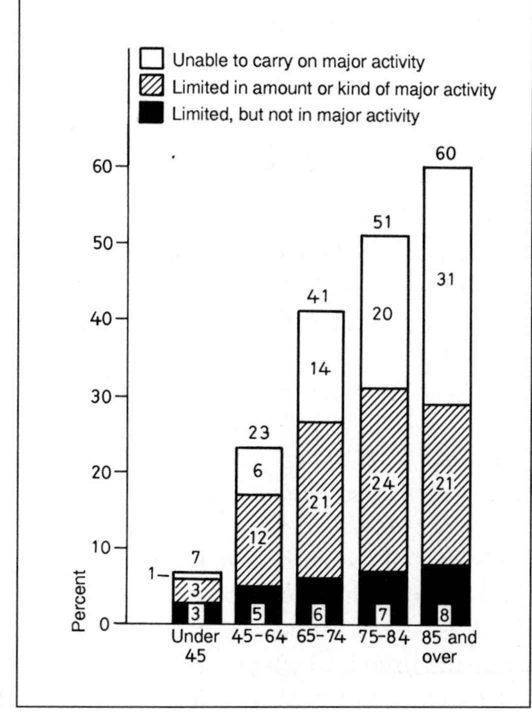

2

Fig. 1. Survival curves for various human populations. Reproduced with permission from *Principles of Mammalian Aging*, second edition, Kohn RR, Prentice-Hall, 1978.

Fig. 2. Percentage of elderly with limitation of activity due to chronic condition, by age group and type of limitation, excluding elderly in institutions. Reproduced with permission from *Essentials of Clinical Geriatrics*, Robert L. Kane, Joseph G. Ouslander, Itamar B. Abrass, McGraw-Hill, 1976.

Table 2. United States population aged 65 and older, by degree of dependence, 1987 (n = 25 million)

Degree of dependence	Residence	Approximate number
Fully independent	community	14 million
Impaired independent	community	6 million
Dependent ('homebound')	community	3–4 million
Dependent	nursing homes	1.4 million

The elderly population can be subdivided not only on the basis of age (table 1), but also according to their degree of impairment and dependency and their place of residence (table 2). These subgroups differ profoundly with regard to prevalence of chronic disease, dietary intakes and nutritional status [16, 19–22].

The high prevalence of disability in the elderly is thought to reflect both the universal aging process and the escalating incidence of age-related diseases [23]. Figure 3 summarizes the work of Shock and colleagues [24]. After the prime of life, age 20–30, many physiologic functions and anatomic structures show a regression for the remainder of the life span at about 5% per decade. While disuse atrophy and adverse environmental factors may contribute to these losses, in large part they appear to reflect a universal aging process.

Another manifestation of the aging process is the marked increase, during the last third of life, of age-related pathologies [23]. One example, shown in figure 4, is the neoplastic diseases. Other examples are cataracts, sensorineural hearing loss, Alzheimer's disease, Parkinson's disease and cerebrovascular accidents.

PUFA and the Health of the Elderly: Two Types of Questions; Observations in Tube-Fed Nursing Home Patients

The environmental improvements of industrialized society have permitted and encouraged the survival of a huge elderly population with marked clinical heterogeneity because of the variety and degree of disability ensuing from the aging process and from the high prevalence of chronic diseases. Now to address the authors' assignment: In what way can the

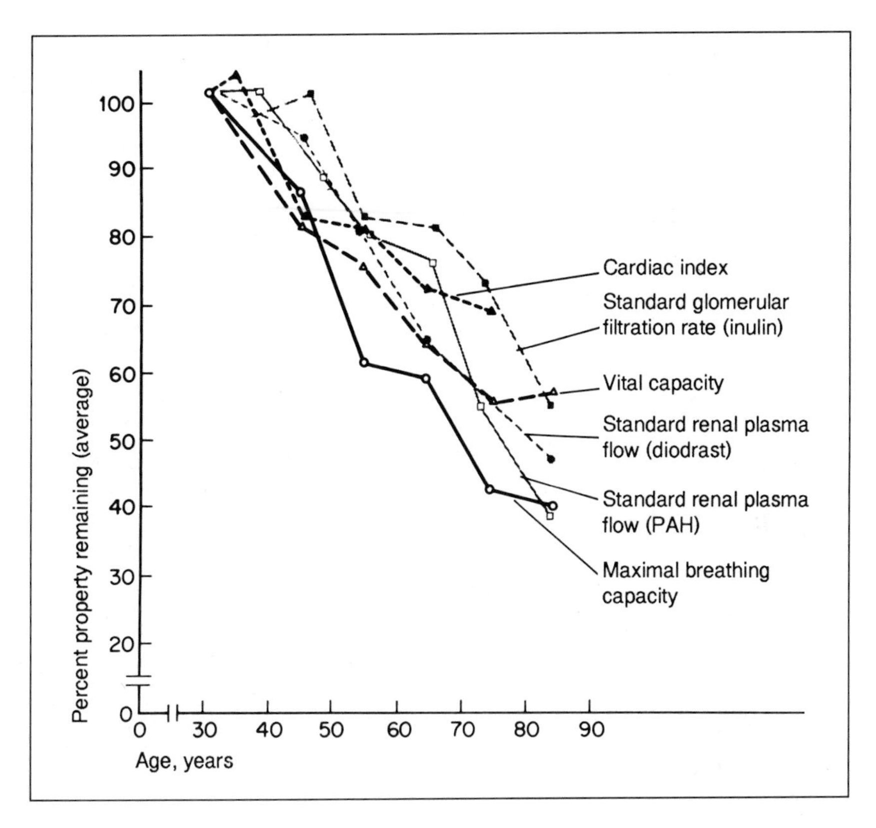

Fig. 3. Efficiency of human physiological mechanisms as a function of age. Level at 30 years is assigned a value of 100%. Reproduced with permission from *Principles of Mammalian Aging,* second edition, Kohn RR, Prentice-Hall, 1978.

modern knowledge about PUFA be usefully applied to improve the health status of this aged sub-population?

Two types of questions call for inquiry. First, consider the new administration of a PUFA-rich diet to elderly subjects in order to achieve a goal such as lowering serum triglycerides or blood pressure, or improving arthritis or psoriasis [7–14]. Physicians will need to know, if the various subgroups of the elderly (table 1 and 2) digest, absorb, metabolize and utilize PUFA in the same manner as young people, and if they respond similarly to the PUFA-derived eicosanoids. The second type of question concerns the potential impact of habitual PUFA intake during the younger years on the later incidence of the geriatric diseases.

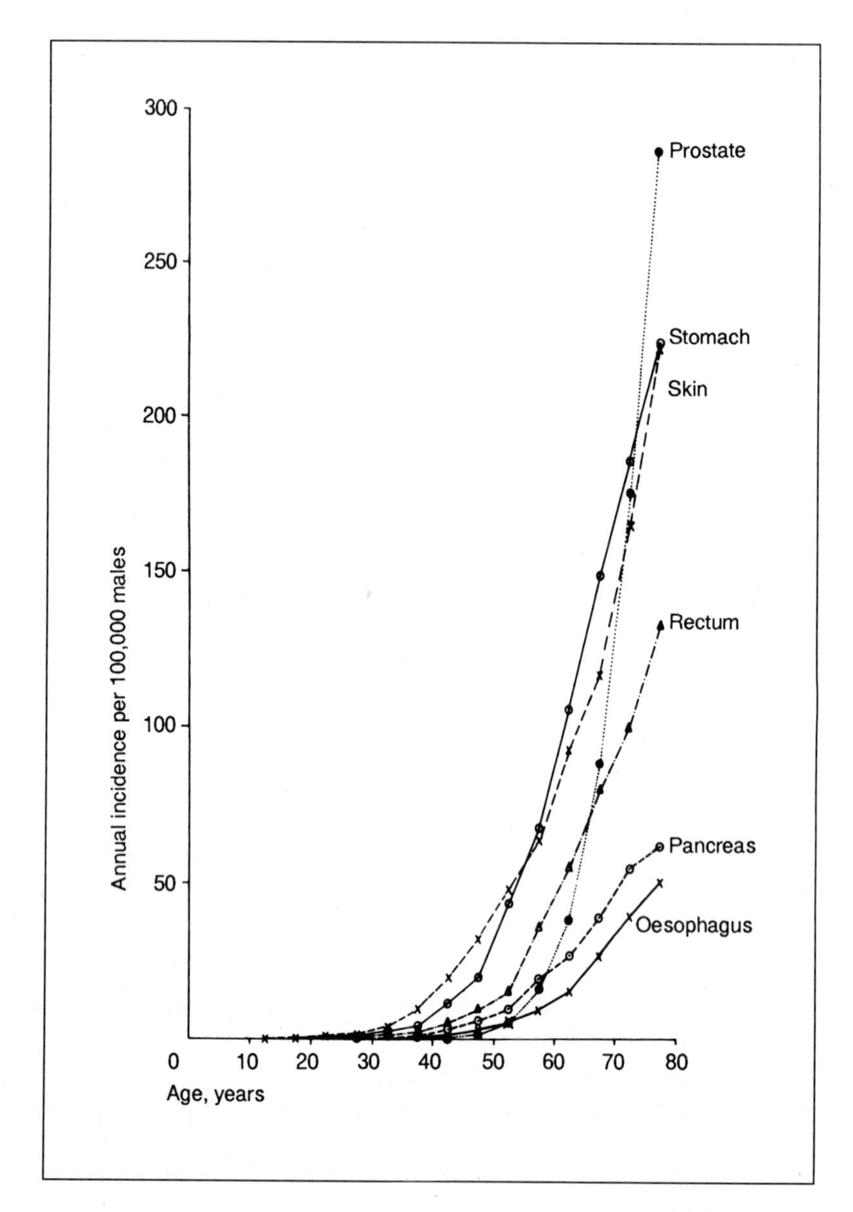

Fig. 4. Age-related incidence of various carcinomas illustrating the steep rise characteristic of the digestive tract and prostate. Reproduced with permission from *Clinical Geriatrics,* second edition, Rossman I, Lippincott, 1979.

Little information is available on the physiology of PUFA in old age. As a first step to gather information in this area, the authors utilized an opportunity which was provided by the tube-fed nursing home population [25]. These individuals constitute about 10% of the 1.5 million individuals residing in the 20,000 nursing homes of our nation (table 2). Their advanced organic brain syndrome has caused inability to eat, even with assistance, so that tube-feeding is required for survival. There are about 60 liquid diets currently on the market for this purpose. The most popular in the Veterans Affairs nursing homes of the authors' region has been 'Isocal'.

Table 3. Fatty acid profile of Isocal

Fatty acid	Concentration, mg/ml	Percent of total
12:0	0.02	0.05
14:0	0.04	0.10
15:0	0.02	0.05
16:0 (palmitic)	4.37	10.50
16:1 ω7	undetectable	undetectable
16:1 ω9	0.04	0.10
17:0	0.04	0.10
18:0 (stearic)	0.54	1.30
18:1 ω9 (oleic)	11.27	27.10
18:2 ω6 (linoleic)	21.62	52.00
18:2 ω3	0.12	0.30
18:3 ω6 (gamma linolenic)	undetectable	undetectable
18:3 ω3 (alpha linolenic)	2.95	7.10
19:0	0.04	0.10
20:0	0.12	0.30
20:1 ω9	0.08	0.20
20:3 ω9 (eicosatrienoic)	undetectable	undetectable
20:3 ω6	undetectable	undetectable
20:4 ω6 (arachidonic)	undetectable	undetectable
20:5 ω3 (eicosapentaenoic)	undetectable	undetectable
21:0	0.01	0.02
22:0	0.12	0.30
22:5 ω6	undetectable	undetectable
22:5 ω3	undetectable	undetectable
22:6 ω3 (docosahexaenoic)	undetectable	undetectable
Other	0.17	0.40
Total	41.57	100.00

Table 4. Serum fatty acids as percent of total fatty acids in healthy young men (group 1) and tube-fed nursing home men (group 2)

	Group 1 (mean ± SD)	Group 2 (mean ± SD)
12:0	0.04 ± 0.05	0.04 ± 0.03
14:0	1.25 ± 0.53	0.75 ± 0.25
16:0 (palmitic)	23.8 ± 2.78	19.3 ± 1.67[1]
16:1 ω7	3.39 ± 1.06	2.50 ± 0.17
18:0 (stearic)	5.83 ± 2.75	3.79 ± 2.70
18:1 ω9 (oleic)	26.4 ± 4.28	15.5 ± 5.28[1]
18:2 ω6 (linoleic)	24.9 ± 4.81	40.9 ± 7.17[1]
18:3 ω6 (gamma-linolenic)	0.67 ± 0.61	1.16 ± 0.49[1]
18:3 ω3 (alpha linolenic)	0.48 ± 0.14	2.07 ± 0.96[1]
20:3 ω9 (eicosatrienoic)	0.19 ± 0.05	0.25 ± 0.18
20:3 ω6	0.96 ± 0.34	0.43 ± 0.29
20:4 ω6 (arachidonic)	5.26 ± 1.60	6.73 ± 1.26
20:5 ω3 (eicosapentaenoic)	0.24 ± 0.07	0.79 ± 0.62
22:5 ω6	0.54 ± 0.24	0.38 ± 0.13
22:5 ω3	0.32 ± 0.19	0.50 ± 0.30
22:6 ω3 (docosahexaenoic)	0.83 ± 0.51	0.63 ± 0.26

[1] Independent sample t test with two-tail $p < 0.02$.

Like many other current enteral formulas, the fat component of this formula is a mixture of safflower and soy oils and provides the ω6 and ω3 fatty acids almost exclusively as linoleic and alpha-linolenic acids (table 3). The assumption is that the patient, through the action of elongase and desaturase enzymes, will generate the correct amounts of arachidonic acid, eicosapentaenoic acid (EPA) and docosahexaenoic acid (DHA). It has been questioned, however, whether old age might compromise these metabolic conversions [26, 27]. To investigate this question, the authors selected 6 nursing home men with advanced organic brain syndrome who had been nourished for over 1 year with Isocal as the sole source of nutrition, and measured the serum total fatty acid profile. The results were compared with the corresponding data from healthy young men, 20–30 years old, who were eating a mixed diet, and also with literature references on healthy 70-year-old men eating mixed foods. The Isocal-nourished nursing home men had the same levels of arachidonic acid, EPA and DHA as did the two control groups [25] (table 4). These data showed that linoleic and alpha-linolenic acids satisfied the requirements for ω6 and ω3 fatty acids, respectively, in the elderly men studied.

Is There a Relationship of PUFA Intake during the First Two Thirds of Life and the Health Status in Old Age?

The question merits consideration because prototype relationships have been described, demonstrated or proposed for other nutrients [28]. (a) Fluoride intake during childhood influences dental health in old age. (b) Protein intake during mid-adulthood may influence renal status in old age. (c) Caloric intake during the first half of life influences the rate of the aging process in rodents. (d) Calcium intake during mid-adulthood influences the degree of osteoporosis and the risk of fractures in old age. (e) Carotenoid intake during early and mid-adulthood may influence the incidence of head, neck and lung cancer in old age [29].

With regard to PUFA nutrition, the following potential relationships seem worth considering.

Alzheimer's Disease

Dementia is defined as a global cognitive impairment involving memory, problem solving, performance of learned perceptual-motor skills, and social skills. Alzheimer's disease accounts for 50–60% of all dementing illness. Although it may begin as mild short-term memory impairment, it is a progressive debilitating illness.

The clinical diagnosis of probable Alzheimer's disease is made by the exclusion of other known causes of dementia. At autopsy, the characteristic findings in the Alzheimer's brain include cerebral atrophy, neuritic plaques and neurofibrillary tangles in the hippocampus and neocortex [30].

It is estimated that up to 16% of persons over 65 suffer from a dementing illness. The prevalence rates of Alzheimer's disease for those over 80 years old are higher and range from 10 to 50%. A recent epidemiologic study of the prevalence of probable Alzheimer's disease in a community population found a 47% prevalence of Alzheimer's disease in persons over 85 years old [31]. This prevalence is higher than in previous studies and is important to the health care system because the 85+ elderly are one of the most rapidly growing segments of the American population. By a conservative estimate, more than 2 million Americans are affected by this disease.

Is it possible that the life-long intake of PUFA could influence an individual's susceptibility to Alzheimer's disease? The rationale for considering this suggestion is as follows. Brain tissue has a high concentration of PUFA [32, 33] and the proportions of different fatty acids in the brain phospholipids reflect their proportions in the diet [34–39]. The rapidity and the degree of this influence is greatest during prenatal and postnatal brain growth but is still detectable thereafter. The maternal dietary fat composition influences the fatty acid intake of breast-fed infants and therefore the composition of their brain fatty acid mixture [40–42]. PUFA in brain tissue are subject to lipoperoxidation [17]. The lipoperoxides then tend to react with proteins and nucleic acids to produce cross-linked, denatured products [43]. The several types of PUFA differ in their susceptibility to this process. The tendency for neurons in Alzheimer's disease to deteriorate, in a process which often includes the precipitation of intracellular proteins, could reflect in part the tendency to lipoperoxidation of neuronal PUFA. Since the mix of neuronal fatty acids is dependent on dietary fatty acids especially in early life, the childhood intake of fatty acids could be one of the 'risk factors' determining the susceptibility of neurons to these late degenerative events.

How can this hypothesis be studied? Since there is at present no animal model of Alzheimer's disease, epidemiologic studies of human populations consuming different fatty acid mixes is the alternative method. Is the age-specific incidence of Alzheimer's disease different between such populations? Incidence rather than prevalence needs to be studied, because the latter reflects not only the incidence but also the duration of survival after onset [44]. Length of survival in turn reflects quality of care, which varies widely [44]. A second caveat is that in such surveys Alzheimer's disease ('neurodegenerative' dementia) must be carefully dinstinguished from the other major type of geriatric dementia, 'multi-infarct' or 'vascular' dementia. As described below, multi-infarct dementia theoretically could also be influenced by diet PUFA, but not necessarily in the same direction as Alzheimer's disease.

Age-Related or Senile Macular Degeneration

Age-related macular degeneration (ARMD) is the leading cause of permanent central visual loss in the elderly. It causes vision impairment through loss of cone photoreceptors in the macula. Although the exact pathogenesis is not known, ARMD appears to be the consequence of the progressive inability of the retinal pigmented epithelium to digest phagocytosed materials. These materials, called drusen, damage the epithelial cells. In addition, new blood vessels which grow from the choroid into the subretinal pigmented epithelial space may leak or bleed, causing detachment of the retinal pigmented epithelium. The prevalence of ARMD increases from 1.6% in persons aged 52–64, to 11% for ages 65–74, and 28% for ages 75–85 [45].

What is the rationale for a possible influence of life-long PUFA intake on an individual's susceptibility to ARMD? The involved photoreceptor cells have a very high content of ω6 and ω3 PUFA ranging in chain length from 20 to 36 carbons [32, 46, 47]. Moreover, the proportions of various fatty acid species in these light-sensitive cells can be radically altered by changing the fatty acid content of the diet [39, 48]. Theoretically, lipoperoxidation of PUFA is one of the mechanisms that causes degeneration and death of postmitotic cells with advancing age [32, 49–51]. The lipoperoxidation process can generate intracellular cross-linked, precipitated lipid or protein products with physical properties similar to those in the degenerating photoreceptor cells of ARMD [51]. The question arises, therefore, whether the susceptibility to late death of these cells may be conditioned by their life-long content of highly unsaturated fatty acids, a content which is strongly influenced by diet.

How can the theory be tested? In the absence of an animal model for ARMD, investigators must seek epidemiologic data on the incidence and prevalance of this condition in populations with life-long differences in fatty acid intake. It is notable, therefore, that the incidence of ARMD in the Greenland population has been reported to be substantially higher than in that of the USA [52].

Intracranial Hemorrhage

The incidence of stroke has declined by approximately 25% in the past 30–35 years. This reduced incidence has been attributed to control of hypertension and a decrease in embolism from heart disease. Age, however, remains a major risk factor for stroke. The incidence of stroke is five-fold greater for ages 70–74, and 10-fold greater for ages over 75, than it is for ages 55–59 [53].

Strokes can be either hemorrhagic or non-hemorrhagic. The ratio of infarcts to hemorrhages is approximately 4:1; however, the in-hospital mortality remains higher for hemorrhagic strokes [54]. The Harvard Cooperative Stroke Registry further subdivides these categories and lists the frequency of the major types of cerebrovascular disease as: thrombosis 32%, embolism 32%, hypertensive hemorrhage 11%, ruptured aneurysms and vascular malformations 4.5%, and lacunar infarcts 18% [55].

The reason for postulating a potential diet PUFA effect on the incidence of hemorrhagic stroke is that the shift from ω6 to ω3 fatty acids in the diet reduces the tendency of platelets to aggregate and lengthens the bleeding time [56–59]. The susceptibility to thrombotic events theoretically would be reduced by such a change in hemostasis but the risk of hemorrhagic events could be increased. In Greenland, one epidemiologic study showed a low rate of coronary heart disease and a high rate of stroke, type not stated [60].

We thus may hypothesize that the administration of high ω3 diets to the elderly could be a risk factor for hemorrhagic strokes, while diminishing the rate of thrombotic strokes. How can the hypothesis be tested? Epidemiologic surveys are needed on the incidence and types of stroke in elderly populations subsisting on different mixtures of fatty acids. Two cautions need to be kept in mind. First, to distinguish the type of stroke requires brain imaging; second, other risk factors for hemorrhagic stroke, like blood pressure level, could also be influenced by diet [61].

The Rate of Aging

In the rat, with a 3-year life span, advancing age causes a series of adverse structural and functional changes which in sum affect nearly all organs in the body [62–68]. These changes are considered to reflect the underlying aging process. If the rodent's diet is restricted to 60% of that consumed ad lib, the rate of progression of many of the adverse aging events is slowed substantially and average longevity is increased by about 50% [65].

Restriction of calories, with the ratio of fat and protein calories kept unchanged, accomplishes the same effect [65]. At isocaloric levels of the diet reduction of the quantity of fat or of protein has little beneficial effect (except that the latter restriction may slow geriatric kidney deterioration).

However, it has not yet been reported whether manipulation of the quality of the dietary fatty acids influences the aging processes in the rodent. The possibility has credibility because of the thesis advanced by Tappel [51, 68] and others [49, 50, 69, 70]. They have proposed that free radicals continuously produced from O_2 as an inevitable consequence of aerobic metabolism react non-enzymatically with PUFA to yield lipoperoxides. An adverse cascade of events then leads, hypothetically, to the accumulation of insoluble or cross-linked lipid, protein and nucleic acid products. According to Tappel, lipoperoxidation of PUFA is involved in the pathogenesis of aging which culminates in body-wide senescence [51, 68].

Different PUFA differ in their susceptibility to lipoperoxidation [43]. Some current synthetic diets for the elderly contain a ten times higher proportion of PUFA than do conventional American diets [71]. Thus it would be relevant to examine in the rat model [65] whether the fatty acid composition of the diet influences the rate of aging, for which several biomarkers are now recognized [61]. Similar use could be made of the nonhuman primate with an approximate 15-year life span, in which a series of aging markers has also been defined [72].

Carcinogenesis

Over half of all cancers occur in the 11% of the US population over age 65. The risk of developing any of the four leading cancers: lung, colon, breast, and prostate, increases with age (fig. 4). Also, the age-adjusted death rates from all cancers increases from age 40 to 80 in both sexes [73, 74].

While old age is the commonest time of onset of clinical cancer, many of these neoplasms have evolved during mid-adulthood through a series of pre-neoplastic stages. There are reasons to suspect that the natural history of the pre-neoplastic lesions could be influenced by the life-long intake of PUFA.

One of the first clues came from the work of Dayton [75] in the 1960s. He conducted a 2-year diet intervention wherein the proportion of ω6 PUFA in the diet was increased in a population of 1,000 men. The death rate from coronary heart disease decreased. The overall death rate, however, remained unchanged because of a simultaneous increase in cancer mortality. Conversely, the age-specific death rate in Greenland from certain types of cancer has been reported to be relatively low; this population consumes a high proportion of ω3 fatty acids [60].

More recently several in vivo and in vitro rodent models of cancer have been studied with regard to the influence of exogenous PUFA [76–79]. Under selected conditions, ω6 fatty acids promote and ω3 fatty acids inhibit tumorigenesis.

Thus the hypothesis is now being tested both epidemiologically and experimentally that the ω6/ω3 PUFA ratio in the diet during early and mid-life may be a risk factor for the occurrence of some types of clinical cancer in old age.

Conclusion

Natural populations exist with radically different PUFA intakes. Moreover, the PUFA content of the diet is now being manipulated in sick people for therapeutic reasons, or is being varied in the general population because of new information on fatty acid physiology which was gained largely in young or middle-aged subjects.

It is possible that the risk of several common health problems in old age could be influenced by PUFA intake during the first two thirds of the life span, as well as during the geriatric period itself. These conditions include Alzheimer's disease, age-related macular degeneration, cancer, intracranial hemorrhage and the generalized aging process which leads ultimately to senescence. These possible 'risk factor' roles of PUFA can be investigated by cross-sectional epidemiologic surveys in natural populations, and by experiments conducted in the rat with an approximate 3-year life span and in nonhuman primates with an approximate 15-year life span.

References

1 Lin DS, Conner WE: Are the n–3 fatty acids from dietary fish oil deposited in the triglyceride stores of adipose tissue? Am J Clin Nutr 1990;51:535–539.
2 Salem N, Kim Hee-Yong, Yergey JA: Docosahexaenoic acid: Membrane function and metabolism; in Simopoulos AP, Kifer RR, Martin RE (eds): Health Effects of Polyunsaturated Fatty Acids in Seafoods. Orlando, Academic Press, 1986, pp 263–317.
3 Sinclair AJ: Incorporation of radioactive polyunsaturated fatty acids into liver and brain of developing rat. Lipids 1975;10:175–184.
4 Mohrhauer H, Holman RT: The effect of dietary essential fatty acids upon composition of polyunsaturated fatty acids in depot fat and erythrocytes of the rat. J Lipid Res 1963;4:346–350.
5 Spielmann D: Metabolism of unsaturated fatty acids. Role of n3 and n6 fatty acids in clinical nutrition; in Kinney JM, Borum PR (eds): Perspectives in Clinical Nutrition. Baltimore, Urban & Schwarzenberg, 1989.
6 Simopoulos AP: Summary of the NATO advanced research workshop on dietary

omega-3 and omega-6 fatty acids: Biological effects and nutritional essentiality. J Nutr 1989;119:521–528.

7 Shekelle RB, Missell L, Paul O, et al: Fish consumption and mortality from coronary heart disease. N Engl J Med 1985;313:820–824.

8 Iacono JM, Dougherty RM, Paska P: Dietary fats and the management of hypertension. Can J Physiol Pharmacol 1986;64:856–862.

9 Kremer JM, Jubiz W, Michalek A, et al: Fish oil fatty acid supplementation in active rheumatoid arthritis. A double-blinded, controlled, crossover study. Ann Intern Med 1987;106:497–503.

10 Black KL, Culp B, Madison D, et al. The protective effects of dietary fish oil on focal cerebral infarction. Prostaglandins Med 1979;3:257–268.

11 Harris WS: Fish oils and plasma lipid and lipoprotein metabolism in humans: A critical review. J Lipid Res 1989;30:785–807.

12 Rudman D, Mattson DE, Feller AG: Serum fatty acid profile of elderly tube-fed men in a nursing home. J Am Geriatr Soc 1989;37:229–234.

13 Horrobin DF: Polyunsaturated oils of marine and plant origins and their uses in clinical medicine; in Galli C, Simopoulos AP (eds): Dietary ω3 and ω6 Fatty Acids: Biological Effects and Nutritional Essentiality. New York, Plenum Publishing, 1989, pp 297–308.

14 Kremer JM, Lawrence DA, Jubiz W: Different doses of fish oil fatty acid ingestion in active rheumatoid arthritis: A prospective study of clinical and immunological parameters; in Galli C, Simopoulos AP (eds): Dietary ω3 and ω6 Fatty Acids: Biological Effects and Nutritional Essentiality. New York, Plenum Publishing, 1989, pp 343–350.

15 Fries JF: Aging, natural death, and the compression of morbidity. N Engl J Med 1980;303:130–135.

16 Kane RL, Solomon DH, Beck JC, et al: Geriatrics in the United States: Manpower Projections and Training Considerations. Lexington, Heath, 1981.

17 Cornoni-Huntley JC, Foley DJ, White LR, et al: Epidemiology of disability in the oldest old: Methodologic issues and preliminary findings. Milbank Mem Fund Q Health Soc 1985;63:350–376.

18 Katz S, Branch LG, Branson MH: Active life expectancy. N Engl J Med 1983;309:1218–1224.

19 Brickner PW, Teresita D, Kaufman A, et al. The homebound aged. A medically unreached group. Ann Intern Med 1975;82:1–6.

20 Allan CA, Brotman H (compilers): Chartbook on aging in America. The 1981 White House Conference on Aging, 1981.

21 Fernandes G: Modulation of immune functions and aging by omega-3 fatty acids and/or calorie restriction; in Ingram DK (ed): Nutritional Modulation of Aging Process. Bridgeport, Food & Nutrition Press, in press.

22 Federal Council on the Aging: The need for long-term care: A chartbook of the federal council on the aging. OHDS 81–20704, Government Printing Office, 1981.

23 Kohn RR: Human aging and disease. J Chronic Dis 1963;16:5–21.

24 Shock NW: Aging of physiological systems. J Chronic Dis 1983;36:137–142.

25 Rudman D, Mattson DE; Feller AG: Serum fatty acid profile of elderly tube-fed men in a nursing home. J Am Geriatr Soc 1989;37:229–234.

26 Horrobin DF: Loss of delta-6-desaturase activity as a key factor in aging. Med Hypotheses 1981;7:1211–1220.
27 Bordoni A, Biagi PL, Turchetto E, et al. Aging influence on delta-6-desaturase activity and fatty acid composition of rat liver microsomes. Biochem Int 1988;17:1001–1009.
28 Rudman D: Nutrition and fitness in elderly people. Am J Clin Nutr 1989;49(suppl): 1090–1098.
29 Willett WC: Vitamin A and lung cancer. Nutr Rev 1990;48:201–211.
30 Cassel CK, Walsh JR (eds): Geriatric Medicine. New York, Springer, 1984, vol 1, pp 43–49.
31 Evans DA, Funkenstein HH, Albert MS, et al: Prevalence of Alzheimer's disease in a community population of older persons. Higher than previously reported. J Am Med Ass 1989;262:2551–2556.
32 Neuringer M, Anderson GJ, Connor WE: The essentiality of n–3 fatty acids for the development and function of the retina and brain. Annu Rev Nutr 1988;8:517–541.
33 O'Brien JS, Fillerup DL, Mean JF: Quantification of fatty acid and fatty aldehyde composition of ethanolamine, choline and serine glycerophosphatides in human cerebral grey and white matter. J Lipid Res 1964;5:329–338.
34 Martinez M, Ballabriga A: Effects of parenteral nutrition with high doses of linoleate on the developing human liver and brain. Lipids 1987;22:133–138.
35 Mohrhauer H, Holman RT: Alteration of the fatty acid composition of brain lipids by varying levels of dietary essential fatty acids. J Neurochem 1963;10:523–530.
36 Foot M, Cruz TF, Clandinin MT: Influence of dietary fat on the lipid composition of rat brain synaptosomal and microsomal membranes. Biochem J 1982;208:631–640.
37 Matheson DF, Oei R, Roots BI: Effect of dietary lipid on the acyl group composition of glycerophospholipids of brain endothelial cells in the developing rat. J Neurochem 1981;36:2073–2079.
38 Tahin QS, Blum M, Carafoli E: The fatty acid composition of subcellular membranes of rat liver, heart, and brain: Diet-induced modifications. Eur J Biochem 1981;121:5–13.
39 Connor WE, Neuringer M: The effects of n–3 fatty acid deficiency and repletion upon the fatty acid composition and function of the brain and retina. Prog Clin Biol Res 1988;282:275–294.
40 Innis SM: Sources of ω3 fatty acids in arctic diets and their effects on red cell and breast milk fatty acids in Canadian Inuit; in Galli C, Simopoulos AP (eds): Dietary ω3 and ω6 Fatty Acids: Biological Effects and Nutritional Essentiality. New York, Plenum Publishing, 1989, pp 135–146.
41 Carlson SE, Rhodes PG, Ferguson MG: Docosahexaenoic acid status of preterm infants at birth and following feeding with human milk or formula. Am J Clin Nutr 1986;44:798–804.
42 Harris WS, Connor WE, Lindsey S: Will dietary ω3 fatty acids change the composition of human milk? Am J Clin Nutr 1984;40:780–785.
43 Kappus H: Lipid peroxidation: Mechanisms, analysis, enzymology and biological relevance. Oxidative Stress, 1985, chapter 12, pp 273–310.
44 Gruenburg EM: The epidemiology of senile dementia; in Schoenberg BS (ed): Neu-

rological Epidemiology: Principles and Clinical Application. New York, Raven Press, 1978, pp 4357–455.
45 Ferris FL III: Senile macular degeneration: Review of epidemiologic features. Am J Epidemiol 1983;118:132–151.
46 Anderson RE: Lipids of ocular tissues. IV. A comparison of the phospholipids from retina of six mammalian species. Exp Eye Res 1970;10:339–344.
47 Aveldano MI; Sprecher H: Very long chain (C_{24} to C_{36}) polyenoic fatty acids of the n–3 and n–6 series in dipolyunsaturated phosphatidylcholines from bovine retina. J Biol Chem 1987;262:1180–1186.
48 Anderson GJ, Connor WE, Corliss JD: Rapid modulation of the n–3 docosahexaenoic acid levels in the brain and retina of the newly hatched chick. J Lipid Res 1989; 30:433–441.
49 Miquel J, Economos AC, Fleming J, et al: Mitochondrial role in cell aging. Exp Gerontol 1980;15:575–591.
50 Roberts J, Adelman RC, Cristofalo VJ: Pharmacological intervention in the aging process. Adv Exp Med Biol 1978;97:231–241.
51 Tappel AL: Lipid peroxidation damage to cell components. Fed Proc 1973;32:1870–1874.
52 Rosenberg T: Prevalence of blindness caused by senile macular degeneration in Greenland. Arctic Med Res 1987;46:64–70.
53 Cassel CK, Walsh JR (eds): Geriatric Medicine. New York, Springer, 1984, vol 1, pp 25–27.
54 Adams RD, Victor M (eds): Principles of neurology, 4th edition. New York, McGraw-Hill, 1989, pp 617–692.
55 Mohr JP, Caplan LR, Melski JW, et al. The Harvard Cooperative Stroke Registry: A prospective registry. Neurology 1978;28:754–762.
56 Lagarde M, Croset M, Harjarine M: In vitro studies on docosahexaenoic acid in human platelets; in Galli C, Simopoulos AP (eds): Dietary ω3 and ω6 Fatty Acids: Biological Effects and Nutritional Essentiality. New York, Plenum Publishing, 1989, pp 91–96.
57 Renaud S: Alpha-linolenic acid, platelet lipids and function; In Galli C, Simopoulos AP (eds): Dietary ω3 and ω6 Fatty Acids: Biological Effects and Nutritional Essentiality. New York, Plenum Publishing, 1989, pp 263–272.
58 Knapp HR, Reilly IA, Alessandrini P, et al: In vivo indexes of platelet and vascular function during fish-oil administration in patients with atherosclerosis. N Engl J Med 1986;314:937–942.
59 Bang HO, Dyerberg J: The bleeding tendency in Greenland Eskimos. Dan Med Bull 1980;27:202–205.
60 Kromann N, Green A: Epidemiological studies in the Upernavik district, Greenland. Incidence of some chronic diseases 1950–1974. Acta Med Scand 1980;208: 401–406.
61 Knapp HR, FitzGerald GA: The antihypertensive effects of fish oil. A controlled study of polyunsaturated fatty acid supplements in essential hypertension. N Engl J Med 1989;320:1037–1043.
62 Weindruch R, Walford RL: Dietary restriction in mice beginning at 1 year of age: Effect on life-span and spontaneous cancer incidence. Science 1982;215:1415–1418.

63 Weindruch R, Gottesman SR, Walford RL: Modification of age-related immune decline in mice dietarily restricted from or after midadulthood. Proc Natl Acad Sci USA 1982;79:898–902.
64 Davis TA, Bales CW, Beauchene RE. Differential effects of dietary caloric and protein restriction in the aging rat. Exp Gerontol. 1983;18:427–435.
65 Masoro EJ: Nutrition as a modulator of the aging process. Physiologist 1984;27: 98–101.
66 Yu BP, Masoro EJ, McMahan CA. Nutritional influences on aging of Fischer 344 rats. I. Physical, metabolic, and longevity characteristics. J Gerontol 1985;40:657–670.
67 Masoro EJ: Physiological system markers of aging. Exp Gerontol 1988;23:391–397.
68 Tappel AL: Free radical lipid peroxidation damage. Department of food science and technology, University of California. Protection against free radial lipid peroxidation reactions, pp 111–131.
69 McArthur MC, Sohal RS: Relationship between metabolic rate, aging, lipid peroxidation, and fluorescent age pigment in milkweed bug, *Oncopeltus fasciatus* (Hemiptera). J Gerontol 1982;37:268–274.
70 Naeim F, Walford RL: Aging and cell membrane complexes. The lipid bilayer, integral proteins, and cytoskeleton. Molec Biol, pp 272–282.
71 Letson A: Nutritional product chart number 2. Enteral formulas. Nutr Support Services 1987;7:12–21.
72 Ershler WB, Coe CL, Laughlin N: Aging and immunity in non-human primates. II. Lymphocyte response in thymosin treated middle-aged monkeys. J Gerontol 1988; 43:B142–146.
73 Devita VT, Hellman S, Rosenberg SA: Chancer: principles and practice of oncology, 3rd ed. Philadelphia, Lippincott, 1989.
74 Crawford J, Cohen HJ: Relationship of cancer and aging. Clin Geriatr Med 1987;3: 419–432.
75 Dayton S: Cholesterol-lowering in man. Adv Exp Med Biol 1972;26:245–254.
76 Karmali RA: Dietary ω3 and ω6 fatty acids in cancer; in Galli C, Simopoulos AP (eds): Dietary ω3 and ω6 Fatty Acids: Biological Effects and Nutritional Essentiality. New York, Plenum Publishing, 1989, pp 351–360.
77 Simopoulos AP: Nutritional cancer risks derived from energy and fat. Med Oncol Tumor Pharmacother. 1987;4:227–239.
78 Nelson RL, Tanure JC, Andrianopoulos G, et al: A comparison of dietary fish oil and corn oil in experimental colorectal carcinogenesis. Nutr Cancer 1988;11:215–220.
79 Cameron RG, Armstrong D, Clandinin MT, et al. Changes in lymphoma development in females SJL/J mice as a function of the ratio in low polyunsaturated/high polyunsaturated fat diet. Cancer Lett 1986;30:175–180.

Daniel Rudman, MD, Division of Geriatrics, Department of Medicine,
Medical College of Wisconsin and the Clement J. Zablocki Veterans Affairs Medical Center, Milwaukee, WI 53295 (USA)

Simopoulos AP, Kifer RR, Martin RE, Barlow SM (eds): Health Effects of ω3 Polyunsaturated Fatty Acids in Seafoods. World Rev Nutr Diet. Basel, Karger, 1991, vol 66, pp 161–164

Nordic Recommended Dietary Allowances for ω3 and ω6 Fatty Acids

Jan I. Pedersen

Institute for Nutrition Research, School of Medicine, University of Oslo, Norway

The evidence now available clearly indicates that α-linolenic acid and the ω3 family of longer-chain polyunsaturated fatty acids are essential for man as for animals. So far this knowledge has not been reflected in any official national dietary recommendations. The argument has been that we do not have enough data to propose recommended dietary allowances (RDAs) for essential fatty acids (EFA) in general and for ω3 fatty acids in particular.

There are reasons to challenge this view and to raise the question if time is now ripe to reach consensus for the establishment of RDAs for both ω6 and ω3 fatty acids.

It should be reminded that RDAs (RDIs or recommended dietary intakes may be a better term [1]) are not equivalent to 'requirements'. RDAs for the different nutrients are based on our knowledge of minimum requirement to avoid clinical symptoms. To this an addition is made that takes into account several factors such as variability in requirement, safety margins, requirement for specific functions or according to selected criteria, biological availability, etc. Recommendations on dietary intakes of fat must also take into consideration our present knowledge about the effects of dietary fat on blood lipids and other factors related to certain chronic diseases. Such knowledge is at the basis of 'Dietary guidelines' formulated by a number of national and international expert groups in order to prevent coronary heart disease, cancer and other chronic diseases. Some would claim that dietary guidelines should be strictly separated from RDAs [2]. The panel responsible for the second edition of the Nordic Dietary Recommendations [3] recognised that recommendations for fat intake occupies a position at the interface between Dietary guidelines and RDAs and should be included in official dietary recommendations.

The view that α-linolenic as well as linoleic acid should be considered as essential for man was expressed already about 10 years ago in a FAO/WHO report on fat in human nutrition, in which estimates for recommended intakes were presented [4]. Unfortunately this report has not received the attention it deserves and has not had much impact in the formulation of national RDAs. The reason may be the special status and influence of the RDAs for the United States in the formulation of dietary recommendations around the world, and up to now these RDAs have expressed a rather conservative view in regard to both ω6 and ω3 fatty acids.

In the 9th edition (1980) of the RDAs for the USA [5] it was stated that linolenic acid has essential properties in animals but that the role of linolenic acid in human nutrition was not yet clear. Doubt was expressed about the dietary need of linolenic acid for man and no figures for requirement or allowance were given. In the recently published 10th edition (1989) [6], the essentiality of linolenic acid is recognised and it is stated that the role of linolenic acid in human nutrition is becoming clarified. Still no RDAs were given, but it is stated that the possibility of establishing RDAs for EFA should be considered. Reference is also given to the proposal of Neuringer and coworkers [7] that the intake of ω3 fatty acids in humans should be 10–25% that of linoleic acid. The reason given for not presenting any figures is that essential fatty acid deficiency has been observed exclusively in patients with medical problems affecting fat intake or absorption. This argument can be turned around: because no RDAs for essential fatty acids have been established, the liquid formulas used in the treatment of these patients have been deficient in such acids.

In fact, RDAs are used i.e. for planning therapeutic and formula diets, and if essential nutrients are omitted, nutritional deficiencies may result. All published cases of ω3 fatty acid deficiency in humans have thus resulted from inadequate planning of therapeutic diets. Because of the absence of recommendations in our national RDAs the codex alimentarius, or food regulations, do not contain any requirement for linolenic acid in neither infant milk formulas nor preparations used for parenteral and tube feeding. For such reasons the committee responsible for the preparation of the second edition of the Nordic Nutrition Recommendations took the position that we now have enough knowledge to give estimates of minimally adequate intakes both for ω3 and ω6 fatty acids. The proposed minimal intakes were primarily based on data presented in the above mentioned FAO/WHO report [4]. The figures from the FAO/WHO report are shown in table 1 and those from the Nordic Nutrition Recommendations

Table 1. Recommended dietary intakes of essential fatty acids (FAO/WHO Report 1977)

	Percent of energy
Infants (linoleic acid)	3
Adults (total EFA)	3
Pregnancy (total EFA)	4.5
Lactation (total EFA)	5–7

Ratio of linoleic to linolenic acid in the diet is 5–7 to 1 for all groups.

Table 2. Nordic Nutrition Recommendations, 2nd Edition, 1988

	Advisable intake, % of energy	
	minimum	maximum
Essential fatty acids (total ω6 + ω3)		
All age groups	3	10
Pregnancy	4.5	10
Lactation	6	10
α-Linolenic and very-long-chain ω3 fatty acids		
All age groups	0.5	

in table 2. Essential fatty acids (sum of ω6 and ω3) should provide at least 3% of the energy intake. In the diet of pregnant women these fatty acids should contribute 4.5%, and for lactating women at least 6% of the energy. Linolenic acid and long-chain polyunsaturated fatty acids of the ω3 family should provide at least 0.5% of the energy. Total intake of polyunsaturated fatty acids should not exceed 10% of the energy.

The ratio between ω6 and ω3 fatty acids (5–7 to 1) proposed in the FAO/WHO report was based on the ratio found in fetal and neonatal structural lipids in the human (5–7 to 1) as well as the ratio found in human milk (5–6 to 1) [4]. The more recent data by Bjerve [7] were not available at the time of preparation of the Nordic recommendations. His experiments showed that normalization of the concentration of ω3 fatty acids in plasma and red cell lipids in ω3 fatty acid deficiency required an intake of linolenic acid between 0.54 and 1.2% of the energy. If these data are to be taken into consideration in future revision of the recommendations, a reasonable proposal would be that linoleic acid should provide 5% and linolenic acid and other ω3 fatty acids 1% of the energy intake.

Dietary requirements of both ω6 and ω3 fatty acids have traditionally been discussed in terms of percent of energy intake. Bjerve has recently proposed to use absolute amounts, e.g. milligrams per kilogram body weight [8]. Fatty acids function both as energy source and as stuctural components. It is not known to what extent the requirement for EFA is related to energy expenditure. It is also uncertain whether the relatively limited number of data available on EFA requirement in humans are suitable to be related to body weight. From a practical point of view, however, whether the requirement is related to body weight or to energy expenditure may not be critical as long as the energy requirement is met.

To conclude, with the large amount of knowledge now available on the role of essential fatty acids, both of the ω6 and the ω3 family in human nutrition, an initiative should be taken to reach consensus on recommended dietary intakes for essential fatty acids to be used in official dietary recommendations.

References

1 Truswell AS, Irwin T, Beaton GH, et al: Recommended dietary intakes around the world. A report by committee 1/5 of the International Union of Nutritional Sciences. Nutr Abstr Rev 1983;53:939–1119.
2 Harper AE: Evolution of recommended dietary allowances – New directions? Ann Rev Nutr 1987;7:509–537.
3 Nordic Council of Ministers: Nordic nutrition recommendations, 2nd edition. Report 1989;2 (English version). Copenhagen, 1989.
4 FAO/WHO: Dietary fats and oils in human nutrition: a joint FAO/WHO report. Rome, FAO, 1978.
5 Food and Nutrition Board: Recommended dietary allowances. 9th ed. Washington, Natl. Acad. Sci./Natl. Res. Council, 1980.
6 Food and Nutrition Board: Recommended dietary allowances. 10th ed. Washington, Natl. Acad. Sci./Natl. Res Council, National Academy Press, 1989.
7 Neuringer M, Anderson GJ, Connor WE: The essentiality of n–3 fatty acids for the development and function of the retina and brain. Ann Rev Nutr 1988;8:517–541.
8 Bjerve KS: n–3 Fatty acid deficiency in man: Pathogenetic mechanisms and dietary requirements. n–3 News 1989;4:1–4.

Jan I. Pedersen, MD, Institute for Nutrition Research, University of Oslo,
P.O. Box 1046, N–0316 Oslo 3 (Norway)

Biochemistry and Physiology

Simopoulos AP, Kifer RR, Martin RE, Barlow SM (eds): Health Effects of ω3 Polyunsaturated Fatty Acids in Seafoods. World Rev Nutr Diet. Basel, Karger, 1991, vol 66, pp 166–176

Enzyme Activities Affecting Tissue Lipid Fatty Acid Composition[1]

Howard Sprecher

Department of Physiological Chemistry, The Ohio State University, Columbus, Ohio, USA

Introduction

Numerous compositional studies show that dietary ω3 fatty acids modify membrane lipid fatty acid composition. These important descriptive studies have, in some cases, been experimentally correlated with altered eicosanoid synthesis. The exciting prospect that dietary ω3 fatty acids may modify physiological processes in a beneficial way is tempered by the realization that we still understand relatively little about what regulates membrane lipid fatty acid composition. The objective of this review is to evaluate critically how the numerous pathways in unsaturated fatty acid and phospholipid biosynthesis interact to control membrane lipid fatty acid composition and thus function.

Metabolism of Lipids in Liver

Fatty Acid Desaturation

The types and amounts of fatty acids in membrane lipids must in part depend on whether there is selective uptake of specific acids by different cells. Intracellular fatty acid metabolism then defines what fatty acids are

[1] These studies were supported by NIH grants DK20387 and DK18844.

at least potentially available for acylation. Rosenthal [1] points out that various cells have different abilities to metabolize fatty acids by desaturation and chain elongation. It thus appears that liver plays a major role in modifying dietary fatty acids, not only for synthesis of its own membrane lipids but also for export. The activity of the rate-limiting 6-desaturase as well as the 5- and 4-desaturases in liver are frequently cited as playing a pivotal role in defining what fatty acids are made available for subsequent acylation into membrane lipids. This assumption may be true but it is imperative to stress that much of our basic knowledge about these three position-specific desaturases may be extrapolated information based on studies with the 9-desaturase. The activity of this lipogenic enzyme is elevated by dietary carbohydrate and insulin [2]. It has a half-life of 4 h in situ [3]. In 1974 the desaturase was purified to homogeneity and the requirement for cytochrome b_5 and cytochrome b_5 reductase, in a reconstituted system, was established [4]. The enhanced activity of the 9-desaturase, in response to dietary carbohydrate, correlates with a 50-fold increase in the mRNA which codes for this enzyme [5]. This mRNA has now been used to construct the cDNA [6].

During the last 10 years there has been little progress in defining what regulates the activities or levels of the 6-, 5- and 4-desaturases or how they are arranged in the membrane to interact with the alternating chain elongation enzymes. In 1981 the 6-desaturase was partially purified and its requirement for cytochrome b_5 and b_5 reductase established [7]. The 5-desaturase can be assayed by incubating an appropriate acyl-CoA with microsomes [8]. Choline phosphoglycerides containing 20:3 ω6 at the *sn*-2 position are also directly desaturated to give an arachidonic acid-containing phospholipid [9]. In liver 18:2 ω6 is converted to 20:4 ω6 only through 18:3 ω6. The product of 18:2 ω6 chain elongation – i.e. 11,14-20:2 – is desaturated to 5,11,14-20:3 and not to 8,11,14-20:3 [10]. The 5-desaturase has never been solubilized nor has it been established whether a single protein can use an acyl-CoA derivative and a phospholipid as substrate and in addition introduce a double bond in 11,14-20:2 in a nonskipped pattern of unsaturation. Little is known about the putative 4-desaturase or its distribution in various tissues. The lack of appropriately labeled substrates even precludes doing the types of rate studies for the 6- and 5-desaturases. Indeed, additional experiments are required to document what regulates the 4-desaturase. For example, liver microsomes from chow-fed rats do not desaturate 22:4 ω6 to 22:5 ω6 [11] but microsomes from rats fed a fat-free diet made small amounts of 22:5 ω6 [8].

Table 2. Percent distribution of radioactive fatty acids in phospholipids and triglycerides when hepatocytes from chow-fed and essential fatty acid-deficient rats were incubated with [3-^{14}C]22:4 ω6 and [3-^{14}C]22:5 ω3 for 120 min

	Fatty acid	Metabolite	Chow-fed	Fat-free
Triglycerides	[3-^{14}C]22:4 ω6	22:4 ω6	72	78
		20:4 ω6	28	10
		22:5 ω6	–	12
	[3-^{14}C]22:5 ω3	22:5 ω3	69	69
		20:5 ω3	31	14
		22:6 ω3	–	17
Phospholipids	[3-^{14}C]22:4 ω6	22:4 ω6	21	20
		20:4 ω6	79	70
		22:5 ω6	–	10
	[3-^{14}C]22:5 ω3	22:5 ω3	57	39
		20:5 ω3	43	33
		22:6 ω3	–	27

20:4 ω6 followed by its acylation into lipids [17]. Collectively, these findings allow us to formulate a model suggesting that microsomes may metabolize some 18:2 ω6 directly to 20:4 ω6. The 18:3 ω6 and 20:3 ω6 which are acylated enter a very labile phospholipid pool. They are cleaved from phospholipids, perhaps by phospholipase A_2, and converted to 20:4 ω6 followed by its acylation into a more stable pool of phosphoglycerides. Cook and Spence [18] have proposed a similar model when they studied the coordinate control of unsaturated fatty acid and phospholipid biosynthesis in neuroblastoma and glial cells.

Retroconversion

Communication between intracellular organelles may be particularly important in defining what regulates the synthesis and acylation of 22-carbon-unsaturated fatty acids into membrane lipids. Both 20:4 ω6 and 20:5 ω3 are chain-elongated at about the same rate (unpublished results) and no rate studies have compared how 22:4 ω6 and 22:5 ω3 are desaturated by the putative 4-desaturase. We incubated [3-^{14}C]-labeled 22:4 ω6 and 22:5 ω3 with hepatocytes from rats fed a chow or fat-free diet to compare their catabolism via the retro-conversion pathway (fig. 1) [19] versus desaturation at position-4. In hepatocytes from chow-fed rats both

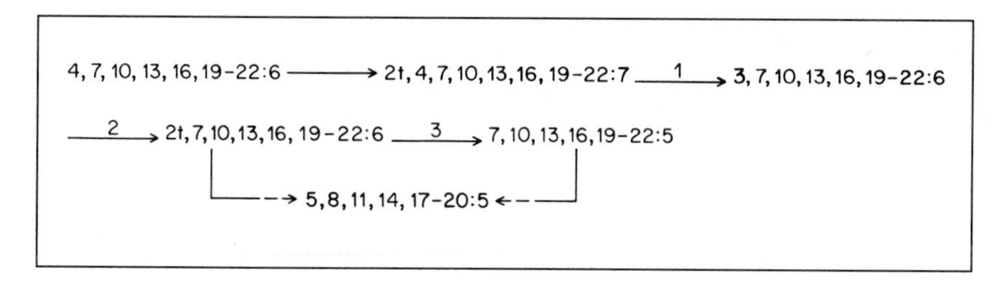

Fig. 1. Pathway for the β-oxidation of unsaturated fatty acids showing the use of (1) 2,4-dienoyl-CoA reductase, (2) Δ^3-*cis*-Δ^2-*trans*-enoyl-CoA isomerase and (3) intramitochondrial nucleotide dependent reduction.

substrates were only acylated or retroconverted followed by acylation (table 2, unpublished results). The percent distribution among metabolites in triglycerides was similar for both [3-^{14}C]-labeled substrates. Metabolite profiles in phospholipids showed that retroconversion of 22:4 ω6 to 20:4 ω6 followed by its acylation was more efficient than conversion of 22:5 ω3 to 20:5 ω3 followed by its esterification. A different labeling pattern was found in hepatocyte lipids from rats raised on a fat-free diet. These cells also desaturated both substrates. Again, the acylation pattern in triglycerides for the two substrates was quite similar. In phospholipids only 20% of the radioactivity was unmetabolized 22:4 ω6, while 70% was converted to arachidonate followed by acylation. The remaining 10% of the substrate was desaturated to 22:5 ω6 followed by acylation. When [3-^{14}C]22:5 ω3 was the substrate a greater percentage of the radioactivity was associated with the desaturated product – i.e. 22:6 ω3 and smaller amounts with the product of retroconversion – i.e. 20:5 ω3. These findings suggest that the regulation of 22-carbon fatty acid synthesis is a complex substrate-specific process involving coordinate control between their synthesis from 20-carbon precursors and the partial degradation of 22-carbon acids back to 20:4 ω6 and 20:5 ω3 as well as the possible dietary regulation of the 4-desaturase. In addition, it must be noted that as yet we do not know in detail how dietary fish oils regulate the activity of enzymes used in synthesizing 22-carbon acids or in their partial or complete oxidation. It seems likely that this type of dietary fat change has the potential of modifying the activities or levels particularly of both Δ^3-*cis*-Δ^2-*trans*-enoyl-CoA isomerase and 2-*trans*-4-*cis*-dienoyl-CoA reductase. In this regard it is interesting to note that a possible inborn error of metabolism may be due to insufficient 2-*trans*-4-*cis*-dienoyl-CoA reductase [20].

Lipid Biosynthesis in Extrahepatic Cells

Platelets, neutrophils and myocytes have only a limited ability to make polyunsaturated fatty acids from 18-carbon dietary ω6 and ω3 precursors. Platelets do not desaturate 18:2 ω6 [21] and only chain-elongate fatty acids at a slow rate [22]. Neutrophils do not metabolize 18:2 ω6 to 20:4 ω6 [23]. Heart microsomes desaturate small amounts of linoleate to 18:3 ω6, but the cellular origin of this activity was not defined [24]. However, we have been unable to show that heart myocytes either desaturate or chain-elongate fatty acids [25]. The results in table 3 show that fish oils modify the fatty acid composition of heart, neutrophil and platelet phospholipids in different ways. The high level of 20:5 ω3 in platelet phospholipids may in part be due to the presence of an activating enzyme specific for 20:4 ω6 and structural analogs such as 20:5 ω3 [26]. Once 20:4 ω6 is initially acylated it is moved among phospholipids via both a substrate-specific CoA-ATP-independent pathway and a CoA-dependent ATP-independent pathway [27]. However, none of these studies explain why 20:5 ω3 is readily incorporated into inositol phosphoglycerides when it is incubated with platelets but not when fish oils are included in the diet. Platelet phospholipids contain small amounts of 22-carbon unsaturated acids the composition of which is altered by dietary fat change. All four 22-carbon ω3 and ω6 acids are metabolized by platelet lipoxygenase(s) or cyclooxygenase. However, evidence for their release from phospholipids in response to agonists has not been convincingly documented [28]. Platelet function thus may primarily be regulated by defining what controls acylation and release of 20:4 ω6 and 20:5 ω3 from phospholipids.

In our studies smaller amounts of 20:5 ω3 were incorporated into neutrophil than platelet phospholipids (table 3). Exogenous 20:4 ω6 [29] like 20:5 ω3 [30] is initially acylated into diacyl-GPC followed by a time-dependent transfer to ether and plasmalogenic phospholipids. When fish oil is fed to rats the 20:5 ω3 pairs with the same fatty chains and in the same ratio as does 20:4 ω6 [31]. Exogenously added platelet-activating factor is deactivated by deacetylation followed by acylation with 20:4 ω6. Neutrophils from fish oil-fed animals deactivate platelet-activating factor and its lyso analog by the transfer of both 20:4 ω6 and 20:5 ω3 both of which pair with the same alkyl groups in about the same ratio [32, unpublished results]. Collectively these studies show that once 20:4 ω6 and 20:5 ω3 are incorporated into neutrophil phospholipids they are metabolized in similar ways. However, it must be emphasized that we still do not under-

Table 3. The influence of dietary fat change on the unsaturated fatty acid composition of total ethanolamine phosphoglycerides from rat platelets and of the 1,2-diacylethanolamine phosphoglycerides from neutrophils and heart. Results are expressed as weight percent.

	Fatty acid						
	18:2 ω6	20:4 ω6	20:5 ω3	22:4 ω6	22:5 ω6	22:5 ω3	22:6 ω3
Platelets[1]	4.0	20.9	10.4	1.3	0.9	4.2	1.6
Neutrophils[1]	5.2	15.0	2.1	2.1	–	4.7	1.4
Heart[1]	7.7	11.8	1.8	0.2	0.4	3.6	22.7
Heart[2]	4.0	22.9	–	2.6	10.9	1.2	8.8

[1] Weanling male rats were fed 2.5% corn oil and 2.5% fish oil for six weeks.
[2] Weanling male rats were fed 5% corn oil for six weeks.

stand what factors define initial acylation specificity. For example, why don't neutrophil and platelet phospholipids contain high levels of 22-carbon-unsaturated fatty acids?

Heart phospholipids, unlike those from neutrophils and platelets, contain high levels of 22-carbon acids. When rats are fed only corn oil the heart phospholipids contain high levels of 22-carbon ω6 acids (table 3). When fish oil is fed the decline in 20:4 ω6 is not accompanied by acylation of large amounts of 22:5 ω3 but rather by increases in 22-carbon ω3 acids. It must be stressed that these animals were fed a relatively low level of fish oil as well as 2.5% corn oil. Adequate 18:2 ω6 was thus available for synthesis of 20- and 22-carbon ω6 acids, however, there is clearly a specificity for incorporating 22-carbon ω3 acids. Fatty acid uptake by myocytes is carrier-mediated but it is not known if it is fatty acid-specific [33]. The results in figure 2 compare the acylation of 20:4 ω6, 20:5 ω3 and 22:6 ω3 into myocyte lipids. All three fatty acids were incorporated into phospholipids. The uptake and acylation of 20:5 ω3 was greater than for 22:6 ω3. These results are thus opposite to what would be expected from compositional analyses. The high level of esterified 22:5 ω3 and 22:6 ω3 cannot be explained by intracellular metabolism of 20:5 ω3. Neither 20:4 ω6 nor 20:5 ω3 was chain-elongated. Both 22:4 ω6 and 22:5 ω3 were substrates for retroconversion but not for desaturation of position-4. These studies suggest that in vivo myocytes must obtain their fatty acids by selective uptake. These cells have the ability to convert 22-carbon acids to 20-carbon com-

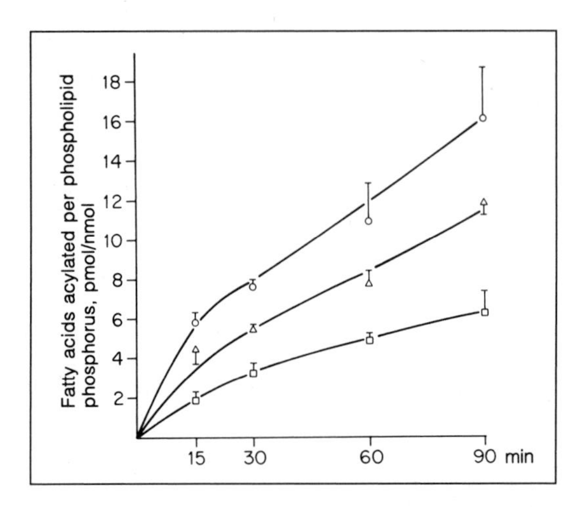

Fig. 2. Time-dependent incorporation of 20:4 ω6 (○), 20:5 ω3 (△) and 22:6 ω3 (□) into rat heart myocyte phospholipids.

pounds but apparently cannot synthesize 20-carbon acids from appropriate precursors. These studies further show that major differences exist in defining what regulates membrane lipid fatty acid composition in three types of cells which have, at best, only a limited ability to make unsaturated fatty acids from 18-carbon precursors.

References

1 Rosenthal MD: Fatty acid metabolism of isolated mammalian cells. Prog Lipid Res 1987;26:87–124.
2 Gellhorn A, Benjamin W: The intracellular localization of an enzymatic defect of lipid metabolism in diabetic rats. Biochim Biophys Acta 1964;84:167–175.
3 Oshino N, Sato R: The dietary control of the microsomal control of the microsomal stearyl CoA desaturation enzyme system in rat liver. Arch Biochem Biophys 1972; 149:369–377.
4 Strittmatter P, Spatz L, Corcoran O, et al: Purification and properties of rat liver microsomal stearyl coenzyme A desaturase. Proc Natl Acad Sci USA 1974;71:4565–4569.
5 Thiede MA, Strittmatter P: The induction and characterization of rat liver stearyl-CoA desaturase mRNA. J Biol Chem 1985;260:14459–14463.
6 Thiede MA, Ozols J, Strittmatter P: Construction and sequence of cDNA for rat liver stearyl coenzyme A desaturase. J Biol Chem 1986;261:13230–13235.
7 Okayasu T, Nagao M, Ishibashi T, et al: Purification and partial characterization of

linoleoyl-CoA desaturase from rat liver microsomes. Arch Biochem Biophys 1981; 206:21–28.

8 Bernet JT, Sprecher H: Studies to determine the role rates of chain elongation and desaturation play in regulating the unsaturated fatty acid composition of rat liver lipids. Biochim Biophys Acta 1975;398:354–363.

9 Pugh EL, Kates M: Direct desaturation of eicosatrienoyl lecithin to arachidonoyl lecithin by rat liver microsomes. J Biol Chem 1977;252:68–72.

10 Sprecher H, Lee C-J: The absence of an 8-desaturase in rat liver: A reevaluation of optional pathways for the metabolism of linoleic and linolenic acids. Biochim Biophys Acta 1975;388:113–125.

11 Ayala S, Gaspar G, Brenner RR, et al: Fate of linoleic, arachidonic and docosa-7,10,13,16-tetraenoic acids in rat testicles. J Lipid Res 1973;14:296–305.

12 Bernert JT, Sprecher H: The isolation of acyl-CoA derivatives as products of partial reactions in the microsomal chain elongation of fatty acids. Biochim Biophys Acta 1979;573:436–442.

13 Bernert JT, Sprecher H: An analysis of partial reactions in the overall chain elongation of saturated and unsaturated fatty acids by rat liver microsomes. J Biol Chem 1977;252:6736–6744.

14 Prasad MR, Nagai MN, Ghesquier D, et al: Evidence for multiple condensing enzymes in rat hepatic microsomes catalyzing the condensation of saturated, monounsaturated, and polyunsaturated acyl coenzyme A. J Biol Chem 1986;261:8213–8217.

15 Lands WEM, Inoue M, Sugiura Y, et al: Selective incorporation of polyunsaturated fatty acids into phosphatidylcholine by rat liver microsomes. J Biol Chem 1982;257: 14968–14972.

16 Voss AC, Sprecher H: Regulation of the metabolism of linoleic acid to arachidonic acid in rat hepatocytes. Lipids 1988;23:660–665.

17 Voss AC, Sprecher H: Metabolism of 6,9,12-octadecatrienoic acid and 6,9,12,15-octadecatetraenoic acid by rat hepatocytes. Biochim Biophys Acta 1988;958:153–162.

18 Cook HW, Spence MW: Studies of the modulation of essential fatty acid metabolism by fatty acids in cultured neuroblastoma and glial cells. Biochim Biophys Acta 1987; 918:217–229.

19 Schulz H, Kunau W-H: Beta-oxidation of unsaturated fatty acids: A revised pathway. Trends Biochim Sci 1987;12:403–406.

20 Roe CR, Millington DS, Norwood DL, et al: 2,4-Dienoyl-coenzyme A reductase deficiency: A possible new disorder of fatty acid oxidation. J Clin Invest 1990;85: 1703–1707.

21 Needleman S, Spector AA, Hoak JC: Enrichment of human platelet phospholipids with linoleic acid diminishes thromboxane release. Prostaglandins 1982;24:607–622.

22 Weiner TW, Sprecher H: 22-Carbon acids. Incorporation into platelet phospholipids and the synthesis of these acids from 20-carbon polyenoic acid precursors by intact platelets. J Biol Chem 1985;260:6032–6038.

23 Cook HW, Clarke JTP, Spence MW: Inability of rabbit polymorphonuclear leukocytes to synthesize arachidonic acid from linoleic acid. Prostaglandins Leukotrienes Med 1983;10:39–52.

24 Brenner RR: The desaturation step in the animal biosynthesis of polyunsaturated fatty acids. Lipids 1971;6:567–575.
25 Hagve T-A, Sprecher H: Metabolism of long-chain polyunsaturated fatty acids in isolated cardiac myocytes. Biochim Biophys Acta 1989;1001:338–344.
26 Neufeld EJ, Sprecher H, Evans RW, et al: Fatty acid structural requirements for activity of arachidonyl-CoA synthetase. J Lipid Res 1984;25:288–293.
27 Nakagawa Y, Waku K: The metabolism of glycerophospholipids and its regulation in monocytes and macrophages. Prog Lipid Res 1989;28:205–243.
28 Careaga-Houck M, Sprecher H: The effect of a fish oil diet on the fatty acid composition of individual phospholipids and eicosanoid production by rat platelets. Lipids 1989;24:477–481.
29 Chilton FH, Murphy RL: Remodeling of arachidonate-containing phosphoglycerides within the human neutrophil. J Biol Chem 1986;261:7771–7777.
30 MacDonald JIS, Sprecher H: Studies on the incorporation and transacylation of various fatty acids in choline and ethanolamine-containing phosphoacylglycerol subclasses in human neutrophils. Biochim Biophys Acta 1989;1004:151–157.
31 Careaga-Houck M, Sprecher H: Effect of a fish oil diet on the composition of rat neutrophil lipids and the molecular species of choline and ethanolamine glycerophospholipids. J Lipid Res 1989;30:77–87.
32 Chabot ML, Schmitt JD, Bullock BC, et al: Reacylation of platelet activating factor with eicosapentaenoic acid in fish oil-enriched monkey neutrophils. Biochim Biophys Acta 1987;922:214–220.
33 Stremmel W: Fatty acid uptake by isolated rat heart myocytes represents a carrier-mediated transport process. J Clin Invest 1988;81:844–852.

Howard Sprecher, PhD, Department of Physiological Chemistry,
The Ohio State University, 333 W. 10th Avenue, Columbus, OH 43210 (USA)

Simopoulos AP, Kifer RR, Martin RE, Barlow SM (eds): Health Effects of ω3 Polyunsaturated Fatty Acids in Seafoods. World Rev Nutr Diet. Basel, Karger, 1991, vol 66, pp 177–194

Dose-Response Relationships for ω3/ω6 Effects

William E.M. Lands

Department of Biological Chemistry, University of Illinois at Chicago, Chicago, Ill., USA

Introduction

Cholesterol and arachidonate are two important lipids that serve as precursors of hormones, and both are retained tenaciously by mammalian tissues. Deficiencies of these two lipids are very hard to obtain, but epidemiological studies lead to the conclusion that excessive participation of both of these lipids in cellular processes occurs commonly associated with cardiovascular diseases in people of western industrial societies. To interpret the health status of individuals with respect to arachidonate (20:4 ω6), we need better criteria for the threshold between adequacy and inadequacy (below which a deficiency state exists) and the threshold between adequacy and excess (above which hyperresponsiveness occurs). For these criteria, we need physiological measurements that can indicate when either of these two lipids is participating in a rate-limiting manner during specific disease-related events (fig. 1). Whereas epidemiologists frequently attempt to correlate diet and death, the task of biomedical researchers is to develop more mechanistic evidence linking diet to the capacity to express tissue mediators and to the subsequent pathophysiological processes. Difficulties in accurately measuring nutrient intake and then predicting relationships between intake and tissue lipid levels have led investigators to monitor plasma lipids as a more reliable indicator of an individual's risk. To approach a systematic interpretation of possible risks, we initiated a study to evaluate quantitatively the influence of dietary essential fatty acids on the levels of eicosanoid precursors and antagonists that are maintained in tissues. Information in this presentation is grouped in relation to the two

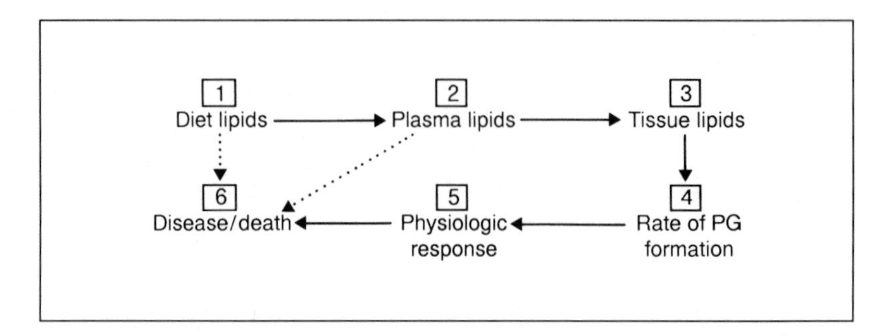

Fig. 1. Functional links between diet and dysfunction for essential fatty acids.

thresholds mentioned above: the lower one related to inadequate actions of essential fatty acids and the upper one related to excessive actions of essential fatty acids.

Valuable insight into the mechanism of action of essential fatty acids derives from the observation that either oral linoleate or injected chorionic gonadotropin could restore to normal the weight of testes, prostate and seminal vesicles in animals made deficient in essential fatty acids [1]. Essential fatty acids can act via prostaglandins to modulate hypothalamic function [2, 3] by stimulating GH release from pituitary, by stimulating GRF release from hypothalamus, by mediating the release of ACTH from the pituitary (perhaps by increasing CRF from the hypothalamus), by enhancing the response of thyroid tissue to TSH, promoting PRL release by decreasing the inhibitory (PIF) and increasing the stimulatory (PRF) factors from the hypothalamus, and by stimulating gonadotropin (LH and FSH), release by stimulating LHRH release from the hypothalamus (although PGI_2 decreased LHRH and PGE_2 increased LHRH [4]). Essential fatty acids may also act by way of leukotrienes C4, D4 and E4, which are synthesized in appreciable amounts in the hypothalamus and median eminence [5]. The synthesizing cells have significant LHRH contents, and anterior pituitary cells in culture released LH in response to added leukotriene C4 [6]. Thus selective eicosanoid modulations of hypothalamic-pituitary function may occur with either prostaglandins or leukotrienes that may be synthesized from either the ω3 or ω6 fatty acids.

Recognition of the essential role of the hypothalamus in modulating the release of pituitary hormones and in influencing behavior such as hunger, sex and aggression places great importance on our choices of dietary supplies of the ω3 and ω6 essential fatty acids due to their subsequent influence upon the quality of human life. Experimenters that wish to relate quantitatively the dietary ω3/ω6 intake to the powerful physiological effects of eicosanoid action will need to be aware of the subtle pleiotropic effects of ω3/ω6 acids upon the hypothalamic-pituitary axis.

Transition from Inadequacy to Adequacy

An influence of dietary ω3 and ω6 fats upon tissue fatty acid patterns and upon eicosanoid-mediated physiology is likely to occur in the range of intake that borders the lower threshold. Thus, we designed experiments with rats on different controlled diets with 18:2 ω6 contents ranging from 0% of calories (0 en%) to 12.2 en% and 18:3 ω3 from 0 to 10.7 en%. We found it useful to classify the fatty acids in three groups based on general differences in their metabolic selectivities for esterification to glycerolipids: saturated (SFA); C-16 and C-18 unsaturated (UFA); and C-20 and C-22 highly unsaturated (HUFA). The saturated acids tend to esterify the sn-1 position in glycerolipids, the unsaturated acids tend to esterify the sn-2 position during de novo synthesis, and the HUFA tend to esterify the sn-2 position during retailoring reactions [7]. These general selectivities are similar for rats and humans, and they result in similar typical values reported world-wide [8]. For example, irrespective of the relative amounts of ω9, ω6, and ω3 fatty acids, human plasma phospholipids have about 44% SFA, 37% UFA and 19% HUFA, and triglycerides have about 31% SFA, 60% UFA, and 9% HUFA. Increased dietary 18:2 ω6 and 18:3 ω3 gave linearly increased contents in tissue triglycerides with the slope for 18:3 ω3 being one half that for 18:2 ω6. The 2-fold faster rate of mitochondrial oxidation of 18:3 ω3 [9] may cause this difference. In contrast, to the changes in triglycerides, the content of ω3 and ω6 HUFA in rat tissues phospholipids increased with increased amounts of dietary precursors until a 'saturation' level occurred. Quantitative comparisons of the HUFA composition showed competitive interactions that fitted the traditional hyperbolic Michaelis-Menten relationship:

$$\text{Response} = V_{max}/[1 + (K_m/S)\,(1 + 1/K_i)].$$

As we developed more precise fits of experimental data to the hyperbolic equations, we found it useful to include also a term (C0) for other

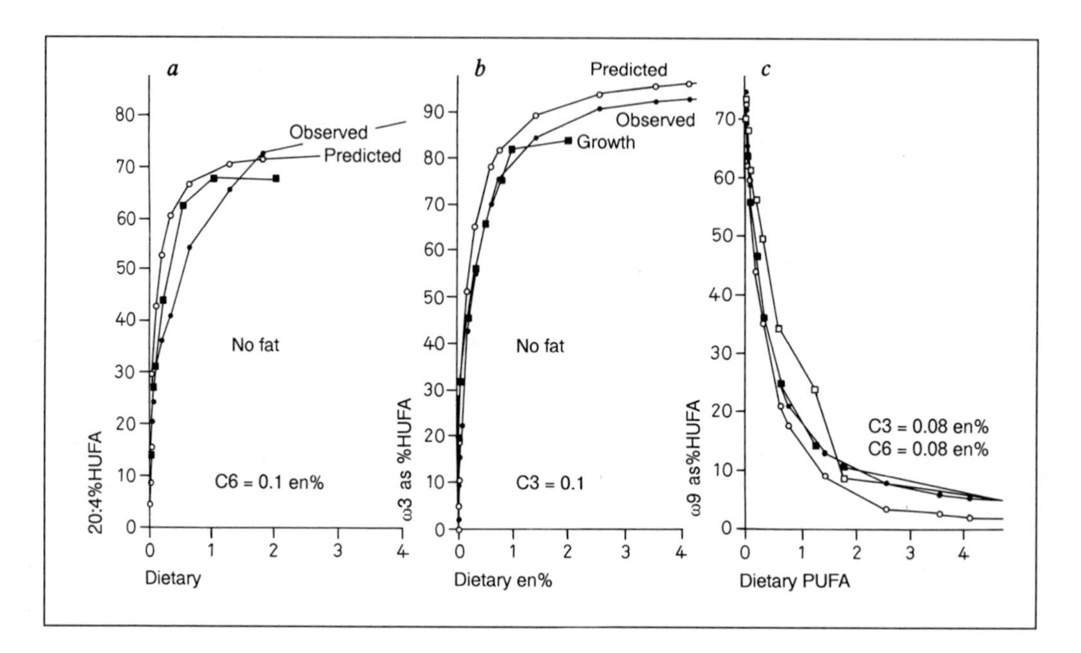

Fig. 2. The hyperbolic relationship between dietary fatty acids and highly unsaturated fatty acids in liver. The observed percentages of total rat liver HUFA as (*a*) ω6, (*b*) ω3 and (*c*) ω9 are derived from data with 42 different diets [10, 11]. The predicted values are based on Michaelis-Menten-like equations and constants shown in each panel.

dietary fatty acids (e.g., 16:0 and 18:1, etc.) and a shape-modifying term (KS) so that the final equation had the general form:

$$\omega 6 \text{ as } \%\text{HUFA} = 100/(1 + C6(1 + \text{en}\%\ \omega 3/C3 + \text{en}\%0/C0 + \text{en}\%6/KS)/\text{en}\%6) \quad (1)$$

The rat tissues maintained levels of ω3 and ω6 HUFA in their phospholipids that closely reflected the dietary supplies of their precursors (expressed as percent of dietary calories; i.e. en%) with an apparent C6 value for dietary 18:2 ω6 of 0.07 en% and an apparent C3 value for dietary 18:3 ω3 of 0.07 en%. These values, which are 10–100 times lower than typical dietary levels, were unexpected. However, subsequent calculations showed that the low values could have been predicted from earlier published results!

For example, rats on 18 different diets [10] and on 24 different diets [11] maintained HUFA compositions in total liver lipids for which the C6

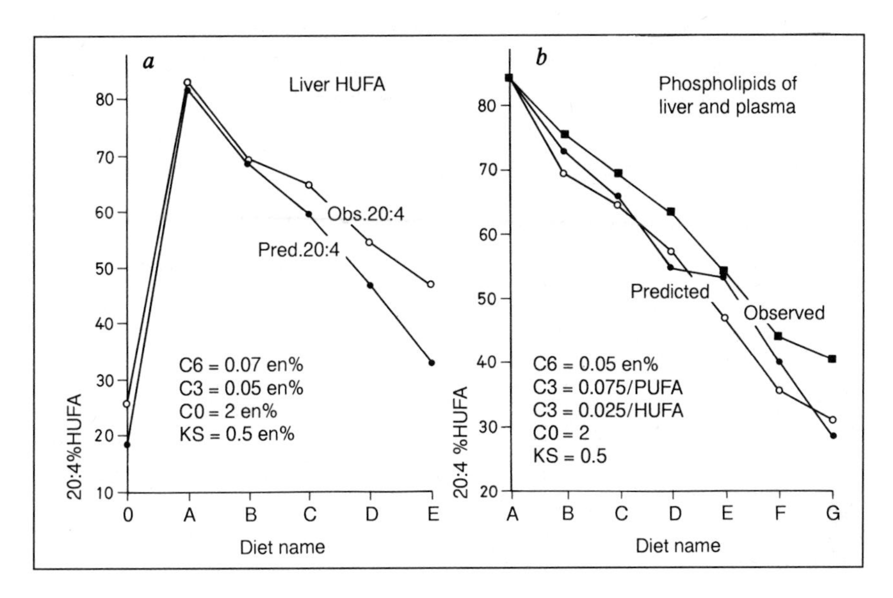

Fig. 3. Confirmation of the hyperbolic relationship for liver and plasma. The observed percentages of 20:4 ω6 in total HUFA are derived from data reported by Hwang et al. [12, 13]. The predicted values are based on Equation 1 using the constants shown in each panel. *a* Diet 0 was fat-free, whereas diets A through E contained 2 en% 18:2 ω6 with increasing amounts of 18:3 ω3. *b* Diets A through D used 18:3 ω3 (C3 = 0.075 en%), whereas E, F and G used fish oil containing ω3 HUGA (C3 = 0.025 en%).

of dietary 18:2 ω6 was about 0.1 en% (fig. 2a) and the C3 of dietary 18:3 ω3 was similar (fig. 2b). In addition, the competitive displacement of 20:3 ω9 by ω3 and ω6 fatty acids reflected a similar hyperbolic relationship to dietary acids (fig. 2c). Thus, dietary levels of ω3 and ω6 polyunsaturated fatty acids play a predictable role in maintaining tissue levels of HUFA. Furthermore, the transition from inadequate to adequate tissue HUFA levels is achieved with a dietary supply of about 0.1 en%. Tissue supplies of HUFA seem closely related to the growth responses described [10] which are superimposed on the compositional data in fig. 2. Clearly, this physiological parameter undergoes a transition from inadequate to adequate with dietary supplies of 18:2 ω6 and 18:3 ω3 near 0.1 en%.

Recent data [12, 13] confirmed the competitive hyperbolic relationship between ω3 and ω6 acids. The data fit values of a C6 (for 18:2 ω6) and C3 (for 18:3 ω3) near 0.1 en% (fig. 3a, b). The levels of 20:4 ω6 maintained

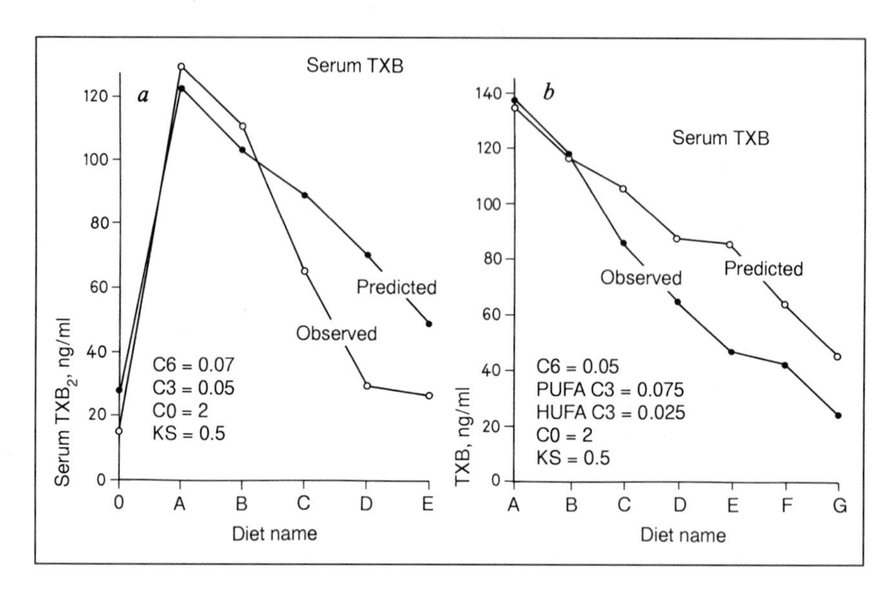

Fig. 4. Observed and predicted capacity for ex vivo thromboxane synthesis. The amount of thromboxane formed in serum from animals on different diets were reported by Hwang et al. [12, 13]. The predicted values are based on Equation 1 using the constants shown in each panel and maximum values for ng TXB/ml of 150 in *a* and 160 in *b*. Diets corresponded to those for figure 3.

in the tissue HUFA were successfully predicted using the hyperbolic relationship (Eq. 1 above) and constants very similar to those that fit data from both Holman's lab and our lab. Again, these results indicated that levels of eicosanoid precursors and antagonists are maintained in tissue lipids in a manner that can be regulated by dietary supplies of linoleate (18:2 ω6). Furthermore, the production of thromboxane by activated platelets in serum from animals on these diets also seemed limited by the amount of 20:4 ω6 as %HUFA since it also fits the predicted hyperbolic competitive interactions (fig. 4a, b). The relationships between diet and tissue fatty acid composition in figures 2 and 3 anticipate a capacity for eicosanoid biosynthesis as outlined in figure 1, and the ex vivo results in figure 4 support the concept that a hyperbolic transition from inadequate to adequate status also fits the capacity for eicosanoid biosynthesis. These data illustrate the quantitative way in which dietary ω6 fats increase thrombotic tendencies, whereas ω3 fats antagonize that effect. A more physiological confirmation of the hyperbolic relationship between dietary

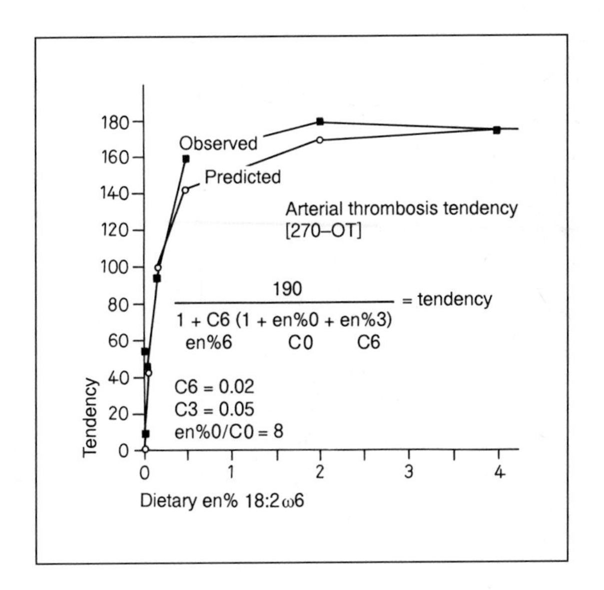

Fig. 5. Observed and predicted in vivo thrombotic tendency. The thrombotic tendency was estimated by subtracting from 270 the occlusion times reported by Hornstra [14]. The predicted values are based on the equation and constants shown.

18:2 ω6 and thrombotic tendency is illustrated in figure 5, which presents results [14] on occlusion times in a system closer to an in vivo model. In this model, the continuous flow of blood through a loop attached to the animal eventually ceases as a thrombus forms.

The excellent series of 8 diets described [15] also showed an enhancement by dietary 18:2 ω6 of in vivo tumor proliferation in rats which had an apparent C6 near 0.1 wt% (fig. 6). The observed tumor abundances fit curves for the predicted arachidonate abundance in HUFA based upon dietary 18:2 ω6. These data support the concepts of Kollmorgen et al. [16] and others that prostaglandin synthesis (from dietary ω6 acids) can enhance breast tumor proliferation. More importantly, the data illustrate that the transition from inadequate to adequate status is very similar for tumor proliferation (fig. 6), for thrombotic tendency (fig. 5), and for growth and abundance of 20:4 ω6 in tissue HUFA (fig. 2, 3). The overall cumulative evidence indicates that competitive interactions between ω3 and ω6 dietary polyunsaturated fatty acids can be demonstrated and quantitatively predicted and that the dose-response range for dietary 18:2 ω6 as

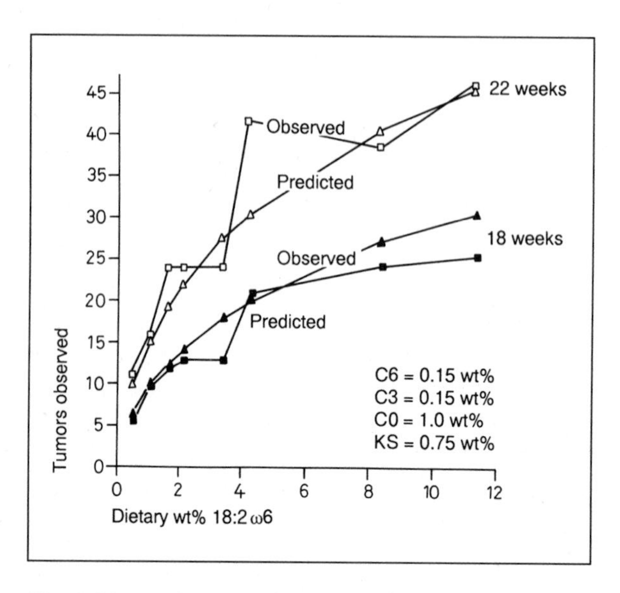

Fig. 6. Observed and predicted in vivo tumor abundance. The number of tumors reported by Ip et al. [15] is compared with those predicted by Equation 1 using the values for constants shown. All values for dietary 18:2 ω6 are as weight percent (wt%) as given in the orginal paper. The maximum value was set at 1.8 times that observed with the diet containing 11.5 wt% 18:2 ω6 (rather than 100% used in the general equation).

a precursor for eicosanoids may be much lower than has been commonly discussed.

The mid-point for dietary 18:2 ω6 to prevent deficiency states in both rats and humans is about 0.1 en%. Clinical observation of 71 infants ingesting one of three diets containing either 0.01, 0.5 or 4.5 en% 18:2 ω6 for a period of 40 days showed no unusual skin changes and no significant difference in weight gain [17]. The serum lipids, however, showed clear differences reflecting the dietary supply of 18:2 ω6 with a ratio of about 1 for triene/tetraene acids maintained in serum HUFA on diets of 0.5 en% 18:2 ω6. As the dietary 18:2 ω6 rose, the dienoic and tetraenoic acids also rose: from 0.01 (6 and 4%) to 0.5 (11 and 5%) to 4.5 en% (23 and 12%). Thus, a wide range of fatty acid compositions may occur in individuals with apparently normal clinical status, and we need more physiologic measurements to determine if the typical ('normal') levels are the preferred levels. Combes et al. [17] noted that diets containing from 0.4 to 0.9 en%

of dietary essential fatty acids had produced thousands of well-grown infants without any clinical evidence of nutritional deficiency.

The difficulty in obtaining a mixture of common foods that has such a low level of 18:2 ω6 (ca. 0.2 en%) makes it unlikely that the threshold to deficiency occurs in free-living adults eating typical foods. Nevertheless, recent efforts to antagonise ω6 eicosanoid functions by using dietary ω3 acids make it important to define physiological endpoints that can be used to monitor carefully the transition to a lower state of function of ω6 eicosanoids. Growth is one physiological criterion commonly used to indicate whether supplies of an essential fatty acid in the diet are above the lower threshold of adequacy. Growth curves for rats indicated that 18:2 ω6, 18:3 ω3 and 20:4 ω3 were all effective [18], and even 22:6 ω3 [19] and the ω3 acids from tuna oil [20] improved weight gain. Other physiological indices of inadequate dietary supplies of essential fatty acids include: scaly skin, necrosis of the tail, increased water consumption that was not evident in excreted urine volumes, renal degeneration that was made worse by increased dietary protein, irregular ovulation and parturition with small litters, and infertile males [21, 22]. An excellent review of these physiologic phenomena was provided by Aaes-Jorgensen [23].

The ω3 fatty acids support growth, development and gestation in the rat [24], but it is important to note that they are inadequate in preventing skin abnormalities and inadequate for normal parturition. Thus any bioassay based on either parturition or skin integrity would discriminate in favor of the ω6 acids over the ω3 acids. For example, a modified method for quantitating essential fatty acids was developed with a water rationing apparatus to limit the available drinking water [25]. This method measures growth in a way that places great importance on the integrity of the skin in preventing water loss, and therefore it selectively permits only the ω6 fatty acids to be effective 'essential fatty acids' [26]. As a result, many beneficial (and essential) actions of ω3 fatty acids have not been fully explored since these acids are unable to support *all* of the functions that are supported by ω6 acids. However, as reasearch on eicosanoids progressed, evidence accumulated to indicate that many pathological processes associated with excessive production and function of ω6 eicosanoids might be diminished by dietary ω3 fatty acids. The prevelance of these eicosanoid-mediated disorders (thrombosis, vasospasm, arthritis, asthma, inflammatory and immune dysfunctions, headache, dysmenorrhea and some metastatic events) suggest that many individuals may be manifesting a hyperresponsiveness due to excessive function of ω6 essential fatty acids.

Transition from Adequacy to Excess

General awareness of certain deficiency symptoms makes it unlikely that a diet-induced state of essential fatty acid deficiency in humans will ever occur in ethical controlled conditions. Demonstration of deficiency symptoms in rats provided the principal type of experimental evidence to support the concept of a requirement of 'essential fatty acids' by mammals. Fortunately, the results with rats have many close parallels with those from humans. However, in 1958, the Food and Nutrition Board of the NAS/NRC [28] declared that 'judging from animal experiments', 1 or 2 en% 18:2 ω6 'should be adequate'. That level was clearly above the dermal and growth threshold levels indicated by animal studies, but it was presumably regarded to be 'safely' above the threshold for a diet-induced deficiency. The recommendation was made years before any information on eicosanoids and their role in pathology was available. Now it is time to recognize and discuss carefully the fact that there are hyperrresponsive conditions associated with excessive mobilization of essential fatty acids. These conditions seem likely to occur in individuals ingesting 18:2 ω6 in amounts 10- to 50-fold greater than the lower threshold. We now are faced with an important question. What will we regard as evidence that a diet-induced state of essential fatty acid hyperfunction is occurring in humans?

Eicosanoids are formed in very small amounts by cells in response to a stimulus, and they are rapidly inactivated by catabolic enzymes. Thus there is little opportunity for prolonged occupancy by active eicosanoid of the receptors that mediate the eicosanoid function. Physiological signalling by essential fatty acids is believed to be accentuated by a transient stimulated hydrolytic release of esterified HUFA precursors (esp. 20:4 ω6) [29] and by transient increased levels of hydroperoxide activator of the biosynthetic oxygenases [30] (fig. 7). Catabolic metabolism has often been overlooked in evaluating the capacity of a tissue to express an eicosanoid-mediated function. A very slow basal rate of synthesis would permit catabolism to prevent any significant occupancy of receptors by active material, and physiological responses would not be evident even while significant amounts of catabolite are formed and accumulated in the urine in a 24-hour period (fig. 7). The ω3 HUFA slow the rate of synthesis of the ω6 eicosanoids, allowing the catabolic enzymes to readily diminish eicosanoid signalling.

The lower threshold of dietary essential fatty acids reflects the availability of adequate cellular supplies of eicosanoid precursor, which con-

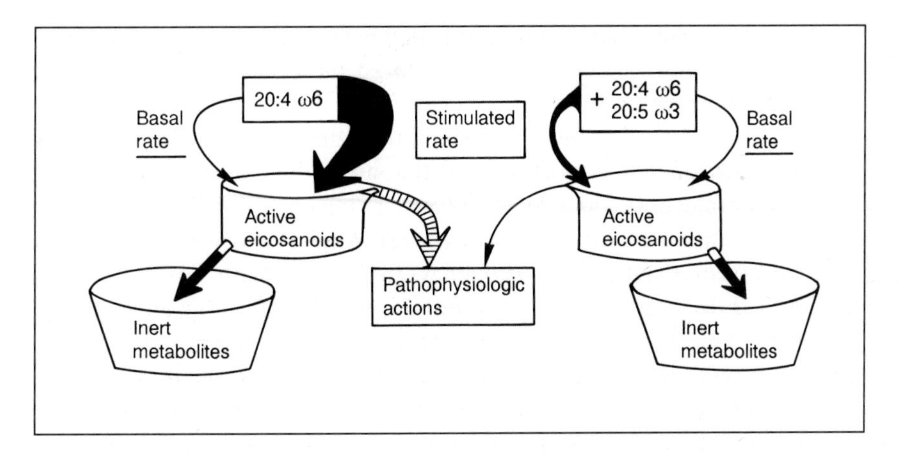

Fig. 7. The importance of eicosanoid synthetic rates in pathophysiology.

trols the capacity to maintain eicosanoid-mediated physiological events. Unfortunately, stimulation can sometimes produce more eicosanoid than is customarily accommodated by the immediate receptors and catabolic enzymes. In this situation, the excess active eicosanoid may diffuse to adjacent cells and tissues leading to pathophysiological actions (fig. 7). Strategies for avoiding this hyperrresponsive condition include attempts at blocking eicosanoid receptors, increasing catabolic disposal of the active eicosanoids, decreasing activity of the synthetic oxygenases with competitive and non-competitive inhibitors, inhibiting the supply of hydroperoxide activators, inhibiting the release of nonesterified substrate from tissue esters, or diminishing the amount of esterified precursor in tissues by limiting the dietary supply. This section examines biochemical and physiological indices to monitor the latter strategy.

One biochemical indication of high dietary levels of ω3 and ω6 fatty acids is the occurrence of almost undetectable levels of 20:3 ω9 in tissue HUFA (fig. 2c). At the low threshold range from physiologic inadequacy to adequacy this acid is present at levels nearly equal to 20:4 ω6 in tissue HUFA.

This was shown in a comprehensive report of the human need for essential fatty acids which involved 428 infants in a 4-year period with different levels of linoleate: 0.04, 0.07, 1.3, 2.8, and 7.3 en% 18:2 ω6 [31]. All of the full-term infants receiving only 0.04 en% 18:2 ω6 showed dryness of skin, desquamation and redness, whereas less than one-half (40%) of those receiving 0.07 en% 18:2 ω6 showed these signs. The latter group had an average serum ratio for triene/tetraene acids of about 1.4, whereas infants free of deficiency symptoms and ingesting 1.3 en% 18:2 ω6 had an average ratio of 0.3. Unfortunately, the study included no diets in the important range of 0.1–0.5 en% 18:2 ω6 leaving unexplored the important transition range with 20:3 ω9/20:4 ω6 ratios near 1 in which no further signs of essential fatty acid deficiency are evident with rats [10, 11] or infants [17]. In a series of 18 diets containing 0.08, 0.3 or 0.6 en% 18:2 ω6 [11], ratios of 20:3 ω9/20:4 ω6 in liver lipids for the animals on the marginal diets (with mild dermatitis) ranged from 0.8 to 1.9, whereas ratios for the animals free of deficiency symptoms ranged from 0.18 to 0.94 [11]. It seems evident from the data that the threshold condition for the most sensitive indicator of the lower threshold to physiological deficiency in rats and humans (i.e., dermal defects), was associated only with very prolonged conditions of about 0.2 en% dietary 18:2 ω6 and with a sustained ratio of about 1 for 20:3 ω9/20:4 ω6 in plasma lipids. The very low level of 20:3 ω9 (0.8–5% of HUFA) and the high level of 20:4 ω6 (50–60% of HUFA) in typical Americans [32, 33] may provide a useful opportunity for attempting to define the upper threshold (between adequacy and hyperresponsiveness).

The hyperbolic relationship that correlates the dietary supply of 18:2 ω6 and 18:3 ω3 with the amounts of ω3 and ω6 HUFA maintained in tissue lipids is also useful for evaluating the impact of dietary fat on the capacity for eicosanoid biosynthesis. Figure 8 indicates that when we use the constants derived from studies with rats, the actual level of 20:4 ω6 observed in plasma HUFA of typical Americans (about 68% HUFA) fits well with the content predicted by Equation 1 using information about the typical American diet (14.5 g 18:2 ω6; 1.7 g 18:3 ω3; 2,255 total cal.). The close fit of observed and predicted values confirms the concept (noted earlier) that the general selectivities for the biosynthesis of fatty acids and glycerolipids are very similar in rats and humans.

The hyperbolic relationship predicts a set of curves for different levels of ω3 rats in the diet (fig. 8). This set of curves is useful for comparing the possible impact of different diets that are characteristic of different countries. For example, in Japan, the typical intake of lipid nutrients differs considerably from that in the USA [34], with a consequently lower value for the level of ω6 acids in the HUFA of plasma phospholipids (about 45% rather than 68%). As we discuss national dietary strategies

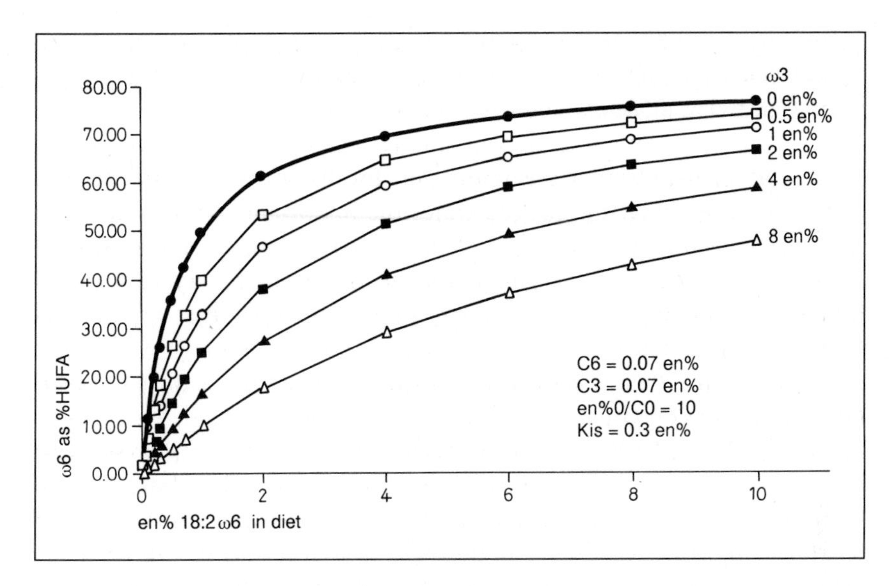

Fig. 8. A general relationship between the amount of 20:4 ω6 in plasma phospholipids and dietary supplies of ω3 and ω6 fatty acids. The curves were derived from Equation 1 using the constants shown in the figure.

that could diminish the frequency and severity of eicosanoid-mediated disorders, we can use the relationships in figure 8 for preliminary estimates of how much of a dietary change might be needed to obtain a desired change in ω6 HUFA of tissue phospholipids. When diets contain little ω3 fat, a 4-fold lowering of the dietary ω6 fat from 8 to 2 en% has little predicted effect on the proportion of ω6 HUFA. In contrast, a 4-fold increase in dietary ω3 fat is predicted to markedly reduce the proportion of ω6 HUFA. Further studies will demonstrate whether or not the level of eicosanoid precursors can be lowered as predicted by the preliminary estimates.

Non-Esterified Fatty Acid Composition and Eicosanoid Function

Pharmacological intervention designed to diminish prostaglandin-mediated pathophysiology has successfully focussed upon decreasing the activity of the synthetic oxygenase with competitive and non-competitive inhibitors. As noted above, merely decreasing the rate of biosynthesis can permit the catabolic enzymes to suppress pathological signalling even when significant amounts of eicosanoid are synthesized slowly over a 24-hour period. Arylacetic acid analogs of fatty acids (e.g., ibuprofen) competitively and reversibly block the binding site for the fatty acid substrate [35], and their commercial success illustrates the importance of this tactic for slowing the synthetic process. Since the ω3 fatty acids have competitive inhibitor constants [36] that are similar to those for the arylacetic acids [35], we can predict a slowing of the synthesis and a decrease in the function of ω6 eicosanoids when ω3 HUFA are present in the NEFA pool of a responding cell. Thus, it becomes important to be aware of the sources of fatty acids that constitute the cellular NEFA pool.

Extracellular NEFA in plasma originates from three major sources: release from adipose tissue following the action of cellular lipase, release by plasma lipoprotein lipase action on dietary chylomicra derived from the intestine; and release by lipoprotein lipase action on lipoproteins derived from the liver. All three sources provide albumin-bound NEFA in plasma which is rapidly introduced into the intracellular NEFA pool (fig. 9). The composition of substrates and antagonists in this intracellular pool has a major influence on the *rate* of biosynthesis of eicosanoids during a transient cellular response. It is clear that exogenous fatty acid analogs like ibuprofen can effectively interfere with the flow of endogenous 20:4 ω6 from esterified precursors through the cellular NEFA on the route to forming prostaglandins. In a similar manner, we must expect the ω3 HUFA entering the cellular NEFA from dietary sources (fig. 9) to slow the rate of conversion of endogenous 20:4 ω6 into prostaglandins.

Dietary information has now been used successfully with Equation 1 to estimate the proportion of ω3, ω6 and ω9 that will be maintained in the HUFA of tissue phospholipids. Also, the equation can indicate the overall capacity of platelets to form an ω6 eicosanoid. Such predictions can be very useful in estimating the impact of dietary selections upon the potential for eicosanoid-mediated responses. However, the vigor of eicosanoid-mediated responses will depend upon the *rates* by which eicosanoid syntheses occur in vivo. These rates will be strongly influenced by the relative abundances of the ω6 precursor and the ω3 antagonists in the NEFA. Some encouraging results from studies described in figures 5 and 6 indicate that the eicosanoid synthetic potential estimated by Equation 1 may be proportionately expressed in vivo. However, earlier studies have demonstrated

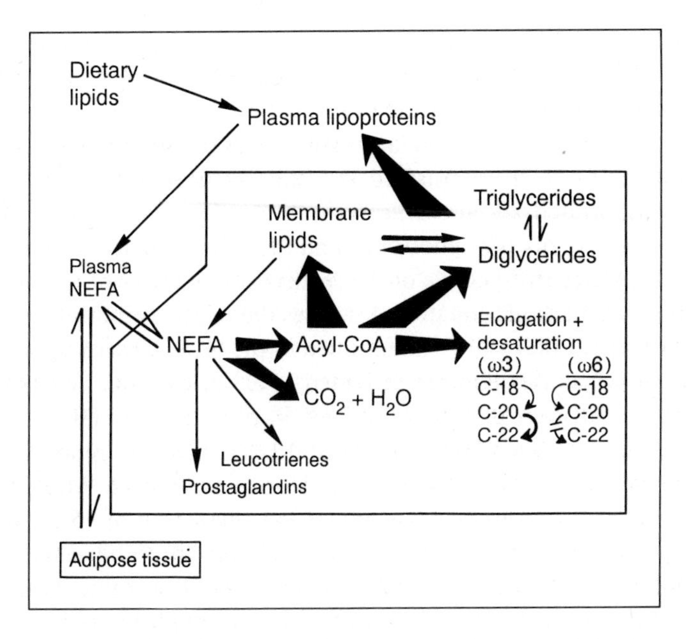

Fig. 9. The pivotal role of cellular non-esterified fatty acids in eicosanoid function.

that the steady-state ω3/ω6 ratios in the HUFA of NEFA may be 6-fold greater than in the HUFA of phospholipids [8, 37]. Thus the potential for eicosanoid synthesis estimated from the predicted composition of phospholipids, must be interpreted further in terms of antagonists in the NEFA. When significant amounts of ω3 HUFA are maintained in tissue phospholipids, we will need further data to estimate the degree of moderation by the ω3 HUFA in the NEFA on the potential rate of synthesis (and thereby the function) of the ω6 eicosanoids. To obtain preliminary data, we examined the effect of dietary fish oil on the ω3 HUFA in human plasma NEFA. Initially, the ratio of 20:5 ω3/20:4 ω6 in plasma phospholipids was 0.5, and it rose to 0.8 at 5 h and 0.9 at 8 h. In contrast, the ratios in the plasma NEFA changed from 0.6 to 4.6 to 3.5 as the entering chylomicrons were cleaved with lipoprotein lipase during lipid transport (fig. 9). As a result of the influx of these dietary NEFA, it is likely that there was a prolonged inhibition of the rate of synthesis of ω6 eicosanoids from the cellular NEFA pools which would exceed that predicted from the slight change in phospholipid.

Conclusions

The general selectivity during the esterification of fatty acids into glycerolipids tends to maintain in plasma phospholipids of humans about 44% saturated fatty acids, 37% unsaturated 16- and 18-carbon fatty acids, and 19% highly unsaturated 20- and 22-carbon fatty acids. Similar levels are maintained in rat plasma phospholipids, and these levels tend to be maintained in both species irrespective of the relative abundances of ω3, ω6 or ω9 fatty acids. A hyperbolic equation describes the relative amounts of ω3, ω6 and ω9 fatty acids that will be maintained in the HUFA based upon the percent of dietary calories that are provided as ω3 or ω6 fatty acids. The mid-point for the efficacy of dietary 18:2 ω6 in providing ω6 HUFA and for 18:3 ω3 in providing ω3 HUFA is about 0.1% of calories; much lower than the typical dietary level of 7–10% of calories as 18:2 ω6. Mid-point values near 0.1% of calories were calculated for the action of 18:2 ω6 in supporting growth, thromboxane formation in serum, thrombosis tendency, and proliferation of mammary tumors. This very low value makes a strategy noted earlier [27] of merely reducing dietary linoleate to diminish eicosanoid-mediated disorders seem infeasible. Also, with the current dietary level of linoleate already 10- to 60-fold greater than the midpoint value, a strategy of increasing it to further 'megadose' levels [27] seems inadvisable. An alternate strategy that seems advisable includes *both* decreasing dietary ω6 acids and increasing dietary ω3 acids. The hyperbolic equation helps estimate the interactions of dietary fatty acids on the capacity of synthesis of ω6 eicosanoids, and it seems useful in evaluating the amount of change in ω3 and ω6 fatty acid ingestion that could be expected to decrease the risk of eicosanoid-related diseases for different individuals. The shape of the curves indicates that an increase in ω3 fats may be more effective than a decrease in ω6 fats.

References

1 Greenberg SM, Ershoff BH: Effects of chorionic gonadotropin on sex organs of male rats deficient in essential fatty acids. Proc Soc Exp Biol Med 1951;78:552–554.

2 Ojeda SR, Naor Z, Negro-Vilar A: The role of prostaglandins in the control of gonadotropin and prolactin secretion. Prostagl Med 1979;5:249–275.

3 Ojeda SR, Negro-Vilar A, McCann SM: Role of prostaglandins in the control of pituitary hormone secretion; in Soto RJ, et al (eds): Physiopathology of Endocrine Diseases and Mechanisms of Hormone Action. New York, Liss, 1981, pp 229–247.

4 Ottlecz A, McCann SM: Concomitant inhibition of pulsatile luteinizing hormone and stimulation of prolactin release by prostacyclin in ovariectomized conscious rats. Life Sci 1988;43:2077–2085.
5 Lindgren JA, Hokfelt T, Dahlen S-E, et al: Leukotrienes in the rat central nervous system. Proc Natl Acad Sci USA 1984;81:6212–6216.
6 Hulting A-L, Lindgren JA, Hokfelt T, et al: Leukotriene C4 as a mediator of luteinizing hormone release from rat anterior pituitary cells. Proc Natl Acad Sci USA 1985;82:3834–3838.
7 Lands WEM, Crawford CG: Enzymes of membrane phospholipid metabolism in animals; in Martinosi (ed): Membrane Bound Enzymes. New York, Plenum Press, 1977, pp 3–85.
8 Lands WEM: ω3 fatty acids as precursors for active metabolic substances: Dissonance between expected and observed events. J Int Med 1989;225(suppl 1);11–20.
9 Clouet P, Niot I, Bezard J: Pathway of α-linolenic acid through the outer mitochondrial membrane in the rat liver and influence on the rate of oxidation. Biochem J 1989;263:867–873.
10 Mohrhauer H, Holman RT: The effect of dose level of essential fatty acids upon fatty acid composition of the rat liver. J Lipid Res 1963;4:151–159.
11 Mohrhauer H, Holman RT: Effect of linolenic acid upon the metabolism of linoleic acid. J Nutr 1963;81:67–74.
12 Hwang DH, Boudreau M, Chanmugam P: Dietary linolenic acid and longer chain ω3 fatty acids: Comparison of effects on arachidonic acid metabolism in rats. J Nutr 1988;118:427–437.
13 Hwang DH, Carroll AE: Decreased formation of prostaglandins derived from arachidonic acid by dietary linoleate in rats. Am J Clin Nutr 1980;33:590–597.
14 Hornstra G: Dietary fats and arterial thrombosis; PhD thesis, 1980, Univ. Limburg, Maastricht.
15 Ip C, Carter CA, Ip MM: Requirement of essential fatty acid for mammary tumorigenesis in the rat. Cancer Res 1985;45:1997–2001.
16 Kollmorgen GM, King MM, Kosanke SD, et al: Influence of dietary fat and indomethacin on the growth of transplantable mammary tumors in rats. Cancer Res 1983;43:4714–4719.
17 Combes MA, Pratt EL, Wiese HF: Essential fatty acids in premature infant feeding. Pediatrics 1962;30:136–144.
18 Turpeinen O: Further studies on the unsaturated fatty acids essential in nutrition. J Nutr 1937;15:351–366.
19 Hume EM, Nunn LC, Smedley-Maclean I, et al: Studies of the essential unsaturated fatty acids in their relation to the fat-deficiency disease of rats. Biochem J 1938;32: 2162–2177.
20 Privett OS, Aaes-Jorgensen E, Holman RT, et al: The effect of concentrates of polyunsaturated acids from tuna oil upon essential fatty acid deficiency. J Nutr 1958;67: 423–432.
21 Burr GO, Burr MM. A new deficiency disease produced by rigid exclusion of fat from the diet. J Biol Chem 1929;82:345.
22 Burr GO, Burr MM: On the nature and role of the fatty acids essential in nutrition. J Biol Chem 1930;86:587:629,
23 Aaes-Jorgensen E: Essential fatty acids. Physiol Rev 1961;41:2–46.

24 Leat WMF, Northrop CA: Effect of linolenic acid on gestation and parturition in the rat. Prog Lipid Res 1981;20:819–821.
25 Thomasson HJ: Biological standardization of essential fatty acids (a new method). Int Rev Vitamin Res 1953;25:62–82.
26 Thomasson HJ: Essential fatty acids. Nature 1962;194:973.
27 Lands WEM: Fish and Human Health. Orlando, Academic Press, 1986, pp 1–186.
28 Food and Nutrition Board Report. 1958. Nat. Acad. Sci./Nat. Res. Council Publ. 575.
29 Lands WEM, Samuelsson B: Phospholipid precursors of prostaglandin. Biochim Biophys Acta 1968;164:426–429.
30 Marshall PJ, Kulmacz RJ, Lands WEM: Constraints on prostaglandin synthesis in tissues. J Biol Chem 1987;262:3510–3517.
31 Hansen AE, Wiese HF, Boelsche AN, et al: Role of linoleic acid in infant nutrition. Clinical and chemical study of 428 infants fed on milk mixtures varying in kind and amount of fat. Pediatrics 1963;31:171–192.
32 Holman RT, Smythe L, Johnson S: Effect of sex and age on fatty acid composition of human serum lipids. Am J Clin Nutr 1979;32:2390–2399.
33 Siguel EN, Chee KM, Gong J, et al: Criteria for essential fatty acid deficiency in plasma as assessed by capillary column gas-liquid chromatography. Clin Chem 1987;33:1869–1873.
34 Lands WEM, Hamazaki T, Yamakazaki K, et al: A story of changing dietary patterns. Am J Clin Nutr 1990;51:991–993.
35 Rome LH, Lands WEM: Structural requirements for the time-dependent inhibition of prostaglandin biosynthesis by anti-inflammatory drugs. Proc Natl Acad Sci USA 1975;72:4863–4865.
36 Lands WEM, LeTellier PR, Rome LH, et al: Inhibition of prostaglandin biosynthesis. Adv Biosci 1973;9:15–27.
37 Prasad MR, Culp B, Lands WEM: Alteration of the acyl chain composition of free fatty acids, acyl coenzyme A and other lipids by dietary polyunsaturated fats. J Biosci 1987;11:443–453.

William E.M. Lands, PhD, Department of Biological Chemistry,
University of Illinois at Chicago, 1853 West Polk Street, Chicago, IL 60612 (USA)

Simopoulos AP, Kifer RR, Martin RE, Barlow SM (eds): Health Effects of ω3 Polyunsaturated Fatty Acids in Seafoods. World Rev Nutr Diet. Basel, Karger, 1991, vol 66, pp 195–204

Effects of Dietary Oils with Extreme ω3/ω6 Ratios – Selective Incorporation and Differential Catabolism

Harumi Okuyama, Keiko Sakai

Faculty of Pharmaceutical Sciences, Nagoya City University, Mizuhoku, Nagoya, Japan

Among commonly used vegetable oils, safflower oil contains the highest proportion of linoleic acid (18:2 ω6) (~75%) but very little α-linolenic acid (18:3 ω3) (~0.01%). In contrast, seeds of *Perilla frutescens* (beefsteak plant), which have long been consumed in northern parts of Asia, provide an oil containing linoleate (~15%) and α-linolenate (~65%). In animal experiments, we have shown that perilla oil is more beneficial than safflower oil and soybean oil (containing 52% linoleate and 8% α-linolenate) for the prevention of chronic diseases such as cancer, stroke, thrombosis, allergy and aging [1–7], some of which are known to be prevented also by fish oil supplements [8]. Furthermore, perilla oil-fed animals exhibited higher learning ability and retinal function than those fed a safflower diet or a soybean diet [9–11].

In this communication, we present data which show the fate of dietary linoleate and α-linolenate in animal bodies. This represents an effort to reveal the biochemical bases for the differential effects of these vegetable oils on the severity of chronic diseases and behavior.

Animals and Diets

A semi-purified diet (Nihon Clea Co., Ltd., Tokyo) supplemented either with safflower oil, soybean oil or perilla oil (10%) was fed to rats through 2 generations or to mice for 5 months after weaning.

Table 1. Organ-specific responses to dietary changes in ω3/ω6 ratios

Organ	Enriched ω3 and ω6 acids in perilla and safflower groups					
	phospholipid		triacylglycerol		cholesterol ester	
	ω3	ω6	ω3	ω6	ω3	ω6
Liver	20:5 22:5	20:4	18:3 20:5 (22:5) (22:6)	18:2 20:4	18:3	18:2 20:4
Plasma	20:5 22:5	20:4	18:3 20:5	18:2 20:4	(18:3) 20:5	(18:2) 20:4
Red cells	20:5 22:5	20:4 22:4 22:5	nd		nd	
PMN	20:5 22:5 (22:6)	20:4 22:4	nd		nd	
Platelets	20:5 22:5	20:4 22:4	nd		nd	
Kidney	20:5 22:5	20:4 22:4	nd		nd	
Adrenal	20:5 22:5	20:4 22:4	18:3	18:2	20:5 22:5 22:6	20:4 22:4 22:5
Retina and brain	22:6	20:4 22:5	nd		nd	

ω3 Acids enriched in the perilla group and ω6 acids enriched in the safflower group are listed in the ω3 and ω6 columns, respectively.
nd = not determined; PMN = polymorphonuclear leukocytes

Incorporation of ω3/ω6 Acids into Lipid Classes in Different Organs

Dietary differences in the ω3/ω6 ratios (safflower oil vs. perilla oil) were reflected in the ratios of 20:5 ω3/20:4 ω6, 22:5 ω3/22:4 ω6 and 22:6 ω3/22:5 ω6 in the body lipids. As noted by many researchers, however, the responses to diets were different among different organs and among differ-

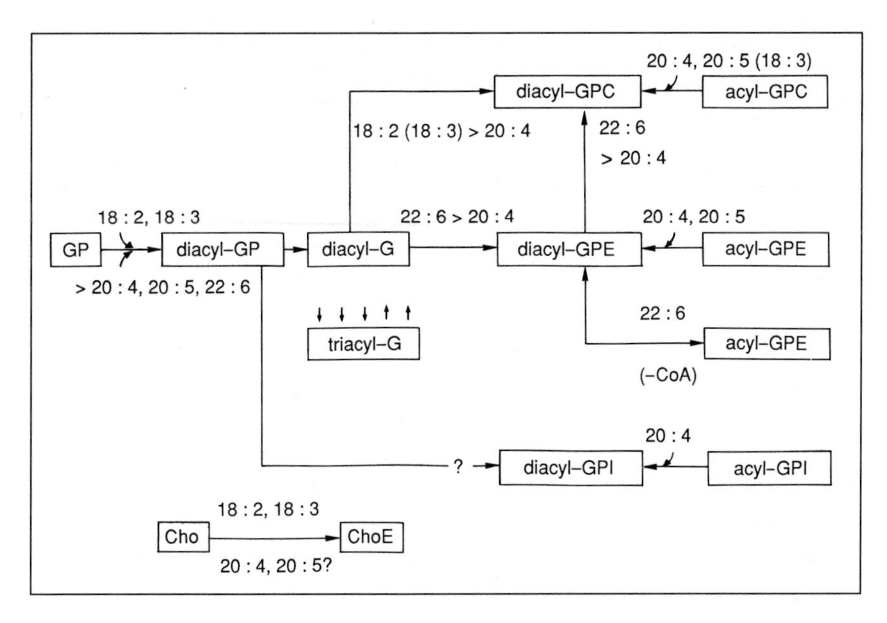

Fig. 1. Selectivities for acyl-CoAs or molecular species containing ω3 and ω6 acids in the synthesis of lipid classes in liver. The 18:2 and 20:4 are ω6 type while 18:3, 20:5 and 22:6 are ω3 type. G = glycerol; GP = *sn*-glycerol 3-phosphate; GPC = *sn*-glycerol 3-phosphocholine; GPE = *sn*-glycerol 3-phosphoethanolamine; GPI = *sn*-glycerol 3-phosphoinositol; Cho = cholesterol; ChoE = cholesterol ester.

ent lipid classes of given organ (table 1). These results raise two major questions; (i) Why do different lipid classes of a given organ respond differently to dietary changes?, and (ii) Why does a given lipid class respond to dietary changes differently in different organs?

As to the question (i), the mechanisms which regulate the fatty acid patterns of different lipid classes have been studied extensively in liver [12] and are summarized in figure 1. Different enzymes involved in the synthesis of individual lipid classes have different selectivities for acyl donors and molecular species of substrates. The manner in which ω3 and ω6 fatty acids are treated in vitro by these enzymes (fig. 1) fits fairly well with the observed fatty acid compositions.

The de novo phosphatidate synthetic pathway and triacylglycerol synthetic pathway have a relatively broad specificity for various acyl donors with a slight preference for 18-carbon polyunsaturated fatty acids over 20- and 22-carbon highly unsaturated fatty acids [13–15]. On the other hand, the acyl-CoA:1-acyl-GPI (glycerophosphoinositol) acyltransferase system is highly selective for arachidonate (20:4 ω6) but both 20:5 ω3 and 18:3 ω3 can compete fairly well with 20:4 ω6 in the acyl-CoA:1-acyl-GPC (glycerophosphocholine) acyltransferase system [16–18]. Diacylglycerol species containing 22:6 ω3 is selectively utilized for the synthesis of phosphatidylethanolamine [13, 14]. These properties of the enzymes determined in vitro fit fairly well with the observed fatty acid compositions of phospholipids and triacylglycerol. The question (ii) is considered in the following paragraph.

Organ Specificities of Enzymes Involved in Lipid Synthesis

Fatty acid compositions of phospholipids are known to be significantly different among different organs (e.g., liver and lung). Fatty acids of triacylglycerol and cholesterol ester also differ significantly in different organs (table 1). The enzymatic mechanisms producing these organ-specific fatty acid patterns have been partly revealed but mostly remain unsolved.

Acyl-CoA: Phospholipid Acyltransferase System

Although these enzymes have not been purified, several lines of evidence indicate that more than two enzymes exist in the acylation of 1-acyl-GPC in liver microsomes [19, 20]. Furthermore, the difference in the selectivities for acyl donors has been noted for the acyl-CoA:1-acyl-GPC acyltransferase systems in different organs. These data have accounted for the difference in the fatty acid composition of phosphatidylcholine in different organs (e.g. liver and lung) [18, 21]. Acyl-CoA:1-acylglycerophosphate acyltransferase systems in different organs also exhibit different acyl selectivities [20].

Desaturation-Elongation System

The perilla and safflower diets induced major increases in the proportions of 22:6 ω3 and 22:5 ω6, respectively, in the phospholipids of retina and brain; whereas in other organs or cells, the proportions of 20:5 ω3 and 20:4 ω6 as well as those of 22:5 ω3 and 22:4 ω6 differed mainly between the two dietary groups (table 1). These results suggest that relative activities for the elongation and desaturation system may be different among different organs. Otherwise, different mechanisms such as those described below must be considered.

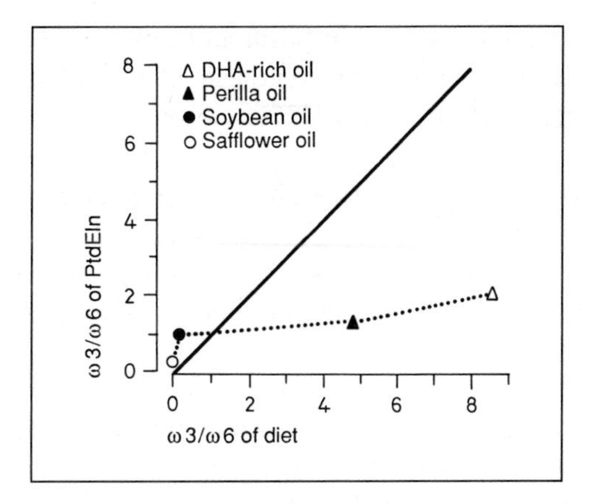

Fig. 2. Effect of dietary ω3/ω6 ratios on the ω3/ω6 ratios of phosphatidylethanolamine in brain. Diet supplemented with either DHA-rich fish oil, perilla oil, soybean oil or safflower oil was fed to rats through 2 generations and the fatty acids of brain phosphatidylethanolamine (PtdEln) were analyzed. The ω3/ω6 ratios of the brain total lipids showed a similar pattern.

Organ-Specificity in the Uptake and Turnover of ω3 and ω6 Fatty Acids?

The organ-specific fatty acid patterns of lipids might originate from differential uptake of ω3 and ω6 acids from blood stream and/or from differential turnover of these acids in the various organs. Very little is known on the possible role of these steps in the regulation of fatty acid patterns of lipid classes in different organs.

Mechanisms Which Keep ω3/ω6 Ratios of Lipids within Limited Ranges

In spite of a very large variation in dietary ω3/ω6 ratios, the ω3/ω6 ratios of tissue phospholipids were kept within limited ranges (fig. 2).

Even when the safflower oil diet with 0.05% of the total fatty acids as ω3 was fed to rats through 2 generations, the proportion of 22:6 ω3 in brain phosphatidylethanolamine was more than 40% of that of the perilla group, whereas the arachidonate content was kept essentially the same in brain phosphatidylinositol under these dietary conditions. Thus, both the ω3 and ω6 fatty acids are highly conserved in brain and retina. The high selectivity for hexaenoic diacylglycerol in the synthesis of phosphatidylethanolamine as well as a transacylase reaction [22], and the high selectivity for 20:4 ω6 in the 1-acyl-GPI acyltransferase system [17, 18] would work to conserve these acids in the respective phospholipids even when their supply from the diet was very limited.

When a diet supplemented with a fish oil rich in 22:6 ω3 was fed to rats through two generations, the 22:6 ω3 content of phosphatidylethanolamine in this group (19%) was not much different from that of the perilla group (17%), indicating the presence of a mechanism to keep the upper limit for the incorporation of ω3 acid. Similarly, there appears to be an upper limit for the incorporation of ω6 acid. It is understood that the levels of the responsible enzymes in given organs are regulated so as to limit the relative rates of incorporations of various fatty acids into phospholipids. The enzymes which synthesize triacylglycerol and cholesterol ester appear to be more variable depending on the dietary conditions, and the triacylglycerol synthesizing enzymes have bigger pool sizes so as to accept excess fatty acids to form depot fats. However, even in these neutral lipid fractions, the ω3/ω6 ratios were not direct reflections of the ratios of diets.

The ω3/ω6 Ratios of Diets Are Not Reflected in the ω3/ω6 Ratios of Whole Bodies

Mice were fed the safflower diet or the perilla diet for 5 months after weaning. For the first month after weaning, mice grew rapidly. Even in this rapidly growing period, approximately 20% of the ingested ω3 and ω6 acids was conserved in the body of the safflower group. However, in the perilla group 9% of the dietary ω3 and 16% of dietary ω6 were conserved in the body. The total fat contents were similar in the two dietary groups.

At 5 months after weaning, the body weights changed relatively little, and these animals are considered to be under a steady-state condition. Therefore, the amounts of ω3 and ω6 acids ingested are considered to be roughly the same as the amounts consumed by the body. At this age, the pool size of ω6 acids in the safflower group was 5.4 g/100 g body weight, which was much larger than the pool size of ω3 in the perilla group (1.8 g/100 g b.w.; fig. 3).

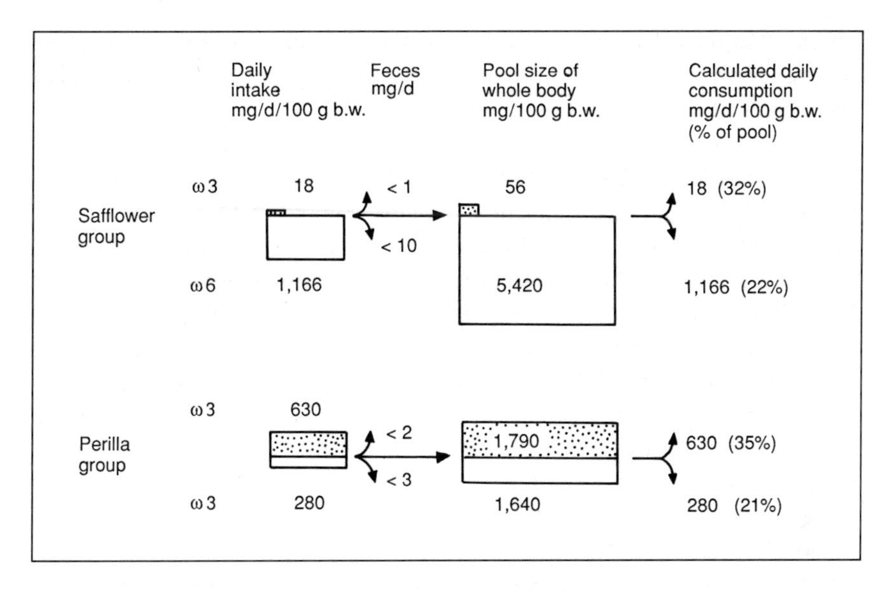

Fig. 3. Pool sizes and turnovers of ω3 and ω6 acids in whole bodies of mice. Mice (ICR) were fed either the safflower diet or the perilla diet for 5 months after weaning, and the fatty acids ingested daily, excreted into feces daily and stored in whole body were determined.

Then, the daily consumption of ω3 in the perilla group is calcualted to be larger (35% of the pool) than that of ω6 (21–22% of the pool) in the two dietary groups. The faster turnover of ω3 as compared with ω6 in the perilla group, at least in part, account for the lower ω3/ω6 ratios of whole body compared to those of diets in the perilla group. In the safflower group, the ω3/ω6 ratios of whole body were similar to that of diet.

Possible Pathways Responsible for the Differential Catabolism of ω3 and ω6 Acids

Because 18:3 ω3 is known to be several-fold more easily oxidizable than 18:2 ω6, the contents of lipid peroxide (conjugated diene) were compared in the lipids extracted from freshly prepared brains of rats fed the two diets through 2 generations. The rates of NADPH-induced lipid peroxidation were also compared. However, no significant differences were observed in these parameters of the two dietary groups. Lipofuscin, a putative age pigment, was determined in brains, and again very little difference was observed. Ethane and pentane are assumed to evolve mostly from ω3 and ω6 acids, respectively, but the total amounts of these two hydrocarbons evolved were below 1/1,000 the amounts of ω3 and ω6 acids ingested. Because the amounts of ω3 and ω6 acids converted to eicosanoids are assumed to be very small, we hypothesize that the majority of ingested ω3 and ω6 acids are catabolized through β-oxidation pathway under these conditions and that this pathway may be responsible for the differential catabolism of ω3 and ω6 acids. This idea is consistent with an in vitro observation [23] that 18:3 ω3 is oxidized faster than 18:2 ω6 in mitochondria.

Conclusion

There are mechanisms which keep ω3/ω6 ratios of tissue lipids within limited ranges, despite a large variation in dietary ω3/ω6 ratios. The mechanisms to conserve ω3 and ω6 acids in membrane phospholipids were discussed. The pool size for ω3 acids in whole body was significantly smaller than that for ω6 acids. The dietary 18:3 ω3 was catabolized faster than 18:2 ω6, possibly through β-oxidation.

Acknowledgment

We are grateful to Prof. Tsuneko Sato, Aichi University School of Medicine, Aichi, Dr. Masaru Sagai, National Institute of Environmental Sciences, Tsukuba, and Dr. Ronald C. Reitz, University of Nevada, Reno, for their helpful discussions.

References

1 Okuyama H: Abura, Kono Oisikute Huannamono (in Japanese) (meaning Oils, very delicious but in anxiety). Tokyo, Nobunkyo, 1989.
2 Hori T, Moriuchi A, Okuyama H, et al: Effect of dietary essential fatty acids on pulmonary metastasis of ascites tumor cells in rats. Chem Pharm Bull 1987;35: 3925–3927.
3 Shimokawa T, Moriuchi A, Hori T, et al: Effect of dietary alpha-linolenate/linoleate balance on mean survival time, incidence of stroke and blood pressure of spontaneously hypertensive rats. Life Sci 1988;43:2067–2075.

4 Hashimoto A, Katagiri M, Torii S, et al: Effect of the dietary α-linolenate/linoleate balance on leukotriene production and histamine release in rats. Prostaglandins 1988;36:3–16.
5 Watanabe S, Suzuki E, Kojima R, et al: Effect of dietary α-linolenate/linoleate balance on collagen-induced platelet aggregation and serotonin release in rats. Chem Pharm Bull 1989;37:1572–1575.
6 Kamano K, Okuyama H, Konishi R, et al: Effects of a high-linoleate and a high-a-linolenate diets on spontaneous mammary tumorigenesis in mice. Anticancer Res 1990;9:1903–1908.
7 Yamamoto N, Okaniwa Y, Mori S, et al: Effects of a high-linoleate and a high-a-linolenate diets on the learning ability of aged rats: Evidence against an autoxidation-related lipid peroxide theory of aging. J Gerontol 1990 (in press).
8 Lands WEM: Fish and Human Health. Orland, Academic Press, 1986.
9 Yamamoto N, Saitoh M, Moriuchi A, et al: Effect of dietary α-linolenate/linoleate balance on brain lipid compositions and learning ability of rats. J Lipid Res 1987;28: 144–151.
10 Watanabe I, Kato M, Aonuma H, et al: Effect of dietary alpha-linolenate/linoleate balance on the lipid composition and electroretinographic responses in rats. Adv Biosci 1987;62:563–570.
11 Yamamoto N, Hashimoto A, Takemoto Y, et al: Effect of the dietary α-linolenate/linoleate balance on lipid compositions and learning ability of rats. II. Discrimination process, extinction process, and glycolipid compositions. J Lipid Res 1988;29:1013–1021.
12 Lands WEM, Crawford CG: Enyzmes of membrane phospholipid metabolism in animals; in Martonosi A (ed): The Enzymes of Biological Membranes. New York, Plenum Press, 1976, pp 3–85.
13 Hill EE, Husbands DR, Lands WEM: The selective incorporation of ^{14}C-glycerol into different species of phosphatidic acid, phosphatidylethanolamine, and phosphatidylcholine. J Biol Chem 1968;243:4440–4445.
14 Akesson B, Elovson J, Arvidson G: Initial incorporation into rat liver glycerophospholipids of intraportally injected [^{3}H] glycerol. Biochim Biophys Acta 1970;210: 15–27.
15 Okuyama H, Lands WEM: Variable selectivities of acyl coenzyme A:monoacylglycerophosphate acyltransferases in rat liver. J Biol Chem 1972;247:1414–1423.
16 Lands WEM, Inoue M, Sugiura Y, et al: Selective incorporation of polyunsaturated fatty acids into phosphatidylcholine by rat liver microsomes. J Biol Chem 1982;257: 14968–14972.
17 Baker RR, Thompson W: Selective acylation of 1-acylglycerophosphorylinositol by rat liver microsomes. J Biol Chem 1973;248:7060–7065.
18 Murase S, Yamada K, Okuyama H: Kinetic parameters of lysophospholipid acyltransferase systems in diet-induced modifications of platelet phospholipid acyl chains. Chem Pharm Bull 1988;36:2109–2117.
19 Reitz RC, El Sheikh M, Lands WEM, et al: Effects of ethylenic bond position upon acyltransferase activity with isomeric *cis*-octadecenoyl coenzyme A thiol esters. Biochim Biophys Acta 1969;176:480–490.

20 Okuyama H, Yamada K, Ikezawa H: Acceptor concentration effect in the selectivity of acyl coenzyme A:1-acylglycerylphosphorylcholine acyltransferase system in rat liver. J Biol Chem 1975;250:1710–1713.
21 Okuyama H, Yamada K, Miyagawa T, et al: Enzymatic basis for the formation of pulmonary surfactant lipids by acyltransferase systems. Arch Biochem Biophys 1983;221:99–107.
22 Masuzawa Y, Okano S, Nakagawa A, et al: Selective acylation of alkyl-lysophospholipids by docosahexaenoic acid in Ehrlich ascites cells. Biochim Biophys Acta 1986; 876:80–90.
23 Clouet P, Niot I, Bezard J: Pathway of α-linolenic acid through the mitochondrial outer membrane in the rat liver and influence on the rate of oxidation. Biochem J 1989;263:867–873.

Harumi Okuyama, MD, Faculty of Pharmaceutical Sciences,
Nagoya City University, 3–1 Tanabedori, Mizuhoku, Nagoya 467 (Japan)

Simopoulos AP, Kifer RR, Martin RE, Barlow SM (eds): Health Effects of ω3 Polyunsaturated Fatty Acids in Seafoods. World Rev Nutr Diet. Basel, Karger, 1991, vol 66, pp 205–216

Dietary Polyunsaturated Fatty Acids and Mortality in the Multiple Risk Factor Intervention Trial (MRFIT)

Therese A. Dolecek, Greg Grandits

Department of Public Health Sciences, Bowman Gray School of Medicine, Wake Forest University, Winston-Salem, N.C. USA, and
The Multiple Risk Factor Intervention Trial Research Group

Introduction

Dietary polyunsaturated fatty acids (PUFA) have been shown to have effects on biochemical and physiological mechanisms which relate to numerous disease occurrences. Perhaps of greatest concern in PUFA biochemistry is the regulation of eicosanoids, which act in diverse ways to influence biological processes, both positively and adversely. The quality, quantity, and balance of these fatty acids appear to play a significant role in determining their actions [1]. Unfortunately, experimental evidence currently available is not sufficient to even suggest optimal dietary PUFA intake levels. Understanding these relationships as they apply to humans could have profound implications for dietary recommendations aimed at achievement of health promotion and disease prevention.

Given that disease patterns as well as dietary practices differ considerably among populations, the examination of dietary PUFA intakes of large populations would be valuable. Findings from such evaluations in combination with results of experimental studies might help to clarify some of the inconsistencies that currently exist in the literature and provide direction for further research in this very important area.

Among the few large population studies that have estimated dietary intake of specific fatty acids is the Multiple Risk Factor Intervention Trial (MRFIT). This report describes dietary PUFA intakes by middle-aged

American men who participated in this trial from 1973 through early 1982. Estimates of PUFA intakes are available using nutrient data from 24-hour dietary recall interviews obtained during the study. Associations between PUFA intakes and mortality are examined.

Methods

Background, Organization, and Design

A detailed description of the MRFIT is provided elsewhere [2]. Briefly, MRFIT was a multi-center, clinical trial in the primary prevention of coronary heart disease (CHD) supported by the National Heart, Lung, and Blood Institute. The study population included men aged 35–57 years at entry who were determined to be at high risk of developing CHD based upon smoking status, diastolic blood pressure, and serum cholesterol levels. The trial was conducted at 22 clinical centers where the men were followed for 6–8 years. From 361,662 screenees, 12,866 men were selected to be participants during three screening visits. Approximately half were randomized to each of two study groups. The special intervention (SI) group received interventions to reduce smoking, blood pressure, and blood cholesterol, while the usual care (UC) group participants were referred to their usual source of medical care and returned annually for examination at their respective clinical centers.

Principal Dietary Data Collection Method

The principal dietary method chosen for the MRFIT was the 24-hour dietary recall, since it is appropriate for measuring the dietary intake of groups [3]. Highly standardized dietary recall interviews were conducted at the baseline third screen visit for all participants and at follow-up years 1, 2, 3, 5, and 6 for the SI men and at years 1, 2, 3, and 6 for participants in the UC group. Data generated from 24-hour recall interviews have been used to describe the MRFIT population at baseline and to monitor the dietary patterns and food selection trends by both SI and UC groups over the follow-up period [4, 5].

While the 24-hour recall is generally not intended to characterize usual intake by individuals considered desirable when studying diet-disease relationships, multiple recalls on the same individual improve accuracy in terms of establishing more reliable usual intake estimates. On this premise, data from recalls obtained at baseline and at follow-up years were used for this evaluation.

Mortality Ascertainment

Mortality was ascertained from the beginning of the trial and continues to be monitored by the MRFIT coordinating center in Minneapolis, Minnesota. Clinics assumed responsibility for follow-up while the trial was in progress until February 28, 1982 at which time the National Death Index became the primary mortality follow-up method. The data presented include deaths ascertained through December 31, 1985. Cause-specific mortality assignments are based on the 9th revision of the International Classification of Diseases. All death certificates and supporting records have been independently coded by two nosologists without knowledge of treatment group and where differences exist adjudication is achieved by a third.

Statistical Analysis

Only data on the UC group were analyzed for this evaluation to avoid the analytical complexities introduced by the multi-intervention effects on the SI group. MRFIT recalls were reanalyzed in 1985 using the University of Minnesota Nutrition Coordinating Center Food Table version number 11 which contains very complete information on individual fatty acids thereby making possible an evaluation of PUFA intake [6]. Mean values available from recall data analysis at baseline and at the first three follow-up years were calculated for each PUFA estimate under study. PUFA intake estimates in grams were established for total PUFA, 18:2 ω6, 18:3 ω3, 18:4 ω6, 20:4 ω6, 20:5 ω3, 22:5 ω3, 22:6 ω3. The sum of 20:5 ω3, 22:5 ω3, and 22:6 ω3 as well as ratios, 18:3 ω3/18:2 ω6 and total ω3/total ω6, were also calculated for analytic purposes.

Participants were divided into quintiles based on their average intake of each PUFA estimate. Relative risks of mortality were calculated relative to the first quintile using proportional hazards regression analysis. Adjusted relative risks were also determined by including age, race and baseline values of diastolic blood pressure, cigarettes smoked per day, high density lipoprotein-cholesterol, and low density lipoprotein-cholesterol in the model. Each PUFA estimate was also entered into the model as a continuous variable using log transformations as appropriate with the regression coefficient and significance level given. Deaths and documented clinical myocardial infarctions during the first 3 years of follow-up were excluded from the analysis. The mortality categories selected for evaluation are broad, since overall death rates were low. Four groups were established including CHD, all cardiovascular diseases (CVD), all causes, and all cancers.

Results

Dietary Polyunsaturated Fatty Acid Distribution

Table 1 shows the distribution of mean dietary PUFA intake among the MRFIT UC participants based on the average of four 24-hour dietary recall measures. The predominant PUFA was linoleic acid (18:2 ω6) contributing approximaely 87% of the mean total PUFA intake. Linolenic acid (18:3 ω3) contributed about 10%. Stearodonic acid (18:4 ω6) appeared in very low concentrations with 65% of participants reporting no intake of the fatty acid. Mean arachidonic acid (20:4 ω6), the precursor of eisosanoids, was reported to be consumed in relatively low amounts in relation to total PUFA intake. The long-chain ω3 fatty acids found in fish oils – eicosapentaenoic acid (20:5), docosapentaenoic acid (22:5), and docosahexaenoic acid (22:6) – were reported to be consumed in very small quantities with the mean sum equaling about 175 mg/day. It should be noted that the reported intake of these fish fatty acids demonstrated considerable variability and that about 20% of the group reported zero intake. The mean intake ratios of 18:3 ω3/18:2 ω6 and total ω3/total ω6 were approximately 0.12 and 0.13, respectively.

Table 1. Reported dietary polyunsaturated fatty acid (PUFA) intake[1] distributions in grams for usual care participants in the Multiple Risk Factor Intervention Trial (MRFIT)

	Mean	Standard deviation	Quartile 1	Median	Quartile 3	% >0
Total PUFA	16.828	7.663	11.740	15.470	20.420	100
18:2 ω6	14.603	6.957	9.990	13.352	17.868	100
18:3 ω3	1.688	0.736	1.189	1.572	2.028	100
18:4 ω6	0.008	0.027	0.000	0.000	0.005	35
20:4 ω6	0.222	0.107	0.151	0.205	0.272	100
20:5 ω3	0.069	0.155	0.000	0.011	0.070	71
22:5 ω3	0.024	0.057	0.000	0.008	0.021	74
22:6 ω3	0.082	0.193	0.000	0.013	0.089	59
20:5+22:5+22:6	0.175	0.184	0.004	0.043	0.187	79
18:3/18:2	0.122	0.034	0.101	0.119	0.139	100
(18:3+20:5+22:5+22:6)/ (18:2+20:4)	0.133	0.051	0.107	0.127	0.149	100

[1] Mean of baseline, follow-up years 1, 2, and 3 dietary recall nutrient data.

Mortality Findings

Tables 2–6 show the death rates by quintile and the results of proportional hazards regression analyses of selected PUFAs, combined PUFAs, and PUFA ratios on mortality outcome groups including CHD, CVD, all cause, and cancer. Tables 2–4 show findings from analyses of independent variables 18:2 ω6, 18:3 ω3 and sum of 20:5 ω3, 22:5 ω3, and 22:6 ω3 expressed in grams. Tables 5 and 6 display results from analyses for 18:3 ω3/18:2 ω6 and total ω3/total ω6 on the four mortality outcomes.

No significant associations were detected for linoleic acid (18:2 ω6) on any mortality group as shown in table 2. Table 3 presents results using 18:3 ω3 as the independent variable in the analysis. The same negative pattern of adjusted relative risks by quintile of intake was apparent for 18:3 ω3 on CHD, CVD, and all cause mortality. However, the inverse relationship was only significant for all cause mortality and marginally significant for CVD. A weaker association was observed for cancer deaths and was not significant. Analysis of the combined fatty acids predominantly found in fish shown in table 4 demonstrated significant inverse associations with CHD, CVD, and all cause mortality groups but not for cancers. The benefit appeared to be in the largest intake quintile with a mean ingestion of about 664 mg/day.

Table 2. Estimated relative risk for dietary 18:2 ω6 intake and mortality in MRFIT usual care participants from proportional hazards regression analyses[1]

Quintile	Mean, g	n	Deaths	Dead, %		RR	ADJ. RR
CHD mortality							
I	7.037	1,251	44	3.52		1.00	1.00
II	10.646	1,252	31	2.48		0.70	0.74
III	13.387	1,252	27	2.16		0.60	0.65
IV	16.839	1,252	46	3.67		1.04	1.09
V	25.065	1,251	27	2.16		0.61	0.65
					Slope	-0.0150	-0.0128
					Z slope	-1.2610	-1.0909
					p value	0.2077	0.2746
CVD mortality							
I	7.037	1,251	60	4.80		1.00	1.00
II	10.646	1,252	37	2.96		0.61	0.65
III	13.387	1,252	40	3.19		0.65	0.71
IV	16.839	1,252	60	4.79		1.00	1.06
V	25.065	1,251	35	2.80		0.58	0.63
					Slope	-0.0121	-0.0089
					Z slope	-1.1874	-0.8888
					p value	0.2347	0.3737
All cause mortality							
I	7.037	1,251	104	8.31		1.00	1.00
II	10.646	1,252	79	6.31		0.75	0.80
III	13.387	1,252	78	6.23		0.73	0.79
IV	16.839	1,252	105	8.39		1.01	1.10
V	25.065	1,251	73	5.84		0.69	0.77
					Slope	-0.0135	-0.0095
					Z slope	-1.8166	-1.2923
					p value	0.0694	0.1957
Cancer mortality							
I	7.037	1.251	18	1.44		1.00	1.00
II	10.646	1,252	34	2.72		1.87	1.98
III	13.387	1,252	25	2.00		1.36	1.44
IV	16.839	1,252	28	2.24		1.55	1.70
V	25.065	1,251	27	2.16		1.48	1.65
					Slope	-0.0001	0.0033
					Z slope	0.0000	0.2646
					p value	0.9953	0.7876

[1] Mean of baseline, follow-up years 1, 2, and 3 dietary recall nutrient data. Deaths and clinical myocardial infarctions in first 3 years excluded. Adjusted for age, race, and baseline smoking, diastolic blood pressure, high density and low density lipoprotein levels.

Table 3. Estimated relative risk for dietary 18:3 ω3 intake and mortality in MRFIT usual care participants from proportional hazards regression analysis[1]

Quintile	Mean, g	n	Deaths	Dead, %		RR	ADJ. RR
CHD mortality							
I	0.873	1.251	43	3.44		1.00	1.00
II	1.273	1.253	40	3.19		0.93	0.98
III	1.577	1.251	24	1.92		0.55	0.57
IV	1.926	1.251	40	3.20		0.92	0.98
V	2.802	1.252	28	2.24		0.64	0.68
					Slope	−0.1897	−0.1657
					Z slope	−1.6703	−1.4560
					p value	0.0951	0.1458
CVD mortality							
I	0.873	1,251	58	4.64		1.00	1.00
II	1.273	1,253	52	4.15		0.89	0.94
III	1.577	1,251	38	3.04		0.64	0.67
IV	1.926	1,251	49	3.92		0.83	0.90
V	2.802	1,252	35	2.80		0.60	0.63
					Slope	−0.2130	−0.1832
					Z slope	−2.1331	−1.8330
					p value	0.0329	0.0667
All cause mortality							
I	0.873	1,251	105	8.39		1.00	1.00
II	1.273	1,253	99	7.90		0.94	0.96
III	1.577	1,251	73	5.84		0.68	0.69
IV	1.926	1,251	91	7.27		0.86	0.89
V	2.802	1,252	71	5.67		0.67	0.69
					Slope	−0.1982	−0.1784
					Z slope	−2.7514	−2.4556
					p value	0.0059	0.0141
Cancer mortality							
I	0.873	1,251	25	2.00		1.00	1.00
II	1.273	1,253	34	2.71		1.36	1.35
III	1.577	1,251	22	1.76		0.86	0.85
IV	1.926	1,251	29	2.32		1.15	1.14
V	2.802	1,252	22	1.76		0.87	0.87
					Slope	−0.1003	−0.0985
					Z slope	−0.8000	−0.7746
					p value	0.4235	0.4392

[1] Mean of baseline, follow-up years 1, 2, and 3 dietary recall nutrient data. Deaths and clinical myocardial infarctions in first 3 years excluded. Adjusted for age, race, and baseline smoking, diastolic blood pressure, high density and low density lipoprotein levels.

Table 4. Estimated relative risk for sum of dietary 20:5, 22:5, and 22:6 ω3 and mortality in MRFIT usual care participants from proportional hazards regression anaylses[1]

Quintile	Mean, g	n	Deaths	Dead, %		RR	ADJ. RR
CHD mortality							
I	0.000	1.307	42	3.21		1.00	1.00
II	0.009	1,197	39	3.26		1.01	1.08
III	0.046	1,251	35	2.80		0.87	0.91
IV	0.153	1,252	35	2.80		0.87	0.88
V	0.664	1,251	24	1.92		0.59	0.60
					Slope	-0.2803	-0.2877
					Z slope	-2.3707	-2.4310
					p value	0.0178	0.0150
CVD mortality							
I	0.000	1.307	55	4.21		1.00	1.00
II	0.009	1,197	51	4.26		1.01	1.06
III	0.046	1,251	47	3.76		0.90	0.92
IV	0.153	1,252	48	3.83		0.91	0.92
V	0.664	1,251	31	2.48		0.58	0.59
					Slope	-0.2857	-0.2936
					Z slope	-2.7749	-2.8460
					p value	0.0055	0.0044
All cause mortality							
I	0.000	1,307	99	7.57		1.00	1.00
II1	0.009	1,197	96	8.02		1.06	1.09
III	0.046	1,251	93	7.43		0.98	1.02
IV	1.153	1,252	80	6.39		0.84	0.85
V	0.664	1,251	71	5.68		0.74	0.76
					Slope	-0.1799	-0.1826
					Z slope	-2.5338	-2.5671
					p value	0.0113	0.0102
Cancer mortality							
I	0.000	1,307	28	2.14		1.00	1.00
II	0.009	1,197	31	2.59		1.21	1.24
III	0.046	1,251	29	2.32		1.09	1.16
IV	0.153	1,252	19	1.52		0.71	0.73
V	0.664	1,251	25	2.00		0.92	0.97
					Slope	-0.1090	0.0985
					Z slope	-0.8718	0.7874
					p value	0.3836	0.4309

[1] Mean of baseline, follow-up years 1, 2, and 3 dietary recall nutrient data (log +0.1). Deaths and clinical myocardial infarctions in first 3 years excluded. Adjusted for age, race, and baseline smoking, diastolic blood pressure, high density and low density lipoprotein levels.

Table 5. Estimated relative risk for dietary 18:3 ω3/18:2 ω6 intake and mortality in MRFIT usual care participants from proportional hazards regression analyses[1]

Quintile	Mean ratio	n	Deaths	Dead, %		RR	ADJ. RR
CHD mortality							
I	0.080	1,251	41	3.28		1.00	1.00
II	0.105	1,252	29	2.32		0.70	0.70
III	0.120	1,252	32	2.56		0.77	0.81
IV	0.135	1,252	33	2.64		0.79	0.81
V	0.170	1,251	40	3.20		0.97	0.95
					Slope	0.2854	0.1846
					Z slope	0.1414	0.1000
					p value	0.8983	0.9331
CVD mortality							
I	0.080	1,251	56	4.48		1.00	1.00
II	0.105	1,252	40	3.19		0.70	0.70
III	0.120	1,252	42	3.35		0.74	0.77
IV	0.135	1,252	44	3.51		0.77	0.78
V	0.170	1,251	50	4.00		0.89	0.85
					Slope	−0.1776	−0.4419
					Z slope	−0.1000	−0.2236
					p value	0.9278	0.8200
All cause mortality							
I	0.080	1,251	101	8.07		1.00	1.00
II	0.105	1,252	79	6.31		0.77	0.76
III	0.120	1,252	86	6.87		0.84	0.83
IV	0.135	1,252	82	6.55		0.80	0.78
V	0.170	1,251	91	7.27		0.89	0.82
					Slope	−0.2824	−1.0739
					Z slope	−0.2000	−0.7483
					p value	0.8432	0.4539
Cancer mortality							
I	0.080	1,251	31	2.48		1.00	1.00
II	0.105	1,252	26	2.08		0.83	0.81
III	0.120	1,252	27	2.16		0.86	0.81
IV	0.135	1,252	24	1.92		0.76	0.71
V	0.170	1,251	24	1.92		0.77	0.67
					Slope	−4.0296	−5.5186
					Z slope	−1.4595	−1.9824
					p value	0.1442	0.0475

[1] Mean of baseline, follow-up years 1, 2, and 3 dietary recall nutrient data. Deaths and clinical myocardial infarctions in first 3 years excluded. Adjusted for age, race, and baseline smoking, diastolic blood pressure, high density and low density lipoprotein levels.

Table 6. Estimated relative risk for dietary total ω3/ω6 and mortality in MRFIT usual care participants from proportional hazards regression analysis[1]

Quintile	Mean ratio	n	Deaths	Dead, %		RR	ADJ. RR
CHD mortality							
I	0.086	1,251	41	3.28		1.00	1.00
II	0.111	1,252	31	2.48		0.76	0.76
III	0.127	1,252	34	2.72		0.83	0.87
IV	0.145	1,252	31	2.48		0.75	0.76
V	0.199	1,251	38	3.04		0.92	0.89
					Slope	−0.5508	−0.5659
					Z slope	−1.2083	−1.2490
					p value	0.2262	0.2114
CVD mortality							
I	0.086	1,251	55	4.40		1.00	1.00
II	0.111	1,252	45	3.59		0.82	0.82
III	0.127	1,252	43	3.43		0.78	0.81
IV	0.145	1,252	40	3.19		0.72	0.73
V	0.199	1,251	49	3.92		0.89	0.84
					Slope	−0.6571	−0.7062
					Z slope	−1.6492	−1.7776
					p value	0.0091	0.0754
All cause mortality							
I	0.086	1,251	97	7.75		1.00	1.00
II	0.111	1,252	92	7.35		0.95	0.93
III	0.127	1,252	88	7.03		0.90	0.90
IV	0.145	1,252	72	5.75		0.74	0.72
V	0.199	1,251	90	7.19		0.92	0.85
					Slope	−0.3814	−0.5205
					Z slope	−1.3454	−1.8193
					p value	0.1783	0.0690
Cancer mortality							
I	0.086	1,251	31	2.48		1.00	1.00
II	0.111	1,252	33	2.64		1.06	1.02
III	0.127	1,252	26	2.08		0.84	0.80
IV	0.145	1,252	20	1.60		0.64	0.61
V	0.199	1,251	22	1.76		0.70	0.62
					Slope	−0.9513	−1.1848
					Z slope	−1.7664	−2.1726
					p value	0.0771	0.0299

[1] Mean of baseline, follow-up years 1, 2, and 3 dietary recall nutrient data (log +0.1). Deaths and clinical myocardial infarctions in first 3 years excluded. Adjusted for age, race, and baseline smoking, diastolic blood pressure, high density and low density lipoprotein levels.

When compared with zero intake, mortality from CHD, CVD, and all cause mortality was 40, 41 and 24% lower, respectively.

Findings from 18:3 ω3/18:2 ω6 ratios showed no association with CHD, CVD, or all cause mortality but an inverse relationship was observed between the ratio and cancer mortality. 33% less cancer deaths occurred in the highest intake quintile when compared with the lowest. The pattern of relative risk across quintiles generally showed a smooth trend and gradual decline in cancer mortality as the 18:3 ω3/18:2 ω6 ratio increased. An inverse relationship with cancer mortality was also demonstrated when total ω3/total ω6 ratio was the independent variable in the analysis (table 6). A reduction of 38% cancer deaths was apparent when the highest intake quintile was compared with the lowest. Marginally significant inverse associations were observed between the total ω3/total ω6 ratio and mortality from CVD and all cause mortality groups but not for CHD.

Discussion

The mean total PUFA intake of 16.83 g reported by the MRFIT UC group represents approximately 6.5% of the average total kilocalories consumed [5]. This intake is consistent with a reported 6.72% by men in the Lipids Research Clinics Prevalence study conducted from 1972 to 1976 [7]. The MRFIT UC PUFA intake was slightly greater than that reported by men inteviewed during NHANES I (3.92%) from 1971 to 1974 and during NHANES II (4.96%) from 1976 to 1980 [unpublished data].

Information from the Surgeon General's Report estimated an average PUFA intake of 8% total kilocalories for the US population during the late 1980s [8]. It would appear that consumption of PUFA in the US population has increased slightly since the 1970s. Since few reports express PUFA in terms of fatty acids, it is not possible to compare trends in specific PUFA intake among these studies.

Data on Japanese dietary PUFA intake during 1975 show very interesting differences when compared with the MRFIT data [personal communication with Dr. Harumi Okuyama]. The Japanese consumed less total fat but about the same amount of total PUFA as that reported by the MRFIT men. The composition of PUFA, however, was strikingly different. The Japanese consume considerably more fish oil fatty acids (20:5 ω3, 22:5 ω3, 22:6 ω3). While the Japanese intake was approximately 1.5 g/day, MRFIT

UC participants consumed an average of 0.18 g/day. The intake of linoleic acid (18:2 ω6) was about the same for the Japanese comparison with the MRFIT reported intake. Linolenic acid (18:3 ω3) intake was slightly greater for Japanese than MRFIT participants. Overall, the Japanese had a greater ω3 to ω6 fatty acid intake ratio, 0.26 compared with 0.13 in the MRFIT group. Given that the disease patterns between Japan and the US differ considerably, it would seem that dietary PUFA composition may play a role in establishing and changing those patterns over time [9].

The findings of this evaluation which show a protective effect of the long-chain fish fatty acids on CHD, CVD, and all cause mortality are consistent with other reports in the literature [10]. It is known that populations consuming large amounts of marine and seafoods such as the fishing villagers of Japan, Greenland Eskimos and Alaskan natives have remarkably low rates of acute myocardial infarction [11–13]. Moreover, several epidemiologic studies of Western industrialized populations have also reported inverse associations between fish consumption and death especially from coronary heart disease [14, 15].

The inverse associations between ω3/ω6 ratios and cancer mortality are intriguing. Although some evidence exists that ω3 fatty acids have antioncogenic properties, there is a need to evaluate other databases to determine if these results can be duplicated. Interpretation is further complicated by the fact that cancer mortality includes deaths from approximately 30 different forms of cancer, each having its own etiologic factors and pathologic uniqueness. Likewise, interpretation of the negative association between dietary 18:3 ω3 intake and all cause mortality is difficult. Whether the critical factors involved in these complex processes will reveal a benefit from qualitative and/or quantitative changes in the PUFA composition of diets remains to be seen.

Conclusions

The results of this evaluation support the hypothesis that fatty acids found primarily in fish oils protect against cardiovascular disease. They also suggest that the composition and balance of PUFA in the diet may influence mortality from cardiovascular disease and possibly various forms of cancer. Further research is needed to define the optimal level and balance of polyunsaturated fatty acids in the diets of humans to promote health and prevent disease.

References

1 Lands WEM: n–3 fatty acids as precursors for active metabolic substances: Dissonance between expected and observed events. J Int Med 1989;225(suppl. 1):11–20.
2 Multiple Risk Factor Intervention Trial Research Group: Multiple Risk Factor Intervention Trial: Risk factor changes and mortality results. J Am Med Ass 1982; 248:1465–1477.
3 Young CM: Dietary methodology; in Committee on Food Consumption Patterns, Food and Nutrition Board: Assessing Changing Food Consumption Patterns. Washington, National Academy Press, 1981.
4 Tillotson JL, Gorder DD, Kassim N: Nutrition data collection in the Multiple Risk Factor Intervention Trial (MRFIT): Description of baseline nutrient intake of randomized population. J Am Diet Ass 1981;78:235–240.
5 Gorder DD, Dolecek TA, Coleman GG, Tillotson JL, Brown HB, Lenz-Litzow K, Bartsch GE, Grandits G: Dietary intake in the Multiple Risk Factor Intervention Trial (MRFIT): Nutrient and food group changes over 6 years. J Am Diet Ass 1986; 86:744–751.
6 Sievert YA, Shakel SF, Buzzard IM: Maintenance of a nutrient database for clinical trials. Controlled Clin Trials 1989;10:416–425.
7 The Lipid Research Clinics Populations Studies Data Book, Volume II: The prevalence study nutrient intake. Lipid Metabolism Atherogenesis Branch, Division of Heart and Vascular Diseases, National Heart, Lung, and Blood Institute. US Dept Health and Human Services. Public Health Service, NIH. NIH Publication No. 82–2014, 1982.
8 The Surgeon General's Report on Nutrition and Health; DHHS (PHS) 1988; Publication 88-5021.
9 Lands WEM, Hamazaki T, Yamazaki K, Okuyama H, Sakai K, Goto Y, Hubbard VS: A study of changing dietary patterns. Am J Clin Nutr 1990;51:991–993.
10 Simopoulos AP, Kifer RR, Martin RE (eds): Health Effects of Polyunsaturated Fatty Acids in Seafoods. Orlando, Academic Press, 1986.
11 Bang HO, Dyerberg J: Lipid metabolism and ischemic heart disease in Greenland Eskimos. Adv Nutr Res 1980;3:1–40.
12 Gottmann AW: A report of 103 autopsies on Alaskan native. Arch Path 1960;70: 117–124.
13 Keys A: Coronary heart disease in seven countries. Circulation 1970;41(suppl I): 162–179.
14 Kromhout D, Bosschieter EB, deLezenne-Coulander C: The inverse relation between fish consumption and 20-year mortality from coronary heart disease. N Engl J Med 1985;312:1205–1209.
15 Shekelle RB, Paul O, Shryock AM, Stamler J: Fish consumption and mortality from coronary heart disease. N Engl J Med 1985;313:820.

Therese A. Dolecek, PhD, RD, Department of Public Health Sciences,
Bowman Gray School of Medicine, Wake Forest University,
300 South Hawthorne Road, Winston-Salem, NC 27103 (USA)

Effects on Diseases

Cardiovascular I: Cell-Vessel Wall Interactions

Simopoulos AP, Kifer RR, Martin RE, Barlow SM (eds): Health Effects of ω3 Polyunsaturated Fatty Acids in Seafoods. World Rev Nutr Diet. Basel, Karger, 1991, vol 66, pp 218–232

Cardiovascular Effects of ω3 Fatty Acids

Atherosclerosis Risk Factor Modification by ω3 Fatty Acids[1]

P.C. Weber[a], *A. Leaf*[b]

[a] Institut für Prophylaxe und Epidemiologie der Kreislaufkrankheiten, Universität München, München, FRG;
[b] Harvard Medical School, Department of Preventive Medicine, Massachusetts General Hospital, Boston, Mass., USA

Introduction

The original observations made in Eskimos on the association between high fish intake and low cardiovascular mortality has led to a number of epidemiological evaluations of this intriguing relationship in industrialized and Westernized populations in Japan, Europe and the United States. As indicated in table 1, both in Japan and in 4 out of 6 surveys it was demonstrated during the past five years that a high intake of fish is associated with a lower rate of cardiovascular death [1–5]. In this brief review, we concentrate primarily on studies published in 1985 and later, the year when the first large survey on the biological actions of ω3 fatty acids was published. We furthermore refer to several recently published reviews on this topic [6–10].

Functional Effects of ω3 Fatty Acids in the Cardiovascular System

Epidemiological studies cannot prove a causal relationship. However, a number of human and animal studies have, during the past five years, demonstrated a multiplicity of functional effects of ω3 fatty acids in the

[1] Some studies reported in this publication were supported by National Institutes of Health Grant DK38165 and DFG We 681.

Table 1. Epidemiological findings on the association between high ω3 fatty acid intake and reduced cardiovascular mortality

Reduced cardiovascular mortality in Eskimos
Reduced myocardial and atherothrombotic cerebral infarction in japanese
Reduced cardiovascular mortality in four out of six large population studies in Holland (Zutphen), Sweden and the United States (Western Electric, MRFIT)

Table 2. Functional effects of ω3 fatty acids in the cardiovascular system

	References
Reduce platelet aggregation	9, 47
Increase platelet survival	46
Increase bleeding time	9, 47
Reduce blood pressure	45, 47, 52
Reduce vasospastic response to vasoconstrictors	47
Increase vascular (arterial) compliance	37, 43
Decrease blood viscosity	17, 18
Reduce albumin leakage in type 1 diabetes mellitus	38
Reduce cardiac arrhythmias	48, 50
Increase cardiac β-receptor function	49
Increase postischemic coronary blood flow	42
Reduce vascular intimal hyperplasia	32–34

cardiovascular system that may contribute to its beneficial effects suggested in those epidemiological findings. Table 2 summarizes the major observations made in these studies.

These range from a reduction of platelet aggregability to a lowering of blood pressure and vascular resistance in normotensives and hypertensives. They include a reduction of cardiac arrhythmias and a partial normalization of increased microvascular albumin leakage in type 1 diabetic patients.

The exact mechanisms underlying the diverse functional effects of ω3 fatty acids in the cardiovascular system are not yet fully understood, despite the increasing number of laboratory studies revealing many diverse biochemical effects of fish oils. In the following we attempt to summarize briefly these effects to which the putative antiatherogenic and anti-inflammatory actions of fish oils are attributable.

It is now generally agreed that these potential health benefits of fish oils are due to their content of ω3 long-chain polyunsaturated fatty acids, specifically eicosapentaenoic acid (EPA, C20:5 ω3) and docosahexaenoic acid (DHA, C22:6 ω3).

Biochemical Effects of ω3 Fatty Acids

Table 3 lists factors thought to play a role in the development of atherosclerosis and in inflammatory and immune responses of human beings. The actions of each are tersely summarized and the effects of feeding fish oils or of incorporating EPA and/or DHA into the phospholipids of responsive cells are succinctly indicated by an arrow.

Arachidonic Acid

In the mammalian organism, linoleic acid (LA, C18:2 ω6), the parent fatty acid of the ω6 fatty acid family, is slowly desaturated and elongated to arachidonic acid (AA, C20:4 ω6), the dominant eicosanoid precursor fatty acid under our Western dietary conditions. AA is a potent aggregator of platelets in vitro. After infusion or dietary intake of AA, in vivo animal and human studies show that platelets are more sticky and lead to thrombus formation. Recently, a new role of AA and some of its metabolites in the intracellular signalling pathway associated with cell proliferation and gene expression has been suggested [11, 12]. Such a function would place AA into a strategic position to modulate key events in atherosclerotic and chronic inflammatory disorders.

ω3 Fatty acids not only replace AA and LA in certain cellular phospholipids (especially PC, PE) but they redistribute these ω6 fatty acids from the phospholipid fraction into cholesterol esters and triacylglycerols [13]. Together with the high affinity of ω3 fatty acids to the ether lipid fraction of phospholipids this may have important implications for the formation of lipid mediators formed from AA or EPA [9, 10].

Thromboxane

Due to the early findings on fish oils and platelet aggregation and the potential role of EPA as eicosanoid precursor, the effects of fish oil on eicosanoids were the first to be studied. Eicosanoids are 20-carbon molecules, biologically active factors, i.e. prostanoids and leukotrienes, derived from the 20-carbon fatty acids AA of the ω6 family or EPA of

Table 3. Effects of ω3 fatty acids on factors involved in the pathophysiology of atherosclerosis and inflammation

Factor	Function	Effect of ω3 fatty acids	References
Arachidonic acid	eicosanoid precursor; aggregates platelets; stimulates white blood cells	↓	9, 10, 47
Thromboxane	platelet aggregation; vasoconstriction; increase of i.c. Ca^{++}	↓	14
Prostacyclin ($PGI_{2/3}$)	prevent platelet aggregation; vasodilation; increase cAMP	↑	14
Leukotriene (LTB_4)	neutrophil chemoattractant; increase of i.c. Ca^{++}	↓	14
Tissue plasminogen activator	increase endogenous fibrinolysis	↑	17
Fibrinogen	blood clotting factor	↓	18
PAF	activates platelets and white blood cells	↓	19
PDGF	chemoattractant and mitogen for smooth muscles and macrophages	↓	20
Oxygen free radicals	cellular damage; enhance LDL uptake via scavenger pathway; stimulate arachidonic acid metabolism	↓	21
Lipid hydroperoxides	stimulate eicosanoid formation	↓	22
Interleukin 1 and tumor necrosis factor	stimulate neutrophil O_2 free radical formation; stimulate lymphocyte proliferation; stimulate SMC proliferation; stimulate PAF; express intercellular adhesion molecule-1 on endothelial cells; inhibit plasminogen activator – thus procoagulants	↓	23
EDRF	reduces arterial vasoconstrictor response	↑	25
VLDL	related to LDL and HDL level	↓	7

the ω3 family. At least two prostanoids, thromboxane and prostacyclin, are involved in atherogenesis. Thromboxane, produced in platelets from AA, is of the 2-series (TXA_2) and promotes platelet aggregation, an important early stage in blood clot formation and in atherogenesis. It also is a potent constrictor of arteries. ω3 fatty acids inhibit the formation of TXA_2 and lead to the formation of small quantities of inactive TXA_3 [14].

Prostacyclin

Prostacyclin, produced by endothelial cells – the innermost lining cells of blood vessels – has opposite actions. It prevents platelets from aggregating and dilates arteries; its actions are antiatherogenic. When fish oils are ingested and EPA displaces AA from cell membrane and plasma phospholipids and redistributes it into e.g. plasma triacylglycerols, the AA cascade is altered and, in addition, prostanoids of a different series, 3-series, are produced. The thromboxane formed from EPA, unlike that from AA, has very little physiologic activity, whereas the prostacyclin (PGI_3) formed from EPA – and DHA [6] after retroconversion to EPA – is fully active, increasing total antiatherogenic prostacyclin activity. Thus, the presence of EPA from fish oil in the phospholipids of membranes of platelets and endothelial cells and in plasma tips the balance in the blood vessels away from atherosclerosis [14].

Leukotrienes

Leukotrienes are potent proinflammatory and immunoactive factors. LTB_4, produced from ω6 AA, is a chemoattractant for neutrophils and monocytes/macrophages which are circulating white blood cells that are important in inflammation and familiar as major cellular constituents of early atherosclerotic lesions. LTB_4 also causes neutrophils and macrophages to become active. LTB_5 produced from EPA, on the other hand, has little physiologic effect, so that when EPA replaces AA in cell membrane phospholipids, the inflammatory component of atherosclerosis is diminished [14].

Disturbances of the cell signalling process are thought to play a major role in diseases like atherosclerosis and chronic inflammation. Arachidonic acid and its metabolites formed via the cyclooxygenase, lipoxygenase or the cytochrome P450 pathways are important modulators of this process [10–12]. The mechanisms involved include also the formation or release of messenger molecules such as inositolphosphates (e.g. IP3), Ca^{++} and diglycerides. ω3 fatty acids have been found to interfere at several sites in this signalling process [10, 15, 16].

Tissue Plasminogen Activator

Much medical and public attention has been attracted to thrombolytic agents: tissue plasminogen activator (TPA), streptokinase, and urokinase which can dissolve clots within blood vessels and, thus, prove useful in treating heart attacks acutely. However, eating fish oil supplements has been shown to increase endogenous TPA and to inhibit its inhibitors [17]. This action should be a major deterrent to the development of atherosclerosis, and also to the development of blood clots at the site of atheromas in coronary arteries which are usually the terminal events blocking blood flow to the heart muscle and causing heart attacks.

Fibrinogen

Fibrinogen, another factor in the blood clotting cascade, has been identified as a cardiac risk factor when its blood level is increased. It is also suppressed by dietary fish oil [18].

Platelet Activating Factor

Platelet activating factor (PAF) is a phospholipid-like molecule now known to possess many widespread physiological effects, largely adverse. At least some of these, the activation of platelets (as its name indicates) and of monocytes/macrophages, contribute to atherogenesis, as described above. PAF in macrophages can be synthesized by several different cell types. It has been shown in humans that the production of PAF is markedly inhibited when the ω3 fatty acids are incorporated in their membrane phospholipids replacing AA as the precursor molecules for PAF formation [19].

Platelet-Derived Growth Factor

Once platelets aggregate, as they will at a site of injury to the endothelial cells lining arteries, they release several factors including platelet-derived growth factor (PDGF), a potent chemoattractant and mitogen responsible, in part, for the migration of smooth muscle cells from deeper layers of the arterial wall to the site of endothelial injury or dysfunction where they multiply. It, along with other factors, causes circulating monocytes, another type of white blood cell, to migrate also to the site of endothelial injury, to multiply and to change into macrophages which are scavenger cells in the developing atherosclerotic process.

Not only platelets, but the other cell types that participate in the development of atherosclerosis, i.e., endothelial cells, monocytes/macrophages, and smooth muscle cells, produce PDGF-like growth factors which stimulate cell growth in the arterial wall. When the low density lipoprotein (LDL) cholesterol level in the blood is elevated and oxidized after trapping in the vascular intima, it is the macrophages and smooth muscle cells that ingest the cholesterol to form the foam cells which contribute to the atherosclerotic plaques. The latter gradually increase in size, narrowing the blood vessels in the heart, brain and periphery, constituting the atherosclerotic disease which, through heart attacks and strokes, is the leading cause of death in all Western industrialized countries.

Recently, it has been shown that incorporating EPA into cultured aortic endothelial cells markedly reduces their production of PDGF-like protein [20].

Oxygen Free Radicals

An important means by which neutrophils and monocytes fight infection is to attack invading bacteria or foreign cells with lethal oxygen free radicals. These highly reactive oxygen species, however, can damage normal cells and, in fact, are thought to be responsible for the cellular debris component of the atherosclerotic plaque. It has been demonstrated that feeding fish oils results in a large reduction in oxygen radical formation by neutrophils and monocytes when these cells are activated [21].

Furthermore, the formation of lipid hydroperoxides in plasma is reduced after dietary fish oil [22].

Interleukin-1 and Tumor Necrosis Factor

Interleukin-1 (IL-1), a peptide molecule produced by several of the cell types discussed here that are incriminated in the development of atherosclerosis, has been shown to cause most of the effects in humans associated with infections: fever, malaise, sleepiness, etc. It also, together with another peptide factor, tumor necrosis factor (TNF), is responsible for several effects that are atherogenic: stimulates synthesis of adhesion molecules (proteins which cause monocytes to adhere to endothelial cells), stimulates production of cytotoxic oxygen free radicals by neutrophils and monocytes, and activates platelets, neutrophils and monocytes. Again, it has been shown that feeding fish oil supplements to humans reduces the production of both IL-1 and TNF [23].

These actions of fish oils may be protective not only against atherosclerosis, but they may have beneficial effects on modulating the excessive or misdirected activity of the immune system as well, which we think is responsible for the so-called autoimmune diseases like lupus, rheumatoid arthritis [24], and, perhaps, even the aging process. So far, there has been no report suggesting that such damping down of the immune system by fish oils creates an increased risk of infections or cancer.

Endothelial Derived Relaxation Factor

Endothelial cell-derived relaxing factor, EDRF, as its name implies, promotes relaxation of arterial smooth muscle cells and diminished constriction of arteries when exposed to several physiologic and pharmacologic vasoconstrictor agents. These actions of EDRF are enhanced by supplementing diets of experimental animals with fish oils [25]. In addition, it has been found that these potentially antiatherosclerotic actions persist even after exposure of arteries to anoxic conditions, as may occur with decreased blood flow during and following a heart attack [26].

Blood Lipids

The causal role in atherogenesis of elevated LDL cholesterol levels in plasma is well established. The most consistent effects of fish oils on serum lipids and lipoproteins have been reductions in levels of serum triglycerides and very low density lipoprotein (VLDL) [27]. The effects, however, on LDL cholesterol have been variable [7]. We have seen no significant effect of fish oil consumption (2–3 g/day of EPA and DHA) on cholesterol levels of individuals with single LDL cholesterol elevations (phenotype 2A) and increases in LDL cholesterol and Apo-B with fish oil have been reported.

Other effects of ω3 fatty acids on the lipid transport and metabolic systems may exert antiatheromatous actions.

Thus, in primates dietary ω3 fatty acids have a disordering effect on the core cholesteryl esters of LDL, resulting in a depression of LDL transition temperature, making cholesteryl esters more available for cholesteryl efflux from arterial plaques [28]. The reduction of lipoprotein lipase activity in swine fed ω3 fatty acids contributes to the reduced secretion of VLDL with subsequent changes in the cholesterol content of IDL, LDL and HDL fractions [29].

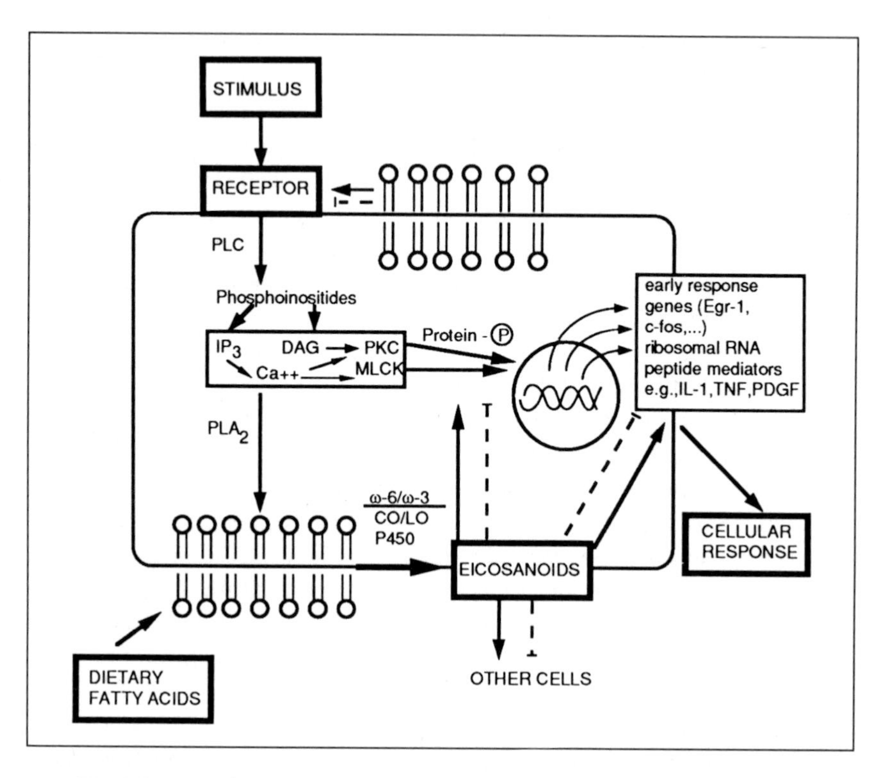

Fig. 1. Scheme of the putative role of ω3 and ω6 eicosanoid precursor fatty acids in cell membrane phospholipids as determined by dietary intake for the modulation of stimulus response coupling. PLC = Phospholipase C; IP_3 = inositol triphosphate; DAG = 1,2 diacylglycerol; PKC = protein kinase C; MLCK = myosine light chain kinase; PLA_2 = phospholipase A_2; CO = cyclooxygenase; LO = lipoxygenase; Protein-P = phosphorylated proteins; Peptide mediators, e.g., = IL-1, TNF, PDGF [adapted from ref. 10].

Whether individual characteristics in the cholesterol transport and reverse transport systems are responsible for the heterogeneity of LDL responses in hypertriglyceridemic subjects [30] is a possibility that needs to be tested.

One's first reaction is to be startled by the diversity of effects of dietary supplementation with fish oil fatty acids on cellular functions. However, one needs to realize the multiple points of action of the ω6 and ω3 fatty acids and their derivatives in the modulation of the stimulus-response coupling at several sites in the cell. This is shown schematically in figure 1.

Adverse Effects

It has been reported that ω3 fatty acids added to the diet will inhibit intimal hyperplasia and atherosclerosis in animals made hypercholesterolemic by diets high in saturated fat and cholesterol. This has been clearly demonstrated in rabbits [31], dogs [32], swine [33] and monkeys [34]. In rabbits [35] and in rats [36], however, increased intimal foam cells and increased monocyte adhesion to endothelial surfaces have also been found.

In diabetic subjects ω3 fatty acids have been found to improve vascular compliance [37] and to normalize the increased microvascular albumin leakage [38] which is a reliable prognostic parameter. In one study ω3 fatty acids have been reported to cause insulin resistance with elevation of blood glucose levels [39]. Further observations should determine how frequent and serious this problem may be and whether it outweights the beneficial vascular effects in diabetes cited above.

Clinical Benefit

Despite all the laboratory, human, animal, and epidemiologic studies suggesting an anti-atheromatous action of ω3 fatty acids, we have been lacking adequate clinical trials which will determine in prospective, placebo-controlled, randomized studies whether all the above experimental and epidemiologic evidence adds up to a demonstrable effect of fish oils to prevent atherosclerosis, e.g. coronary heart disease in humans at high risk for heart attacks. At present, the most direct and quickest conditions in which to seek the answer to this important clinical question is to learn if fish oil supplements can prevent the high rate of restenosis of coronary arteries following dilatation of the narrowed, atheromatous artery by coronary angioplasty, since the 25–40% restenosis rate occurs within 6 months of the angioplasty procedure and pathologically represents an accelerated atherosclerotic process. So far there have been five reports, four fully published, which are summarized in table 4.

The results of these studies are encouraging. They support the hypothesis that ω3 fatty acids may be protective against restenosis following coronary angioplasty. A critique of these studies would include: (1) The populations studied were quite different; (2) different end points were used; (3) different doses and preparations of ω3 fatty acids were used; (4) data on

Table 4. Effect of ω3 fatty acids to prevent restenosis post coronary angioplasty

Study	n	End point	Beneficial	Follow-up months	Dose ω3 FA, g/day	p
Dehmer [53]	82	angio	yes	3–4	5.6	0.007
Slack [57]	138	EST	yes	6	2–3	*
Milner [55]	143	cl + angio	yes	6	4.5	0.03
Grigg [54]	104	angio	no	4	3.0	NS
Reis [56]	133	EST + angio	no	6	6.0	NS

angio = angiography; EST = exercise stress test; cl = clinical symptoms.
* p value not given but beneficial results claimed.

other concomitant therapies are incomplete; (5) time of commencing intervention varied; (6) studies are small.

A large, placebo-controlled multicenter study with fish oil in patients undergoing PTCA is under way in the USA.

Importantly, in addition to these studies, in a recently published large, controlled prospective trial on dietary intervention in 2,033 men who had recovered from myocardial infarction, the subjects advised to eat fatty fish had a 29% reduction in 2-year all-cause mortality ($p < 0.05$) compared with those not so advised [40]. The advice on fat reduction plus increase of the P/S ratio, or the advice to increase fibre intake were ineffective in this trial. This study is the first prospective dietary intervention trial for secondary prevention of coronary heart disease that demonstrates clinical benefit (quo ad vitam) using a feasible approach.

In conjunction with the epidemiological, biochemical and functional data on ω3 fatty acids it seems safe to explore this approach in further secondary and primary prevention studies of atherosclerosis.

References

1 Curb JD, Reed DM: Fish consumption and mortality from coronary heart disease. N Engl J Med 1985;313:821.

2 Kromhout D, Bosschieter EB, de Lezenne CC: The inverse relation between fish consumption and 20-year mortality from coronary heart disease. N Engl J Med 1985;312:1205–1209.

3 Norell SE, Ahlbom A, Feychting M, Pedersen NL: Fish consumption and mortality from coronary heart disease. Br Med J 1986;293:426.

4 Shekelle RB, Missell LV, Paul O, Shryock AM, Stamler J: Fish consumption and mortality from coronary heart disease. N Engl J Med 1985;313:820.
5 Vollset SE, Heuch I, Bjelke E: Fish consumption and mortality from coronary heart disease. N Engl J Med 1985;313:820–821.
6 Fischer S, Vischer A, Praec-Mursic V, Weber PC: Dietary docosahexaenoic acid is retroconverted in man to eicosapentaenoic acid, which can be quickly transformed to prostaglandin I_3. Prostaglandins 1987;34:367–373
7 Harris WS: Fish oils and plasma lipid and lipoprotein metabolism in humans: a critical review. J Lipid Res 1989;30:785–807.
8 V. Schacky C: Prophylaxis of atherosclerosis with marine omega-3 fatty acids. Ann Intern Med 1987;107:890–899.
9 Weber PC: Clinical studies on the effects of n-3 fatty acids on cells and eicosanoids in the cardiovascular system. J Intern Med 1989;225 (suppl 1):61–68.
10 Weber PC: Membrane phospholipid modification by dietary n-3 fatty acids: effects on eicosanoid formation and cell function; in Karnovski, Leaf, Bolis (eds): Aberrations in Membrane Structure and Function Alan R. Liss., Inc., New York, 1988, pp 263–274.
11 Kacich RL, Williams LT, Coughlin SR: Arachidonic acid and cyclic adenosine monophosphate stimulation of c-fos expression by a pathway independent of phorbol ester-sensitive protein kinase C. Mol Endocrinol 1988;2:73–77.
12 Sellmayer A, Weber PC, Bonventre JV: Endogenous arachidonic acid metabolites modulate growth and mRNA levels of immediate early genes in rat mesangial cells (abstract). Kidney Int 1990;37:350.
13 Garg ML, Wierzbicki AA, Thomson AB, Clandinin MT: Omega-3 fatty acids increase the arachidonic acid content of liver cholesterol ester and plasma triacylglycerol fractions in the rat. Biochem J 1989;261:11–15.
14 Weber PC, Fischer S, v. Schacky C, Lorenz R, Strasser T: The conversion of dietary eicosapentaenoic acid to prostanoids and leukotrienes in man. Prog Lipid Res 1986; 25:273–276.
15 Bankey PE, Billiar TR, Wang WY, Carlson A, Holman RT, Cerra FB: Modulation of Kupffer cell membrane phospholipid function by n-3 polyunsaturated fatty acids. J Surg Res 1989;46:439–444.
16 Locher R, Sachinidis A, Steiner A, Vogt E, Vetter W: Fish oil affects phosphoinositide turnover and thromboxane A metabolism in cultured vascular muscle cells. Biochim Biophys Acta 1989;1012:279–283.
17 Barcelli U, Glas-Greenwalt P, Pollack VE: Enhancing effect of dietary supplement with n-3 fatty acids on plasma fibrinolysis in normal subjects. Thromb Res 1985;39: 307–312.
18 Hostmark AT, Bjerkedal T, Kierulf P, Flaten H, Ulshagen K: Fish oil and plasma fibrinogen. Br Med J 1988;297:180–181.
19 Sperling RI, Robin JL, Kylander KA, Lee TH, Lewis RA, Austen KF: The effects of n-3 polyunsaturated fatty acids on the generation of platelet-activating factor-acether by human monocytes. J Immunol 1987;139:4186–4191.
20 Fox PL, DiCorleto PE: Fish oil inhibits endothelial cell production of platelet-derived growth factor-like protein. Science 1988;241:453–456.
21 Fisher M, Upchurch KS, Levine PH, Johnson MH, Vaudreuil CH, Natale A, Hoo-

gasian JJ: Effects of dietary fish oil supplementation on polymorphonuclear leukocyte inflammatory potential. Inflammation 1986;10:387–392.
22 Lands WE, Miller JF, Rich S: Influence of dietary fish oil on plasma lipid hydroperoxides; in Samuelsson, Paoletti, Ramwell (eds): Prostaglandins and Related Compounds. Advances in Prostaglandin, Thromboxane, and Leukotriene Research. New York, Raven Press, 1987, vol. 17, pp 876–879.
23 Endres S, Ghorbani R, Kelley VE, Georgilis K, Lonnemann G, van der Meer JW, Cannon JG, Rogers TS, Klempner MS, Weber PC, Schaefer EJ, Wolff SM, Dinarello CA: The effect of dietary supplementation with n-3 polyunsaturated fatty acids on the synthesis of interleukin-1 and tumor necrosis factor by mononuclear cells. N Engl J Med 1989;320:265–271.
24 Clark WF, Parbtani A, Huff MW, Reid B, Holub BJ, Falardeau P: Omega-3 fatty acid dietary supplementation in systemic lupus erythematosus. Kidney International 1989;36:653–660.
25 Shimokawa H, Lam JY, Chesebro JH, Bowie EJW, Vanhoutte PM: Effects of dietary supplementation with cod-liver oil on endothelium-dependent responses on porcine coronary arteries. Circulation 1987;76:898–905.
26 Malis C, Varadarajan GS, Force T, Weber PC, Leaf A, Bonventre JV: Effects of dietary fish oils on vascular contractility subsequent to anoxia (abstract). Circulation 1988;78:(suppl II):II-216.
27 Weintraub MS, Zechner R, Brown A, Eisenberg S, Breslow JL: Dietary polyunsaturated fats of the w-6 and w-3 series reduce postprandial lipoprotein levels. J Clin Invest 1988;82:1884–1893.
28 Parks JS, Bullock BC: Effect of fish oil versus lard diets on the chemical and physical properties of low density lipoproteins of nonhuman primates. J Lipid Res 1987;28:173–182.
29 Groot PH, Scheek LM, Dubelaar M-L, Verdouw PD, Hartog JM, Lamers JM: Effects of diets supplemented with lard fat or mackerel oil on plasma lipoprotein lipid concentrations and lipoprotein lipase activities in domestic swine. Atherosclerosis 1989;77:1–6.
30 Schectman G, Kaul S, Kissebah A: Heterogeneity of low density lipoprotein response to fish-oil supplementation in hypertriglyceridemic subjects. Arteriosclerosis 1989;9:345–354.
31 Zhu B-Q, Smith DL, Sievers RE, Isenberg WM, Smith DL, Parmeley WW: Inhibition of atherosclerosis by fish oil in cholesterol-fed rabbit. J Am Coll Cardiol 1988;12:7073–7078.
32 Landymore RW, MacAulay M, Sheridan B, Cameron C: Comparison of cod-liver oil and aspirin-dipyridamole for the prevention of intimal hyperplasia in autologous vein grafts. Ann Thorac Surg 1986;41:54–57.
33 Weiner BH, Ockene IS, Levine PH, et al: Inhibition of atherosclerosis by cod-liver oil in a hyperlipidemic swine model. N Engl J Med 1986;15:841–846.
34 Davis HR, Bridenstine RT, Vesselinovitch D, Wissler RW: Fish oil inhibits the development of atherosclerosis in rhesus monkeys. Arteriosclerosis 1987;7:441.
35 Thiery J, Seidel D: Fish oil feeding results in enhancement of cholesterol-induced atherosclerosis in rabbits. Atherosclerosis 1987;63:53–56.
36 Rogers KA, Karnovasky MJ: Dietary fish oil enhances monocyte adhesion and fatty streak formation in the hypercholesterolemic rat. Am J Path 1988;132:382–388.

37 Wahlqvist ML, Lo ChS, Myers KA: Fish intake and arterial wall characteristics in healthy people and diabetic patients. Lancet 1989;i:944–946.

38 Jensen T, Stender S, Goldstein K, Holmer G, Deckert T: Partial normalization by dietary cod-liver oil of increased microvascula albumin leakage in patients with insulin-dependent diabetes and albuminuria. N Engl J Med 1989;321:1572–1577.

39 Glauber H, Wallace P, Griver K, Brechtel G: Adverse metabolic effect of omega-3 fatty acids in non-insulin-dependent diabetes mellitus. Ann Int Med 1988;108:663–668.

40 Burr ML, Fehily AM, Gilbert JF, Rogers S, Holliday RM, Sweetnam PM, Elwood PC, Deadman NM: Effects of changes in fat, fish, and fibre intakes on death and myocardial reinfarction: Diet and reinfarction trial (DART). Lancet 1989;ii:757–761.

41 Leaf A, Weber PC: Cardiovascular effects of n-3 fatty acids. N Engl J Med 1988;318: 549–557.

42 Force T, Malis CD, Guerrero JL, Varadarajan GS, Bonventre JV, Weber PC, Leaf A: n-3 Fatty acids increase postischemic blood flow but do not reduce myocardial necrosis. Am J Physiol 1989;257:H1204–H1210.

43 Hamazaki T, Urakaze M, Sawazaki S, Yamazaki K, Taki H, Yano S: Comparison of pulse wave velocity of the aorta between inhabitants of fishing and farming villages in Japan. Atherosclerosis 1988;73:157–160.

44 Hock CE, Holahan MA, Reibel DK: Effect of dietary fish oil on myocardial phospholipids and myocardial ischemic damage. Am J Physiol 1987;252:H554.

45 Knapp HR, Fitzgerald GA: The antihypertensive effects of fish oil. N Engl J Med 1989;320:1037–1043.

46 Levine PH, Fisher M, Schneider PB, Whitten RH, Weiner BH, Ockene IS, Johnson BF, Johnson MH, Doyle EM, Riendeau PA, Hoogasian JJ: Dietary Supplementation with omega-3 fatty acids prolongs platelet survival in hyperlipidemic patients with atherosclerosis. Arch Intern Med 1989;149:1113–1116.

47 Lorenz R, Spengler U, Fischer S, Duhm J, Weber PC: Platelet function, thromboxane formation and blood pressure control during supplementation of the western diet with cod liver oil. Circulation 1983;67:504–511.

48 McLennan PL, Abeywardena NY, Charnock JS: Dietary fish oil prevents ventricular fibrillation following coronary artery occlusion and reperfusion. Am Heart J 1988; 116:709–717.

49 Patten GS, Rinaldi JA, McMurchie EJ: Effects of dietary eicosapentaenoate 20:5 n-3 on cardiac beta-adrenergic receptor activity in the marmoset monkey. Biochem Biophys Res Commun 1989;162:686–693.

50 Riemersma RA, Sargent CA: Dietary fish oil and ischaemic arrhythmias. J Intern Med 1989;225 (suppl 1):111–116.

51 Sarris GE, Fann JI, Sokoloff MH, Smith DL, Loveday M, Kosek C, Stephens RJ, Cooper AD, May K, Willis AL, Miller DC: Mechanisms responsible for inhibition of vein-graft arteriosclerosis by fish oil. Circulation 1989;80 (suppl 1):I-109–I-123.

52 Singer P, Berger I, Luck K, Taube C, Naumann E, Gödicke W: Long-term effect of mackerel diet on blood pressure, serum lipids and thromboxane formation in patients with mild essential hypertension. Atherosclerosis 1986;62:259–265.

53 Dehmer GJ, Popma JJ, Van den Berg EK: Reduction in the rate of early restenosis after coronary angioplasty by a diet supplemented with n-3 fatty acids. N Engl J Med 1988;319:733–740.

54 Grigg LE, Kay IWH, Valentine PA, et al: Determinants of restenosis and lack of effect of dietary supplementation with eicosapentaenoic acid on the incidence of coronary artery restenosis after angioplasty. J Am Coll Cardiol 1989;13:665–672.
55 Milner MR, Gallino RA, Leffingwell A, Pichard AD, Brooks-Robinson S, Rosenberg J, Little T, Lindsay J: Usefulness of fish oil supplements in preventing clinical evidence of restenosis after percutaneous transluminal coronary angioplasty. Am J Cardiol 1989;64:294–298.
56 Reis GJ, Boucher TM, Sipperly ME, Silverman DI, McCabe CH, Baim DS, Sacks FM, Grossman W, Pasternak RC: Randomised trial of fish oil for prevention of restenosis after coronary angioplasty. Lancet 1989;ii:177–181.
57 Slack JD, Pinkerton CA, Van Tassel J, et al: Can oral fish oil supplement minimize re-stenosis after percutaneous transluminal coronary angioplasty. J Am Coll Cardiol 1987;9 (suppl):64a.

Dr. P.C. Weber, Institut für die Prophylaxe und Epidemiologie der Kreislaufkrankheiten, Universität München, Pettenkofer Strasse 9, D-8000 München 2 (FRG) imopoulos AP, Kifer RR, Martin RE, Barlow SM (eds): Health Effects of ω3 Polyunsaturated Fatty Acids in Seafoods. World Rev Nutr Diet. Basel, Karger, 1991, vol 66, pp 233–244

Fish Oil and the Platelet-Blood Vessel Wall Interaction

Paul M. Vanhoutte, Hiroaki Shimokawa, Chantal Boulanger

Center for Experimental Therapeutics, Baylor College of Medicine, Houston, Tex., USA

Introduction

Diets rich in marine oils may prevent coronary artery disease and lower the occurrence of atherosclerosis [1]. The major components of fish oil are the ω3 unsaturated fatty acids, eicosapentaenoic and docosahexaenoic acid, and they most likely are responsible for the protective effects of marine oils. This essay summarizes work performed in the author's laboratory that may help to understand the cardioprotective effects of fish oils [2].

ω3 Unsaturated Fatty Acids and Metabolism of Arachidonic Acid

When ω3 fatty acids are included in the diet, a competition with arachidonic acid occurs at several levels: (a) at the level of the precursors of eicosapentaenoic acid and arachidonic acid which compete for the elongases and desaturases enzymes, so that a decreased formation of arachidonic acid is observed [3]; (b) eicosapentaenoic acid competes with arachidonic acid for the 2-acyl position on the membrane phospholipids [4], and (c) eicosapentaenoic acid acts as a substrate for the cyclooxygenase enzyme [5, 6]. In endothelial cells, eicosapentaenoic acid does not inhibit the production of prostacyclin and leads to the production of prostaglandin I_3; the latter has the same biological activity as the vasodilator prostaglandin I_2 [5]. In platelets, the ω3 fatty acids inhibit the production of thromboxane A_2, which induces vasoconstriction, platelet aggregation and inhibition of the production of prostacyclin by endothelial cells; they lead to the formation of the physiologically inactive thromboxane A_3 [5–7]. The ω3 fatty acids also compete with arachidonic acid for lipoxygenases and thus, inhibit the formation of leukotriene B_4, a potent chemoattractant for monocytes and polymorphonuclear leukocytes; eicosapentaenoic acid leads to the formation of leukotriene B_5, which is less active than, but competes for the same receptors as leukotriene B_4 [8]. These various interactions of ω3 unsaturated fatty acids with the metabolism of arachidonic acid concur to reduce the amount of its metabolites accelerating platelet-aggregation and vasoconstriction, while favoring the formation of vasodilator and antiaggregating prostanoids. This obviously can contribute to the protective effects of the fish oils [2, 7]. However, it appears that other actions of the ω3 unsaturated fatty acids on the endothelial cells contribute as well.

Endothelium-Dependent Responses

The endothelium plays a pivotal role in the local control of hemostasis and vascular tone [9–14]. Dysfunction of the endothelial cells may contribute to the etiology of vasospasm and the pathogenesis of vascular diseases such as atherosclerosis and hypertension [11–15].

Endothelium-Derived Relaxing Factors

The endothelial cells act as a sensor for changes in the composition and the mechanical properties of the blood. Hence, endothelial cells release prostacyclin and several non-prostanoid relaxing factors when exposed to stimuli such as increased shear stress, acetylcholine, bradykinin, thrombin, or platelets producing adenosine diphosphate and serotonin (fig. 1) [9, 10, 12].

The major endothelium-derived relaxing factor (EDRF) released from cultured endothelial cells is nitric oxide or a nitrosothiol which yields nitric oxide [16–19]. Nitric oxide is derived enzymatically from *L*-arginine [20]. It causes relaxation of vascular smooth muscle by activation of soluble guanylate cyclase, with a resulting accumulation of cyclic guanosine monophosphate (cyclic GMP) (fig. 2). Endothelium-derived relaxing factor also inhibits the adhesion of platelets, and acts synergistically with prostacyclin to inhibit platelet aggregation [21–25].

The endothelial cells release another non-prostanoid relaxing factor which hyperpolarizes the vascular smooth muscle (endothelium-derived hyperpolarizing factor, EDHF) (fig. 2) [26, 27]. The release of EDHF pre-

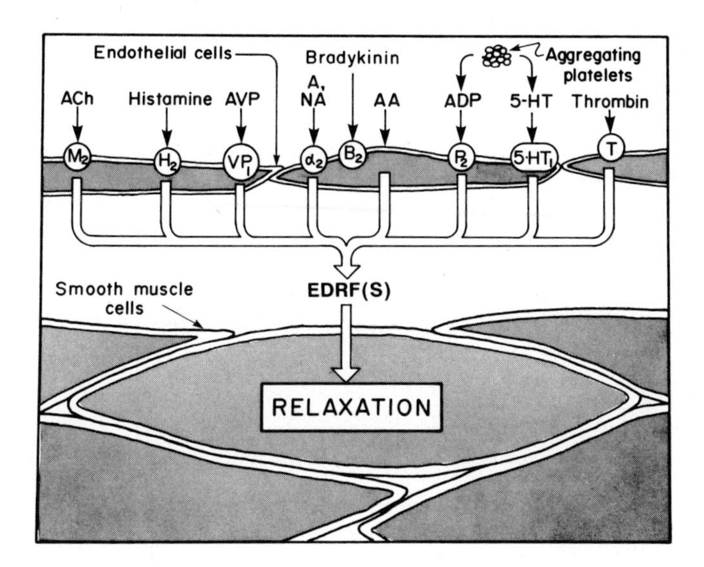

Fig. 1. Neurohumoral mediators which cause the release of endothelium-derived relaxing factor (EDRF) through activation of specific endothelial receptors (indicated by circles). In addition, EDRF can be released independent of receptor-operated mechanisms by the calcium ionophore A23187 (not shown). A = Adrenaline (epinephrine), AA = arachidonic acid, ACh = acetylcholine, ADP = adenosine diphosphate, α = alpha-adrenergic receptor, AVP = arginine vasopressin, B = kinin receptor, H = histaminergic receptor, 5-HT = serotonin (5-hydroxytryptamine), serotonergic receptor, M = muscarinic receptor, NA = noradrenaline (norepinephrine), P = purinergic receptor, T = thrombin receptor, and VP = vasopressinergic receptor. [From ref. 14, by permission.]

sumably explains why the perfusate through canine femoral arteries with endothelium relaxes bioassay rings without endothelium when exposed to shear stress or to acetylcholine (but not to bradykinin); these relaxations are inhibited by ouabain [28]. Likewise, under basal conditions and during stimulation with adenosine diphosphate, cultured porcine endothelial cells release a relaxing factor (different from nitric oxide), the action of which can be prevented by ouabain (fig. 3) [29].

ω3 Unsaturated Fatty Acids and Endothelium-Dependent Relaxations

When isolated blood vessels are exposed to aggregating platelets, they relax in an endothelium-dependent manner, because the adenosine diphosphate and serotonin from the platelets trigger the release of EDRF (fig. 4)

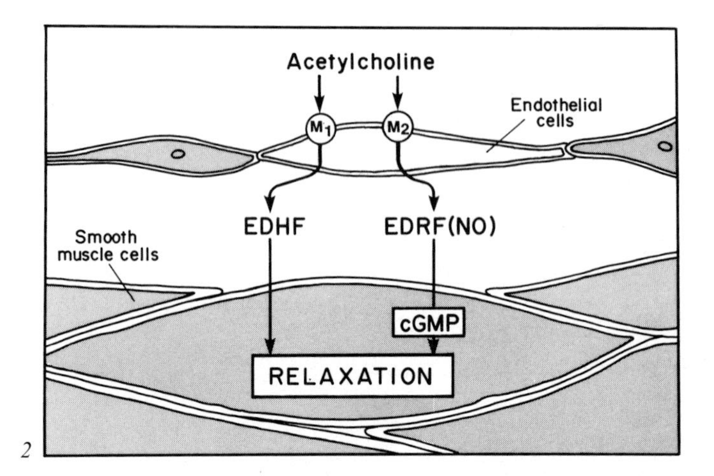

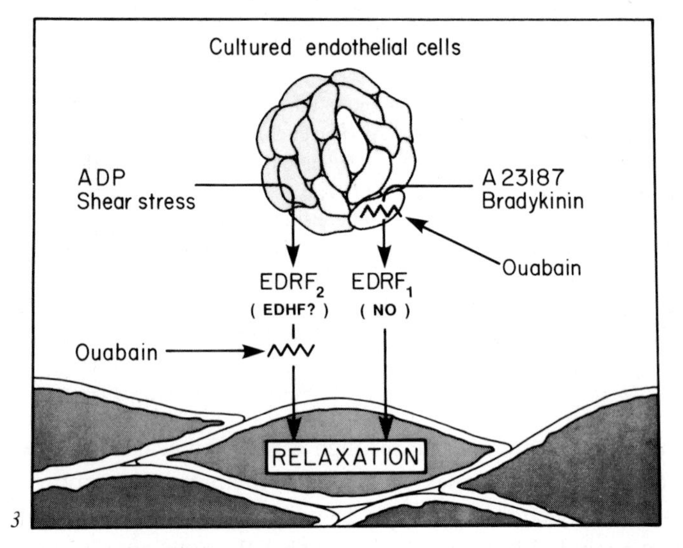

Fig. 2. The endothelial cells, when exposed to acetylcholine, release two relaxing factors. EDHF hyperpolarizes the cell membrane, thus initiating the relaxation and/or making the vascular smooth muscle more sensitive to the action of EDRF which presumably is NO. The latter sustains the relaxation by entering the cell and activating soluble guanylate cyclase which leads to an accumulation of cyclic GMP (cGMP). The muscarinic receptors (M) on the endothelial cell membrane triggering the release of the two factors do not belong to the same subtype. [From ref. 14, by permission.]

Fig. 3. Cultured porcine endothelial cells release two endothelium-derived relaxing factors. EDRF = Endothelium-derived relaxing factor; EDHF = endothelium-derived hyperpolarizing factor; NO = nitric oxid; ADP = adenosine diphosphate. [From ref. 14, by permission.]

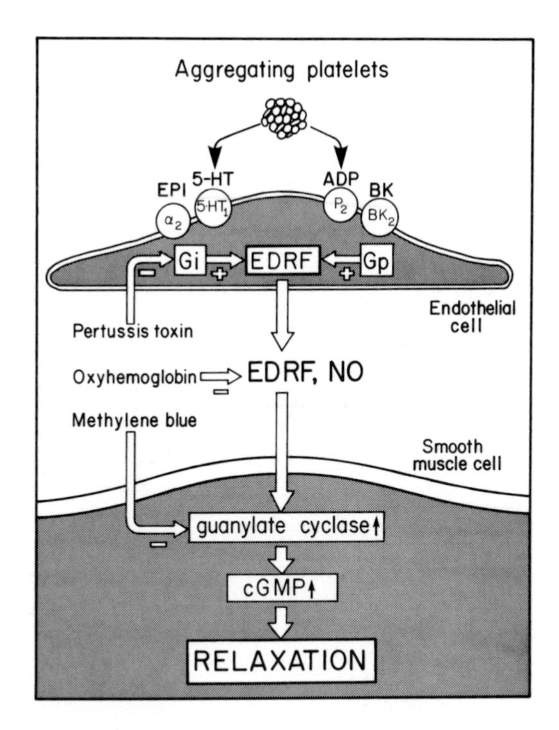

Fig. 4. The release of EDRF by endothelial cells involves at least two types of signal-transducing G-proteins, one of which is sensitive to pertussis toxin. Agonists at the endothelial cell membrane can use one and/or the other pathway. $\alpha_2 = \alpha_2$-Adrenoceptor, ADP = adenosine diphosphate, BK = bradykinin, kinin receptor, cGMP = cyclic GMP, EDRF = endothelium-derived relaxing factor, Gi and Gp = Gi and Gp proteins, 5-HT = serotonin, serotonin receptor, NO = nitric oxide, $P_2 = P_{2}y$-purinoceptor. [From ref. 41, by permission.]

[30–32]. Dietary supplementation of pigs with cod liver oil augments the endothelium-dependent relaxations to aggregating platelets in their coronary arteries (fig. 5) [33]. These relaxations are not mediated by prostacyclin since they are insensitive to indomethacin. In quiescent rings without endothelium, aggregating platelets elicit contractions which are not different between the control and fish oil-treated groups; however, the presence of the endothelium markedly reduces the contractions evoked by aggregating platelets, an effect that is considerably reinforced by dietary supplementation of the donor animals with cod liver oil [30–33]. The endothelium-dependent relaxations to adenosine diphosphate and serotonin are augmented also in blood vessels taken from animals treated with cod liver

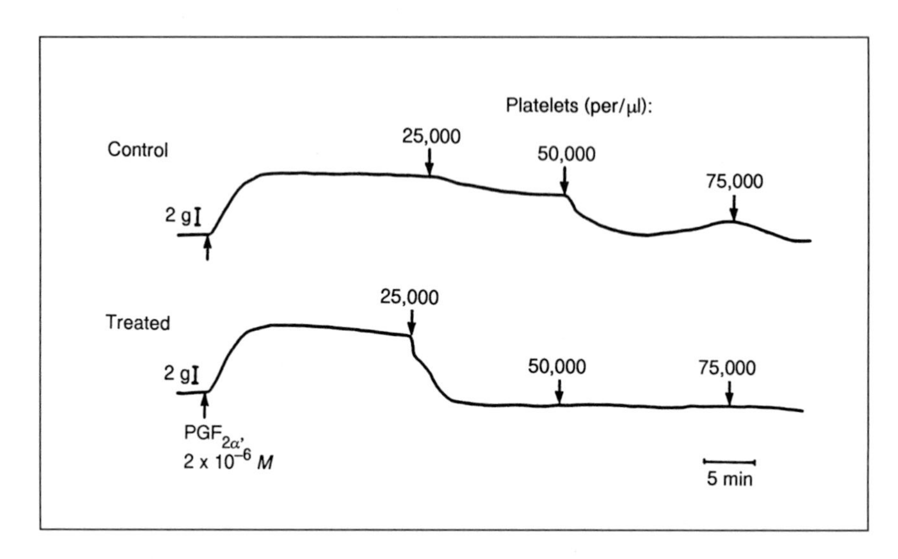

Fig. 5. In coronary arteries of the pig with endothelium, aggregating platelets cause relaxations (upper). Treatment of the pigs with cod liver oil (30 g/kg for 4 weeks) markedly augments the relaxation. [Data from ref. 33; from ref. 2, by permission.]

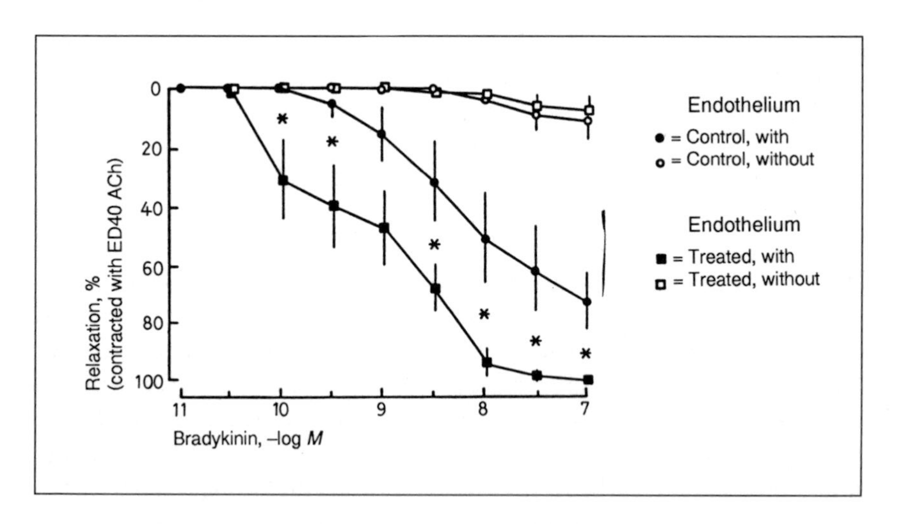

Fig. 6. Cumulative concentration-response curve to bradykinin during contractions evoked by the individual EC_{40} of acetylcholine in coronary microvessels of control and treated pigs treated with ω3 unsaturated fatty acids in the presence of indomethacin (10^{-5} *M*). [From ref. 35, by permission.]

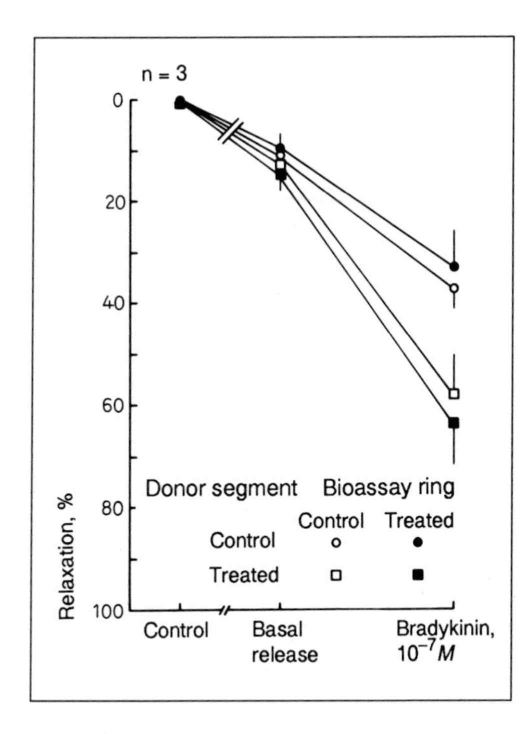

Fig. 7. Cumulative concentration-response curves to bradykinin during contraction evoked by $PGF_{2\alpha}$ (2×10^{-6} *M*) in bioassay experiments. Responses expressed as percent change in tension from contraction under endothelial superfusion. Data shown as means ± SE. * Statistically significant difference ($p < 0.01$) noted between control and treated donor segments at three different concentrations of bradykinin, regardless of whether bioassay ring was taken from control or treated pig. † Statistically significant difference ($p < 0.01$) between control and treated bioassay rings. n = Number of animals. [From ref. 34, by permission.]

oil, as are those evoked by bradykinin [33]. However, the endothelium-dependent relaxations evoked by the calcium ionophore A23187 (which does not activate receptors on the cell membrane of endothelial cells) [14] are not different in arteries from control and fish oil-treated animals, indicating that the basic process leading to the release of endothelium-derived relaxing factor(s) is not altered fundamentally by the chronic exposure to cod liver oil.

When eicosapentaenoic and docosahexaenoic acid, rather than fish oil is included in the diet, the endothelium-dependent relaxations to adenosine diphosphate, serotonin, bradykinin and aggregating platelets are facilitated to the same extent as after chronic exposure of the donor animals to cod liver oil [34]. This is also the case in coronary microvessels (fig. 6) [35]. Bioassay experiments demonstrate that bradykinin releases endothelium-relaxing factor(s) to a greater extent in blood vessels obtained from animals treated with the ω3 unsaturated fatty acid (fig. 7) [34]. The potentiating effect of the ω3 unsaturated fatty acids can also be demonstrated in hypercholesterolemic and atherosclerotic arteries (fig. 8) [36].

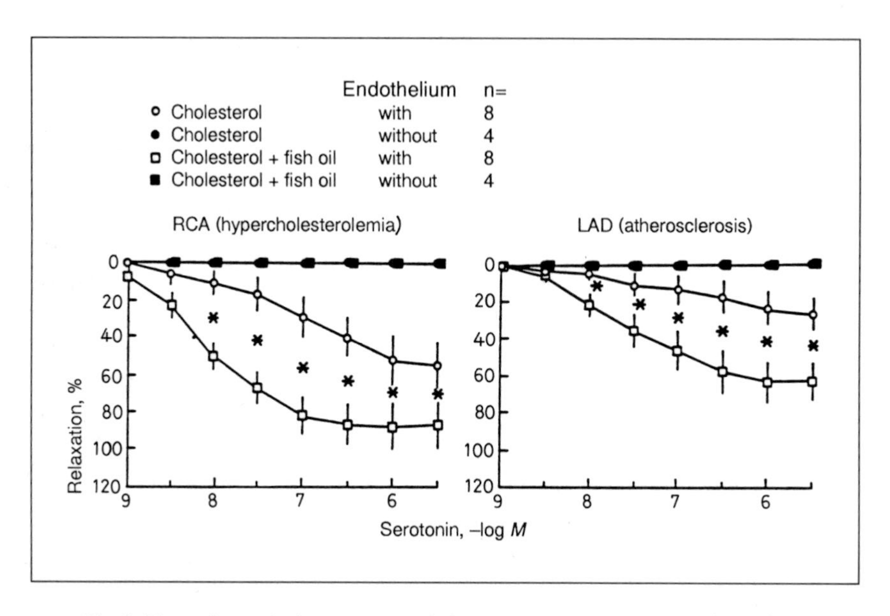

Fig. 8. Plots of cumulative concentration-response curves to serotonin during a contraction evoked by prostaglandin $F_{2\alpha}$ (2×10^{-6} *M*) in the presence of indomethacin (10^{-5} *M*) and ketanserin (10^{-6} *M*). Relaxations are expressed as percent decrease in tension from the contraction evoked by prostaglandin $F_{2\alpha}$. Data are mean ± SEM. Asterisk denotes a statistically significant difference ($p < 0.05$) between cholesterol- and cod liver oil-fed groups. RCA = Right coronary artery, LAD = left anterior descending coronary artery. [From ref. 36, by permission.]

In cultured porcine endothelial cells, incubation for 8–10 days with eicosapentaenoic acid augments the release of endothelium-derived relaxing factors mainly during stimulation with adenosine diphosphate (fig. 9) but minimally with bradykinin. However, the culture with eicosapentaenoic acid does not affect the release of relaxing factor(s) during exposure to the calcium ionophore A23187 [37]; it does not affect the production of cyclic GMP by the cultured endothelial cells (taken as a measure of the production of nitric oxide by these cells) [38, 39] when they are stimulated by adenosine diphosphate, bradykinin or nitric oxide [37]. These results suggest that marine fish oils and ω3 unsaturated fatty acids potentiate endothelium-dependent relaxations in isolated blood vessels by directly stimulating receptor-mediated release of relaxing factor(s) from the endothelial cells, and that the factor that differs from nitric oxide (fig. 3) may play a major role in the potentiation.

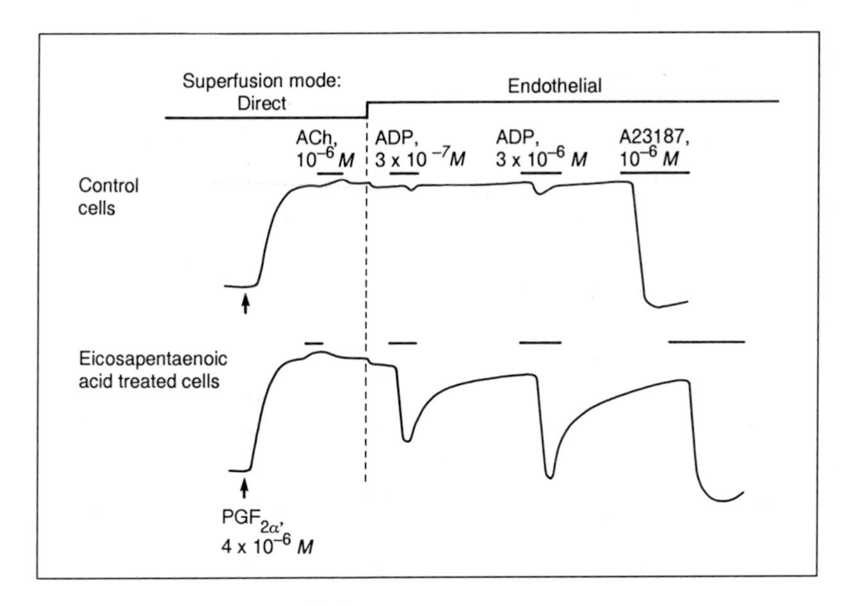

Fig. 9. Isometric tension recording of a bioassay ring contracted with prostaglandin $F_{2\alpha}$ ($PGF_{2\alpha}$) superfused with solution flowing over microcarrier beads covered with porcine aortic endothelial cells. Endothelium-derived relaxing factor was either released upon stimulation with adenosine diphosphate (ADP) from control cells (top trace) or from endothelial cells chronically exposed to eicosapentaenoic acid (2.5×10^{-5} *M*) (bottom trace). [From ref. 37, by permission.]

ω3 Unsaturated Fatty Acids, Endothelium, and Atherosclerosis

The endothelium-dependent responses to aggregating platelets, and the platelet-products, serotonin and adenosine diphosphate, are impaired in coronary arteries of hypercholesterolemic pigs; the responses are blunted further in atherosclerotic blood vessels [40]. The impaired relaxation reflects a reduced release of endothelium-derived relaxing factor(s) [40]. The release of relaxing factors in response to platelet aggregation is a major component in the antispastic and antiaggregatory protective role exerted by the endothelium [11, 13, 14]. A reduced interaction between platelets and atherosclerotic blood vessels favors platelet aggregation and platelet-induced contractions of vascular smooth muscle which could lead to vasospasm and thrombosis. In addition, ongoing constriction and platelet-activation favors stimulation of growth in the intima. Dietary supplementation with ω3 unsaturated fatty acids delays the impairment of endothelium-dependent relaxations in hypercholesterolemia and in atherosclerosis, partly by improving the release of endothelium-derived relaxing factor(s) and thus amplifying the protective role of the endothelium. This endothelial effect of the ω3 unsaturated fatty acids concurs with their other antiatheromatous actions [2].

References

1 Bang HO, Dyerberg J: Lipid metabolism and ischemic heart disease in Greenland Eskimos. Adv Nutr Res 1980;3:1–22.
2 Boulanger C, Shimokawa H, Schini VB, et al: Vascular endothelium and n–3 unsaturated fatty acids; in Rubanyi GM, Vanhoutte PM (eds): Endothelium-Derived Contracting Factors. Basel, Karger, 1990, pp 169–177.
3 Holman RT: Nutritional and metabolic interrelationships between fatty acids. Fed Proc 1964;23:1062–1067.
4 Goodnight SH Jr, Harris WS, Connor WE: The effects of dietary omega–3 fatty acids on platelet composition and function in man: A prospective, controlled study. Blood 1981;58:880–885.
5 Fisher S, Weber PC: Thromboxane A_3 is formed in human platelets after dietary eicosapentaenoic acid. Biochem Biophys Res Comm 1983;116:1091–1099.
6 Needleman P, Raz A, Minkes MS, et al: Triene prostaglandins: Prostacyclin and thromboxane biosynthesis and unique biological properties. Proc Natl Acad Sci USA 1979;76:944–948.
7 Knapp HR, Reilly IAG, Allesandrini P, et al: In vivo indexes of platelet and vascular function during fish-oil administration in patients with atherosclerosis. N Eng J Med 1986;314:937–942.
8 Goldman DW, Pickett WC, Goetzl EJ: Human neutrophil chemotactic and degranulating activities of leukotriene B_5 (TB_5) derived from eicosapentaenoic acid. Biochem Biophys Res Comm 1983;117:282–288.
9 Furchgott RF, Zawadzki JV: The obligatory role of endothelial cells in the relaxation of arterial smooth muscle by acetylcholine. Nature 1980;286:373–375.
10 Furchgott RF, Vanhoutte PM: Endothelium-derived relaxing and contracting factors. The FASEB J 1989;3:2007–2018.
11 Vanhoutte PM, Houston DS: Platelets, endothelium and vasospasm. Circ 1985;72: 728–734.
12 Vanhoutte PM: Endothelium and the control of vascular tissue. NIPS 1987;2:18–22.
13 Vanhoutte PM, Shimokawa H: Endothelium-derived relaxing factor and coronary vasospasm. Circulation 1989;80:1–9.
14 Lüscher TF, Vanhoutte PM: The endothelium: Modulator of cardiovascular function. Boca Raton, CRC Press, 1990, pp 1–228.
15 Ross R: The pathogenesis of atherosclerosis – an update. N Engl J Med 1986;314: 488–500.

16 Furchgott RF: Studies on relaxation of rabbit aorta by sodium nitrite: The basis for the proposal that acid-activatable inhibitory factor from bovine retractor penis is inorganic nitrite and the endothelium-derived relaxing factor is nitric oxide; in Vanhoutte PM (ed): Vasodilatation: Vascular Smooth Muscle, Peptides, Autonomic Nerves and Endothelium. New York, Raven Press, 1988, pp 401–414.
17 Ignarro LJ, Byrns RE, Wood KS: Biochemical and pharmacological properties of endothelium-derived relaxing factor and its similarity to nitric oxide radical; in Vanhoutte PM (ed): Vasodilatation: Vascular Smooth Muscle, Peptides, Autonomic Nerves and Endothelium. New York, Raven Press, 1988, pp 427–436.
18 Palmer RMJ, Ferrige AG, Moncada S: Nitric oxide release accounts for the biological activity of endothelium-derived relaxing factor. Nature 1987;327:524–526.
19 Myers PR, Minor Jr RL, Guerra Jr R, et al: Vasorelaxant properties of the endothelium-derived relaxing factor more closely resemble S-nitrosocysteine than NO. Nature 1990;345:161–163.
20 Palmer RMJ, Rees DD, Ashton DS, et al. *L*-Arginine is the physiological precursor for the formation of nitric oxide in endothelium-dependent relaxation. Biochem Biophys Res Comm 1988;153:1251–1256.
21 Moncada S, Vane JR: Pharmacology and endogenous roles of prostaglandin endoperoxides, thromboxane A_2 and prostacyclin. Pharm Rev 1979;30:293–331.
22 Busse R, Luckhoff A, Bassenge E: Endothelium-derived relaxant factor inhibits platelet activation. Naunyn-Schmiedeberg's Arch Pharm 1987;336:566–571.
23 Radomski MW, Palmer RMJ, Moncada S. The anti-aggregating properties of vascular endothelium: Interactions between prostacyclin and nitric oxide. Br J Pharm 1987;92:639–646.
24 Radomski MW, Palmer RMJ, Moncada S: Endogenous nitric oxide inhibitis human platelet adhesion to vascular endothelium. Lancet 1987b;ii:1057–1068.
25 Sneddon JM, Vane JR: Endothelium-derived relaxing factor reduces platelet adhesion to bovine endothelial cells. Proc Natl Acad Sci USA 1988;85:2800–2804.
26 Feletou M, Vanhoutte PM: Endothelium-dependent hyperpolarization of canine coronary smooth muscle. Br J Pharm 1988;93:515–524.
27 Komori K, Lorenz RR, Vanhoutte PM: Nitric oxide, acetylcholine, and electrical and mechanical properties of canine arterial smooth muscle. Am J Physiol 1989; 255:H207–H212.
28 Hoeffner U, Flavahan NA, Vanhoutte PM: Release of different relaxing factors from the endothelium of the canine femoral artery. Am J Physiol 1989;257:H330–H333.
29 Boulanger C, Hendrickson HH, Lorenz RR, et al: Release of different relaxing factors by cultured porcine endothelial cells. Circ Res 1989;64:1070–1078.
30 Cohen RA, Shepherd JT, Vanhoutte PM: Inhibitory role of the endothelium in the response of isolated coronary arteries to platelets. Science 1983;221:273–274.
31 Houston DS, Shepherd JT, Vanhoutte PM: Adenine nucleotides, serotonin and endothelium-dependent relaxations to platelets. Am J Physiol 1985;248:H389–H395.
32 Flavahan NA, Shimokawa H, Vanhoutte PM: Pertussis toxin inhibitis endothelium-dependent relaxations to certain agonists in porcine coronary arteries. J Physiol 1989;408:549–560.

33 Shimokawa H, Lam JYT, Chesebro JH, et al: Effects of dietary supplementation with cod-liver oil on endothelium-dependent responses in porcine coronary arteries. Circulation 1987;76:898–905.
34 Shimokawa H, Vanhoutte PM: Dietary omega–3 polyunsaturated fatty acids and endothelium-dependent relaxations in porcine coronary arteries. Am J Physiol 1988;256:H968–H973.
35 Shimokawa H, Aarhus LL, Vanhoutte PM: Dietary omega–3 polyunsaturated fatty acids augment endothelium-dependent relaxation to bradykinin in coronary microvessels of the pig. Br J Pharm 1988;95:1197–1203.
36 Shimokawa H, Vanhoutte PM: Dietary cod-liver oil improves endothelium-dependent responses in hypercholesterolemic and atherosclerotic porcine coronary arteries. Circulation 1988;78:1421–1430.
37 Boulanger C, Schini VB, Hendrickson H, et al: Chronic exposure of cultured endothelial cells to eicosapentaenoic acid potentiates the release of endothelium-derived relaxing factor(s). Br J Pharm 1990;99:176–180.
38 Schini VB, Moncada S, Vanhoutte PM: N^G-monomethyl-*L*-arginine inhibits thrombin-stimulated production of cyclic GMP in cultured porcine aortic endothelial cells. Amsterdam, Elsevier Science Publishers (Biomedical Division), 1990, vol 50, pp 1–6.
39 Boulanger C, Schini VB, Moncada S, et al: Stimulation of cyclic GMP production in cultured porcine endothelial cells by bradykinin, adenosine diphosphate, calcium ionophore A23187 and nitric oxide. Br J Pharm (in press).
40 Shimokawa H, Vanhoutte PM: Impaired endothelium-dependent relaxation to aggregating platelets and related vasoactive substances in porcine coronary arteries in hypercholesterolemia and in atherosclerosis. Circ Res 1989;64:900–914.
41 Vanhoutte PM: Endothelium-derived vasoactive factors. Hypertension Annual, in press.

Paul M. Vanhoutte, MD, PhD, Center for Experimental Therapeutics,
Baylor College of Medicine, One Baylor Plaza, Room 802E,
Houston, TX 77030 (USA)

Simopoulos AP, Kifer RR, Martin RE, Barlow SM (eds): Health Effects of ω3 Polyunsaturated Fatty Acids in Seafoods. World Rev Nutr Diet. Basel, Karger, 1991, vol 66, pp 245–249

Effects of Dietary ω3 Fatty Acid Supplementation on Leukocyte Free Radical Production

Marc Fisher, Peter H. Levine

Departments of Neurology, Medicine and Blood Research Laboratory, The Medical Center of Central Massachusetts – Memorial, and Departments of Neurology and Medicine, University of Massachusetts Medical School, Worcester, Mass., USA

The effects of ω3 fatty acids upon cells and their products has been an area of increasing interest in trying to understand how these compounds influence physiologic and pathologic processes. Regarding atherogenesis, ω3 fatty acids have been observed to reduce production of such diverse cellular factors as interleukin-1, tumor necrosis factor, leukotriene B_4, platelet-derived growth factor and platelet-activating factor [1]. Impeding the generation of these and other interesting compounds by the cellular participants in atherogenesis such as monocytes, endothelial cells and platelets may explain in part the beneficial effects of ω3 fatty acids upon experimental and human atherosclerosis. Additionally, these cellular effects of ω3 fatty acids may also be important in their favorable influence on human and experimental autoimmune disorders. Free radicals are oxygen metabolites with potentially highly toxic properties produced during the inflammatory response of monocytes and polymorphonuclear leukocytes (PMN) [2]. These leukocytes are key contributors to the cellular pathophysiology of autoimmune disorders such as rheumatoid arthritis and systemic lupus (SLE) [3]. Monocyte-derived tissue macrophages also play an important role in the initiation and propagation of atherosclerotic plaques [4]. Therefore, we assessed the effects of dietary ω3 fatty acid supplementation on free radical production by stimulated human PMN and monocytes.

Methods

Two groups of 6 healthy volunteers were given 30 ml of cod liver oil (CLO) containing 3.6 g of eicosapentaenoic acid (EPA) and 2.4 g of docosahexaenoic acid (DHA) daily for 6 weeks. Fasting blood specimens were drawn for the following studies just prior to starting the dietary ω3 fatty acid supplementation, and were repeated 6 weeks later.

Whole blood was drawn into glass tubes containing an anticoagulant and then centrifuged. The PMN were harvested by a dextran sedimentation technique and the monocytes by a density gradient technique. The PMN preparation contained 66% PMN and the monocyte preparation was 80% pure. Leukocyte fatty acids were analyzed by gas chromotography.

Chemiluminescence was measured in the rear photomultiplier tube of an ambient liquid scintillation counter. An aliquot of cells, 0.5 ml of PMN containing 1×10^7 cells/ml or 1 ml of monocytes containing 1×10^6 cells/ml was activated by the addition of latex particles. All samples were run in duplicate and the results are expressed in counts/5 min/10^3 leukocytes. Superoxide production was measured from PMN or monocytes stimulated by zymosan as a function of cytochrome c reduction. Aliquots of 0.2 ml of monocytes or PMN were incubated for 30 min.

Cytochrome c reduction was determined by measuring the difference in absorbance of the incubated supernatant at 550 nm in a Beckman spectrophotometer. Control samples with and without stimulus plus superoxide dismutase were run simultaneously. The final results of superoxide production in nmoles/30 min/10^6 leukocytes represents an average of duplicate trials for the stimulated sample adjusted for control production.

The data was analyzed by paired t tests comparing the pre and post ω3 fatty acid supplementation results.

Results

Chemiluminescence generation by stimulated human monocytes and PMN declined significantly after 6 weeks of dietary ω3 fatty acid supplementation as outlined in table 1. The effect on chemiluminescence appeared to be greater in monocytes than in PMN. Superoxide production was also reduced by dietary ω3 fatty acid supplementation in both leukocyte species and was inhibited to a similar extent (table 1). The reduction in chemiluminescence and superoxide measurements in conjunction with the consumption of ω3 fatty acids for 6 weeks was associated with substantial changes in the cellular fatty acid composition (table 2).

Arachidonic acid levels declined significantly and EPA levels rose significantly in both cells. Stearic acid levels declined significantly in the monocytes but not the PMN. DHA levels rose in the monocytes and were unfortunately not measured in the PMN.

Table 1. Chemiluminescence generation and superoxide production by stimulated monocytes and PMN before and after 6 weeks of daily dietary ω3 fatty acid supplementation (mean 3b SEM) in 6 healthy human subjects

	Chemiluminescence counts/5 min/1×10^3 cells		Superoxide nmol/30 min/l/10^6 cells	
	before	after	before	after
Monocytes	941 ± 47	506 ± 52***	15.2 ± 2.9	6.3 ± 1.5**
PMN	30 ± 4.4	22 ± 3.2*	33.3 ± 3	12.5 ± 0.8**

* $p < 0.05$; ** $p < 0.01$; *** $p < 0.001$.

Table 2. Leukocyte fatty acid content as a percentage of the total before and after ω3 fatty acid supplementation (mean ± SEM)

	Monocytes		PMN	
	before	after	before	after
Palmitic	18.8 ± 1.7	13.6 ± 2.4	14.7 ± 1.2	14.5 ± 0.6
Stearic	21.1 ± 0.6	16.5 ± 1.3*	18.8 ± 0.8	19.0 ± 1.4
Oleic	16.7 ± 0.5	16.8 ± 1.2	21.5 ± 0.9	20.3 ± 2.0
Linoleic	8.8 ± 0.5	7.4 ± 1.4	11.8 ± 2.0	10.8 ± 0.8
Arachidonic	22.6 ± 0.8	15.0 ± 2.3	18.9 ± 1.3	12.2 ± 1.5*
EPA	0.04 ± 0.4	2.1 ± 0.5*	0	3.3 ± 1.3*
DHA	0.7 ± 0.4	2.2 ± 0.4*		

* $p < 0.005$.

Discussion

These results indicate that dietary ω3 fatty acid supplementation can reduce free radical production by stimulated monocytes and PMN. Chemiluminescence is a general marker for the generation of excited oxygen metabolites such as superoxide, the hydroxyl radical and singlet oxygen [5]. Superoxide is an oxygen metabolite which is produced by the univalent reduction of molecular oxygen. The superoxide radical can accept an additional electron and hydrogen to form hydrogen peroxide, a compound of limited toxicity but which can diffuse across membranes [6]. With the aid of transition metals, hydrogen peroxide can accept an additional electron to generate the hydroxyl radical, a highly toxic molecule which can peroxidize lipids, proteins and carbohydrates [7]. Stimulated PMN and monocytes can generate these free radical species by the respiratory burst mechanism. This involves increased NADPH production which reacts with the oxygen to yield superoxide and the other free radical species as described. Additionally, the activation of leukocyte myeloperoxidase generates hypochlorous acid, another highly toxic compound, from hydrogen peroxidase and chloride [8]. The leukocyte respiratory burst can be associated with tissue destruction if it occurs in approximation to susceptible cells. It is tempting to speculate that ω3 fatty acids may suppress the respiratory burst of stimulated leukocytes, as reflected by the reduced production of free radicals.

The beneficial effects of ω3 fatty acids upon experimental atherogenesis has generated substantial interest [9]. This effect appears to occur independently of lipid lowering and is probably related to cellular effects of ω3 fatty acids. Monocytes are recruited to atherosclerotic plaques early in their development. They are transformed into macrophages, which imbibe lipid, mainly LDL-cholesterol, to become foam cells. Native LDL is taken up slowly by macrophages, but peroxidized LDL is rapidly absorbed by the scavenger receptor [10]. LDL peroxidation can be induced by monocytes/macrophages, endothelial cells or smooth muscle cells in association with the respiratory burst [11]. Additionally, peroxidized LDL can promote further plaque development because it is toxic for endothelial cells, has the ability to recruit circulating monocytes into the arterial wall and can reduce the egress of macrophages from this locale. Therefore, our observation that ω3 fatty acids reduce leukocyte production of toxic oxygen metabolites has potentially wide-ranging implications concerning the cellular aspects of atherosclerotic plaque development.

Monocytes and PMN are thought to directly contribute to the tissue destruction in autoimmune disorders such as rheumatoid arthritis and SLE. It has been observed that ω3 fatty acids favorably affect these conditions presumably at the cellular level [12]. The reduced production of toxic oxygen products which we have shown may also explain in part how ω3 fatty acids are effective in autoimmune disorders. Further studies are needed which directly evaluate the effects of ω3 fatty acids on leukocytes and other cells in patients with autoimmune disorders. Additional studies should be performed in such patients to ascertain if free radical mediated tissue destruction can be directly ameliorated by ω3 fatty acids. Studies of the cellular consequences of ω3 fatty acids offer a new route to an improved understanding of disease processes as well as the potential for novel therapeutic interventions. Therapy with these agents will require much more in the way of basic and applied research studies, however.

References

1 Fisher M, Leaf A, Levine PH: ω3 fatty acids and cellular aspects of atherosclerosis. Arch Int Med 1989;149:1726–1728.
2 Southorn PA, Powis G: Free radicals in medicine. I. Chemical nature and biologic reactions. Mayo Clin Proc 1988;63:381–389.
3 Nwia Y, Sakane T, Shingo M, Yokomaya BB: Effects of stimulated neutrophils synovial fluid of patients with rheumatoid arthritis on lymphocytes. J Clin Immunol 1983;3:228–240.
4 Watanabe T, Hirata M, Yoshikawa Y, et al: Role of macrophages in atherosclerosis. Sequential observation of cholesterol-induced rabbit aortic lesions by immunoperoxidase techniques using antimacrophage antibody. Lab Invest 1985;53:80–90.
5 Trush MA, Wilson ME, VanDyke, K: The generation of chemiluminescence by phagocytic cells. Methods Enzymol 1978;57:642–647.
6 Cohen MV: Free radicals in ischemic and reperfusion myocardial injury. Ann Int Med 1989;111:918–931.
7 Blake DR, Allen RE, Lunce J: Free radicals in biologic systems – a review oriented to inflammatory processes. Br Med Bull 1987;43:371–385.
8 Weiss SJ: Tissue destruction by neutrophils. N Engl J Med 1989;320:365–376.
9 Goodnight S, Fisher M, Fitzgerald GA, Levine PH: An assessment of the therapeutic use of dietary fish oil in atherosclerotic vascular disease. Chest 1989;95:19S–25S.
10 Goldstein JL, Ho YK, Basu SK, Brown MJ: Binding site on macrophages that mediates uptake and degradation of acetylated low density lipoprotein producing massive cholesterol deposition. Proc Natl Acad Sci USA 1979;76:333–337.
11 Steinberg D, Parthasarathy S, Carew TE, Khoo JC, Witztum JL: Beyond cholesterol: Modifications of low density lipoprotein that increase its atherogenicity. N Engl J Med 1989;320:915–924.
12 Kremer JM, Jubiz W, Michalek A, et al: Fish oil fatty acid supplementation in active rheumatoid arthitis, a double-binded, controlled crossower study. Ann Int Med 1987;106:497–503.

Marc Fisher, MD, The Medical Center of Central Massachusetts – Memorial, 119 Belmont Street, Worcester, MA 01605 (USA)

Simopoulos AP, Kifer RR, Martin RE, Barlow SM (eds): Health Effects of ω3 Polyunsaturated Fatty Acids in Seafoods. World Rev Nutr Diet. Basel, Karger, 1991, vol 66, pp 250–254

Effects of ω3 and ω6 Polyunsaturated Fatty Acids on the Action of Cardiac Glycosides on Cultured Rat Myocardial Cells[1]

Haifa Hallaq, Alexander Leaf

Department of Preventive Medicine, Harvard Medical School,
Massachusetts General Hospital, Boston, Mass., USA

There is increasing evidence that fish oil fatty acids, eicosapentaenoic acid (EPA) and docosahexaenoic acid (DHA) have effects on myocardial function and the stability of heart action. Gudbjarnason et al. [1] have for some years pointed to a relationship between cardiac disease mortality and the ω3 fatty acid content of phospholipids of cardiac myocytes. Recently, MaLennan et al. [2, 3] have demonstrated convincingly in rats that dietary saturated fat increases ventricular arrhythmias induced by temporary or permanent coronary occlusion in aging rats, whereas dietary vegetable oils diminished ventricular arrhythmias and fish oil feedings essentially abolished fatal arrhythmias during ischemia and reflow. These effects indicate a stabilizing action of ω3 fatty acids on cardiac function.

We have examined another aspect of stabilization of cardiac myocyte function by ω3 fatty acids, namely a preventive effect on toxicity from cardiac glycosides – another arrhythmogenic stress. We have examined the effects of incorporating purified EPA into isolated neonatal cardiac myocytes on the responses of the myocytes to ouabain. Cardiac myocytes were isolated from hearts of 1-day-old rats and cultured in suitable media containing no additives (control), arachidonic acid (AA) (5 μ*M*) or EPA (5 μ*M*) for 3–5 days at 37 °C.

[1] This work was supported in part by the National Institutes of Health research grant DK 38165. Dr. Hallaq was a Fellow of the Massachusetts Affiliate, Inc. of the American Heart Association during the period of this study.

Effects on Heart Rate and Contractions

A comparison of the responses of the beating rate and amplitude of contraction of single cardiac myocytes on exposure to 10^{-4} *M* ouabain was made. The toxic effects of this high dose of cardiac glycoside were clearly seen in the increased spontaneous beating rate of the myocytes and the decreased amplitude of the contractions in the control myocyte and arachidonic acid-enriched cells. The elevation of the baseline as the toxic effects develop represents the decreased diastolic relaxation of the myocytes as they proceed toward a state of lethal contracture. By contrast, the response of similar myocytes which had been enriched with EPA show a typical salutary cardiotonic response to cardiac glycosides with slowing of the beating rate, increase in amplitude of contractions, including an increased relaxation during diastole.

Thus, EPA, an ω3 polyunsaturated fatty acid from fish oil, when present during 3–5 days in the culture media of the isolated cardiac myocytes clearly protects the myocytes from toxicity from high concentrations of ouabain (10^{-4} *M*) incorporation of ω3 and ω6 fatty acids on membrane composition.

The effects on membrane phospholipid fatty acid composition of enriching media with AA or EPA as compared to control cells with no additives in the culture media were examined. Isolated cardiac myocytes from 1-day-old rats were incubated for 3–5 days with no additives, or with AA (5 μ*M*), or EPA (5 μ*M*), in the culture media. Phospholipids were extracted, hydrolyzed to liberate their free fatty acids and the latter were methylated and quantified by gas chromatography. The results are expressed as μg of fatty acids per mg protein in the samples.

AA, C20:4 ω6, was not increased in the fatty acids of membrane phospholipids by enrichment of the media with AA, but was slightly reduced with EPA enrichment. EPA was present only in small amounts in the phospholipids of neonatal cardiac myocytes and showed a small, but significant increase with EPA enrichment. Interestingly, there was no detectable increase in the elongation/desaturation products of either the ω6 AA or the ω3 EPA. Since the myocytes were carefully washed following removal from the 3- to 5-day incubation prior to exposure to ouabain, we may conclude that the resistance to the toxic effects of high dose of ouabain results from the modest changes found in the fatty acids of the cardiac myocytes.

Effects on Cytosolic Free Calcium

Since intracellular calcium concentrations are important in causing contraction of muscle cells, we examined the cytosolic free calcium concentrations within the isolated cardiac myocytes before and 10 min following addition of ouabain. In the absence of ouabain the cytosolic free calcium levels were the same for the control cells, those enriched with AA and those enriched with EPA, approximately 140 n*M*. With exposure to 10^{-6} *M* ouabain there was a large increase in cytosolic free calcium in all preparations to approximately 200 n*M*, but less of an increase in the myocytes enriched with EPA. At the truly toxic levels of ouabain, 10^{-4} *M*, there was a large further increase in cytosolic calcium in the control and AA-enriched myocytes, but no further increase in the EPA-enriched cells. This physiologic level of cytosolic calcium accounts for the preserved normal contractility of the EPA-enriched myocytes in the presence of 10^{-4} *M* ouabain.

Inhibition of Na,K-ATPase

To determine the reason for the lower calcium levels and continued normal – even enhanced – contractility of the EPA-enriched myocytes, we thought perhaps the EPA had somehow prevented ouabain from inhibiting the Na,K-ATPase in the myocyte membrane. This hypothesis was tested by assaying Na,K-ATPase by two independent methods. First, the rate of influx of ^{86}Rb into the neonatal myocytes was determined as a measure of potassium influx. The rate of the total Rb uptake expressed as µmol/mg of myocyte protein/minute was the same in all preparations. The ouabain inhibitable component of the Rb influx was also the same, as was that portion inhibitable by bumetanide. The bumetanide is an inhibitor of the facilitated co-transport pathway for Na+K+2Cl and these values indicate that this pathway is also unaffected by enrichment with AA or EPA.

Second, we used the NADH-coupled enzyme assay to determine the rate of ATP phosphorolyses by the Na,K-ATPase, a more direct assessment of enzyme activity. The total myocyte ATPase activity was found to be the same in all three preparations. Following exposure to ouabain 10^{-4} *M*, a reduction of some 40% in tissue ATPase activity was found, but no significant differences were found among the three preparations – control cells, cells enriched with AA, or those enriched with EPA. To be certain that there would be equal inhibition of Na,K-ATPase in all three preparations, further ATPase assays were performed in the presence of 10^{-3} *M* ouabain. Though this very high concentration of ouabain did produce a further inhibition of tissue ATPase, presumably of the Na,K-ATPase, there were still no detectable differences between the three preparations.

From these determinations we conclude that differences in cytosolic free calcium levels could not be explained by incomplete inhibition of the Na,K-ATPase in the EPA-enriched myocytes by the concentrations of ouabain used. Incomplete inhibition of the Na,K-ATPase in the EPA-enriched myocytes seemed a possibility as it would have explained the lower levels of cytosolic calcium and the continued normal contractions of the EPA-enriched myocytes. The results, however, exclude this possibility.

Effects on Calcium Uptake by Myocytes

Another possible explanation for the lower cytosolic calcium levels in the EPA-enriched myocytes, of course, could be a slower rate of calcium uptake by these cells. Calcium uptake was, therefore, measured as the initial influx slope of ^{45}Ca. ^{45}Ca uptake was slightly, but not significantly increased in the control and AA-enriched myocytes on exposures to 10^{-4} *M* ouabain, but significantly reduced in EPA-enriched myocytes at this concentration of ouabain.

Comments

Thus, a reduced rate of calcium influx (efflux has not been measured) produced somehow by the enrichment of membrane phospholipids with EPA, could account for the lower cytosolic calcium levels in these myocytes. There are, of course, several calcium channels in cardiac myocytes and we cannot at this time say which channels are inhibited by EPA-enriched phospholipids. Since the action of ouabain is to inhibit Na,K-ATPase, the sodium pump, it results in elevated intracellular sodium concentrations which in turn depolarize the cell membrane potential. The elevated cytosolic sodium levels and the depolarized membrane potential favor influx of calcium via the Na-Ca antiport channels. This is the usual explanation for the elevation of cytosolic calcium levels with exposure of cardiac myocytes to ouabain and the toxicity of cardiac glycosides when calcium levels exceed physiologic tolerance. If the EPA is inhibiting the

Na-Ca antiport channels, it could by this means prevent ouabain toxicity. This we would speculate is the mechanism by which EPA, when incorporated into the phospholipids of the sarcoplasm of cardiac cells, protects these cells from ouabain toxicity and probably more generally from stresses which can lead to calcium overload.

References

1 Gudbjarnason S, Benediktsdottir VE, Skuladottir G: Effects of n–3 polyunsaturated fatty acids on coronary heart disease. Bibl Nutr Dieta 1989;43:1–12.

2 McLennan PL, Abeywardena MY, Charnock JS: Dietary fish oil prevents ventricular fibrillation following coronary artery occlusion and reperfusion Am Heart J 1988;116:709–717.

3 McLennan PL, Abeywardena MY, Charnock JS: The influence of age and dietary fat in an animal model of sudden cardiac death. Aust NZ J Med 1989;19:1–5.

Haifa Hallaq, PhD, Department of Preventive Medicine,
Massachusetts General Hospital, Boston, MA 02114 (USA)

Simopoulos AP, Kifer RR, Martin RE, Barlow SM (eds): Health Effects of ω3 Polyunsaturated Fatty Acids in Seafoods. World Rev Nutr Diet. Basel, Karger, 1991, vol 66, pp 255–267

Involvement of Fish Oils and Growth Factors in Atherogenesis[1]

Paul L. Fox, Paul E. DiCorleto

Department of Vascular Cell Biology and Atherosclerosis Research, Cleveland Clinic Research Institute, FF4–15, Cleveland, Ohio, USA

Introduction

The observation that diets enriched in fish and fish oils are correlated with a reduced incidence of cardiovascular disease in man has stimulated two decades of investigations aimed at elucidating the underlying mechanisms. These studies, whether in cell culture, in experimental animal models, or in clinical trials, have focused primarily on two aspects of vascular biology: (1) lipoprotein metabolism, specifically the cholesterol- and triglyceride-lowering effects of fish oils, and (2) coagulation, specifically the effects of ω3 fatty acids on eicosanoid metabolism and platelet aggregation. Recently, we have suggested an alternative mechanism through which fish oils could mediate their beneficial effects, i.e., by inhibiting the release of specific growth factors that stimulate the growth of vascular smooth muscle cells (SMC). This cell is the primary proliferative cell involved in intimal thickening, a hallmark of the occlusive atherosclerotic lesion. In this report, the results from our laboratories and others that have led to the formulation of this hypothesis will be described.

[1] Supported in part by research grants HL 40352 (to PLF) and HL29582 (to PED) from the National Heart, Lung and Blood Institute of the National Institutes of Health.

The Reduction of Intimal Hyperplasia by Fish Oils

In vivo studies of atherosclerosis are greatly hampered by the extremely slow rate of progression of the disease. Several animal models of atherosclerosis have been developed to overcome this impediment, and it is from these models that much of the information on the cellular events involved in atherogenesis is derived. The first report that fish oils reduced intimal hyperplasia was that by Landymore et al. [1] who observed that cod liver oil markedly reduced intimal thickening in autogenous vein grafts implanted as arterial bypasses in cholesterol-fed dogs. Interestingly, they found that fish oils did not alter any platelet parameters, including platelet counts and prothrombin time, suggesting that the observed vessel response to the diet did not depend on interactions with platelets. Cahill et al. [2] confirmed this result using a similar animal model; they showed that MaxEPA, like cod liver oil, also prevented intimal thickening in vein grafts implanted in dogs. Like Landymore's group, they also found that the fish oil did not affect coagulation parameters, nor did it alter vessel prostacyclin production, serum lipid levels, or the number of hepatic low-density lipoprotein (LDL) receptors. They did, however, report a small but significant reduction in serum thromboxane levels. Sarris et al. [3], using a similar canine vein graft model, recently showed that MaxEPA, with or without aspirin, was significantly more effective than aspirin alone at reducing intimal thickening. In a related animal model, Casali et al. [4] demonstrated that a diet of fish and fish oils improved the patency of synthetic vascular grafts in dogs. They did not, however, direcly measure intimal thickening in these grafts. The effect of fish oils on atherosclerosis in a rat model of cardiac transplantation was studied by Sarris et al. [5]. Lewis rats that received (in the presence of cyclosporine) heterotopic hearts from Brown-Norway donors were found to develop severe coronary atherosclerosis that was mitigated by Super MaxEPA but not by safflower oil, a polyunsaturated lipid that is rich in ω6 fatty acids.

The effects of fish oils on intimal thickening of native vessels, rather than implanted grafts, have also been investigated. Luminal encroachment in coronary arteries of cholesterol-fed pigs was substantially inhibited by the addition of cod liver oil to the diet [6]. This result was confirmed for multiple coronary arteries as well as several regions of the aorta [7]. Studies on the effects of fish oils on the development of atherosclerosis and intimal hyperplasia in rabbit models have led to conflicting results. Fish oil (Protochol) was shown to reduce significantly the area of the involved plaque region in cholesterol-fed rabbits, however, the effects on intimal thickening were not determined [8]. Hearn et al. [9] reported mixed results with respect to the effect of MaxEPA on intimal hyperplasia in rabbits who underwent balloon de-endothelialization of the aorta and iliac arteries while consuming a cholesterol- and peanut oil-enriched diet. They reported a mild sparing of the lumen diameter in the iliacs, but thickening of the aorta wall (although without reduction of the overall luminal area). Thiery and Seidel [10] reported that MaxEPA enhances lesion area in cholesterol-fed rabbits (but lumen diameters were not measured). Finally, two groups have reported that fish oils (MaxEPA or menhaden oil) do not alter the surface extent of atherosclerosis or lesion thickness in Watanabe heritable hyperlipidemic rabbits [11, 12]. Using the hypercholesterolemic rat model of atherosclerosis, Rogers and Karnovsky [13] have shown that MaxEPA enhances monocyte adhesion to the vessel wall, and augments fatty streak formation. The effect of fish oil on atherosclerosis in nonhuman primates has been reported in a single study. Davis et al. [14] showed that menhaden oil reduced lesion area in the aorta and carotid artery of cholesterol-fed rhesus monkeys. The control treatment consisted of equivalent amounts of coconut oil, which is known to accelerate atherogenesis, and thus it is not clear if the observed results were due to a specific beneficial property of the fish oil, or simply from a decrease in the saturated fatty acid content in the diet.

The apparent discrepancies in several of the animal studies have not yet been resolved, but there are clearly many variables that must be considered. Some of these critical variables are: the specific animal model of atherosclerosis (and especially the physical disruption and/or diet used to induce lesions), the dose and type of fish oil used (the quantity of ω3 fatty acids, anti-oxidants, and minor unidentified components, may all contribute to the activity), the duration of the fish oil treatment, the methods employed for quantitation of atherosclerotic lesions, and the nature of the control treatment to which the fish oil treatment is compared. It may be significant that in both studies in which atherosclerosis was reported to be enhanced by fish oils, parameters other than intimal thickening were measured [10, 13].

The effect of fish oils on restenosis of coronary arteries after percutaneous transluminal angioplasty was examined in several recent clinical trials. This procedure has a very high initial success rate, but the long-term benefit is compromised by recurring restenosis in approximately one third of the vessels, the failure generally occurring within six months. Dehmer et al. [15] reported that MaxEPA reduced the early restenosis rate from 36% of the lesions in the control group to 16% in the treatment group. A nearly identical reduction in lesion restenosis, from 35% in controls to 19% in the treatment group, was reported by Milner et al. [16] in patients receiving Promega. These intriguing results, however, were not confirmed by two other recent reports. Grigg et al. [17] reported similar rates of restenosis between a treatment group that received MaxEPA (29% of lesions) and the control group that received a mixture of olive oil and corn oil (31% of lesions). Likewise, Reis et al. [18] did not observe significant differences in restenosis in a treatment group given either Promega or ethyl esters of fatty acids extracted from fish oils (32% of lesions), and the control group receiving 12 g/day of olive oil (23% of lesions).

The discrepancies in these clinical studies are not likely due to the ancillary therapies since all treatment and control groups received anti-platelet agents and calcium channel blockers (although according to differing regimens). The dose and source of the fish oil also does not readily explain the discrepant results; in terms of the actual amount of ω3 fatty acids administered, the effective treatments contained 5.4 and 4.5 g/day, while the ineffective treatments contained 3.0 and 6.0 g/day. However, one of the two treatment groups in the Reis study received fatty acid ethyl esters whose activity has not been established in the animal models, and the results from each group were not independently reported.

Growth Factors and Atherosclerosis

Balloon catheter injury of the major arteries, in several animal models, initiates a vessel wall response that results, within days, in an occlusive cellular lesion resembling an atherosclerotic plaque. The principal proliferative cell, in both the surgically induced lesion and in the atherosclerotic plaque, is the smooth muscle cell (SMC). This observation, combined with the identification of a serum growth factor derived from platelets that stimulates the proliferation of cultured SMC, has led Ross and Glomset [19] to propose the 'response to injury hypothesis of atherosclerosis'. According to this hypothesis, the key initiating event of atherogenesis is the removal of the endothelium (by physical injury in the animal model systems, or by physical, chemical, or immunological injury in the natural progression). Exposure of the subendothelial surface initiates a healing response that includes platelet adherence to the denuded vessel wall. Subsequent platelet aggregation and degranulation releases growth factors that stimulate proliferation of intimal SMC, leading to the formation of the occlusive lesion. Ross and co-workers have isolated and identified the mitogenic factor stored in the platelet α-granule, platelet-derived growth factor (PDGF). Although platelets contain multiple mitogens, including large amounts of transforming growth factor-β, PDGF is the principal human serum mitogen for vascular smooth muscle cells. Human PDGF is a 30 kD dimer containing two polypeptides, the A-chain and B-chain. It binds to specific, high affinity receptors on the plasma membrane of SMC (as well as on fibroblasts and other cells of mesenchymal origin), initiating a series of intracellular responses that culminates in cell division. In addition to stimulating cell proliferation, PDGF is also a potent SMC chemoattractant [20]. A scenario thus can be imagined in which an elevated concentration of PDGF in the intima stimulates both inward migration and proliferation of SMC.

The significance of the platelet and platelet factors in lesion development is currently being investigated. The experiments of Friedman et al. [21], in which de-endothelialization of the aorta of thrombocytopenic rabbits results in less intimal thickening than in normal rabbits, suggest that platelets may play a crucial role. However, several observations argue against platelet release as a key step in the atherogenic process: (1) endothelial denudation and platelet adhesion are not generally observed in the early stages of atherosclerosis, and (2) Reidy and Silver [22] have shown that a shallow, defined injury to the aorta that does not cause medial injury, results in platelet aggregation but no intimal thickening. A recent experiment by Fingerle et al. [23], using thrombocytopenic rats, may reconcile this apparent discrepancy. They report that platelets are not involved in the earliest proliferative responses of SMC, but that they are likely to be involved in later responses, perhaps by influencing the migration of medial SMC to the intima.

During the last decade much attention has focused on non-platelet sources of PDGF in the arterial environment. DiCorleto and Bowen-Pope [24] have shown that endothelial cells (EC) in vitro secrete multiple mitogens including PDGFc (PDGF-like protein), in amounts sufficient to maximally stimulate the proliferation of cultured SMC. EC-derived PDGFc is biochemically and immunologically indistinguishable from PDGF and the molecules are likely to be either closely related or identical. EC cultured from all major vessels, and from all species examined to date (including human umbilical vein and human aortic EC), secrete PDGFc into their media. The process is constitutive, i.e., exogenous stimulation of the cells is not required; however, factors related to arterial injury further stimulate the release of the growth factor. For example, certain members of the coagulation cascade, including thrombin and Factor Xa, increase the rate of synthesis of PDGFc by cultured EC up to 10-fold [25, 26]. In addition, several agents that mortally injure cultured bovine aortic EC, including bacterial lipopolysaccharide, cause the release of a large burst of PDGFc, presumably from dying cells [27]. EC production of PDGFc is also regulated by the physical structure of the cultured cells. Jaye et al. [28] have shown that certain in vitro conditions stimulate human umbilical vein EC to form tubular structures reminiscent of small diameter vessels. This reorganization is accompanied by decreased expression of mRNA for the B-chain of PDGF. Cultured EC also secrete heparin-like inhibitors of SMC growth [29], but when SMC were co-cultured with EC, the EC-derived growth-promoting activity dominates and the SMC replicate as if in the presence of maximal concentrations of mitogen [30].

The above observations have led us and others to propose that while EC under physiological conditions express multiple anti-atherogenic and anti-thrombotic properties, they may be 'activated' by defined stimuli to express pro-atherogenic and pro-thrombogenic functions. Furthermore, the observation that cells of the vessel wall synthesize and secrete growth factors, combined with the apparent insufficiency of platelets to account for the proliferative response of the vessel, have led to reappraisal and modification of the original 'response to injury' hypothesis. According to this updated hypothesis, injury to the endothelium directly leads to release of growth factors from vessel wall cells, rather than indirectly from circulating platelets (fig. 1). According to this model, it is the presence of the endothelium, rather than its absence, that is critical for the expression of its pro-atherogenic properties.

Direct in vivo measurements of growth factor production by EC or other vascular cells are not easily accomplished due to dilution, turnover, and other technical problems. However, the transcriptional activity of the genes encoding growth factors can be measured in freshly isolated tissue by extracting the mRNA and determining the level of the specific transcripts by Northern analysis. Using this technique, Barrett and Benditt [31] have shown that the expression of PDGF B-chain mRNA is at least 10-fold lower in endothelium freshly scraped from human umbilical vein or bovine aorta than in the comparable cultured EC. Furthermore, B-chain mRNA is present in 5-fold greater amounts in human carotid artery lesions than in normal artery, although the cell responsible for the production of

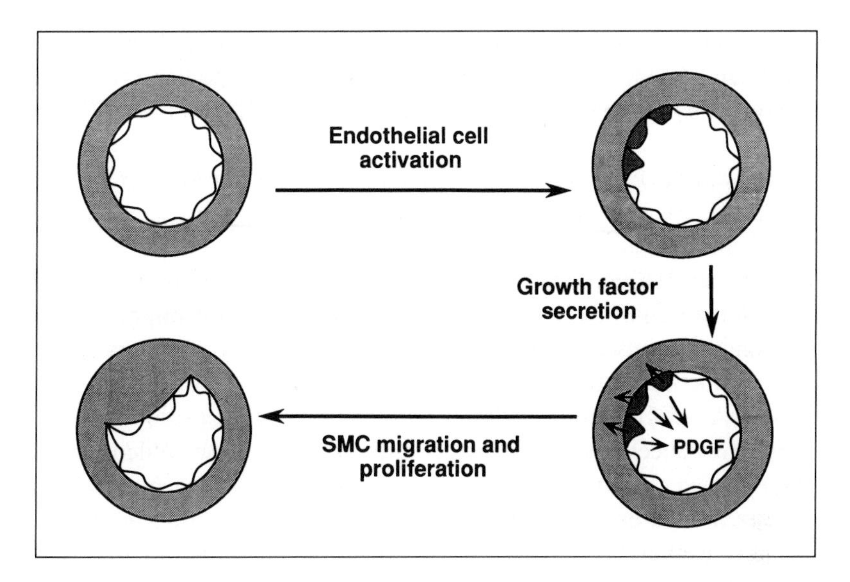

Fig. 1. Modified response to injury hypothesis of atherosclerosis. Schematic showing the proposed relationship of endothelial cell activation to lesion formation. PDGF = platelet-derived growth factor, SMC = smooth muscle cell.

PDGF has not been specifically identified [32]. These results raise the interesting possibility that the culturing process results in EC 'activation', and that EC monolayers in vitro may share properties with endothelium exposed to pathological stimuli in vivo. If this is the case, then studies on the conditions that inactivate cultured EC, i.e. lead to the reduction of growth factor production, may give insights into the physiological conditions required to maintain the endothelium in the 'inactive' or anti-atherogenic state.

Vascular cells other than EC also have been shown to synthesize and secrete PDGFc in vitro. Circulating human blood monocytes, the precursors for tissue macrophages, do not express either of the genes for PDGF, but when activated in vitro by bacterial lipopolysaccharide or by immune complexes, secrete PDGFc [33]. In addition, SMC themselves under certain circumstances produce growth factors. SMC cultured from the aorta of rat pups secrete significantly more PDGF than SMC isolated from adult rats, suggesting that a normal function of PDGF may be related to blood vessel development [34]. Cultured SMC derived from intimal lesions of mechanically injured rat carotid arteries [35] and from human atherosclerotic plaques [36] also produce significantly more PDGF than medial SMC from uninjured, normal arteries. Elevated growth factor production by lesion cells in vivo is supported by the study by Wilcox et al. [37] in which in situ hybridization histochemistry is used to determine the location and identity of the lesion cells expressing the genes for the A-chain and B-chain of PDGF. EC and intimal SMC express both gene transcripts, but medial SMC and macrophages show little of either. While this discussion is focused on PDGF production by vessel wall cells, regulation of the expression of the PDGF receptor on the target SMC is equally important. Recent evidence suggests that the PDGF receptor is a family of dimers containing two distinct subunits [38]. The individual subunits have different affinities for the A- and B-chains of PDGF, thereby providing a second mechanism conferring specificity of the cell proliferative response. Wilcox et al. [37] have shown that mRNA for one of the two PDGF receptor subtypes is localized predominantly in the intima of atherosclerotic lesions. These results together suggest that SMC, by producing growth factors to which they also respond, may stimulate their own growth in an autocrine fashion, similar to the mechanism postulated for the uncontrolled cell proliferation characteristic of certain tumors.

Fish Oils and Growth Factors for Smooth Muscle Cells

Our studies on the effects of fish oils on growth factor production were in part motivated by the studies showing that fish oils reduced intimal thickening in experimental animals, and in part by our previous studies on the effects of plasma lipoproteins on growth factor production by EC. Certain modified lipoproteins, including LDL that is modified by acetylation [39], and LDL that is oxidized by in vitro free radical oxidation [40], inhibit PDGFc production by greater than 75% compared to lipoprotein-free controls. The inhibitory activity of both lipoproteins is directly related to the level of lipid peroxidation as measured as thiobarbituric acid-reactive substances. The solvent-extracted lipids from oxidized LDL also inhibit growth factor production, although the potency of the extract is less than that of the intact lipoprotein, presumably due to the greater uptake of the latter by specific high affinity receptors on the EC surface. Since these results, and those from other laboratories, clearly show that oxidized lipids dramatically influence EC function, we have begun to investigate the interactions with EC of another easily oxidizable lipid mixture, the marine

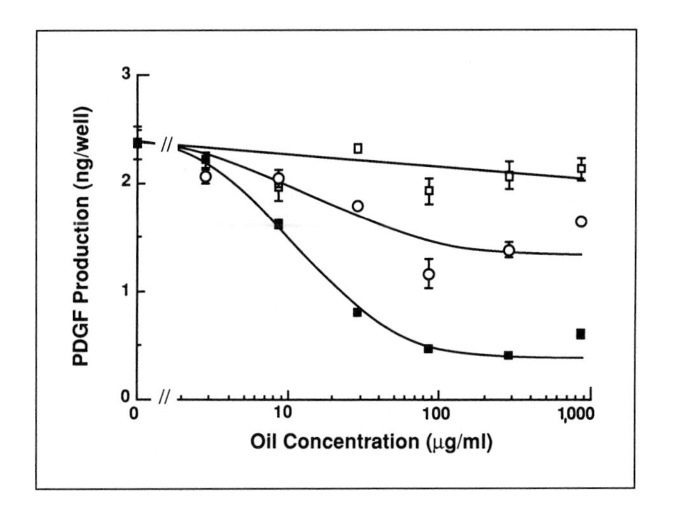

Fig. 2. Inhibition of endothelial cell production of platelet-derived growth factor by fish oil. Emulsions of MaxEPA (■), safflower oil (○), and peanut oil (□) were prepared by sonication with egg phosphatidylcholine. The emulsions were incubated for 72 h with confluent cultures of bovine aortic EC, and PDGFc in the condition medium was assayed by a specific radioreceptor assay.

lipids rich in ω3 fatty acids. In these experiments a stable emulsion of MaxEPA, containing approximately 30% ω3 fatty acids, is prepared by sonication with egg or dimyristoyl phosphatidylcholine [41]. MaxEPA emulsions inhibit the production of PDGF by at least 70% in confluent cultures of bovine aortic EC (fig. 2). Cod liver oil, containing approximately 20% ω3 fatty acids, is nearly as inhibitory. Safflower oil (containing only trace amounts of ω3 fatty acids) and trilinolein also inhibit growth factor production but with a potency 10–20 times less than the fish oils. Peanut oil (containing primarily saturated and monounsaturated fatty acids), trimyristin, and triolein completely lack inhibitory activity. The inhibitory activity of the fish oils is for the most part specific since the rate of overall protein synthesis, as measured by incorporation of radiolabeled leucine into protein, is unaltered. This same result also indicates that the inhibition of growth factor production is not due to a toxic effect of the lipid on the EC cultures.

Cellular uptake of MaxEPA has been measured using emulsion labeled with [^{3}H]cholesteryl oleyl ether, a hydrophobic, non-hydrolyzable marker lipid. Up to 20% of the lipid is taken up by the EC cultures [41]. The

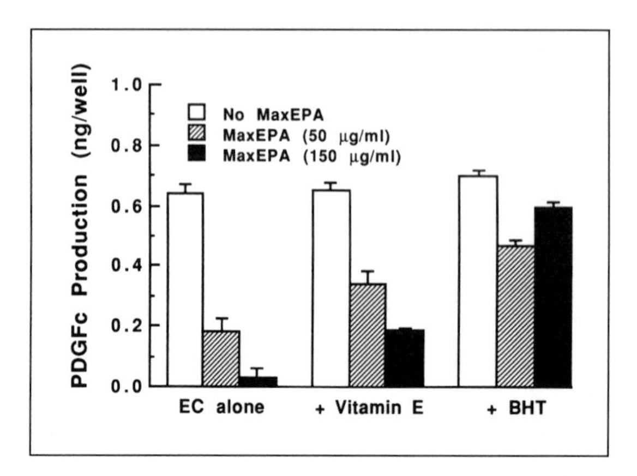

Fig. 3. Effect of antioxidants on the inhibition of PDGFc production by MaxEPA. Bovine aortic EC were preincubated for 24 h with vitamin E (25 μ*M*) or butylated hydroxytoluene (20 μ*M*). After 72 h, the EC-conditioned medium was collected and PDGFc measured by radioreceptor assay.

identity of the lipid(s) in fish oils that specifically inhibits PDGF production has not been determined. A candidate molecule is cholesterol which is a minor component of most fish oil extracts, comprising up to 0.5% of the total mass. Incubation of EC with cholesterol/albumin complexes results in sufficient lipid-loading to increase the cellular cholesterol content by 50%, however, the inhibitory activity of cholesterol is only marginal and cannot account for the activity of fish oils [40]. The inhibitory activity of MaxEPA is suppressed by the addition of antioxidants to the cultures (fig. 3); vitamin E partially reverses the effect of the oil while butylated hydroxytoluene is nearly completely suppressive [41]. That oxidative processes play a critical role in this particular activity of fish oils is evident, but it has not yet been determined if the inhibitor molecule itself is an oxidized lipid, or if oxidative processes within the cell are required for the expression of the inhibitory activity. If the inhibitor is a minor component of fish oil, e.g., an oxidized lipid, then identification of this lipid should result in more potent, and perhaps more tolerable, treatments. Furthermore, it is certainly possible that differences in the level of oxidation of fish oil and the presence of antioxidants in the extract, may contribute to the discrepancies observed in both the animal studies and the clinical trials.

The effects of fish oils on the release of growth factor activity from other cells has been recently reported. Eicosapentaenoic acid and docosahexaenoic acid, ω3 fatty acids that are major constituents of fish oils, inhibit the release of unidentified mitogens from ADP-stimulated platelets and from monocyte-derived macrophages in vitro [42]. In addition, Sarris et al. [3] have reported that implantation of autologous vein grafts in hypercholesterolemic dogs caused increased serum mitogenic activity that was mitigated by dietary fish oils. Although the cellular source of the mitogen is not established with certainty, the authors speculate that the fish oil-mediated inhibition of mitogen release from platelets, and not changes in lipoprotein or eicosanoid metabolism, is responsible for the anti-atherogenic properties of fish oils.

Conclusion

Investigation of the effects of fish oils on the metabolism of vascular cells remains in its infancy. The functional changes induced by fish oils must be better understood at the cellular level, and the ensuing responses, especially the secretion of paracrine agents that alter nearby cells, must be identified. Finally, the identity of the lipid(s) responsible for the cellular responses, and the mechanisms underlying these responses, also must be determined. These investigations at the cellular level should provide important insights into, and suggest new directions for, in vivo studies on the role of fish oils in cardiovascular disease.

References

1 Landymore RW, Kinley CE, Cooper JH, et al: Cod-liver oil in the prevention of intimal hyperplasia in autogenous vein grafts for arterial bypass. J Thorac Cardiovasc Surg 1985;89:351–357.
2 Cahill PD, Sarris GE, Cooper AD, et al: Inhibition of vein graft intimal thickening by eicosapentaenoic acid: Reduced thromboxane production without change in lipoprotein levels or low-density lipoprotein receptor density. J Vasc Surg 1988;7:108–118.
3 Sarris GE, Fann JI, Sokoloff MH, et al: Mechanisms responsible for inhibition of vein-graft arteriosclerosis by fish oil. Circulation 1989;80:I-109–I-123.
4 Casali RE, Hale JA, LeNarz L, et al: Improved graft patency associated with altered platelet function induced by marine fatty acids in dogs. J Surg Res 1986;40:6–12.
5 Sarris GE, Mitchell RS; Billingham ME, et al: Inhibition of accelerated cardiac allograft arteriosclerosis by fish oil: J Thorac Cardiovasc Surg 1989;97:841–855.
6 Weiner BH, Ockene IS, Levine PH, et al: Inhibition of atherosclerosis by cod-liver oil in a hyperlipidemic swine model. N Engl J Med 1986;315:841–846.
7 Kim DN, Ho H-T, Lawrence DA, et al: Modification of lipoprotein patterns and

retardation of atherogenesis by a fish oil supplement to a hyperlipidemic diet for swine. Atherosclerosis 1989;76:35–54.
8 Zhu B-Q, Smith DL, Sievers RE, et al: Inhibition of atherosclerosis by fish oil in cholesterol-fed rabbits. J Am Coll Cardiol 1988;12:1073–1078.
9 Hearn JA, Sgoutas DS, Robinson KA, et al: Marine lipid concentrate and atherosclerosis in the rabbit model. Atherosclerosis 1989;75:39–47.
10 Thiery J, Seidel D: Fish oil feeding results in an enhancement of cholesterol-induced atherosclerosis in rabbits. Atherosclerosis 1987;63:53–56.
11 Clubb FJ, Schmitz JM, Butler MM, et al: Effect of dietary omega-3 fatty acid on serum lipids, platelet function, and atherosclerosis in Watanabe heritable hyperlipidemic rabbits. Arteriosclerosis 1989;9:529–537.
12 Rich S, Miller JF, Charous S, et al: Development of atherosclerosis in genetically hyperlipidemic rabbits during chronic fish-oil ingestion. Arteriosclerosis 1989;9: 189–194.
13 Rogers KA, Karnovsky MJ: Dietary fish oil enhances monocyte adhesion and fatty streak formation in the hypercholesterolemic rat. Am J Pathol 1988;132:382–388.
14 Davis HR, Bridenstine RT, Vesselinovitch D, et al: Fish oil inhibits development of atherosclerosis in rhesus monkeys. Arteriosclerosis 1987;7:441–449.
15 Dehmer GJ, Popma JJ, van den Berg EK, et al: Reduction in the rate of early restenosis after coronary angioplasty by a diet supplemented with n–3 fatty acids. N Engl J Med 1988;319:733–740.
16 Milner MR, Gallino RA, Leffingwell A, et al: Usefulness of fish oil supplements in preventing clinical evidence of restenosis after percutaneous transluminal coronary angioplasty. Am J Cardiol 1989;64:294–299.
17 Grigg LE, Kay TWH, Valentine PA, et al: Determinants of restenosis and lack of effect of dietary supplementation with eicosapentaenoic acid on the incidence of coronary artery restenosis after angioplasty. J Am Coll Cardiol 1989;13:665–672.
18 Reis GJ, Boucher TM, Sipperly ME, et al: Randomised trial of fish oil prevention of restenosis after coronary angioplasty. Lancet 1989;ii:177–181.
19 Ross R, Glomset JA: Atherosclerosis and the arterial smooth muscle cell. Science 1973;180:1332–1339.
20 Grotendorst GR, Chang T, Seppä HEJ, et al: Platelet-derived growth factor is a chemoattractant for vascular smooth muscle cells. J Cell Physiol 1982;113:261–266.
21 Friedman RJ, Stemerman MB, Wenz B, et al: The effect of thrombocytopenia on experimental arteriosclerotic lesion formation in rabbits: Smooth muscle cell proliferation and re-endothelialization. J Clin Invest 1977;60:1191–1201.
22 Reidy MA, Silver M: Endothelial regeneration. VII. Lack of intimal proliferation after defined injury to rat aorta. 1985;118:173–177.
23 Fingerle J, Johnson R, Clowes AW, et al: Role of platelets in smooth muscle cell proliferation and migration after vascular injury in rat carotid artery. Proc Natl Acad Sci USA 1989;86:8412–8416.
24 DiCorleto PE, Bowen-Pope DF: Cultured endothelial cells produce a platelet-derived growth factor-like protein. Proc Natl Acad Sci USA 1983;80:1919–1923.
25 Harlan JM, Thompson PJ, Ross RR, et al: α-Thrombin induces release of platelet-derived growth factor-like molecule(s) by cultured endothelial cells. J Cell Biol 1986; 103:1129–1133.
26 Gajdusek C, Carbon S, Ross R, et al: Activation of coagulation releases endothelial cell mitogens. J Cell Biol 1986;103:419–428.

27 Fox PL, DiCorleto PE: Regulation of production of a platelet-derived growth factor-like protein by cultured bovine aortic endothelial cells. J Cell Physiol 1984;121: 298–308.
28 Jaye M, McConathy E, Drohan W, et al: Modulation of the *sis* gene transcript during endothelial cell differentiation in vitro. Science 1985;228:882–885.
29 Castellot JJ Jr, Addonizio ML, Rosenberg, R: Cultured endothelial cells produce a heparinlike inhibitor of smooth muscle cell growth. J Cell Biol 1981;90:372–379.
30 Gajdusek C, DiCorleto P, Ross R, et al: An endothelial cell-derived growth factor. J Cell Biol 1980;85:467–472.
31 Barrett TB, Gajdusek CM, Schwartz SM, et al: Expression of the *sis* gene by endothelial cells in culture and in vivo. Proc Natl Acad Sci USA 1984;81:6772–6774.
32 Barrett TB, Benditt EP: *sis* (platelet-derived growth factor B chain) gene transcript levels are elevated in human atherosclerotic lesions compared to normal artery. Proc Natl Acad Sci USA 1987;84:1099–1103.
33 Shimokado K, Raines EW, Madtes DK, et al: A significant part of macrophage-derived growth factor consists of at least two forms of PDGF. Cell 1985;43:277–286.
34 Seifert RA, Schwartz AM, Bowen-Pope DF: Developmentally regulated production of platelet-derived growth factor-like molecules. Nature 1984;311:669–671.
35 Walker LN, Bowen-Pope DF, Ross R, et al: Production of PDGF-like molecules by cultured arterial smooth muscle cells accompanies proliferation after arterial injury. Proc Natl Acad Sci USA 1986;83:7311–7315.
36 Libby P, Warner SJC, Salomon RN, et al: Production of platelet-derived growth factor-like mitogen by smooth-muscle cells from human atheroma. N Engl J Med 1988;318:1493–1498.
37 Wilcox JN, Smith KM, Williams LT, et al: Platelet-derived growth factor mRNA detection in human atherosclerotic plaques by in situ hybridization. J Clin Invest 1988;82:1134–1143.
38 Seifert RA, Hart CE, Phillips PE; et al: Two different subunits associate to create isoform-specific platelet-derived growth factor receptors. J Biol Chem 1989;264: 8771–8778.
39 Fox PL, DiCorleto PE: Modified low density lipoproteins suppress production of a platelet-derived growth factor-like protein by cultured endothelial cells. Proc Natl Acad Sci USA 1986;83:4774–4778.
40 Fox PL, Chisolm GM, DiCorleto PE: Lipoprotein-mediated inhibition of endothelial cell production of platelet-derived growth factor-like protein depends on free radical lipid peroxidation. J Biol Chem 1987;262:6046–6054.
41 Fox PL, DiCorleto PE: Fish oils inhibit endothelial cell production of platelet-derived growth factor-like protein. Science 1988;241:453–456.
42 Smith DL, Wills AL, Nguyen N, et al: Eskimo plasma constituents, dihomo-γ-linolenic acid, eicosapentaenoic acid and docosahexaenoic acid inhibit the release of atherogenic mitogens. Lipids 1989;24:70–75.

Paul L. Fox, MD,
Department of Vascular Cell Biology and Atherosclerosis Research,
Cleveland Clinic Research Institute, FF4–15, 9500 Euclid Avenue,
Cleveland, OH 44195 (USA)

Cardiovascular II: Heart

Simopoulos AP, Kifer RR, Martin RE, Barlow SM (eds): Health Effects of ω3 Polyunsaturated Fatty Acids in Seafoods. World Rev Nutr Diet. Basel, Karger, 1991, vol 66, pp 268–277

Review: Fish Oil and Cardiac Function

P.J. Nestel

CSIRO Division of Human Nutrition, Adelaide, S.A., Australia

Introduction

In this chapter, I will review recent evidence which offers possible reasons for the low incidence of cardiac deaths in fish-eating populations. Although the protection relates to eating fish, this has been extrapolated to the ω3 fatty acids in the oil. The explanations sought for the benefits of fish have therefore been approached experimentally mainly by using dietary fish oils. Possibilities include (1) suppression of sudden death from cardiac arrhythmia; (2) protection of the ischemic myocardium; (3) improved myocardial function; (4) reduction of cardiovascular risk.

Reduced Cardiac Mortality

Apart from the well-known low incidence of cardiac deaths in several high fish-eating populations, (though there are exceptions), further support has come from several recent reports.

The eicosapentaenoic acid (EPA) content of blood platelets was found to be inversely related to the prevalence of clinical coronary artery disease in Scottish men [1]. This was not as persuasive as the inverse correlation with adipose linoleic acid but does suggest that even in a low fish-consuming population individuals who eat more fish have less coronary disease. A more direct approach was taken by Gudbjarnason and coworkers [2], who found in men who have died suddenly from cardiac causes, a reduced docosahexaenoic:arachidonic acid ratio in cardiac phospholipids.

Wahlqvist and associates in Melbourne [3] have identified a new cardiovascular benefit related to eating fish. Arterial compliance, measured by ultrasound, was greater in subjects who ate fish regularly. This is an important factor in the hemodynamic control of the circulation and one which might influence arterial pressure and cardiac work.

Intervention studies are urgently required to answer the critical question about the protective effects of fish. Hardarson et al. [4] failed to reduce the incidence of ventricular premature beasts (VPB) in men given cod liver oil during recovery from myocardial infarction. However, ventricular extrasystolic beats are not necessarily related electrophysiologically to ventricular tachycardia and ventricular fibrillation which are the common precursors of sudden cardiac death.

The only intervention study of substance is that recently reported by Burr et al. [5]. Mortality from all causes was lower in those survivors of myocardial infarction who had been advised to eat fatty fish or take fish oil capsules. However, cardiac events were not reduced suggesting that the increased survival reflected fewer sudden cardiac deaths. Only the first two years' experience was reported and later outcomes will clarify whether reinfarction is preventible.

The questions which need the most urgent answers concern (1) reduction of tachyarrhythmia, (2) protection of myocardium, (3) minimization of infarction and (4) reduction of risk factors.

Experimental Myocardial Ischemia

Research objectives in this area are to prevent or at least limit infarct size and preserve sufficient healthy myocardium to reduce the likelihood of later ventricular dysfunction. In the clinical setting, patients who survive an acute myocardial infarction face an increased risk of sudden death over the following 6 months or so. The prognosis is related to ventricular dysfunction and continuing ischemia which can trigger arrhythmias and induce cardiac failure.

Myocardial ischemia need not progress to infarction or death of myocytes. Prolonged ischemia may depress contractile function for varying duration and to varying degrees of severity. These reversible but potentially fatal states are referred to as 'hibernating' or 'stunned' myocardium [6].

Possible mechanisms include abnormal energy utilization, production of oxygen-derived free radicals, abnormal calcium flux, accumulation of leucocytes in the ischemic tissue, formation of leukotrienes, abnormal microvascular flow, and combinations of these. Measurements of these parameters and events have been used as indices of the effectiveness of fish oil feeding. Once death of tissue occurs, infarct size can be visualized morphologically or assessed functionally through the leakage of creatine kinase (CK). Lysophosphoglycerides which accumulate in ischemic myocardium and can affect electrical stability [7] are derived from membrane phospholipid and presumably amenable to changes in dietary fatty acids.

The significance of neutrophil leakage from the increasingly permeable capillaries is becoming more evident. The influx of neutrophils into ischemic myocardium leads to extension of injury and the generation of oxygen-related free radicals [8]. The importance of free radical formation in conditions such as the stunned myocardium is apparent from restoration of contractility through the use of scavengers of oxygen metabolites [9]. Recent experiments using electron paramagnetic resonance spectroscopy have shown directly that reperfusing an ischemic myocardium provokes free radical production.

Some of the benefits claimed for fish oil in this area are: (1) Increased flow through ischemic area; (2) reduced infarct size; (3) reduced CK efflux; (4) reduced neutrophil infiltration; (5) reduced leukotriene production; (6) reduced injury by free radicals; (7) prevention of calcium overload.

A decade ago, Culp et al. [10] showed that prefeeding menhaden oil to dogs significantly reduced infarct size following induction of coronary thrombosis. A more recent study [11] with rats fed fish oil failed to reduce the extent of infarction after prolonged (40 min) left coronary artery occlusion followed by 2-hour reperfusion. Nevertheless postischemic blood flow was improved. (The rat heart does not have the collateral blood flow of the dog.) A report by Hock et al. [12] suggests that rats may be protected by prefeeding menhaden oil, provided the duration of ischemia is shorter (15 min). Sixteen of 21 fish oil-fed rats survived compared with only 9 of 22 fed corn oil. Survival may have been unrelated to infarction since neither CK loss nor leukocyte infiltration into heart muscle was influenced. However, and earlier report from this group did show lessened susceptibility to ischemia in fish oil-fed rats [13].

Both CK leakage and leukocyte infiltration were favourably reduced in dogs given EPA and subsequently subjected to occlusion of the circumflex artery of the heart [14]. Infarct size was halved in another study of fish oil fed dogs [15].

Myocardial function has been investigated in animals fed fish oil, sometimes in response to ischemia. Measurements have included physiological dynamics and biochemical indices of function.

Hartog et al. [16] and Lamers et al. [17] added mackerel oil to the diet of young pigs. Cardiac hemodynamics (echocardiography) were not affected, despite considerable enrichment of cardiac phospholipid with ω3 fatty acid. Earlier reports with isolated rat heart preparations had suggested increased contractility in linoleic acid-fed rats. Following induction of brief periods of ischemia, the hearts of fish oil-fed and lard-fed pigs recovered equally, despite the formation of peroxidative products in the hearts of pigs fed mackerel oil [17].

Swanson et al. [18] have described fish oil-induced changes in calcium flux in the sarcoplasmic reticulum of fish oil-fed mice. They interpret their findings to show that the reduced flux might protect the heart from the large fluctuations in calcium which occur with ischemia and which are partly responsible for disordered contractility and rhythm.

McLennan et al. [19] have also noted that cardiac papillary muscle obtained from rats fed tuna fish oil developed a reduced inotropic response to calcium. By contrast, a diet rich in saturated fat augmented the response. Since indomethacin inhibited these responses, the effect was attributed to the altered nature of eicosanoids.

Cardiac Arrhythmia

Suppression of serious ventricular tachyarrhythmias is emerging as the best documented cardiac outcome of fish oil feeding to experimental animals. Supplementation with linoleic acid-rich oils had been previously shown to be also moderately effective, but the studies by Charnock and McLennan [20–22] suggest that fish oils, at least those rich in docosahexaenoic acid (DHA), are considerably superior in preventing experimental ventricular fibrillation.

The evidence for linoleic acid will first be reviewed briefly. It has been observed in a variety of experimental models. Most persuasive are those in which the animals had been fed for long periods of time [20–22], because, at least in rats, it has become clear that the vulnerability to arrhythmia and protection by polyunsaturated fatty acids (PUFA), increases with age. The tachyarrhythmias inducible with isoproterenol and norepinephrine are also suppressible with PUFA [19, 23, 24], as are those following coronary artery ligation [20–22, 25]. In the marmoset monkey, the opposing effects of dietary saturated fatty acids and of PUFA on isoproterenol-induced arrhythmia was reflected in corresponding changes in adenylate cyclase activity [26].

Research with fish oil preparations is not as extensive. As benefits, resulting in suppression of cardiac arrhythmias, are postulated: (1) Reduced excitability during ischemia, (2) blunted response to adrenergic agonists, (3) reduced eicosanoid production, (4) minimised effect of ageing, (5) altered membrane characteristics. In studies with rats fed tuna fish oil for at least several months, ventricular fibrillation following an occlusion-reperfusion manoeuvre is uncommon [21, 22]. However, in a similar study in young pigs prefed mackerel oil for 2 months, the incidence of serious arrhythmias did not differ from those fed lard [27]. Nor was the complication of ventricular fibrillation reduced in dogs prefed menhaden oil for 5–6 weeks before coronary thrombosis was induced [10].

Interestingly, Reibel et al. [28] have observed a modification by fish oil only of alpha- and not of beta-adrenoreceptor function in rat heart (isoproterenol is a beta agonist). Others have observed enhanced alpha-adrenergic responsiveness in ischemic and reperfused heart [7].

The major mechanism proposed for the beneficial effect of fish oil is related to eicosanoid metabolism. Less 6-keto $PGF_{1\alpha}$ is released from ischemic rat heart when the animal has been prefed cod liver oil [29]. Indomethacin, an inhibitor of eicosanoid synthesis, has also been found to modulate the induction of cardiac arrhythmias [30]. The heart produces both thromboxane and prostacyclin which may have opposing effects; the synthesis of the former is suppressed more by fish oil. The formation of the potentially injurious leukotrienes is also inhibited by ω3 fatty acids, although not equally by EPA and DHA. These compounds may act directly on cardiac electrophysiologic function or indirectly by reducing coronary flow.

It is clear that the effects of catecholamines, eicosanoids, calcium and oxygen-derived free radicals exert interrelated effects on cardiac contractility, cardiac conduction and myocardial perfusion. Changes in any of these or in a combination of these may trigger a fatal tachyarrhythmia.

Reperfusion Injury

This is one of the major issues in current cardiology. The value of fish oils in this area has been tentatively exploratory but is certain to develop rapidly. The metabolic events which follow rapid reinstitution of perfusion after occlusion of the arterial circulation occur within seconds or at most minutes. Whereas total reperfusion is possible under experimental circum-

Table 1. Reperfusion injury

Major disturbances	Common mechanisms
Cardiac arrhythmia	Free radical activity
Myocardial stunning	Damage to membrane function
Microvascular injury	Damage to ion pumps
? Acceleration of cell necrosis	Calcium overload
	Contractile dysfunction

stances but rarely clinically. the findings in experimental reperfusion are probably more severe. Nevertheless the two major complications, cardiac arrhythmia and myocardial stunning, do occur commonly in clinical cardiology.

Cardiac arrhythmias occur after brief or mild ischemia when reversal of injury is possible with reperfusion. Possible mechanisms are listed in table 1.

Free radical scavengers and antioxidants protect or reverse the complication of arrhythmia, supporting the importance of free oxygen-generated radicals. Xanthine oxidase, generated during the preceding anaerobic phase, interacts with oxygen and catalysts when flow is restored and so produces superoxide.

Myocardial stunning is a more subacute process. Free radicals and calcium overload are at least two factors which lead to the contractile dysfunction which may last hours or even days. Free radical scavengers have again proved beneficial under experimental conditions. Creatine kinase is lost from heart cells and ultrastructural changes to myofilaments and mitochondria can develop quickly.

Table 1 summarizes several of the major disturbances induced by post-ischemic reperfusion.

Myocardial Membrane Fatty Acid Changes

The changes in the fatty acid profile of myocytes and especially of membrane phospholipids with fish oil feeding have been well categorized. They will not be reviewed further other than to point out that ω6 and ω3 fatty acid enrichment probably underlie most of the changes discussed above. Some are related to effects which altered fluidity might exert on membrane-bound enzymes, proteins and receptors. Mostly, the changes would influence the profile of eicosanoid production and the probability for peroxidation and generation of free radicals to occur.

Several of the publications cited above contain comprehensive description of myocardial phospholipid fatty acids in fish oil-fed animals.

One further point of interest is the cholesterol:phospholipid ratio in myocyte membranes which is influenced by the nature of dietary fat. The fluidity of membranes is also affected by this ratio which McMurchie et al. [26] have shown is influenced by dietary PUFA.

Modification of Cardiovascular Risk Factors

Dietary fish oils modify most of the major cardiovascular risk factors including plasma lipids, blood pressure and coagulant factors. It is therefore relevant to ask to what extent the reduction in cardiovascular mortality in fish-eating populations reflects overall improvement in risk or to direct effects on the heart as reviewed above. Clearly this is currently not possible. I will briefly touch on two points: (1) changes in the nature of the HDL particle which by influencing reverse cholesterol transport would modify membrane characteristics; (2) the relative effects of fish and fish oil.

Abbey et al. have observed in men taking fish oil supplements significant changes in HDL characteristics [31]. The proportion of HDL_2 rose, increasing the HDL_2:HDL_3 ratio; this was reflected in an increase in the apoprotein A_1:A_2 ratio and was partly due to decreased activity of lipid transfer protein which normally transfers cholesteryl esters from HDL to other lipoproteins. The potential significance of this is a reduction in the redistribution of cholesterol from non-atherogenic to atherogenic lipoproteins. In experiments with marmoset monkeys Abbey et al. have found EPA to correct the maldistribution of cholesterol among lipoproteins which occurs with high saturated fat/cholesterol diets [32]. In the presence of EPA supplement, the bulk of plasma cholesterol was transported in HDL rather than in LDL which characterized the high-fat, high-cholesterol diet.

We have also recently compared the effects on cardiovascular risk of a diet rich in fish and a diet in which similar amounts of ω3 fatty acids (4g/d) were provided as in an oil supplement. The predominant ω3 fatty acids were DHA in the fish and EPA in the supplement.

Table 2. Effects of fish (DHA+) and fish oil (EPA+) on plasma lipids

	Fish	Fish oil	Control
Plasma cholesterol, Δμmol/l	-0.41	-0.10	-0.40
VLDL cholesterol, Δμmol/l	-0.57*	-0.74*	+0.17
HDL cholesterol, Δμmol/l	+0.11*	+0.09*	-0.03
Plasma triglyceride, Δμmol/l	-0.39*	-0.57*	+0.20
Fibrinogen, Δg/l	-0.15*	+0.38	+0.18
Thromboxane B_2, Δng/ml	-12*	+6	+17
Bleeding time, Δsec	+43*	-12	-25
Systolic blood pressure, Δmm Hg	-5*	-4*	-3

* Significantly different from baseline diet.

EPA out-performed DHA (fish) only in terms of plasma triglyceride lowering although both were highly effective. On the other hand, EPA raised LDL cholesterol, relative to DHA (fish). DHA (fish) alone significantly prolonged bleeding time and lowered plasma fibrinogen and thromboxane generation. Blood pressure lowering was similar (table 2).

These differences between the effects of DHA and EPA (assuming no additional influence from fish) demonstrate the need to examine separately these major ω3 fatty acids in studies of cardiac function. This has not yet been done.

Appropriate Human Models

The final consideration of this review is to emphasize the complex pathophysiology of coronary heart disease. Impaired perfusion occurs through structural (atherosclerotic) changes as well as to reversible effects such as coronary spasm. The latter phenomenon is only becoming understood but it is known that ω3 fatty acids modify spasm in experimental situations.

The pathology of diseased heart muscle is heterogeneous. Ventricular dilation, patchy fibrosis (which may involve the conducting system), periodic ischemia and micronecrosis are impossible to emulate experimentally. In such a context it may be difficult to determine from clinical trials alone the mechanisms through which fish and fish oil protect the heart.

This should not, however, inhibit the organization of intervention trials, provided the dissimilarities between the experimental models and the clinical reality are taken into account.

References

1 Wood DA, Riemersma RA, Butler S, et al: Linoleic and eicosapentaenoic acids in adipose tissue and platelets and risk of coronary heart disease. Lancet 1987;i:177–183.
2 Gudbjarnason S: Dynamics of n–3 and n–6 fatty acids in phospholipids of heart muscle. J Intern Med 1989;225:117–128.
3 Wahlqvist ML, Lo CS, Myers KA: Fish intake and arterial wall characteristics in healthy people and diabetic patients. Lancet 1989;ii:944–946.
4 Hardarson T, Kristinsson A, Skuladottir G, et al: Cod liver oil does not reduce ventricular extrasystoles after myocardial infarction. J Intern Med 1989;226:33–37.
5 Burr ML, Fehily AM, Gilbert JF, et al: Effects of changes in fat, fish, and fibre intakes on death and myocardial reinfarction: diet and reinfarction trial (DART). Lancet 1989;ii:757–761.
6 Kloner RA, Przyklenk K, Patel B: Altered myocardial states. The stunned and hibernating myocardium. Am J Med 1989;86:14–22.
7 Corr PB, Pogwizd, SM: Mechanisms contributing to arrhythmogenesis during early myocardial ischemia, subsequent reperfusion, and chronic infarction. Angiology 1988;39:684–689.
8 Mehta JL, Nichols WW, Mehta P: Neutrophils as potential participants in acute myocardial ischemia: relevance to reperfusion. J Am Coll Cardiol 1988;11:1309–1316.
9 Bolli R: Oxygen-derived free radicals and postischemic myocardial dysfunction ('Stunned Myocardium'). J Am Coll Cardiol 1988;12:239–249.
10 Culp BR, Lands WEM, Luccesi BR, et al: The effect of dietary supplementation of fish oil on experimental myocardial infarction. Prostaglandins 1980;20:1021–1031.
11 Force T, Malis CD, Guerrero JL, et al: N-3 fatty acids increase postischemic blood flow but do not reduce myocardial necrosis. Am J Physiol 1989;257:1204–1210.
12 Hock CE, Beck LD, Reibel DK: Effect of dietary menhaden oil (MO) on ischemia-reperfusion injury to the myocardium. Circulation 1989;80(suppl 2):237.
13 Hock CE, Holahan MA, Reibel DK: Effect of dietary fish oil on myocardial phospholipids and myocardial ischemic damage. Am J Physiol 1987;252:554–560.
14 Otsuji S, Hirota H, Akegami H, et al: Dietary eicosapentaenoic acid reduces the extent of ischemic myocardial injury through inhibition of neutrophil infiltration. Circulation 1989;80(suppl 2):33.
15 Oskarsson HJ, Godwin JE, Ferraro ME, et al: Dietary supplementation with fish oil reduces infarct size in dogs. Circulation 1988;78(suppl 2):216.
16 Hartog JM, Lamers JMJ, Montfoort A, et al: Comparison of mackerel-oil and lard-fat enriched diets on plasma lipids, cardiac membrane phospholipids, cardiovascular performance, and morphology in young pigs. Am J Clin Nutr 1987;46:258–266.

17 Lamers JMJ, Hartog JM, Guarnier C, et al: Lipid peroxidation in normoxic and ischemic-reperfused hearts of fish oil and lard fat fed pigs. J Mol Cell Cardiol 1988; 20:605–615.
18 Swanson JE, Lokesh BR, Kinsella JE: Ca^{2+}-Mg^{2+} ATPase of mouse cardiac sarcoplasmic reticulum is affected by membrane n–6 and n–3 polyunsaturated fatty acid content. J Nutr 1989;119:364–372.
19 McLennan PL, Abeywardena MY, Charnock JS: A comparison of the long-term effects of n–3 and n–6 polyunsaturated fatty acid dietary supplements and the action of indomethacin upon the mechanical performance and susceptibility of the rat heart to dysrhythmia. Prostaglandins Leukotrienes Med 1987;27:183–195.
20 McLennan PL, Abeywardena MY, Charnock JS: Influence of dietary lipids on arrhythmias and infarction after coronary artery ligation in rats. Can J Physiol Pharmacol 1985;63:1411–1417.
21 McLennan PL, Abeywardena MY, Charnock JS: Dietary fish oil prevents vetricular fibrillation following coronary artery occlusion and reperfusion. Am Heart J 1988; 116:709–717.
22 McLennan PL, Abeywardena MY, Charnock JS: The influence of age and dietary fat in an animal model of sudden cardiac death. Aust NZ J Med 1989;19:1–5.
23 Gudbjarnasson S, Hallgrimsson J: Prostaglandins and polyunsaturated fatty acids in heart muscle. Acta Biol Med Ger 1976;35:1069–1080.
24 Hoffmann P: Cardiovascular actions of dietary polyunsaturates and related mechanisms. Prostaglandins Leukotrienes Med 1986;21:113–147.
25 Lepran I, Nemecz G, Koltai M, et al: Effect of a linoleic acid-rich diet on the acute phase of coronary occlusion in conscious rats: influence of indomethacin and aspirin. J Cardiovas Pharmacol 1981;3:847–853.
26 McMurchie EJ, Patten GS, McLennan PL, et al: The influence of dietary lipid supplementation on cardiac β-adrenergic receptor adenylate cyclase activity in the marmoset monkey. Biochim Biophys Acta 1988;937:347–358.
27 Hartog JM, Lamers JMJ, Verdouw PD: The effects of dietary mackerel oil on plasma and cell membrane lipids, on hemodynamics and cardiac arryhtmias during recurrent acute ischemia in the pig. Basic Res Cardiol 1986;81:567–580.
28 Reibel DK, Holahan MA, Hock CE: Effects of dietary fish oil on cardiac responsiveness to adrenoceptor stimulation. Am J Physiol 1988;254:494–499.
29 Karmazyn M, Horackova M, Murphy MG: Effects of dietary cod liver oil on fatty acid composition and calcium transport in isolated adult rat ventricular myocytes and on the response of isolated hearts to ischemia and reperfusion. Can J Physiol Pharmacol 1987;65:201–209.
30 Abeywardena MY, McLennan PL, Charnock JS: Dietary fat modulation of rat heart PGI/TXA balance. J Mol Cell Cardiol 1988;20:vi.
31 Abbey M, Clifton P, Kestin M, et al: Lipoproteins, lecithin: cholesterol acyltransferase and lipid transfer protein activity in human subjects consuming n–6 fatty acids and n–3 fatty acids of vegetable and marine origin. Arteriosclerosis (in press).
32 Abbey M, Clifton PM, McMurchie EJ, et al: The effect of a high fat/cholesterol diet with or without eicosapentaenoic acid on plasma lipids, lipoproteins and lipid transfer protein activity in the marmoset. Atherosclerosis (in press).

P.J. Nestel, MD, CSIRO Division of Human Nutrition, P.O. Box 10041,
Gouger Street, Adelaide, S.A. 5000 (Australia)

Simopoulos AP, Kifer RR, Martin RE, Barlow SM (eds): Health Effects of ω3 Polyunsaturated Fatty Acids in Seafoods. World Rev Nutr Diet. Basel, Karger, 1991, vol 66, pp 278–291

Antiarrhythmic Effects of Fish Oils

John S. Charnock

CSIRO Division of Human Nutrition, Glenthorne Laboratory,
O'Halloran Hill, S.A., Australia

Introduction

Despite recent falls in mortality from coronary heart disease in many Western nations [1], sudden cardiac death (SCD) remains the most common mortal event in industrialized societies [2, 3]. Several recent estimates suggest that in the USA alone there are about 400,000 deaths from this cause every year [4]. Apparently, many of these are associated with fatal arrhythmias [5]. It is a matter for some concern that the death rate from SCD is increasing in many South East Asian countries, and that in Singapore, where the rate has been increasing rapidly since the early 1980s, it is now equal to or greater than that in New Zealand or Australia [6].

In their reviews of sudden cardiac death, both Lown [2] and Anderson [7] have drawn attention to the problem that SCD can occur without substantial warning in individuals of apparently good health. When such a medical emergency occurs away from the technology necessary for rapid resuscitation, death frequently occurs within minutes [3, 8]. There is considerable clinical evidence that the primary cause of death is terminal arrhythmia following severe ventricular fibrillation or ventricular arrest. While post-myocardial infarction (MI) patients are clearly a very high risk group, many deaths occur which are not associated with MI [8, 9, 10]. In fact, up to 20% of SCD may not even be associated with significant atherosclerosis [9].

Of the many risk factors for SCD which have been considered, that of the diet is generally accepted as one of major importance [1, 11, 12]. There can be little doubt today that all forms of coronary heart disease are positively associated with the high fat diets which are now common-place in many countries of the world [11, 12]. Conversely, several epidemiological studies have pointed to the marked reduction in premature cardiac deaths amongst populations who regularly consume a high fish diet [13, 14]. It is also known that enforced dietary change which increased the amount of fish eaten in place of meats and dairy products quickly resulted in a marked reduction in the number of cardiac deaths [15, 16]. Very recently, Burr et al. [17] have described the significant fall in the death rate of post-MI patients who increased their regular consumption of fish (or fish oil) over a 2-year period. As the decline in death rate in this latter study was not associated with a change in the incidence of re-infarction or the number of ischemic events, it is reasonable to suppose that the decline was due to a reduction in fatal arrhythmias [Burr, personal communication]. All these population-based or intervention studies strongly suggest that the increased consumption of the ω3 PUFAs of fish or fish oil have a beneficial effect upon the function of the heart and may reduce the death rate from 'heart attacks'.

The following review describes the systematic investigation of this possibility carried out in a series of animal experiments (using both rodents and non-human primates), which were designed specifically to examine the effect of various dietary lipids upon the mechanical performance of the heart and its susceptibility to develop severe arrhythmias both in vitro and in vivo.

Experiments with Isolated Papillary Muscles

Initial experiments with isolated papillary muscles from dietary-manipulated rats clearly indicated that the contractility of these heart muscles was greatly influenced by both the age of the animal and the type of dietary fat that had been consumed over the preceding months [18]. In general the contractility of the papillary muscles increased with the age of the animal, but this change could be greatly attenuated by dietary PUFAs of either vegetable (ω6) or marine (ω3) origin, whereas an increased intake of saturated fats of either animal or plant origin markedly increased the degree of contractility observed in vitro [19]. Under the stimulus of catecholamine stress following the addition of isoprenaline to the tissue bath, a significantly greater incidence of dysrhythmia was observed in the papillary muscles from the hearts of mature rats fed a dietary supplement of

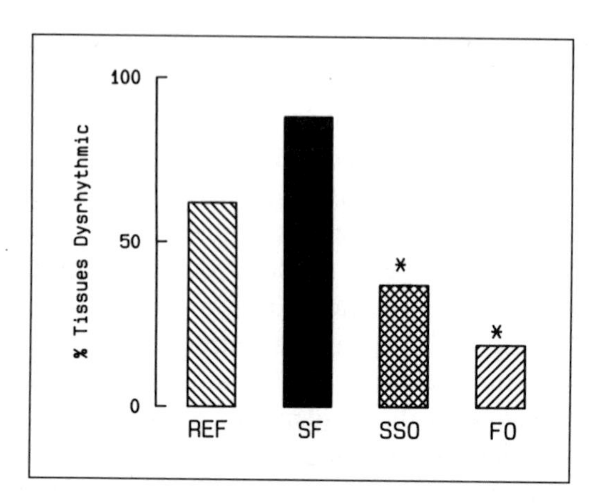

Fig. 1. Effect of different dietary lipid supplements on the incidence I (% tissues dysrhythmic) of isoprenaline-induced arrhythmia in the isolated papillary muscles of rats after 12 months feeding. REF is a commercial rat chow containing 4% fat (w/w) of mixed animal and plant origin and represents a relatively low-fat control diet. SF is the control diet supplemented by the addition of saturated animal fat obtained from sheep perirenal adipose tissue (24% palmitic; 42% stearic; 25% oleic acid). SSO is sunflower seed oil (68% linoleic, 20% oleic acid). FO is fish oil obtained from eviscerated Bluefin Tuna (18% palmitic, 22% oleic, 7% EPA, 25% DHA). All supplements are 12% (w/w). * %I in the SSO and FO groups is significantly less ($p < 0.05$) than in either REF- or SF-supplemented groups ($n = 10$ rats/group; 2 papillary muscles from each animal).

saturated animal fat compared to that which could be induced in the muscles from animals fed PUFA-enriched diets [19, 20] (fig. 1).

Subsequent experiments with the isolated papillary and atrial muscles of dietary-manipulated marmoset monkeys gave similar results to those seen in rats in that the incidence of isoprenaline-induced tachyarrhythmias was reduced by a sunflower seed oil-supplemented diet and increased by a sheep fat diet [21]. More recent (unpublished) observations show a result following ω3 PUFA-enriched diets similar to that seen for the sunflower seed oil diets. These initial results with isolated muscle preparations from both rats and marmosets strongly suggest a *direct* effect of dietary lipids upon the mechanical performance of cardiac muscles which is not dependent upon changes in electrical stimulation to the heart, nor apparently are the effects mediated by the well established cardiovascular actions of dietary lipids on the processes of atherosclerosis or thrombogenesis [22, 23]. On the other hand, abolition of these dietary-induced effects by the introduction of the cyclooxygenase inhibitor indomethacin into the test system in vitro [20] clearly suggests that the action of eicosanoid 'local hormones' may be central to the changes in contractility and susceptibility to arrthythmia which were observed. These in vitro studies have provided the background for more intensive investigation of the effects of long-term dietary lipids in whole animal models of cardia arrhythmia.

Cardiac Arrhythmia in Rats

A model for the study of cardiac arrhythmia in rats was first developed by Selye et al. in the 1960s [24] and has since been refined so that the arrhythmias which follow transient ischemia induced by ligation of the anterior descending branch of the left coronary artery are now widely accepted as a meaningful model of the occlusion-induced arrhythmias encountered in man [25]. When the blood flow is restored to an ischemic region of the heart, the reperfusion arrhythmias which follow are also thought to closely resemble the reperfusion arrhythmias in man which frequently follow myocardial infarction, and the model can therefore be very valuable for the study of factors which influence SCD [26].

Examination of the effect of age in PUFA- versus saturated fat-supplemented animals again revealed a marked effect of age upon the animals' susceptibility to develop serious ventricular fibrillation under ischemic stress [27]. After 20 months feeding, the extent of ventricular fibrillation (VF) in PUFA-fed animals was significantly less than that in either saturated fat-fed animals or even the low fat-fed control animals [27]. More recent studies comparing isocaloric supplements of either ω6 PUFA- or ω3 PUFA-enriched diets in mature animals revealed that while both dietary supplements were effective in animals subjected to a 15-min period of ligation of the coronary artery, there were very marked advantages to ω3 PUFA-enriched diets during reperfusion, where the incidence and duration of both ventricular tachycardia (VT) and VF were significantly less than in ω6 PUFA-fed animals [28] (fig. 2).

It is this apparent advantage of ω3 PUFA-enriched diets over ω6 PUFA-enriched diets which suggests that either fish or fish oil supplements could play a significant role in reducing the incidence of SCD or post-myocardial re-infarction arrhythmias in man [26, 28]. That these very beneficial effects can also be obtained in very old animals which had received high saturated fat diets for at least half their normal life-span

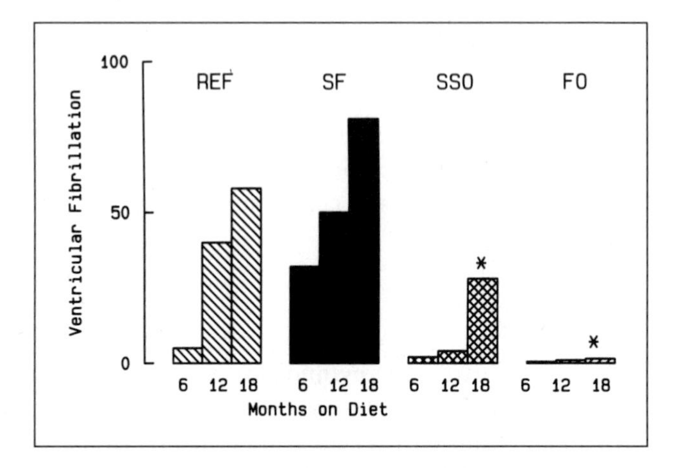

Fig. 2. Influence of age and dietary fat on the duration of ventricular fibrillation (VF) during 15 min occlusion of the left descending coronary artery of anaesthetised rats. Periods of feeding are shown as 6, 8 or 12 months on diets: REF, SF, SSO and FO are as given in figure 1. VF is given as duration in seconds, * is significantly less ($p < 0.05$) in the SSO or FO groups than in either REF- or SF-supplemented animals (n = 25 animals/dietary group).

before being exposed to either ω6 or ω3 PUFA diets [29] is most encouraging (fig. 3), and is in complete agreement with the findings reported from a recent 2-year intervention (DART) trial by Burr et al. [17], with post-myocardial infarction men over 50 years of age.

Effect of Dietary Lipids in a Non-Human Primate

One of the problems facing those investigators who must use animal models in the study of human disease is the degree of confidence with which their results can be translated to man. In this regard, certain non-human primate models are often preferred to those of other more common laboratory animals. Associated work in our Division has already demonstrated that a small non-human primate, the marmoset monkey *(Callithrix jacchus)* has a blood lipid profile similar to that of man [30], and that this profile responds to dietary intervention with changes similar to those seen under atherogenic conditions. Perhaps of even more significance to this present study, a comparative investigation has also revealed that unlike the

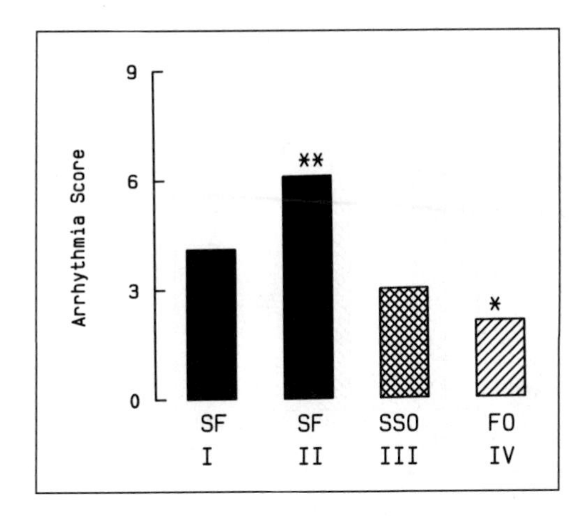

Fig. 3. The severity of cardiac arrhythmia in anaesthetised rats during reperfusion of an ischaemic heart. Severity measured on a hierachical arrhythmia score from 0–9 [see ref. 28 for details]. Diets are as described for figure 1. Feeding regimes were as follows: group I = SF for 9 months; group II = SF for 18 months; group III = SF for 9 months + SSO for 9 months; group IV = SF for 9 months + FO for 9 months (n = 6 animals/dietary group). * The severity of cardiac arrhythmias in the FO group (IV) is significantly ($p < 0.01$) less than in either groups I or II. ** the severity in the long-term SF-fed group (II) is greater than in all other groups ($p < 0.01$).

common laboratory rat, the marmoset monkey has a predominant β-adrenergic receptor system in its heart [31] making this aspect of its cardiac function very similar to that of man.

The quantitative diagnostic procedure of gated blood pool radionuclide angiography has been widely used clinically to access ventricular function in man and the extent of risk of SCD [32, 33]. After miniaturizing this procedure to make it applicable to the small marmoset heart (weighing about 3 g [34], we have been able to compare the effects of different dietary lipid supplements upon heart rate, blood pressure, left ventricular ejection fraction (LVEF), peak emptying and filling rates and the end-diastolic volume (EDV) of surviving marmosets after short and relatively long-term feeding periods [35, 36].

Our initial observations which were made after only 8 months of dietary supplementation demonstrated that in comparison to a diet enriched with ω6 PUFAs, saturated fats produced increases in the heart rate and pressure rate index (an indirect measure of oxygen consumption),

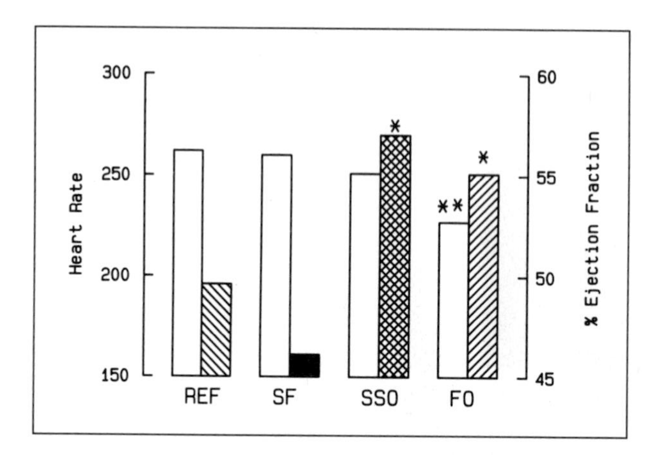

Fig. 4. Effect of feeding different dietary lipid supplements for two years on the cardiac function of adult male marmosets. Ejection fraction (% filled volume) of the left ventricle was measured by radionuclide angiography. Dietary supplements are as shown for figure 1. Open bars = heart rate (beats/min); ** HR in the FO group is significantly less ($p < 0.01$) than in all other dietary groups. Hatched bars = % left ventricular ejection fraction; * LVEF in the SSO and FO groups is significantly greater ($p < 0.01$) than in either the REF or SF groups.

which were accompanied by significant increases in both the peak emptying rate (PER up 71%) and peak filling rate (PFR up 84%). Conversely, and of great potential clinical significance, dietary supplementation with PUFA produced a significant improvement in the left ventricular ejection fraction (LVEF up 23%) which was not accompanied by either an increase in the resting blood pressure or the calculated pressure rate index [35].

Very recently, we have completed a much more detailed set of observations after about 24 months of feeding. The results confirm and extend the observations made after only 8 months of feeding. The LVEF was again elevated by about 20% in marmosets fed either ω6 or ω3 PUFA-enriched diets. In comparison to saturated fat-fed animals, the heart rate and pressure rate index were reduced in the ω3 PUFA-fed group (fig. 4). For the first time we were able to measure both the cardiac output (i.e. the volume of blood pumped per minute) and EDV. Here we also saw a major difference between the ω3 PUFA-fed animals and those receiving saturated fats with this latter group showing a much reduced EDV (0.21 vs. 0.25 $ml \cdot g^{-1}$). That is, there can be no doubt that prolonged administration of a fish oil supplement to male adult marmoset monkeys results in an increased ejec-

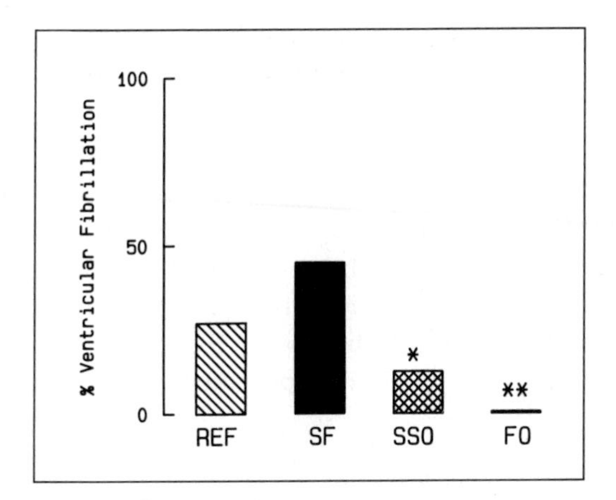

Fig. 5. Effect of feeding different dietary lipid supplements on the stability of electrically stimulated marmoset hearts. Adult male marmosets (n = 8 animals/dietary group) were fed dietary supplements as given in figure 1 for 3 years. % ventricular fibrillation is incidence of animals in each dietary group displaying prolonged VF. * VF in SSO group is significantly less ($p < 0.05$) than that in the SF group. ** VF in the FO group is significantly less ($p < 0.01$) than in all other dietary groups.

tion fraction, increased end diastolic volume and a reduced heart and pressure rate index [36]. It is, therefore, reasonable to conclude that these changes in cardiac function are of great benefit to this small primate as all these parameters of cardiac performance are associated with a reduced risk of cardiac dysfunction in man [32, 33].

A more direct means of evaluation of the vulnerability of the heart to develop arrhythmia can be determined by the clinical test of programmed electrical stimulation [37]. In this mildly invasive procedure, the heart is subjected to a step-wise increase in electrical stimulation until such time as the heart develops a clear pattern of ventricular fibrillation. The level of stimulation that is required to achieve this is expressed as the ventricular fibrillation threshold (VFT), which, like the incidence of sustained VF at a predetermined level of stimulation, is a useful clinical index of the stability of the heart. The less the stability, the greater the risk of death [38, 39]. In a series of recent studies which have just been concluded in our laboratory, we have been able to show that under normoxic conditions, both ω6 and ω3 PUFA-supplemented diets were effective in reducing the vulnerability (VFT) of the heart of the marmoset to develop arrhythmia. This effect was even more marked when the marmosets' hearts were rendered partially ischemic by ligation of a coronary artery [40]. In addition, in mature marmosets (4 years old), the incidence of sustained, and hence clinically significant ventricular fibrillation in saturated fat-fed animals was 45%. This was reduced to only 13% in those animals which had received an ω6 PUFA-rich vegetable oil dietary supplement, but was most dramatically reduced to nil in those animals which had been fed a diet containing 6% (by weight) fish oil supplement for nearly 3 years (fig. 5). To our knowledge, this is the first demonstration that dietary ω3 PUFA supplements can abolish cardiac arrhythmias in a non-human primate model of ventricular fibrillation [40].

Mechanisms of Arrhythmia

Since the pioneering electrophysiological studies of Hoffman more than 20 years ago [41] it has been thought that dysrhythmias result from abnormalities of impulse initiation or impulse conduction or both. That this view still prevails is apparent from two recent reviews of SCD which focus upon the electrophysiological properties of the ischemic myocardium [9, 10]. However, considerable experimental evidence in rats and dogs, with much less in man, has implicated the local eicosanoid hormones thromboxane and prostacyclin in arrhythmia [42–44]. There is a widespread belief that thromboxane (TXA_2) is strongly arrhythmogenic, while physiological concentrations of prostacyclin (PGI_2) may be anti-arrhythmic, and it is the 'balance' between these two opposing actions which determines the extent of arrhythmia which can occur. Certainly under experimental conditions, agents which either reduce the release of thromboxane or promote the local production of prostacyclin during ischemia or reperfusion greatly reduce the severity of arrhythmia in anaesthetised dogs subjected to occlusion of their coronary artery [43, 45]. Nevertheless, the precise cellular mechanism(s) by which these eicosanoid-mediated effects occur have not yet been identified; doubtless the interaction of eicosanoids with mechanisms of impulse initiation or conduction will involve as yet not defined processes in the myocardial membranes. Certainly it would seem logical to suppose that these will involve the influx/efflux of Ca^{2+} ions into or out of the myocardial cell [21].

In collaboration with Fischer and Schweer in Germany, some recent work in our laboratory has demonstrated that like man exposed to a fish or fish oil supplement [46], the marmoset monkey also excretes increased amounts of the urinary metabolites of PGI_2 and PGI_3 [47], which are accompanied by a reduction in the urinary excretion of thromboxane metabolites. These results greatly increase our confidence that the marmoset monkey is a suitable animal model for the study of dietary manipulation of eicosanoid metabolism. We have therefore studied the effects of various dietary lipids upon the biosynthesis of eicosanoids by the isolated myocardium of the marmoset after 24 months feeding either saturated fat, sunflower seed oil or fish oil supplements (6% w/w). While both PUFA diets led to significant reduction in basal myocardial PGI_2 and TXA_2 production in comparison to that found in the hearts of saturated fat-fed animals, the reduction from feeding a fish oil supplement was by far the greatest effect. PGI_2 was reduced by 48% and TXA_2 reduced by 82% – thus shifting the 'balance' of myocardial eicosanoids greatly in favour of the antiarrhythmic effects of PGI_2 yet maintaining the absolute level of eicosanoids at *physiological* rather than pharmacological levels [48, 49]. Concurrent examination of both myocardial phospholipid (membrane) and non-esterified fatty acid (cytosolic) fatty acid composition is in progress and may help to explain the relationship between myocardial membrane composition, eicosanoid release and cardiac arrhythmia [50]. Preliminary findings at this stage suggest that not only is myocardial substrate availability for eicosanoid biosynthesis greatly reduced by feeding fish oil dietary supplements, but that a specific inhibition of thromboxane synthetase may also be occurring in the presence of cellular ω3 PUFAs [49, 51].

Conclusions

These experimental findings from two different animal models and at various levels of myocardial complexity must lend strong support to the numerous epidemiological studies which suggest that a decline in mortality from ischemic heart disease is associated with an increased dietary consumption of fish or fish oil [14, 16, 17, 52]. It seems to us that this effect is more likely to be associated with a direct action of ω3 PUFAs upon the myocardium, than either a coincidental reduction in dietary saturated fatty acids or the addition of some other as yet unidentified component(s) of fish protein [53, 54]. The retrospective finding of the sudden and greatly reduced incidence of death from 'heart attacks' in Norway during the Second World War (1940–1945), which were associated with increased fish consumption, but which rapidly returned to pre-war levels when the imposed dietary restrictions were removed [15, 16] also suggests that dietary interventions can be expected to induce quite rapid changes in cardiac function in free living populations. Apparently extensive clinical trials to test this hypothesis are now in progress [17, 55]. However, it is also apparent that a much better knowledge of the cellular mechanisms which underlie these beneficial dietary-induced changes in cardiac function will also be necessary, as it is only from such an understanding of these mechanisms that appropriate clinical strategies can evolve [56].

Acknowledgements

I wish to express my sincere thanks to my colleagues Dr. Peter McLennan and Dr. Mahinda Abeywardena for their invaluable contribution to this work over many years. Mrs. Thelma Bridle provided outstanding technical assistance and much additional effort during the latter and most critical stages of the project. Mr. Jim Cooper's dedication to the care and welfare of the animals was of the highest standard possible and was invaluable. My thanks are also due to Dr. Leighton Barnden and Dr. Ian Buttfield and the staff of the Department of Nuclear Medicine at the Queen Elizabeth Hospital for their major contribution to the radionuclide angiographic studies reported here. I would also like to thank Dr. B.S. Hetzel, former Chief of the Division of Human Nutrition and Dr. F. de Zwart, Director of the Department of Nuclear Medicine for their encouragement and interest in this study from the outset. The early phase of the study was assisted in part by a grant-in-aid (FIRTA 66/83) from the Department of Primary Industry, Government of Australia.

References

1 Hetzel BS, Charnock JS, Dwyer T, et al: Fall in coronary heart disease mortality in U.S.A. and Australia due to sudden death: Evidence for the role of polyunsaturated fat. J Clin Epidemiol 1989;42:885–893.
2 Lown B: Sudden cardiac death: The major challenge confronting contemporary cardiology. Am J Cardiol 1979;43:313–328.
3 Eisenberg MS, Bergner L, Hallstrom AP, et al: Sudden cardiac death. Sci Am 1986; 254:25–31.
4 American Heart Association: Heart Facts 1986. Publ Am Heart Assoc, Dallas, 1986.
5 Leaf A, Weber PC: Cardiovascular effects of n–3 fatty acids. New Engl J Med 1988; 318:549–557.
6 Beaglehole R, Bonita R, Stewart A: Cardiovascular disease mortality trends in the Western Pacific. New Zealand Med J 1988;101:441–443.

7 Anderson JL: Sudden cardiac death, ventricular arrhythmias and anti-arrhythmic therapy. Aust NZ J Med 1986;16:409–412.
8 Myerburg RJ: Epidemiology of ventricular tachycardia/ventricular fibrillation and sudden cardiac death. PACE 1986;9:1334–1338.
9 Keefe DL, Schwartz J, Somberg JC: The substrate and the trigger: The role of myocardial vulnerability in sudden cardiac death. Am Heart J 1987;113:218–225.
10 Myerburg RJ, Kessler KM, Bassett AL, et al: A biological approach to sudden cardiac death: Structure, function and cause. Am J Cardiol 1989;63:1512–1516.
11 Shaper AG: Coronary heart disease risks and reasons. London, Current Medical Literature, 1989, pp 49–51.
12 Hetzel BS, McMichael T: The life style factor: Lifestyle and health. Penguin Books, Melbourne, Aust, 1987, pp 37–70.
13 Dyerberg J, Bang HO: Dietary fat and thrombosis. Lancet 1978;i:152–153.
14 Kromhout D, Bosschieter EB, Coulander CL: The inverse relationship between fish consumption and 20 year mortality from coronary heart disease. N Engl J Med 1985;312:1205–1209.
15 Bang HO, Dyerberg J: Personal reflections on the incidence of ischemic heart disease in Oslo during the second world war. Acta Med Scand 1981;210:245–248.
16 Shaper AG: National trends in mortality from ischemic heart disease: Implications for prevention. Lancet 1986;i:795.
17 Burr ML, Gilbert JF, Holliday RM, et al: Effects of changes in fat, fish and fibre intakes on death and myocardial reinfarction: Diet and reinfarction trial (DART). Lancet 1989;ii:757–761.
18 Charnock JS, McLennan PL, Abeywardena MY, et al: Diet and cardiac arrhythmia: Effects of lipids on age-related changes in myocardial function in the rat. Ann Nutr Metab 1985;29:306–318.
19 Charnock JS: Dietary fats and cardiac function. Proc Nutr Soc Aust 1985;10:25–34.
20 McLennan PL, Abeywardena MY, Charnock JS: A comparison of the long term effects of n–3 and n–6 PUFA dietary supplements and the action of indomethacin upon the mechanical performance and susceptibility of the rat heart to dysrhythmia. Prostaglandins Leukotrienes Med 1987;27:183–195.
21 McLennan PL, Abeywardena MY, Charnock JS, et al: Dietary lipid modulation of myocardial β-adrenergic mechanisms, Ca^{2+}-dependent automaticity and arrhythmogenesis in the marmoset. J Cardiovasc Pharmacol 1987;10:293–300.
22 Turner J, McLennan PL, Abeywardena MY, Charnock JS: Histopathological sequalae of dietary fats in aorta and coronary artery of the rat; Relationship to eicosanoids and arrhythmia. J Mol Cell Cardiol 1987;19(suppl II):Abstract IX.
23 Turner J, McLennan PL, Abeywardena MY, Charnock JS: Absence of coronary or aortic atherosclerosis in rats having dietary lipid modified vulnerability to cardiac arrhythmias. Atherosclerosis 1990;82:105–112.
24 Selye H, Bajusz E, Grasso S, et al: Simple techniques for surgical occlusion of coronary vessels in the rat. Angiology 1960;11:398–407.
25 Curtis MJ, MacLeod BA, Walker MJA: Models for the study of arrhythmias in myocardial ischemia and infarction: The use of the rat. J Mol Cell Cardiol 1987;19:339–419.

26 McLennan PL, Abeywardena MY, Charnock JS: The influence of age and dietary fat in an animal model of sudden cardiac death. Aust NZ J Med 1989;19:1–5.
27 McLennan PL, Abeywardena MY, Charnock JS: Influence of dietary lipids on arrhythmias and infarction after coronary artery ligation in rats. Can J Physiol Pharmacol 1985;63:1411–1417.
28 McLennan PL, Abeywardena MY, Charnock JS: Dietary fish oil prevents ventricular fibrillation following coronary artery occlusion and reperfusion. Am Heart J 1988;116:709–717.
29 McLennan PL, Abeywardena MY, Charnock JS: Reversal of the arrhythmogenic effects of long-term saturated fatty acid intake by dietary n–3 and n–6 polyunsaturated fatty acids. Am J Clin Nutr 1990;51:53–58.
30 Abbey M, Clifton P, McMurchie EJ, et al: The effect of a high fat/cholesterol diet with or without EPA on plasma lipids, lipoproteins and lipid transfer protein activity in the marmoset. Atherosclerosis 1990;81:163–174.
31 McMurchie EJ, Patten GS, McLennan PL, et al: A comparison of the properties of the cardiac beta-adrenergic receptor adenyl cyclase system in the rat and the marmoset monkey. Comp Biochem Physiol 1987;88B:989–998.
32 Moss AJ, Bigger JT Jr, Case RB, et al: Multicentre postinfarction research group. Risk stratification and survival after myocardial infarction. N Engl J Med 1983;309:331–334.
33 Lesch M, Kehoe RF: Predictability of sudden cardiac death. N Engl J Med 1984;310:255–257.
34 McIntosh GH, Barnden LH, Buttfield IH, et al: Gated blood-pool studies of cardiac function in the rat and marmoset. J Nucl Med 1983;24:728–732.
35 Charnock JS, McLennan PL, McIntosh GH, et al: A radionuclide angiographic study of the influence of dietary lipid supplements upon cardiac function in the marmoset monkey. Cardiovasc Res 1987;21:369–376.
36 McLennan PL, Barnden LR, Charnock JS: Dietary polyunsaturated fatty acids improve cardiac function in the marmoset monkey. J Mol Cell Cardiol 1989;21(suppl II):p 32, Abstract 96.
37 Bigger JT Jr, Reiffel JA, Livelli FD Jr: Sensitivity, specificity, and reproducibility of programmed ventricular stimulation. Circulation 1986;73(suppl II):1173–1179.
38 Swerdlow CD, Winkle RA, Mason JW: Prognostic significance of the number of induced ventricular complexes during assessment of therapy for ventricular tachyarrhythmias. Circulation 1983;68:400–405.
39 Richards DA, Cody DV, Denniss AR, et al: Ventricular electrical instability: A predictor of death after myocardial infarction. Am J Cardiol 1983;51:75–80.
40 McLennan PL, Bridle TM, Abeywardena MY, Charnock JS: Dietary lipid modification of ventricular fibrillation threshold in the marmoset monkey. J Mol Cell Cardiol 1990 (in press).
41 Hoffman BF: The genesis of cardiac arrhythmias. Prog Cardiovasc Dis 1966;8:319–329.
42 Lepran I, Nemecz Gy, Koltai M, et al: Effect of a linoleic acid-rich diet on the acute phase of coronary occlusion in conscious rats: Influence of indomethacin and aspirin. J Cardiovasc Pharmacol 1981;3:847–853.
43 Coker SJ, Parratt JR: Prostacyclin – antiarrhytmic or arrhythmogenic? Comparison of the effects of intravenous and intracoronary prostacyclin and ZK36374 during

coronary artery occlusion and reperfusion in anaesthetised greyhounds. J Cardiovasc Pharmacol 1983;5:557–567.

44 Nowak J, Murray JJ, Fitzgerald GA, et al: Eicosanoids and sudden cardiac death. Advances Prostaglandin, Thromboxane Leukotriene Res 1987;17:20–24.

45 Parratt JR, Coker SJ, Wainwright CL: Eicosanoids and susceptibility to ventricular arrhythmias during myocardial ischaemia and reperfusion. J Mol Cell Cardiol 1987; 5(suppl):55–66.

46 Fischer S, Weber PC: Prostaglandin I_3 is formed in vivo in man after dietary eicosapentaenoic acid. Nature 1984;307:165–168.

47 Abeywardena MY, Fischer S, Schweer H, et al: In vivo formation of metabolites of prostaglandins I_2 and I_3 in the marmoset monkey following dietary supplementation with tuna fish oil. Biochim Biophys Acta 1989;1003:161–166.

48 Abeywardena MY, McLennan PL, Charnock JS: Dietary fat induced changes in cardiac phospholipids and eicosanoids in the primate. Proc Taipei Conf Prostaglandin and Leukotrine Research, 1988, Abstract P48, p 154.

49 Abeywardena MY, McLennan PL, Charnock JS: Differential effects of dietary fish oil on myocardial PGI_2 and TXA_2 production. Am J Physiol 1990 (in press).

50 Abeywardena MY, McLennan PL, Charnock JS: Replacement of endogenous fatty acids of marmoset monkey myocardium by dietary fish or sunflower seed oil supplements. J Mol Cell Cardiol 1990 (in press).

51 Abeywardena MY, McLennan PL, Charnock JS: Selective inhibition of thromboxane A_2 by dietary fish oil. J Mol Cell Cardiol 1990 (in press).

52 Nelson AM: Diet therapy in coronary disease: Effect on mortality of a high protein, high seafood, fat controlled diet. Geriatrics 1972;27:103–116.

53 Goldstein MR: Fish and the heart: Lancet 1989;ii:1450.

54 McLennan PL, Charnock JS, Sundram K, et al: Fish and the heart. Lancet 1989;ii: 1451.

55 Burr ML, Fehily AM: Fish and the heart. Lancet 1989;ii:1451–1452.

56 Charnock JS, Abeywardena MY, McLennan PL: Dietary lipids, membrane composition and cardiac function. Coll INSERM 1989;195:135–144.

John S. Charnock, PhD, DSc, CSIRO Division of Human Nutrition,
Glenthorne Laboratory, Majors Road, O'Halloran Hill, SA 5158 (Australia)

Simopoulos AP, Kifer RR, Martin RE, Barlow SM (eds): Health Effects of ω3 Polyunsaturated Fatty Acids in Seafoods. World Rev Nutr Diet. Basel, Karger, 1991, vol 66, pp 292–305

Balance between ω3 and ω6 Fatty Acids in Heart Muscle in Relation to Diet, Stress and Ageing

Sigmundur Gudbjarnason, V. Edda Benediktsdóttir, Elín Gudmundsdóttir

Science Institute, University of Iceland, Reykjavík, Iceland

Introduction

Epidemiological studies have suggested that the incidence of death from ischemic heart disease may be inversely related to dietary fish intake or ω3 fatty acid consumption [1]. The mechanism of the beneficial effects of ω3 fatty acids is uncertain, and there may be several reasonable mechanisms. The ω3 fatty acids may, for example, influence the development of atherosclerosis [2, 3], they may modify the types and amounts of eicosanoids produced in platelets, neutrophils and endothelial cells and thereby influence blood clotting and thrombosis [4] and the ω3 fatty acids can also modify properties of cell membranes [5, 6].

The role of docosahexaenoic acid (DHA) in neural development and retina has extended the interest in ω3 fatty acids beyond circulating cells, endothelial cells and the liver to biophysical effects in specific membranes [5].

In this paper we examine cardiovascular disease mortality in Nordic countries in relation to fish consumption and in relation to arachidonic acid (AA) levels in plasma phospholipids of the middle aged population in these countries. In experimental studies we examine alterations in the levels of ω3 and ω6 fatty acids in phospholipids of rat heart muscle in relation to various forms of stress, ageing and composition of dietary fat. The influence of dietary ω3 or ω6 fatty acids upon properties of sarcolemmal β_1-receptors and slow calcium channels will also be described and discussed in relation to the incidence of ventricular fibrillation and sudden cardiac death. The results emphasize the importance of a balance between ω3 and ω6 fatty acids in cardiac sarcolemma.

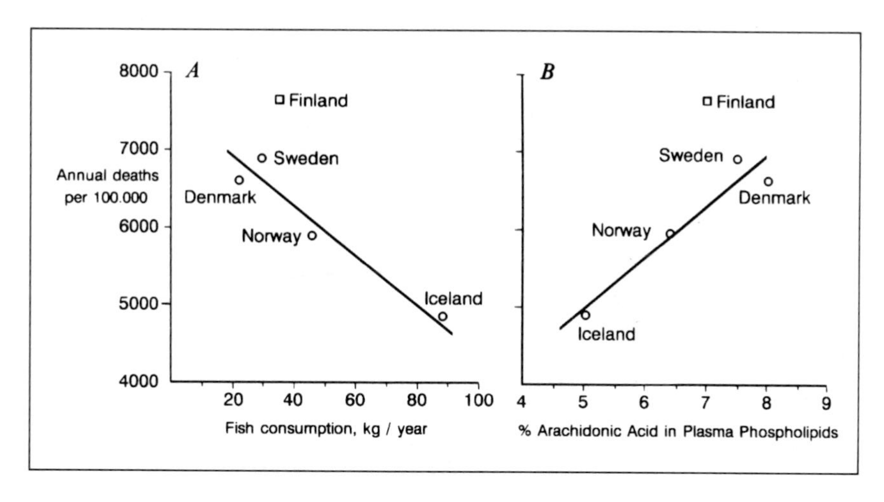

Fig. 1. Death rates from cardiovascular disease at old age (≥ 75 years) in relation to fish consumption (*A*) and the arachidonic acid level in plasma phospholipids. (*B*) in the Nordic countries.

Fish Consumption and Cardiovascular Diseases

Cardiovascular disease mortality in Iceland is lower in the older age groups (≥ 75 years) or the population at risk than in other Nordic countries [7]. Figure 1A shows the cardiovascular disease mortality in Nordic countries in relation to fish consumption [7, 8]. At the age of 75 years or above, the cardiovascular disease mortality decreases with increasing fish consumption. This inverse relationship seems to be a function of both age and fish consumption since this relationship was not observed for the younger age groups. It is also of interest that the Finnish population does not follow this relationship which seems to fit only the Nordic populations with the same genetic background.

Figure 1B shows that the cardiovascular disease mortality in Nordic countries in the older age groups (≥ 75 years) correlates with the level of AA in plasma phospholipids of the normal, healthy population at the age of 40–60 years, i.e. before development of cardiovascular diseases (9–14). Cardiovascular disease mortality seems to be a function of both age and AA levels in plasma phospholipids. These figures suggest that fish consumption and plasma levels of AA have opposite effects on the development of atherosclerosis and cardiovascular disease in these populations.

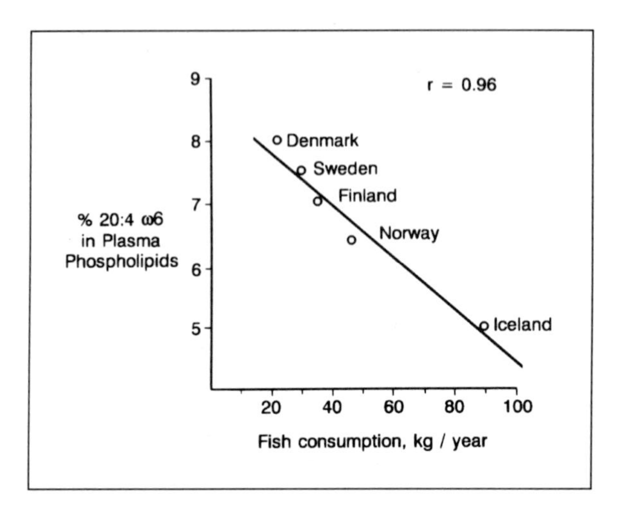

Fig. 2. Relationship between arachidonic acid levels in plasma phospholipids (age 40–60 years) and per capita fish consumption in Nordic countries.

Figure 2 shows a linear relationship between AA levels of plasma phospholipids in the normal middle aged populations and the overall fish consumption (kg/man/year) in these countries. This relationship illustrates the influence of fish consumption upon the availability of AA in plasma and thus in the circulatory system.

Dietary intake of ω3 fatty acids in fish or fish oils leads to a significant incorporation of eicosapentaenoic acid, 20:5 ω3 (EPA), and DHA 22:6 ω3, into cellular phospholipids. These ω3 fatty acids influence the AA or eicosanoid metabolism of different cells differently, influencing thereby numerous biochemical and pathophysiological processes.

The effects on platelets and leukocytes are well known. Increased availability of EPA decreases the AA level in platelets and decreases TXA_2 production. This should lead to reduced platelet aggregation and reduced risk of thrombosis. TXA_2-induced coronary constriction should be diminished, reducing the risk of coronary spasm and ischemia. Increased intake of EPA should decrease the conversion of AA to leukotrienes, thereby reducing coronary constriction and spasm and improve microcirculation by reducing blood viscosity.

The effects of ω3 fatty acids on heart muscle are less well understood. Increased availability of ω3 fatty acids alters the fatty acid composition of membrane phospholipids markedly, increasing the level of 22:6 ω3 and to some extent also 20:5 ω3, but decreasing both 20:4 ω6 and 18:2 ω6. Extensive replacement of ω6 fatty acids by ω3 fatty acids is being studied in relation to cellular functions. Membrane composition can be expected to influence various membrane properties and functions such as ion transport, receptor properties, stability and peroxidation of polyunsaturated fatty acids.

Dietary Fat and Moderate Stress

The influence of dietary fat upon the heart and adaptation to moderate stress induced by epinephrine was examined in rats fed diets containing 10% cod liver oil, butter or corn oil. At the end of a four-month feeding period the various dietary groups were examined before, during and one week after repeated administration of epinephrine [15].

Rats fed diets containing either 10% butter, corn oil or cod liver oil showed markedly different fatty acid composition of phospholipids in cardiac cell membranes such as sarcolemma [16]. Dietary cod liver oil lowered the AA level in phosphatidyl ethanolamine (PE) by 50% compared to butter- or corn oil-fed rats, replacing AA with DHA. Rats fed either butter or corn oil had remarkably similar levels of AA and DHA in heart muscle [15, 16].

Adaptation to moderate stress induced by repeated administration of epinephrine for 15 days resulted in marked but reversible alterations in the fatty acid profile of cardiac phospholipids [15].

Daily injections of epinephrine were well tolerated with no deaths in these groups of rats. The level of AA increased in phosphatidyl choline (PC) in all groups during epinephrine administration and returned to control levels one week after the epinephrine administration was stopped. The level of AA decreased or remained the same in PE during stress (fig. 3) [15].

The levels of linoleic acid (LA) decreased in all groups, both in PC and PE during stress returning to control one week after cessation of epinephrine administration [15]. The DHA levels in PC and PE were highest in rats fed cod liver oil. Epinephrine treatment induced reversible elevation in DHA levels of these phospholipids (fig. 4). The epinephrine-induced changes in PC and PE were specific for each phospholipid class and resembled the changes induced by norepinephrine or isoproterenol.

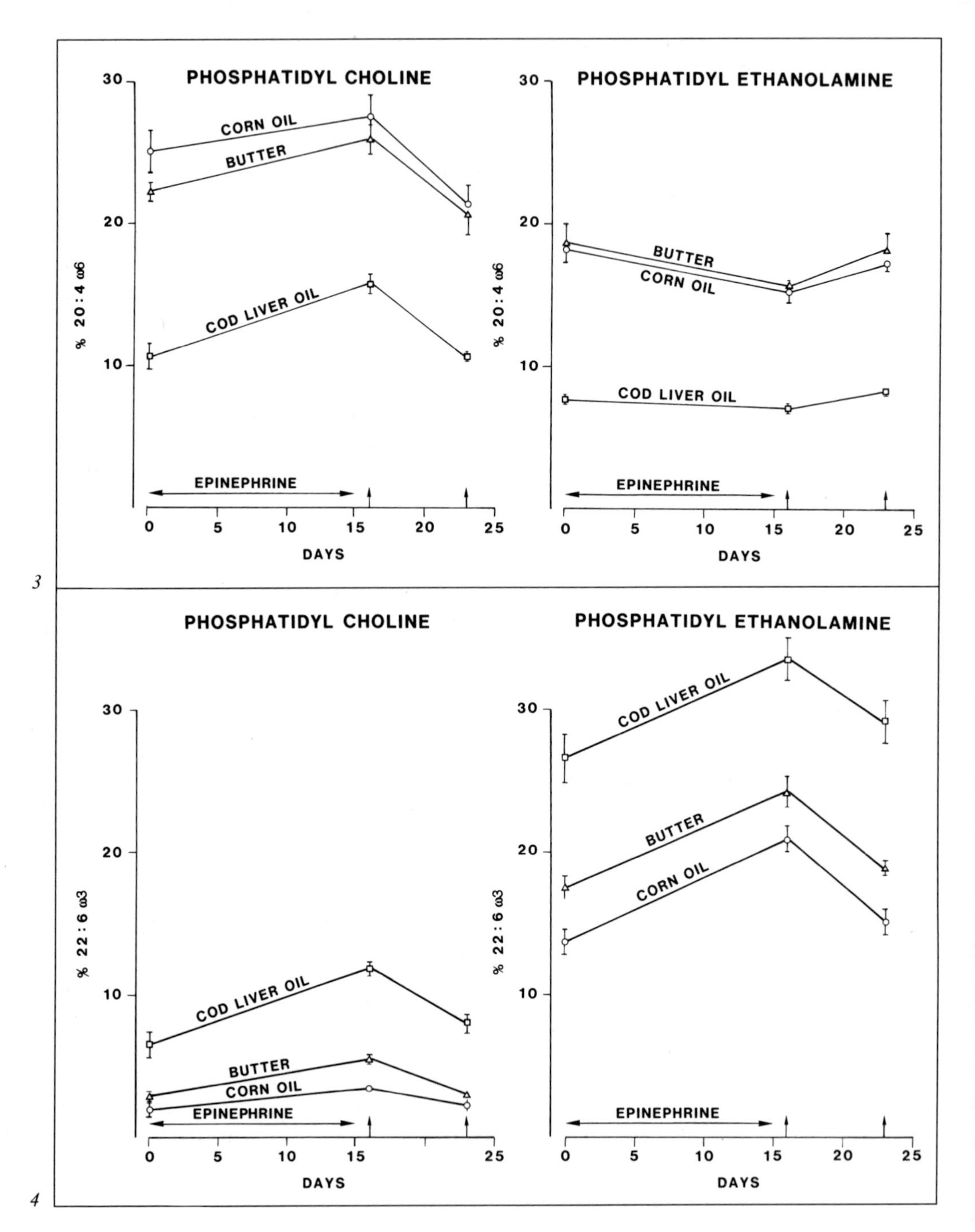

PHOSPHATIDYL CHOLINE
PHOSPHATIDYL ETHANOLAMINE
CORN OIL
BUTTER
COD LIVER OIL
EPINEPHRINE
% 20 : 4 ω6
DAYS
3
PHOSPHATIDYL CHOLINE
PHOSPHATIDYL ETHANOLAMINE
COD LIVER OIL
BUTTER
CORN OIL
EPINEPHRINE
% 22 : 6 ω3
DAYS
4

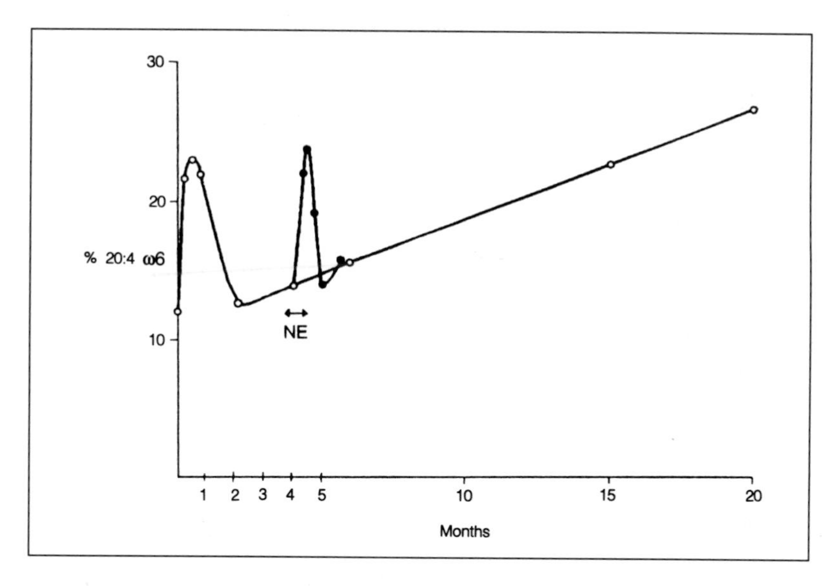

Fig. 5. Alterations in the level of arachidonic acid in phosphatidylcholine of rat heart muscle during neonatal development and ageing, and during and after repeated administration of norepinephrine (NE).

Cardiac Phospholipids during Neonatal Development, Ageing and Severe Stress

The fatty acid composition of individual phospholipids was examined in the rat heart during neonatal development [17] and ageing [18] and during and after repeated administration of norepinephrine, when rats were injected daily for 15 days with norepinephrine bitartrate [19].

Reversible alterations in the level of 20:4 ω6 in PC are shown in figure 5. In the neonatal heart the AA level increased rapidly during the first two weeks of life but after weaning the AA level decreased again. After the age of two months there was a gradual increase in AA during ageing. During repeated administration of norepinephrine the AA level increased

---

Fig. 3. Changes in arachidonic acid levels of phospholipids in rat heart muscle during and after repeated administration of epinephrine.

Fig. 4. Changes in docosahexaenoic acid levels of phospholipids in rat heart muscle during and after repeated administration of epinephrine.

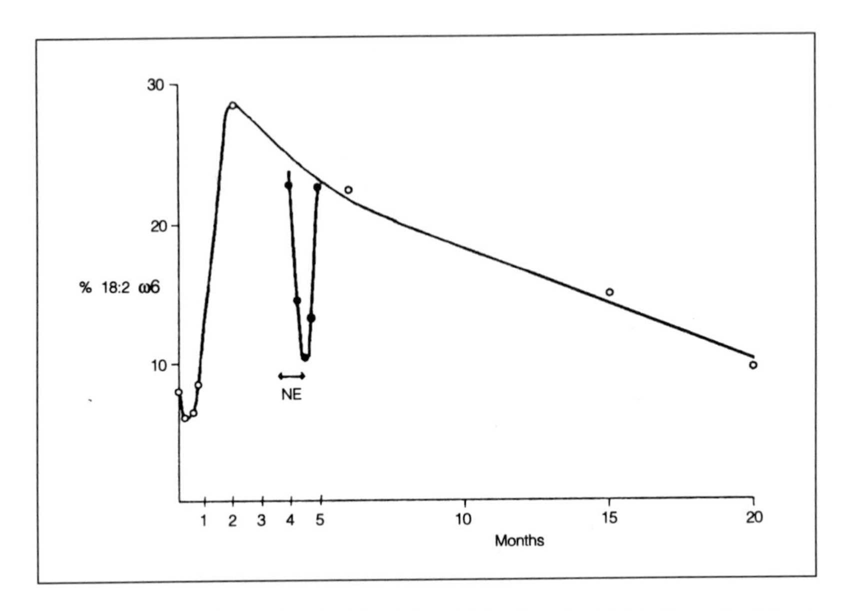

Fig. 6. Alterations in the level of linoleic acid in phosphatidylcholine of rat heart muscle during neonatal development and ageing, and during and after repeated administration of norepinephrine (NE).

markedly in PC but when the administration of norepinephrine was stopped the level of AA returned to normal.

Figure 6 shows the alterations in LA in PC. In the neonatal heart the level of LA was low and remained low during the first three weeks of life, after weaning the level of LA increased markedly. After the age of two months this trend was reversed and LA decreased gradually with age. During repeated aministration of norepinephrine the level of LA decreased rapidly, returning to normal after the norepinephrine administration was stopped.

During adaptation to neonatal stress and ageing and during norepinephrine administration there was an inverse relationship between 22:6 ω3 and 18:2 ω3 in cardiac PE (fig. 7). In the neonatal heart 22:6 ω3 increased markedly during the first three weeks of life but during this period 18:2 ω6 remained low. After weaning and up to the age of two months 18:2 ω6 increased steadily whereas 22:6 ω3 remained unaltered. After the age of two months the level of 22:6 ω3 increased, gradually replacing 18:2 ω6 in cardiac PE. During the administration of norepinephrine the level of 22:6 ω3 increased, markedly decreasing the level of 18:2 ω6.

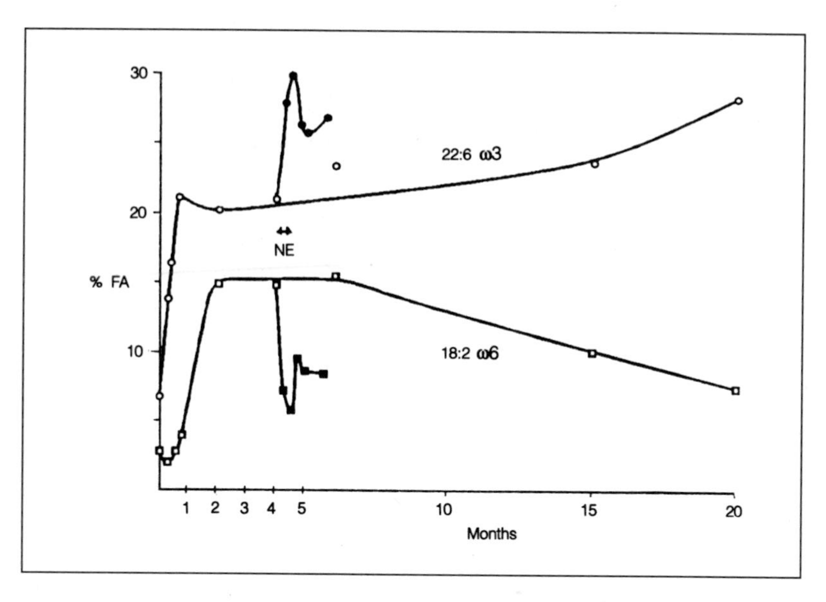

Fig. 7. Alterations in the levels of docosahexaenoic acid and linoleic acid in phosphatidylethanolamine of rat heart muscle during neonatal development and ageing, and during and after repeated administration of norepinephrine (NE).

The results show marked and characteristic changes in the fatty acyl chain composition of cardiac membrane phospholipids during adaptation to various forms of stress. The modification of membrane lipids may be in response to the demands imposed upon the heart during stress with accompanying changes in cardiac function and metabolism. The differences in the response of PC and PE to stress are also noteworthy. The norepinephrine-induced changes in PC are completely reversible, whereas the changes in PE are only partly reversible. The consequences of the stress episode are still observed 5 weeks after cessation of norepinephrine administration in cardiac PE.

Dietary Fat and β_1-Receptors in Cardiac Sarcolemma

The properties of membrane-bound receptors, i.e. β_1-adrenergic receptors and slow calcium channels, were examined in heart muscle in relation to dietary fat and ageing. Both β_1-adrenergic receptors and slow cal-

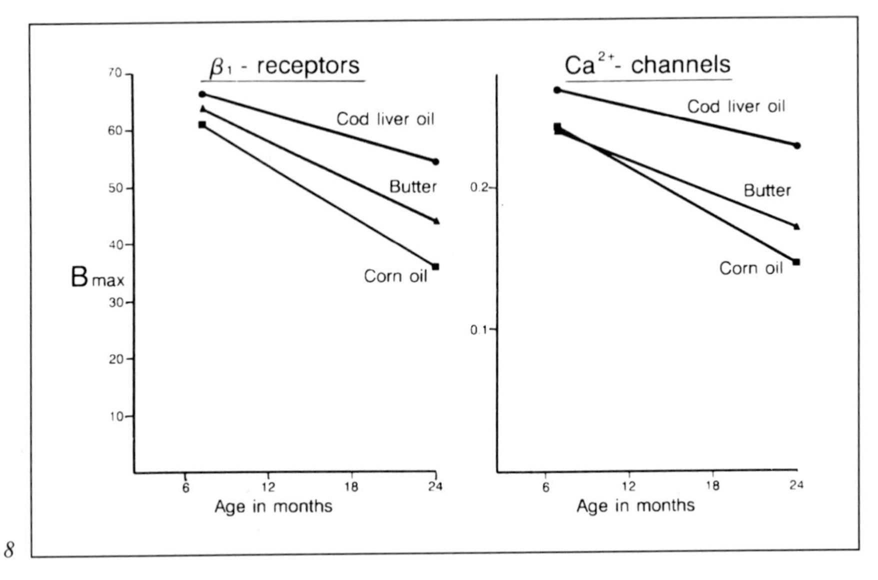

8

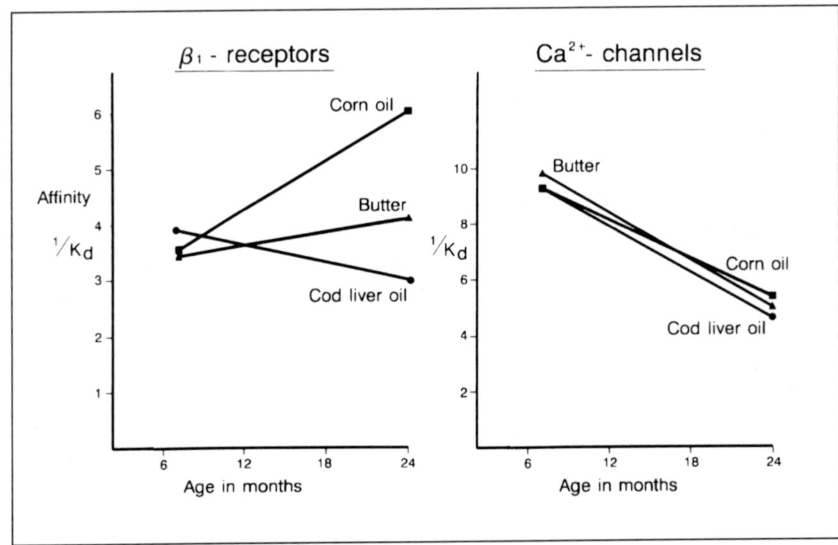

9

Fig. 8. Number or density of binding sites of sarcolemmal β_1-receptors and slow Ca^{2+} channels in relation to age and dietary fat.

Fig. 9. Affinity of β_1-receptors and slow Ca^{2+} channels in sarcolemma in relation to age and dietary fat.

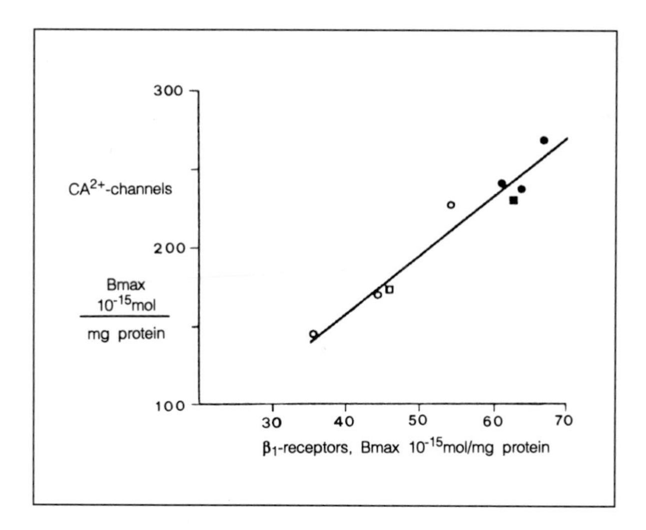

Fig. 10. Relationship between the number or density of binding sites of β_1-receptors and slow Ca^{2+} channels in sarcolemma of rat heart muscle.

cium channels play an important role in the contractile cycle of the mammalian heart. β_1-receptors and calcium channels were studied in sarcolemma of rat ventricular muscle. Two month old rats were fed for 5 or 22 months diets containing either 10% cod liver oil, butter or corn oil as described above. At the age of 7 or 24 months the density and affinity of β_1-adrenergic receptors and slow calcium channels were determined in a sarcolemmal preparation with ^{3}H-dihydroalprenolol and ^{3}H-nitrendipine, respectively.

Density of both β_1-receptors and calcium channels decreased significantly in corn oil-fed rats from the age of 7 to 24 months, to a lesser extent in rats fed butter and even less in rats fed cod liver oil (fig. 8). Affinity of calcium channels for ^{3}H-nitrendipine decreased in all dietary groups with ageing. Affinity of β_1-receptors for ^{3}H-dihydroalprenolol decreased with ageing in rats fed cod liver oil, remained unchanged in rats fed butter but increased markedly in rats fed corn oil (fig. 9) (unpublished data).

The number of sarcolemmal β_1-receptors and calcium channels diminished in the ageing rat dependent upon the composition of dietary fat. The affinity of β_1-receptors was sensitive to the fatty acid composition of dietary fats in the ageing rat. The K_d of ^{3}H-dihydroalprenolol binding increased significantly with an increasing level of DHA in both PC and PE of sarcolemma. The affinity of β_1-receptors decreased thus with increasing levels of DHA in membrane phospholipids of the ageing rat.

The relationship between the density of binding sites of β_1-receptors and slow calcium channels in cardiac sarcolemma illustrates the coregulation of β_1-receptors and calcium channels (fig. 10). Despite considerable variation in density of binding sites depending on age and dietary fat the stoichiometry remains the same, i.e. there are four calcium channels for each β_1-receptor (fig. 10).

Dietary Fat and Ventricular Fibrillation

The functional significance of changes in the level of ω3 and ω6 fatty acids of membrane phospholipids during ageing, stress, or induced by dietary fat may relate to the vulnerability to stress and sudden cardiac death. The importance of these polyunsaturated fatty acids in stress tolerance was examined using isoproterenol as a stressor (which frequently causes ventricular fibrillation and sudden death) in adult rats weighing about 500 g. Isoproterenol, 1 mg/kg body weight injected subcutaneously, induces ventricular fibrillation in adult, heavy rats but not in light rats. Dietary fat and oils were used to modify the fatty acid composition of phospholipids in heart muscle as described above. When these rats were subjected to administration of isoproterenol, the incidence of fatal ventricular fibrillation was 22% in rats fed 10% cod liver oil, 33% in rats fed 10% butter and 54% in rats fed diets containing 10% corn oil. The lowest incidence of ventricular fibrillation was observed in rats fed cod liver oil, with the highest level of DHA and the lowest level of AA in sarcolemmal PC and PE, i.e. the lowest ratio of 20:4 ω6/22:6 ω3. The highest mortality was observed in rats fed corn oil, with the lowest level of DHA and the highest level of AA, i.e. the highest ratio of 20:4 ω6/22:6 ω3 in sarcolemmal phospholipids [6]. Dietary intake of ω3 fatty acids in fish or fish oil such as cod liver oil reduces the ratio of 20:4 ω6/22:6 ω3 and thereby the functional availability of AA, 20:4 ω6.

The level of DHA in phospholipids of mammalian hearts relates in some unknown way to heart rate. There is a semilogarithmic relationship between DHA and heart rate in various species. The whale, with the lowest heart rate (8–10 beats/min), has the lowest level of DHA, about 7% in cardiac PE, whereas the mouse has the highest level of DHA (over 40%) in cardiac PE and the highest heart rate [6].

Balance between ω6 and ω3 Fatty Acids and Sudden Death in Man

Studies on human cardiac autopsy samples also suggest that changes in tissue levels of ω3 and ω6 fatty acids may be associated with sudden cardiac death. Heart muscle samples from ventricular tissue, were obtained from accident victims and if these hearts were without any lesions or abnormalities on macroscopic or microscopic examination, they were considered normal. Autopsy samples were also obtained from heart muscle of men who died a sudden cardiac death. Sudden death is defined as instant death or death within 1 h, usually within minutes of symptoms [20].

Analysis of human autopsy material have to be viewed with caution because of autolytic changes that take place from the time of death until the time of sampling. The autolytic changes in heart muscle are relatively slow compared to those in many other tissues, such as the liver.

The ratio of 20:4 ω6/22:6 ω3 in cardiac phospholipids of autopsy samples from people who died suddenly in accidents changes as a function of age. With advancing age there is an increase in DHA and this ratio decreases with age in the normal heart. The ratio 20:4 ω6/22:6 ω3 was frequently found to be markedly higher in men who died suddenly at home or at their place of work than was expected for that age group [6, 20].

The studies on the rat heart and human heart suggest that the balance between AA and DHA in cardiac phospholipids may play an important role in development of ventricular fibrillation and sudden death. Similar alterations in ω6 and ω3 fatty acids take place in endothelial cells and various blood cells, thereby affecting metabolism of AA and formation of the eicosanoids.

Conclusion

The epidemiological data presented in this paper indicate that increasing fish consumption reduces plasma levels of AA and subsequently also cardiovascular disease mortality in ageing Nordic populations.

The experimental studies illustrate that the levels of ω6 and ω3 fatty acids in cardiac membranes are rapidly modified by stress or dietary fat. The balance between AA and DHA modulates the number and affinity of β_1-adrenergic receptors in ageing sarcolemma and also the incidence of ventricular fibrillation and sudden death.

References

1 Kromhout D: n–3 fatty acids and coronary heart disease: Epidemiology from Eskimos to Western populations. J Int Med 1989;225(suppl 1):47–51.

2 Weber PC: Clinical studies on the effects of n–3 fatty acids on cells and eicosanoids in the cardiovascular system. J Int Med 1989;225(suppl 1):61–68.

3 Hirai A, Terano T, Tamura Y, Yoshida S: Eicosapentaenoic acid and adult diseases in Japan: Epidemiological and clinical aspects. J Int Med 1989;225(suppl 1):69–75.

4 Kristensen SD, Schmidt EB, Dyerberg I: Dietary supplementation with n–3 polyunsaturated fatty acids and human platelet function: A review with particular emphasis on implications for cardiovascular diseases. J Int Med 1989;225(suppl 1):141–150.

5 Neuringer M, Connor WE: Omega–3 fatty acids in the retina; in Galli C, Simopoulos AP (eds): Dietary ω3 and ω6 Fatty Acids. Biological Effects and Nutritional Essentiality. New York, Plenum Press, 1989, pp 177–190.

6 Gudbjarnason S: Dynamics of n–3 and n–6 fatty acids in phospholipids of heart muscle. J Int Med 1989;225(suppl 1):117–128.

7 Yearbook of Nordic Statistics, vol 25, 1986.

8 FAO Fishery Statistics, Rome 1988.

9 Kirkeby K, Ingvaldsen P, Bjerkedal I: Fatty acid composition of serum lipids in men with myocardial infarction. Acta Med Scand 1972;192:513–519.

10 Dyerberg I, Bang HO, Hjörne N: Fatty acid composition of the plasma lipids in Greenland Eskimos. Am J Clin Nutr 1975;28:958–966.

11 Mietinen TA, Naukkarinen V, Huttunon IK, Mattila S, Kumlin T: Fatty acid composition of serum lipids predicts myocardial infarction. Brit Med J 1982;285:993–996.

12 Doughorty RM, Galli BAC, Ferro-Luzzi A, Iacono IM: Lipid and phospholipid fatty acid composition of plasma, red blood cells, and platelets and how they are affected by dietary lipids: A study of normal subjects from Italy, Finland and the USA. Am J Clin Nutr 1987;45:443–455.

13 Boberg M, Croon LB, Gustafsson IB, Vessby B: Platelet fatty acid composition in relation to fatty acid composition in plasma and to serum lipoprotein lipids in healthy subjects with special reference to the linoleic acid pathway. Clin Science 1985;68:581–587.

14 Skúladóttir G, Hardarson Th, Sigfússon N, Oddsson G, Gudbjarnason S: Arachidonic acid levels in serum phospholipids of patients with angina pectoris or fatal myocardial infarction. Acta Med Scand 1985;218:55–58.

15 Benediktsdóttir VE, Gudbjarnason S: Reversible alterations in fatty acid composition of heart muscle membrane phospholipids induced by epinephrine in rats fed different fats. J Lipid Res 1988;29:765–772.

16 Benediktsdóttir VE, Gudbjarnason S: Modification of the fatty acid composition of rat heart sarcolemma with dietary cod liver oil, corn oil or butter. J Mol Cell Cardiol 1988;20:141–147.

17 Gudmundsdóttir A, Gudbjarnason S: Neonatal changes in fatty acid profile of phospholipids in rat heart muscle. Biochim Biophys Acta 1983;752:284–290.

18 Gudbjarnason S, Benediktsdóttir VE: Fatty acid composition of phospholipids in heart muscle modified by age, diet and stress, in Miguel I, Quantanilha AT, Weber H (eds): CRC Handbook of Free Radicals and Antioxidants in Medicine. Boca Raton, CRC Press, 1989; vol III, pp 67–74.
19 Emilsson A, Gudbjarnason S: Reversible alterations in fatty acid profile of glycerophospholipids in rat heart muscle induced by repeated norepinephrine administration. Biochim Biophys Acta 1983;730:1–6.
20 Gudbjarnason S, Hallgrímsson I, Skúladóttir G: Myocardial lipids in relation to coronary artery disease in man; in Peters H, Gresham GA, Paoletti R (eds): Arterial Pollution. New York, Plenum Press, 1983, pp 101–114.

Sigmundur Gudbjarnason, MD, Science Institute, University of Iceland, Dunhaga 3, Reykjavík 107 (Iceland)

Simopoulos AP, Kifer RR, Martin RE, Barlow SM (eds): Health Effects of ω3 Polyunsaturated Fatty Acids in Seafoods. World Rev Nutr Diet. Basel, Karger, 1991, vol 66, pp 306–312

Fatty Fish and Heart Disease: A Randomized Controlled Trial

Michael L. Burr, Ann M. Fehily

MRC Epidemiology Unit, Cardiff, UK

Introduction

The possibility that fish confers protection against ischaemic heart disease (IHD) first arose out of observations on Greenland Eskimos [1]. It was noted that they rarely suffer from IHD and their serum cholesterol and (especially) triglyceride levels are particularly low, which pointed to a dietary factor. Their bleeding time was found to be prolonged and the aggregability of their platelets was low, effects which could be induced by fish oils in experimental studies.

A randomized trial was set up to test the hypothesis that dietary intervention reduces the risk of death in subjects with heart disease. Men who had recently recovered from myocardial infarction (MI) were chosen as a high-risk group, and a factorial design was adopted so as to allow the testing of three dietary factors independently: fat, fish and fibre.

Subjects and Methods

The subjects were non-diabetic men under 70 years of age who were recovering from MI. They were identified while in the wards of 21 hospitals in South Wales and South-West England. On their discharge they were visited by a doctor who measured their blood pressure and took a sample of blood for serum cholesterol estimation. They were then visited by a dietitian who excluded anyone whose diet already incorporated one of the regimens to be tested in the trial. The men who were suitable for the trial were randomly allocated to receive advice or no advice on each of the following three dietary factors (so that a subject could be advised on any combination of these dietary factors or none at all): (1) 'fat advice', designed to reduce fat intake to about 30% total energy and to increase the polyunsaturated:saturated (P:S) ratio to about 1.0; (2) 'fish-advice' – i.e. an intake of about 300 g (two portions) every week of fatty fish (mackerel, herring, pilchard, sardine, salmon or trout), equivalent to 2.4 g eicosapentaenoic acid (EPA); (3) 'fibre advice' – a daily intake of 18 g cereal fibre.

The dietitian visited the subjects again after a month, at three months, and then at three-monthly intervals to reinforce the dietary advice. Men in the fish advice group who could not take sufficient fatty fish were given 'Maxepa' capsules as a partial or total substitute, 3 capsules daily supplying 0.5 g of EPA if no fatty fish was eaten. At 6 months and at 2 years the subjects were seen again; blood was taken, blood pressure measured, and a detailed dietary questionnaire was administered. A random subset of 25% of the subjects was asked to weigh and record everything eaten and drunk for 7 days. Full details of the trial and the dietary advice have already been published [2–4].

End Points of the Trial

It was decided at the inception of the trial that the primary end-point would be all-cause mortality. Since every surviving subject was contacted two years after he entered the trial, at this point we had complete information as to whether each man was alive or dead.

The other main end-point was incidence of IHD events, defined as deaths attributed to IHD plus reinfarctions. Death certificates were obtained from relatives or from the National Health Service central registry. The men were asked whether they had been readmitted to hospital, and in every case where the history suggested the possibility of reinfarction the hospital notes were examined. All the medical interviews, the classifications of death certificates and the decisions regarding diagnosis of MI were undertaken by doctors who did not know the allocation of the subjects within the trial.

Results

During the period of recruitment, 4,371 men were identified in hospital as potentially eligible for the trial. Of these, 2,101 were excluded for various reasons (1,044 because they already intended to eat one of the diets being tested), and a further 237 men died before they could be randomized. The remaining 2,033 men entered the trial at a mean interval of 41 days after MI; their mean age was 56.6 years.

Table 1 shows the effects of the three types of dietary advice on the diets of the subjects, as assessed by questionnaire at 6 months and 2 years. The group given fat advice had a lower mean fat energy intake and a higher mean P:S ratio than the group not given fat advice, though there was considerable overlap in the distributions of fat energy. The advice on fish and fibre produced a substantial rise in EPA and cereal fibre intakes, respectively, in comparison with the groups not so advised. The EPA intake includes some derived from Maxepa capsules, which were taken by 14% of the fish advice group at 6 months.

Table 2 shows the mean daily intakes of certain nutrients and foodstuffs, obtained from the seven-day weighed intakes at 6 months, in subjects advised and not advised to eat fatty fish. There was virtually no difference between the two groups in their intakes of energy, protein, fat, carbohydrate or fibre. The increased intake of fatty fish in the fish advice

Table 1. Effects of dietary advice on intake

	At 6 months			At 2 years		
	subjects	mean	SD	subjects	mean	SD
Fat energy, %						
Fat advice	937	32.1	6.0	869	32.3	5.9
No fat advice	942	35.3	5.9	876	35.0	5.8
P:S ratio						
Fat advice	937	0.78	0.30	869	0.78	0.32
No fat advice	942	0.40	0.23	876	0.44	0.25
EPA, g/week						
Fish advice	947	2.3	1.3	883	2.4	1.4
No fish advice	932	0.7	0.7	862	0.6	0.7
Cereal fibre, g/day						
Fibre advice	926	19	8	849	17	8
No fibre advice	953	9	5	896	9	5

Table 2. Mean daily intakes of certain nutrients and foodstuffs (from weighed intake records at 6 months)

	Fish advice group	No fish advice group
Subjects	220	239
Energy, MJ	7.5	7.5
Protein, g	77	75
Total fat, g	66	66
Carbohydrate, g	216	218
Dietary fibre, g	23	23
Fatty fish, g	32	9
White fish, g	25	21
Meat, g	105	111
Meat products, g	22	23
Cheese and eggs, g	25	28

group was accompanied by small reductions in consumption of meat, meat products, cheese and eggs.

Deaths and IHD events in the paired dietary groups are shown in table 3. All-cause mortality was lower in the subjects given fish advice than in the rest, the difference being attributable to a reduction in IHD deaths. The difference in overall mortality was statistically significant ($p = 0.011$) using the logrank test, which takes account of the relative timing of the deaths in the two groups. Incidence of IHD events was not significantly different in the two groups, there being rather more non-fatal reinfarctions in the fish advice group than among the other subjects. There were no significant differences in mortality or IHD events attributable to fat or fibre advice, and no significant interactions between the effects of the three diets.

Table 4 shows the effect of each type of dietary advice on mortality and IHD incidence, expressed as a relative risk. In each case this has been adjusted for certain variables which were present in slightly different proportions in the randomized groupings and which could conceivably have biased the associations. There were ten such variables: previous history of MI, angina, or hypertension; x-ray evidence of cardiomegaly, pulmonary congestion, or pulmonary oedema; and treatment (at entry into the trial) with β-blockers, other antihypertensives, digoxin/antiarrhythmic agents, or anticoagulants. With the dietary allocations these confounders were entered into a Cox proportional hazards regression model. Fish advice was

Table 3. Deaths and reinfarctions in relation to dietary advice

Diet group	Total subjects	All deaths		IHD deaths		Non fatal MI		IHD events	
		n	%	n	%	n	%	n	%
Fat advice	1,018	111	10.9	97	9.5	35	3.4	132	13.0
No fat advice	1,015	113	11.1	97	9.6	47	4.6	144	14.2
Fish advice	1,015	94*	9.3	78*	7.7	49	4.8	127	12.5
No fish advice	1,018	130*	12.8	116*	11.4	33	3.2	149	14.6
Fibre advice	1,017	123	12.1	109	10.7	41	4.0	150	14.7
No fibre advice	1,016	101	9.9	85	8.4	41	4.0	126	12.4

* $p < 0.05$; ** $p < 0.01$ (logrank test).

Table 4. Effects of dietary intervention on deaths and IHD events (adjusted for confounding variables)

Dietary intervention	Effects on all deaths			Effects on IHD events		
	relative risk	95% CI	Z	relative risk	95% CI	Z
Fat advice	1.00	(0.77–1.30)	–0.01	0.91	(0.72–1.16)	–0.74
Fish advice	0.71	(0.54–0.93)	–2.49*	0.84	(0.66–1.07)	–1.41
Fibre advice	1.27	(0.99–1.65)	1.73	1.23	(0.97–1.57)	1.72

* $p < 0.05$.

associated with a relative risk of 0.71 for all deaths (i.e. a 29% reduction attributable to fish advice); the other relative risks were not significantly different from unity, although there were increased risks in the fibre advice group.

The percentage of EPA in the plasma total fatty acids was measured in a subset of subjects. The geometric mean percentage of EPA was 0.59% (95% CI 0.52–0.67%) in 107 men given fish advice, and 0.46% (0.41–0.51%) in 96 men not given fish advice ($p < 0.01$). Serum cholesterol tended to rise slightly in the fish advice group: a net rise of 2.1% was attributable to fish advice ($p < 0.01$) at 6 months. Fat advice produced a net fall of 3.6% ($p < 0.001$) at 6 months relative to the group not given fat advice. No significant change in serum cholesterol was attributable to the fibre advice. There were no changes in blood pressure attributable to any of the dietary regimens.

Discussion

The men complied well with the fish advice, as shown by the dietary questionnaires, weighed intake records, and plasma fatty acid measurements. There were of course other dietary changes consequent upon the increase in fish intake, viz. minor reductions in the intakes of meat, meat products, cheese and eggs. Larger reductions in these intakes were achieved by the fat advice diet, which produced a fall in serum cholesterol but not in mortality, so it is very unlikely that the effect of fish advice on mortality was attributable to an incidental reduction in saturated fats.

The amount of fatty fish was modest, being on average about 300 g weekly or its equivalent in fish oil, supplying 2.5 g EPA each week. These findings confirm the results of the Zutphen study [5], in which men whose daily fish intake was at least 30 g had less than half the IHD mortality of men who ate no fish. Other cohort studies have had varying results [6–11]; in general it seems that people who eat moderate amounts of fish have a lower IHD mortality than those who eat no fish, but there is little or no additional advantage in eating large amounts.

There have been several controlled trials examining the effect of fish oil on restenosis after angioplasty, but the results have been conflicting: in four there appeared to be a favourable effect [12–15], while two showed no benefit [16, 17]. Animal work suggests that fish oil may reduce the incidence of ischaemia-induced arrhythmias [18], and if this is true for human IHD it could explain the finding of a protection against fatal but not non-fatal disease in this trial.

The fat advice given in the trial did not reduce mortality, possibly because the reduction in serum cholesterol was insufficient or the duration was too short. The subjects given fibre advice had somewhat higher mortality and incidence rates than the men not given this advice, but the differences were not statistically significant and were presumably fortuitous, since cohort studies show a high cereal fibre intake to be associated with reduced mortality.

Conclusion

A moderate intake of fatty fish improves survival in men who have recovered from recent MI.

References

1 Dyerberg J, Bang HO, Stoffersen E, et al: Eicosapentaenoic acid and prevention of thrombosis and atherosclerosis? Lancet 1978;ii:117–119.

2 Burr ML, Fehily AM, Rogers S, et al: Diet and reinfarction trial (DART): Design, recruitment, and compliance. Eur Heart J 1989;10:558–567.

3 Fehily AM, Vaughan-Williams E, Shiels K, et al: The effect of dietary advice on nutrient intakes: Evidence from the diet and reinfarction trial (DART). J Hum Nutr Dietet 1989;2:225–235.

4 Burr ML, Fehily AM, Gilbert JF, et al: Effects of changes in fat, fish and fibre intakes on death and myocardial reinfarction: Diet and reinfarction trial (DART). Lancet 1989;ii:757–761.

5 Kromhout D, Bosschieter EB, Coulander CdeL: The inverse relation between fish consumption and 20-year mortality from coronary heart disease. N Engl J Med 1985;312:1205–1209.
6 Shekelle RB, Missell L, Paul O, et al: Fish consumption and mortality from coronary heart disease. N Engl J Med 1985;313:820.
7 Vollset S, Heuch I, Bjelke E: Fish consumption and mortality from coronary heart disease. N Engl J Med 1985;313:820–821.
8 Curb JD, Reed DM: Fish consumption and mortality from coronary heart disease. N Engl J Med 1985;313:821.
9 Norell SE, Ahlbom A, Feychting M, et al: Fish consumption and mortality from coronary heart disease. Br Med J 1986;293:426.
10 Lapidus L, Andersson H, Bengtsson C, et al: Dietary habits in relation to incidence of cardiovascular disease and death in women: A 12-year follow-up of participants in the population study of women in Gothenburg, Sweden. Am J Clin Nutr 1986;44: 444–448.
11 Dolecek TA, Grandits G, Cagginla AW, et al: Dietary omega 3 fatty acids and mortality in the multiple risk factor intervention trial. Second World Congress of Preventive Cardiology, abstract no. 255, 1989.
12 Slack JD, Pinkerton CA, Van Tassel J, et al: Can oral fish oil supplement minimize restenosis after percutaneous transluminal coronary angioplasty? J Am Coll Cardiol 1987;9:64A.
13 Ilsley CDJ, Nye ER, Sutherland W, et al: Randomised placebo-controlled trial of Maxepa and Aspirin/Persantin after successful coronary angioplasty. Aust NZ J Med 1987;17:559.
14 Milner MR, Gallino RA, Leffingwell A, et al: High dose omega–3 fatty acid supplementation reduces clinical restenosis after coronary angioplasty. Circulation 1988; 78(suppl 2):634.
15 Dehmer GJ, Popma JJ, van den Berg EK, et al: Reduction in the rate of early restenosis after coronary angioplasty by a diet supplemented with n–3 fatty acids. N Engl J Med 1988;319:733–740.
16 Grigg LE, Kay TWH, Valentine PA, et al: Determinants of restenosis and lack of effect of dietary supplementation with eicosapentaenoic acid on the incidence of coronary artery restenosis after angioplasty. J Am Coll Cardiol 1989;13:665–672.
17 Reis GJ, Boucher TM, Sipperly ME, et al: Randomised trial of fish oil for prevention of restenosis after coronary angioplasty. Lancet 1989;ii:177–181.
18 McLennan PL, Abeywardena MY, Charnock JS: Dietary fish oil presents ventricular fibrillation following coronary artery occlusion and reperfusion. Am Heart J 1988; 116:709–717.

Michael L. Burr, MD, MRC Epidemiology Unit, 4 Richmond Road,
Cardiff CF2 3AS (UK)

Cardiovascular III: Circulation – Blood Pressure

Simopoulos AP, Kifer RR, Martin RE, Barlow SM (eds): Health Effects of ω3 Polyunsaturated Fatty Acids in Seafoods. World Rev Nutr Diet. Basel, Karger, 1991, vol 66, pp 313–328

Hypotensive Effects of ω3 Fatty Acids: Mechanistic Aspects[1]

Howard R. Knapp

Division of Clinical Pharmacology, Vanderbilt University, Nashville, Tenn., USA

Introduction

Carefully controlled investigations of a small number of hypertensive patients [1] and a population-based intervention study [2] have both confirmed earlier suggestions [3] that dietary ω3 fatty acids exert antihypertensive effects in humans, but the mechanisms responsible have not been established. Prostaglandins and other eicosanoids have been shown to be involved in numerous processes affecting blood pressure regulation [4] and are made from polyunsaturated fatty acids (PUFA). Because of this, it has been suggested frequently that alterations in blood pressure during dietary supplementation with ω3 or ω6 polyunsaturates are due to changes in the endogenous synthesis of vasoactive eicosanoids, but this hypothesis has rarely been tested [1]. Other possible mechanisms advanced include effects of ω3 fatty acids on renal function, a reduction in vascular responsiveness to systemic vasoconstrictors; and a lowering of blood viscosity [5]. It is conceivable that variable proportions of some or all of these mechanisms could be operative in patients with different origins of their hypertension, but it will be important to elucidate this to help predict those most likely to benefit from ω3 PUFA.

[1] This work was supported in part by a grant from the National Heart, Lung and Blood Institute, HL-35380. Dr. Knapp is an Established Investigator of the American Heart Association.

This article will review the current state of research into the effect of ω3 PUFA on human vascular tone, as well as related areas such as interactions with vasocative drugs and renal function. While the most useful data is that gathered in human studies, some result of experiments in animals will also be presented when providing information that cannot be obtained easily from clinical investigations.

ω3 Fatty Acids and Endogenous Eicosanoids

For a number of years, it has been considered that the assessment of eicosanoid production in vivo is best done by the measurement of urinary metabolites [6]. Although avoiding the artifacts associated with in vitro incubation, tissue sampling or blood drawing, this approach also has recognized limitations. For instance, urinary metabolite excretion provides an integrated index of endogenous synthesis, so phasic changes may be difficult to detect. Also, changes in eicosanoid synthesis by a particular organ might be missed, and if changes in metabolite excretion are found, the site of altered eicosanoid formation in the body usually cannot be assigned unambiguously. Sensitive GC/MS methods recently have been developed to measure enzymatically formed plasma eicosanoid metabolites to circumvent some of these problems [7], but despite its shortcomings, the measurement of urinary metabolites has provided a great deal of insight into the endogenous metabolism of ω3 fatty acids and their effects on that of arachidonic acid.

Although a number of studies on the effects of dietary ω3 fatty acid supplements on endogenous eicosanoids have been published, only one has attempted to correlate the changes induced in urinary eicosanoid metabolites with changes in blood pressure [1]. It has been shown by two research groups that, despite the contrary findings of many in vitro studies [8, 9], dietary EPA is converted to PGI_3 in man and does not suppress formation of PGI_2 from arachidonate [10, 11]. Recently, urinary metabolite data from rats fed fish oil have shown that this apparent disparity may be due to differences between in vivo and in vitro assessment of eicosanoid formation, rather than a species difference [12]. In any case, pharmacologic doses of fish oil administered to males with essential hypertension caused a significant increase in total prostacyclin formation ($PGI_2 + PGI_3$) during the first week of supplementation, but the increase was not maintained as blood pressure fell over the 4-week study period [1]. It is possible that the early surge in prostacyclin production caused changes in other hormone systems regulating blood pressure, but the blood pressure reduction was not associated with a continually increased output of this vasodilator.

The effects of fish oil supplementation on endogenous thromboxane production appear to depend to some extent upon how much thromboxane is being made prior to the intervention. It has been shown that when patients with severe atherosclerosis, who have accelerated platelet destruction and markedly increased thromboxane B_2 metabolite (TxB_2-M) excretion, take large doses of fish oil there is a dramatic reduction in TxB_2-M excretion and appearance of TxB_3-M, along with amelioration of several other indices of platelet-vascular interaction [11]. Two apparently normal subjects with above-average immunoreactive TxB_2-M excretion also were reported to have a lowered rate during fish oil supplementation [13], but there is a very modest reduction in TxB_2 production, nearly matched by the synthesis of (less active) TxB_3, in subjects with only average amounts of thromboxane being synthesized at baseline who ingest fish oil [14]. This suggests that thromboxane synthesis inhibition by ω3 supplements would be most likely to contribute to their hypotensive effects in situations where elevated amounts of thromboxane were being produced. White males with essential hypertension, however, excrete quite normal amounts of this metabolite [1], and have much less of a reduction in thromboxane formation during high dose fish oil supplementation than would be achieved by low-dose aspirin [15]. Although selective thromboxane synthase inhibitors have been found to lower blood pressure in hypertensive rats [16], no reports of a positive result in essential hypertension trials have been published, and the possible role of thromboxane in human essential hypertension has not been established. Doses of aspirin which would markedly reduce thromboxane synthesis have not been found to benefit essential hypertension, and cyclooxygenase inhibitors usually antagonize the actions of antihypertensive drugs [17].

A possible exception to the above, however, may be found in the toxemia of pregnancy syndrome, in which the patients have been shown to produce excessive amounts of thromboxane [18] and to have their hypertension reduced by low-dose aspirin [19]. Animal models of this syndrome have also shown benefit from selective thromboxane synthase inhibition [20]. However, there have been reports that women in the Faroe Islands, where the consumption of fish is high, have a longer gestation time and higher birth weight than women elsewhere in Europe [21]; whether the rates of caesarian section, neonatal mortality and eclampsia [22] may be lower there than in Denmark appears to be controversial [18]. Therefore, while it is possible that ω3 fatty acid supplements would have a beneficial effect in this syndrome, this would obviously have to be studied in a very controlled setting, and the benefits compared against that of once-daily, low-dose aspirin, which is not only convenient but apparently has a high efficacy for blood pressure reduction and a low incidence of adverse effects [19].

Aside from changes in prostacyclin and thromboxane synthesis during the lowering of blood pressure by fish oil, one study has examined the effects on prostaglandin E metabolite (PGE-M) excretion. This whole-body index of PGE_2 (and PGE_1) synthesis showed a non-significant trend towards lower values while the subjects took large doses of fish oil [1], and no metabolite derived from PGE_3 was detected. There are many cells in the body which make PGE_2, including the microvascular endothelium [17], and if ω3 fatty acids generally lowered production of E-type prostaglandins there could be altered responses to physiological stimuli in a number of clinical conditions. One of the most widely recognized of these is impaired renal function, in which renal blood flow may become dependent upon local PGE_2 synthesis [17]. Urinary PGE is considered to largely reflect renal synthesis in women, but the extremely high concentrations of such compounds in seminal fluid contribute a large and highly variable proportion of measured urinary PGE and PGF in men. As a result, it is difficult to interpret the reported non-significant reductions in urinary immunoreactive PGE [23] or lower excretion of PGE compounds (by mass spectrometry) in three of four male subjects ingesting large amounts of ω3 fatty acids [24]. This is particularly true in view of the suppressed formation of PGE and PGF compounds in the seminal fluid of ten normal males taking large doses of fish oil [25]. The question of fish oil effects on renal prostaglandin synthesis will have to be addressed first in normal women and then, coupled with functional studies, in subjects of both sexes with pre-existing renal disease.

While the hypotensive effects of ω3 fatty acids do not clearly relate to altered formation of endogenous cyclooxygenase products, little is known about the role of lipoxygenase products, or fatty acid epoxides formed by cytochrome P-450 enzymes [26], in human vascular function. Also, no currently available methods provide useful, non-invasive indices of endogenous synthesis of such compounds. We have recently found, however, that human subjects taking fish oil supplements excrete diols derived from the P450-mediated epoxidation of eicosapentaenoic acid [27], in addition to the previously described ones from arachidonic acid [26], and further work is under way to determine their possible role in the vascular effects of marine oils.

ω3 Fatty Acids and Vasoconstrictor Responses

There is some evidence that the response of human arteries to vasoconstrictor substances in vivo is lessened by the local synthesis of vasodilator prostaglandins [17]. Therefore, the fact that dietary supplementation with ω3 fatty acids reduces both vasoconstriction of arterial preparations from animals ex vivo [28] as well as the pressor response of normal volunteers to a norepinephrine infusion [23] suggests that they are not acting as cyclooxygenase inhibitors. Unfortunately, the interpretation of these findings is obscured by similar results obtained when animals [29] or human subjects [30] were given ω6 fatty acid supplements. Also, the studies on ω3 supplementation in men [23] had control values obtained 'randomly' before or after the study period, and also had increased urinary sodium at the end of the fish oil period. Other workers found an identical pressor response to a psychophysiological stress test before and after two weeks of mackerel supplements rich in ω3 fatty acids [31], suggesting that the response to endogenous catecholamines was unchanged.

A final study on this point was performed in mildly hypertensive males who had taken large doses of fish oil for 4 weeks [32]. An infusion of phenylephrine, a pure alpha-adrenergic agonist with practically no direct cardiac effects in man, was carried out in six subjects at the end of the supplementation and recovery periods. There was no difference seen on the two occasions in the pressor response to moderate infusion rates producing a 15 mm Hg increase in systolic pressure, about the increase reported to be lessened in the earlier work mentioned above [23]. With still higher infusion rates, however, the subjects appeared to develop reflex bradycardia at a lower systolic pressure on the occasion when they were still taking fish oil. This response would tend to confirm that these volunteers actually did have a reduction in blood pressure and suggests that they had resetting of their baroreceptors in response to this blood pressure reduction. The excretion of prostacyclin metabolites during the two infusion studies was also compared and, interestingly, they made more of this vasodilator on the occasion when they were off fish oil, and achieved higher blood pressures in response to the infusion. Thus, the indirect blunting of the pressor response during fish oil ingestion cannot be attributed to enhanced prostacyclin synthesis, as this eicosanoid has been reported to do in animal studies [33].

While dietary ω3 fatty acids do not appear to cause a direct reduction in the pressor response to catecholamines in man, some interesting findings have been reported in studies of the renin-angiotensin system. Increased plasma renin activity (PRA) has been reported in volunteers eating a mackerel-enriched diet [31, 34], while one of these groups [34] found decreased PRA in subjects taking large doses of fish oil [23]. Other workers, however, did not detect changes in the PRA of a larger number of subjects taking a lower dose of fish oil [35]. None of the subjects in these studies were on controlled diets, but the reported decrease in PRA was associated with increased urinary sodium excretion and a paradoxical non-significant decrease in pressor response to an angiotensin infusion [23]. It has also been suggested that Eskimos have higher PRA, and therefore less angiotensin-responsiveness, than do age- and sex-matched Danes [36], but these data are more difficult to interpret due to both the difference in time between blood sampling and measurement between the groups, as well as the near absence of postural change in renin activity in the Danish controls compared to a doubling with standing in the Eskimo subjects. Obviously, carefully controlled studies of ω3 fatty acid effects in the renin-angiotensin system will provide important information to the understanding of their antihypertensive actions.

ω3 Fatty Acids and the Kidney

The antihypertensive effects of fish oil could be due in part to beneficial alterations in the renal handling of electrolytes, or in the release or response to renin and related hormones. On the other hand, since hypertension is a leading cause of renal failure, fish oil could protect renal function in some situations by lowering arterial pressure. Since fish oil supplements have been reported to have beneficial effects on the blood pressures of patients on hemodialysis with little residual renal function [37], the effects of ω3 fatty acids are not likely to be limited to altered renal function, but must also involve vasoactive hormones or vascular reflexes. The only published investigations of fish oil in human renal disease have been limited to low doses [38], and no controlled studies of fish oil-induced changes in the renal function of normal human subjects have been reported. Eicosanoids may have different roles in the renal physiology of different species, and some aspects of ω3 fatty acid metabolism may be different in rats and man. Therefore, extrapolation of studies in laboratory animals to man may not be entirely valid, but some useful impressions can be obtained.

It is not surprising that there are considerable differences in the reported effects of dietary fish oil in various animal models of renal impairment, since the pathogenic mechanisms of these involve very different processes. Also, if ω3 fatty acids actually do act as cyclooxygenase inhibitors in human organs making large quantities of prostaglandins in vivo [25], then an absence of functional change in normal subjects [39] may not predict the effects on stressed kidneys whose function has become dependent upon local production of prostaglandins. A possibly analogous situation may exist for the use of cyclooxygenase inhibiting drugs, which have negligible effects on the renal function of normal subjects but cause further deterioration of previously impaired renal function in many patients [17]. Therefore, while the lack of renal functional change seen in normal humans [39] or rats [40] given pharmacologic doses of fish oil might be encouraging from the safety standpoint, it may not be predictive of the situation in patients.

The reduced renal mass and function found in patients with hypertensive nephrosclerosis or renovascular atherosclerosis may be reflected to some degree by the changes that take place after partial nephrectomy of rats. Since patients with significant renovascular disease might be candidates for ω3 fatty acid-induced modulation of cardiovascular disease, it is important to determine what effect dietary ω3 enrichment might have on their renal function. Rats subjected to subtotal nephrectomy have been previously found to have their renal function worsened by cyclooxygenase inhibition [41] and ameliorated by thromboxane synthase inhibition [42], administration of prostaglandins [43] or their ω6 fatty acid precursors [44]. When such animals were given dietary fish oil, reduced urinary PGE and renal function, and increased proteinuria and mortality were found [45]. In a study of uninephrectomy, animals fed fish oil again had decreased urinary PGE excretion and an increased kidney weight without a comparable increase in function [46]. It was felt that the accelerated glomerular hyperperfusion and hyperfiltration seen in the fish oil-fed animals would lead to early nephrosclerosis and loss of function. Obviously, this question will have to be addressed in patients in the near future.

The effects of dietary fish oil on immune-mediated renal dysfunction are complex and different from those described above. Several groups have published studies indicating that feeding fish oil prolongs the survival of mice which develop lupus nephritis [47, 48], and results in marked histologic improvement. Recently, other workers have reported much less benefit from feeding fish oil to some of these same strains of mice [49], and two negative clinical studies of fish oil supplements in human lupus nephritis have appeared in abstract form [50, 51]. It is of interest that some of the strains of lupus-prone mice develop serious hypertension, while others do not [52], but the possible role of blood pressure reduction in the beneficial effects of dietary fish oil in these animals has not been discussed. On the other hand, feeding fish oil-supplemented diets did not result in morphologic or functional improvement of rat renal allografts [53], or in lessened histologic changes in the kidneys of rabbits given serum sickness nephritis [54].

Ciclosporin is now being widely used in organ transplantation, and in many individuals causes impaired renal function [55] which has been associated with increased thromboxane formation [56]. It has been shown that using fish oil, instead of olive oil, as the vehicle for its administration to rats caused an attenuation of the ciclosporin nephrotoxicity [57], but a recent report has suggested that this effect may not be linked to thromboxane synthesis reduction [58]. Two pilot studies have examined the interaction of ω3 fatty acids and ciclosporin in patients. In the first, patients with psoriasis who were given ciclosporin were found to have less of a decline in renal function if they also took moderate doses (3.6 g EPA/2.4 g DHA) of ω3 fatty acids per day [59], and it was also noted that the fish oil did not appear to have a major influence upon ciclosporin pharmacokinetics. The other pilot study [60] involved administration of low doses (providing 1.5 g EPA/day) of ω3 fatty acids to renal transplant recipients for six months. The mild increase in blood pressure produced by ciclosporin was not seen in subjects given ω3 supplements, but the between-group difference did not reach statistical significance. A theoretically favorable reduction in thromboxane, but not prostacyclin, was found in the patients taking the ω3 preparation, and no adverse effects attributable to the supplements were seen. Recently, a randomized, controlled study of the effects of fish oil supplements on ciclosporin therapy of renal transplant recipients has been published [61]. The fish oil supplement caused a significant decrease in renal vascular resistance and increased glomerular filtration rates, as well as a lowering of mean arterial pressure. This finding certainly suggests a potential application of ω3 fatty acid therapy that bears further study.

Several papers on the effects of fish oil in human diabetic nephropathy have appeared. Although a reduction in the percent of transcapillary albumin leakage in diabetics has recently been reported [62], there was no decrease in albuminuria found, in agreement with earlier observations [63]. Also, there was an encouraging lack of deterioration seen in the glomerular filtration rate, although these patients received a moderate dose of ω3 fatty acids (2.0 g EPA/2.6 g DHA) daily for 8 weeks. More detailed investigations of fish oil effects in diabetics are needed, since they represent a large proportion of patients with vascular disease who are logical candidates for trials of ω3 fatty acid therapy, and also have a tendency towards deterioration of renal function in response to a variety of metabolic insults (e.g. cyclooxygenase inhibition).

ω3 Fatty Acids and Blood Viscosity

Alteration in the rheological properties of either plasma or blood cells could contribute to changes in blood pressure, thrombosis, bleeding time, and atherosclerosis. Review articles on the effects of fish oil usually mention that such supplements reduce blood viscosity, but the literature presents a less straightforward picture. In fact, several laboratories have reported conflicting results on this point in different papers. Part of the problem may lie in the variability of the different methods used, as well as in the small numbers of both subjects and observations per subject in most studies. A lack of blinded study design is also a major difficulty in studies of this type. Finally, few studies have actually compared the effects of ω3 and ω6 fatty acid supplements to determine whether any differences found from baseline are specific for ω3 fatty acids, or are really a function of increased dietary unsaturated fatty acids per se.

Although there are occasional reports of lowered plasma viscosity [64] or fibrinogen (which is believed to make a major contribution to plasma viscosity) concentrations [65] in subjects receiving fish oil supplements, the majority of workers have found no change in these parameters [66–73]. Increased, rather than decreased, fibrinogen levels have also been reported during fish oil supplementation [63, 74], and a non-significant shortening of clotting time previously prolonged by oral anticoagulants was suggested to be due to a fish oil-induced increase in the production of fibrinogen and

Table 1. Effects of fish oil on plasma rheology

Fibrinogen levels	Plasma viscosity	First author [ref.]
No change		Mortensen [35]
No change		Sanders [66]
No change	no change	Terano [67]
	no change	Hirai [68]
No change	no change	Simons [69]
	no change	Woodcock [70]
No change		Rogers [71]
No change		Dart [72]
Increased		Emeis [73]
Increased		Haines [63]
Increased		Smith [74]
Decreased		Hostmark [65]
	decreased	Bach [64]

Table 2. Effect of fish oil on RBC/whole blood rheology

Lab	Decreased viscosity [ref.]	Lab	No change [ref.]
I	Tamura [75]	I	Hamazaki [37]
	Hirai [68]		(dialysis pt)
	Terano [67]		
II	Popp-Snidjers [76]	II	Popp-Snidjers [78]
III	Cartwright [77]	(other)	Bruckner [79]
	Woodcock [70]		Simons [69]
(Other)	Bach [64]		Rogers [71]

factor VII [74]. The findings in a number of publications have been summarized in table 1. Overall, there appears to be little evidence that there is a sufficient reduction in plasma viscosity by fish oil supplements to account for their antihypertensive effects.

On the question of decreased whole-blood viscosity and increased erythrocyte deformability/membrane fluidity, there is less agreement. As can be seen in table 2, some workers have reported such changes while a similar number of negative reports can be found, and several groups have published both types of data. Although there is a greater number of positive than negative reports, this impression derives from multiple reports from the same laboratories; there are more laboratories finding a lack of effect than a reduction in whole-blood viscosity by fish oil ingestion. It is not clear that the ω3 fatty acid dose or duration of treatment account for the different findings, and further carefully controlled studies using blinded measurements will be useful, particularly if other dietary control groups, such as one receiving ω6 fatty acid supplements, are included.

Conclusion

Many questions about how ω3 fatty acids lower blood pressure remain unanswered, and more carefully controlled studies are needed in a number of areas. Understanding whether fish oil supplements will alter blood viscosity or renal prostaglandin production, for instance, may allow us to expand the indications for their use or avoid potential difficulties in particular patient groups. Although much information can be obtained from in vitro and animal studies, we have already seen how data from those sorts of experiments might not accurately reflect the results of ω3 fatty acid metabolism in man. An increasing number of reports are providing support for the notion that a greater intake of fish or fish oils rich in ω3 fatty acids can have health benefits for people eating a Western diet. Only well-designed clinical trials, however, will define the eventual usefulness of ω3 fatty acid therapy in an expanding number of human diseases.

References

1 Knapp H, FitzGerald GA: The antihypertensive effects of fish oil: A controlled study of polyunsaturated fatty acid supplements in essential hypertension. N Engl J Med 1989;320:1037–1043.
2 Bønaa KH, Bjerve KS, Straume B, et al: Effect of eicosapentaenoic and docosahexaenoic acids on blood pressure in hypertension. N Engl J Med 1990;322:795–801.
3 Knapp HR: Omega-3 fatty acids, endogenous prostaglandins, and blood pressure regulation in humans. Nutr Rev 1989;47:301–313.
4 Cinotti GA, Pugliese F: Prostaglandins and hypertension. Am J Hyp 1989;2:10S–15S.
5 McMillan DE: Antihypertensive effects of fish oil. N Engl J Med 1989;321:1610.
6 Hamberg M, Samuelsson B: On the metabolism of prostaglandin E_1 and E_2 in man. J Biol Chem 1971;246:6713–6721.
7 Lawson JA, Patrono C, Ciabattoni G, et al: Long-lived enzymatic metabolites of thromboxane B_2 in the human circulation. Anal Biochem 1986;155:198–205.

8 Hamazaki T, Hirai A, Terano T, et al: Effects of orally administered ethyl ester of eicosapentaenoic acid (EPA; C20:5, ω-3) on PGI_2-like production by rat aorta. Prostaglandins 1982;23:557–567.
9 Croft KD, Beilin LJ, Legge FM, et al: Effects of diets enriched in eicosapentaenoic or docosahexaenoic acids on prostanoid metabolism in the rat. Lipids 1987;22:647–650.
10 Fischer S, Weber PC: Prostaglandin I_3 is formed in vivo in man after dietary eicosapentaenoic acid. Nature 1984;307:165–168.
11 Knapp HR, Reilly IAG, Alessandrini P, et al: In vivo indexes of platelet and vascular function during fish-oil administration in patients with atherosclerosis. N Engl J Med 1986;314:937–942.
12 Knapp HR, Salem N: Formation of PGI_3 in the rat during dietary fish oil supplementation. Prostaglandins 1989;38:509–523.
13 von Schacky C, Fischer S, Weber PC: Long-term effects of dietary marine omega-3 fatty acids upon plasma and cellular lipids, platelet function, and eicosanoid formation in humans. J Clin Invest 1985;76:1626–1631.
14 Knapp HR: Effect of dietary n-3 fatty acids on platelet turnover and thromboxane synthesis in normal healthy males. Clin Res 1990;38:11A.
15 Knapp HR, Healy C, Lawson J, et al: Effects of low-dose aspirin on endogenous eicosanoid formation in normal and atherosclerotic men. Thromb Res 1988;50: 377–386.
16 Uderman HD, Jackson EK, Puett D, et al: Thromboxane synthetase inhibitor UK38,485 lowers blood pressure in the adult spontaneously hypertensive rat. J Cardiovasc Pharmacol 1984;6:969–972.
17 Oates JA, FitzGerald GA, Branch RA, et al: Clinical implications of prostaglandin and thromboxane A_2 formation. N Engl J Med 1988;319:689–698.
18 Fitzgerald DJ, FitzGerald GA: Eicosanoids in the pathogenesis of preeclampsia; in Laragh JH, Brenner BM (eds): Hypertension: Pathophysiology, Diagnosis and Management. New York, Raven Press, 1990, chapter 111, pp 1789–1807.
19 Lubbe WF: Low-dose aspirin in prevention of toxaemia of pregnancy: Does it have a place? Drugs 1987;34:515–518.
20 Keith JC, Thatcher CD, Schaub RG: Beneficial effects of U-63,557A, a thromboxane synthetase inhibitor, in an ovine model of pregnancy-induced hypertension. Am J Obstet Gynecol 1987;157:199–203.
21 Olsen SF, Hansen HS, Sorensen, TIA, et al: Intake of marine fat, rich in (n-3) polyunsatureated fatty acids, may increase birthweight by prolonging gestation. Lancet 1986;2:367–369.
22 Anderson HJ, Anderson LF, Fuchs AR: Diet, pre-eclampsia, and intrauterine growth retardation. Lancet 1989;2:1146.
23 Lorenz R, Spengler U, Fischer S, et al: Platelet function, thromboxane formation and blood pressure control during supplementation of the Western diet with cod liver oil. Circulation 1983;67:504–511.
24 Fischer S, von Schacky C, Schweer H: Prostaglandins E_3 and $F_{3\alpha}$ are excreted in human urine after ingestion of n-3 polyunsaturated fatty acids. Biochim Biophys Acta 1988;963:501–508.
25 Knapp HR: Prostaglandins in human semen during fish oil ingestion: Evidence for

in vivo cyclooxygenase inhibition and appearance of novel trienoic compounds. Prostaglandins 1990;39:423–439.
26 Toto R, Siddhanta A, Manna S, et al: Arachidonic acid epoxygenase: Detection of epoxyeicosatrienoic acids in human urine. Biochim Biophys Acta 1987;191:132–139.
27 Knapp HR: Excretion of P450-derived diols of eicosapentoaenoic acid during fish oil ingestion by man. Clin Res 1990;38:418A.
28 Lockette WE, Webb RC, Culp BR, et al: Vascular reactivity and high dietary eicosapentaenoic acid. Prostaglandins 1982;24:631–639.
29 Schölkens BA, Gehring D, Schlotte V, et al: Evening primrose oil, a dietary prostaglandin precursor, diminishes vascular reactivity to renin and angiotensin II in rats. Prostaglandins Leukotrienes Med 1982;8:273–285.
30 Pipkin FB, Morrison RA, O'Brien PMS: The effect of dietary supplementation with linoleic and α-linolenic acids on the pressor and biochemical response to exogenous angiotensin II in human pregnancy. Prog Lipid Res 1986;25:425–429.
31 Singer P, Wirth M, Boigt S, et al: Blood pressure and lipid-lowering effect of mackerel and herring diet in patients with mild essential hypertension. Atherosclerosis 1985;56:223–235.
32 Knapp HR, Gregory D, Nolan S: Dietary polyunsaturates, vascular function and prostaglandins; in Galli G, Simopoulos AP (eds): Dietary ω3 and ω6 Fatty Acids. New York, Plenum, 1989, pp 283–295.
33 Panzenbeck MJ, Tan W, Hajdu MA, et al: Intracoronary infusion of prosaglandin I_2 attenuates arterial baroreflex control of heart rate in conscious dogs. Circ Res 1988; 63:860–868.
34 Lorenz R, Spengler U, Siess W, et al: Einfluss veränderter Prostaglandinbildung auf die sympathoadrenerge Aktivität und die Blutdruckregulation. Verh Dt Ges Inn Med 1980;86:692–694.
35 Mortensen JZ; Schmidt EB, Nielsen AH, et al: The effect of n-6 and n-3 polyunsaturated fatty acids on hemostasis, blood lipids and blood pressure. Thromb Haemostasis 1983;50:543–546.
36 Jorgensen KA, Nielsen AH, Dyerberg J: Hemostatic factors and renin in Greenland eskimos on a high eicosapentaenoic acid intake. Acta Med Scand 1986;219:473–479.
37 Hamazaki T, Nakazawa R, Tateno S, et al: Effects of fish oil rich in eicosapentaenoic acid on serum lipid in hyperlipidemic hemodialysis patients. Kidney Int 1984;26: 81–84.
38 Schaap GH, Biol HJG, Popp-Snijders C, et al: Effects of protein intake variation and ω3 polyunsaturated fatty acids on renal function in chronic renal disease. Life Sciences 1987;41:2759–2765.
39 Düsing R, Struck A, Scherf H, et al: Dietary fish oil supplements: Effects on renal hemodynamics and renal excretory function in healthy volunteers. Kidney Int 1987; 31:268.
40 Barcelli UO, Beach DC, Pollak VE: The influence of n-6 and n-3 fatty acids on kidney phospholipid composition and on eicosanoid production in aging rats. Lipids 1988;23:309–312.
41 Logan JL, Lee SM, Benson B: Inhibition of compensatory renal growth by indomethacin. Prostaglandins 1986;31:253–262.

42 Purkerson ML, Joist JH, Yates J, et al: Inhibition of thromboxane synthesis ameliorates the progressive kidney disease of rats with subtotal renal ablation. Proc Natl Acad Sci USA 1985;82:193–197.

43 Mauk RH, Patak RV, Fadem SZ, et al: Effect of prostaglandin E administration in a nephrotoxic and a vasoconstrictor model of acute renal failure. Kidney Int 1977;12: 122.

44 Heifets M, Morrissey JJ, Purkerson ML, et al: Effect of dietary lipids on renal function in rats with subtotal nephrectomy. Kidney Int 1987;32:335–341.

45 Scharschmidt LA, Gibbons NB, McGarry L, et al: Effect of dietary fish oil on renal insufficiency in rats with subtotal nephrectomy. Kidney Int 1987;32:700–709.

46 Logan JL, Michael UF, Benson B: Effects of dietary fish oil on renal growth and function in uninephrectomized rats. Kidney Int 1990;37:57–63.

47 Prickett JK, Robinson DR, Steinberg AD: Dietary enrichment with the polyunsaturated fatty acid eicosapentaenoic acid prevents proteinuria and prolongs survival in NZB × NZW F_1 mice. J Clin Invest 1981;68:556.

48 Kelley VE, Ferretti A, Izui S, et al: A fish oil diet rich in eicosapentaenoic acid reduces cyclooxygenase metabolites, and suppresses lupus in MRL-lpr mice. J Immunol 1985;134:1914–1919.

49 Westberg G, Tarkowski A, Svalander C: Effect of eicosapentaenoic acid rich menhaden oil and MaxEPA on the autoimmune disease of Mrl/1 mice. Int Arch Allergy Appl Immunol 1989;88:454–461.

50 Moore GF, Yarboro C, Sebring NG, et al: Eicosapentaenoic acid (EPA) in the treatment of systemic lupus erythematosus (SLE). Arthritis Rheum 1987;30(4):S33.

51 Westberg G, Tarkowski A: Effect of Max-EPA in patients with SLE: A double blind cross-over study. Kidney Int 1989;35:235.

52 Rudofsky UH, Dilwith RL, Roths JB, et al: Differences in the occurrence of hypertension among (NZB × NZW) F_1, MRL-lpr, and BXSB mice with lupus nephritis. Am J Pathol 1984;116:107–114.

53 Coffman TM, Yohay D, Carr DR, et al: Effect of dietary fish oil supplementation on eicosanoid production by rat renal allografts. Transplantation 1988;45:470–474.

54 Bolton-Smith C, Gibney MJ, Gallagher PJ, et al: Effect of polyunsaturated fatty acids of the n-3 and n-6 series on lipid composition and eicosanoid synthesis of platelets and aorta and on immunological induction of atherosclerosis in rabbits. Atherosclerosis 1988;72:29–35.

55 Myers BD: Cyclosporine nephrotoxicity. Kidney Int 1986;30:964–974.

56 Benigni A, Chiabrando C, Piccinelli A, et al: Increased urinary excretion of thromboxane B_2 and 2,3-dinor-TxB_2 in cyclosporin A nephrotoxicity. Kidney Int 1988;34: 164–174.

57 Elzinga L, Kelley VE, Houghton DC, et al: Modification of experimental nephrotoxicity with fish oil as the venicle for cyclosporine. Transplantation 1987;43:271–274.

58 Walker RJ, Lazzaro VA, Duggin GG, et al: Dietary eicosapentaenoic acid does not modify cyclosporin-induced inhibition of angiotensin II-stimulated prostaglandin synthesis in mesangial cells. Renal Failure 1989;11:125–132.

59 Stoof TJ, Korstanje MJ, Bilo HJG, et al: Does fish oil protect renal function in cyclosporin-treated psoriasis patients? J Intern Med 1989;226:437–441.

60 Urakaze M, Hamazaki T, Kashiwabara H, et al: Favorable effects of fish oil concentrate on risk factors for thrombosis in renal allograft recipients. Nephron 1989;53: 102–109.
61 van der Heide JJH, Bilo HJB, Tegzess AM, et al: The effects of dietary supplementation with fish oil on renal function in cyclosporin-treated renal transplant recipients. Transplantation 1990;49:523–527.
62 Jensen T, Stender S, Goldstein K, et al: Partial normalization by dietary cod-liver oil of increased microvascular albumin leakage in patients with insulin-dependent diabetes and albuminuria. N Engl J Med 1989;321:1572–1577.
63 Haines AP, Sanders TAB, Imeson JD, et al: Effects of a fish oil supplement on platelet function, haemostatic variables and albuminuria in insulin-dependent diabetics. Thromb Res 1986;43:643–655.
64 Bach R, Schmidt U, Jung F, et al: Effects of fish oil capsules in two dosages on blood pressure, platelet functions, haemorheological and clinical chemistry parameters in apparently healthy subjects. Ann Nutr Metab 1989;33:359–367.
65 Hostmark AT, Mjerkedal T, Kierulf P, et al: Fish oil and plasma fibrinogen. Br Med J 1988;297:180–181.
66 Sanders TAB, Vickers M, Haines AP: Effect on blood lipids and haemostasis of a supplement of cod-liver oil, rich in eicosapentaenoic and docosahexaenoic acids, in healthy young men. Clin Sci 1981;61:317–324.
67 Terano T, Hirai A, Hamazaki T, et al: Effect of oral administration of highly purified eicosapentaenoic acid on platelet function, blood viscosity and red cell deformability in healthy human subjects. Atherosclerosis 1983;46:321–331.
68 Hirai A, Terano T, Saito H, et al: Eicosapentaenoic acid and platelet function in Japanese; in Lovenberg W, Yamori Y (eds): Nutritional Prevention of Cardiovascular Disease. New York, Academic Press, 1984, pp 231–239.
69 Simons LA, Hickie JB, Balasubramaniam S: On the effects of dietary n-3 fatty acids (Maxepa) on plasma lipids and lipoproteins in patients with hyperlipidaemia. Atherosclerosis 1985;54:75–88.
70 Woodcock BE, Smith E, Lambert WH, et al: Beneficial effect of fish oil on blood viscosity in peripheral vascular disease. Br Med J 1984;288:592–594.
71 Rogers S, James KS, Butland BK, et al: Effects of a fish oil supplement on serum lipids, blood pressure, bleeding time, haemostatic and rheological variables. Atherosclerosis 1987;63:137–143.
72 Dart AM, Riemersma RA, Oliver MF: Effects of maxepa on serum lipids in hypercholesterolaemic subjects. Atherosclerosis 1989;80:119–124.
73 Emeis JJ, van Houwelingen AC, van den Hoogen CM, et al: A moderate fish intake increases plasminogen activator inhibitor type-1 in human volunteers. Blood 1989; 74:233–237.
74 Smith P, Arnesen H, Opstad T, et al: Influence of highly concentrated n-3 fatty acids on serum lipids and hemostatic variables in survivors of myocardial infarction receiving either oral anticoagulants or matching placebo. Thromb Res 1989;53:467–474.
75 Tamura Y, Hirai A, Terano T, et al: Effect of administration of highly purified eicosapentaenoic acid on platelet and erythrocyte functions in patients with thrombotic disorders; in Lovenberg W, Yamori Y (eds): Nutritional Prevention of Cardiovascular Disease. New York, Academic Press, 1984, pp 323–330.

76 Popp-Snijders C, Schouten JA, Van Der Meer J, et al: Fatty fish-induced changes in membrane lipid composition and viscocity of human erythrocyte suspensions. Scand J Clin Lab Invest 1986;46:253–258.
77 Cartwright IJ, Pockley AG, Galloway JH, et al: The effects of dietary w-3 polyunsaturated fatty acids on erythrocyte membrane phospholipids, erythrocyte deformability and blood viscosity in healthy volunteers. Atherosclerosis 1985;55:267–281.
78 Popp-Snijders C, Schouten JA, van Blitterswijk WJ, et al: Changes in membrane lipid composition of human erythrocytes after dietary supplementation of (n-3) polyunsaturated fatty acids. Maintenance of membrane fluidity. Biochim Biophys Acta 1986;854:31–37.
79 Bruckner G, Webb P, Greenwell L, et al: Fish oil increases peripheral capillary blood cell velocity in humans. Atherosclerosis 1987;66:237–245.

Howard R, Knapp, MD, Division of Clinical Pharmacology,
University of Iowa, Iowa City, IA 52242 (USA)

Simopoulos AP, Kifer RR, Martin RE, Barlow SM (eds): Health Effects of ω3 Polyunsaturated Fatty Acids in Seafoods. World Rev Nutr Diet. Basel, Karger, 1991, vol 66, pp 329–348

Blood Pressure-Lowering Effect of ω3 Polyunsaturated Fatty Acids in Clinical Studies

Peter Singer

Berlin, FRG

Introduction

Patients with even mild hypertension are widely considered to be at higher risk for developing cardiovascular disease [1]. Since no studies thus far have demonstrated a cardiac benefit to result from drug treatment of mild hypertension, non-pharmacological interventions are increasingly being recommeded as an alternative to early drug treatment in this large patient group [2]. Arterial hypertension, and even mild hypertension that can be found in nearly 90% of the hypertensive population, is frequently associated with other cardiovascular risk factors such as lipid disorders [3, 4], and the combination of such risk factors causes a greatly enhanced likelihood of vascular occlusive events. The fact that both elevated blood pressure and hyperlipidemia are favorably influenced by ω3 fatty acid intake makes it a very logical approach to modify the total cardiovascular risk of hypertensive patients. ω3 polyunsaturated fatty acids may also beneficially influence several other disorders (e.g. platelet hyperaggregability, stress hyperreactivity) frequently coexisting with hypertension. Consequently, fish oil can be anticipated to have a variety of possible benefits in hypertensive patients. Despite the relative ease with which one can measure blood pressure, it seems that it has fairly seldom been reported in clinical studies using fish oil supplementation.

Methodological Aspects

Studies of blood pressure changes in either normotensive subjects or in those with borderline elevations must naturally expect to measure the very small real changes which can occur in this setting. To detect such small changes in response to any intervention in a manageable number of subjects requires strictly controlled conditions, and this has certainly been true in the case of ω3 polyunsaturated fatty acids. Carefully designed studies are necessary to avoid the confounding effects of several well-known phenomenoma (table 1). Results from many of the human fish oil supplementation trials are of limited value because of their poor experimental design [5, 6]. For example, it was recently reported that casual blood pressure values taken in the doctor's office were initially higher, but consistently lower than self-recorded values after 4 weeks of fish oil supplementation suggesting an observer error due to the knowledge of the treatment [7]. For a number of reasons, data from animal studies can rarely be extrapolated directly to humans. For instance, spontaneously hypertensive rats (SHR) are dissimilar to man in their fatty acid composition and changes in the eicosanoid system [8, 9]. Nevertheless, the overall impression from the literature is that fish oil exerts a slight, but statistically significant hypotensive effect of fish oil at least in subjects with (mildly) elevated blood pressure.

Clinical Studies in Normotensive Volunteers

In 1981 appeared the first report [10] describing a blood pressure-lowering effect in normotensive volunteers. However, the authors doubted the significance of the results, since the values did not return to the pretreatment level after a washout/recovery period. The decrease of blood pressure was considered to result from the subjects' habituation to the measurements. Similar results were obtained by other authors [11], and although it is possible that their subjects also had failed to reach their true baseline during a run-in (pre-study) period, it cannot be excluded that the long-lasting hypotensive effect might rather be due to the longer-term incorporation of ω3 fatty acids in membranes, since ω3 polyunsaturated fatty acids in platelet and erythrocyte phosphoglycerides remained elevated over pretreatment values after withdrawal of the fish oil [10]. This possibility should be considered and tested in future studies, since there

Table 1. Confounding factors in studies on ω3 polyunsaturated fatty acids and blood pressure (modified after Knapp [5])

Problem	Solution
Relatively weak efficacy	standardized conditions (same day, daytime and room; trained observer, standardized device)
High variability of blood pressure	multiple measurements
Habituation to blood pressure measurement	run-in and recovery periods
Regression to the mean	run-in and recovery periods
Measurement errors and bias	random-zero sphygmomanometer, ambulatory monitors
Multiple effects of dietary changes	adequate controls
Placebo effects	adequate placebo

should be some eventual return to pretreatment blood pressure values unless the subjects have had a permanent lowering of their blood pressure by the fish oil. In general, in the majority of the 8 studies, which included 194 normotensive volunteers, only a weak and insignificant blood pressure-lowering effect in response to ω3 polyunsaturated fatty acids was found (table 2). Therefore, no final conclusion can be drawn from the results obtained in normotensive subjects.

Clinical Studies in Hypertensive Patients

Most of the 8 studies summarized in table 2 were conducted in patients with mild essential hypertension. It is likely that the elevated initial blood pressure level allowed the detection of a significant decrease of systolic and diastolic blood pressure, which was dissimilar to the findings in normotensive volunteers. This is supported by the observation that the decrease of a higher initial systolic blood pressure (160 mm Hg and above) was more pronounced (–17 mm Hg in recumbent position and –24 mm Hg in upright position) when compared with initial values below 160 mm Hg (–7 and –5 mm Hg, respectively) after 2 weeks of mackerel diet providing 5.0 g/day of ω3 polyunsaturated fatty acids (fig. 1).

In the first study reporting the efficacy of fish oil in a small group of 8 patients with mild essential hypertension [19] a significant decline of sys-

Table 2. Effects of fish oil on blood pressure in normotensive volunteers, patients with mild hypertension and patients with several risk factors

First authors	n	Dietary period weeks	Daily dose, g		Blood pressure mm Hg	
			EPA	DHA	systolic	diastolic
Normotensives						
Sanders, 1981 [10]	12	6	1.8	2.2	-13**	-12**
Lorenz, 1983 [12]	8	4	4.4	6.4	-3	-1
Mortensen, 1983 [13]	20	4	0.8	1.3	-4*	-4
Rogers, 1987 [14]	60	3–6	1.7	1.1	-7	-7*
Bruckner, 1987 [15]	11	3	?	?	-2	-3
v. Houwelingen, 1987 [16]	40	6	1.7	3.0	-3	-2
Demke, 1988 [17]	13	4	0.9	0.6	-3	0
Bach, 1989 [18]	30	5	1.1	1.4	-10*	-3
Hypertensives						
Norris, 1986 [19]	8	6	3.0	2.0	-10*	-2
Haller, 1987 [20]	10	6	2.2	1.4	-9*	-10*
Urakaze, 1988 [21]	12	4	1.0	0.5	-9*	-7*
Knapp, 1989 [22]	8	4	1.8	1.2	-1	-0.4
	8	4	9.0	6.0	-7*	-4*
Bønaa, 1989 [23]	157	10	3.3	1.8	-6**	-3*
Steiner, 1989 [7]	14	8	1.1	0.5	-8*	-4*
	14	8	2.2	1.0	-10*	-4*
Künzel, 1989 [24]	118	4	1.5	1.1	-11**	-9**
Schmidt, 1990 [25]	47	4	1.1	0.7	-5*	-5*
	36	12	1.1	0.7	-12**	-7**
Patients with several risk factors						
Hamazaki, 1984 [26]	12 HD	12	1.6	1.0	+2	-7*
Rylance, 1986 [27]	16 HD	8	3.6	2.4	-19**	-8*
Tenschert, 1988 [28]	10 HD	28	0.2	0.1	-16**	-8**
Kasim, 1988 [29]	22 DII	8	1.6	1.1	-14**	-8**
Olivieri, 1988 [30]	20 HLP	8	1.7	1.3	-12**	-9**
Künzel, 1989 [24]	479 HLP	4	1.5	1.1	-5**	-3**
Jensen, 1989 [31]	18 DI	8	2.0	2.6	-7*	-5*

HD = Hemodialysis, DI = diabetes mellitus (IDDM), DII = diabetes mellitus (NIDDM), HLP = hyperlipoproteinemia.
* $p < 0.05$; ** $p < 0.01$.

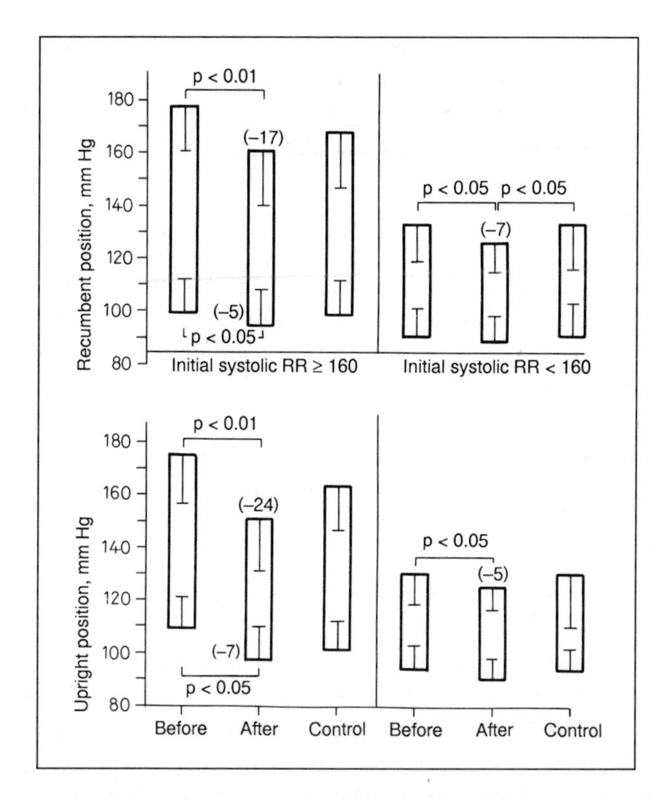

Fig. 1. Blood pressure-lowering effect of mackerel diet (2.2 g of eicosapentaenoic acid and 2.8 g of docosahexaenoic acid per day over 2 weeks) dependent on the initial level of systolic blood pressure (≥ 160 and < 160 mm Hg, respectively). Data taken from Singer et al. [35].

tolic blood pressure was described (table 2). In 6 of 7 studies (358 hypertensive patients) using a dose ranging from 1.5 to 5.1 g/day of fish oil a significant decrease of both systolic and diastolic blood pressure was reported.

One of the studies reporting a negative result in a 4-week trial of 10 ml fish oil per day (3.0 g/day ω3 PUFA) in mild essential hypertension found that extraordinarily high doses (50 ml/day, 15 g/day ω3 PUFA) of fish oil significantly reduced the blood pressure in hypertensive subjects [22]. This carefully designed study, which employed ambulatory blood pressure monitors, long run-in and recovery periods, and several parallel control groups, was the first to describe an increased urinary excretion of prostaglandin

I_3-metabolite and thromboxane A_3-metabolite as well as a depression of thromboxane A_2-metabolite in patients with mild hypertension. Therefore, more strictly controlled studies on dose-time response relationship are warranted from the practical point of view. The hitherto largest study conducted on a population basis [23] revealed a slight, but statistically significant decrease of systolic (–6 mm Hg) and diastolic (–3 mm Hg) blood pressure with fish oil supplementation, as well as a lower blood pressure at baseline in subjects who habitually consumed larger quantities of fish. This is an important result in a population-based study, since it indicates that such supplementation might be of interest from the primary prevention standpoint. From the viewpoint of total risk factor modification, it must be emphasized that more information on concomitant effects of fish oil on associated disorders such as hyperlipidemia, hyperinsulinemia, hyperaggregability, etc., in patients with (mild) hypertension are needed.

Clinical Studies in Patients with Several Risk Factors

Since 6 of 7 studies in patients with several risk factors other than hypertension (table 2) revealed a significant fall of systolic and diastolic blood pressure with fish oil supplementation, it might be reasonable to conclude that subjects with multiple risk factors are also appropriate candidates for fish oil supplementation. The only study performed over a long-term period of 28 weeks [28] was uncontrolled but did show a remarkable reduction in both systolic (–16 mm Hg) and diastolic (–8 mm Hg) blood pressures using a dosage as low as 0.3 g/day of ω3 polyunsaturated fatty acids (equivalent to 1 g/day of fish oil). If this observation is confirmed in a larger, controlled study, it raises the question whether this is only the result of longer-term supplementation of ω3 polyunsaturated fatty acids, or of a high sensitivity to fish oil in patients with renal insufficiency undergoing hemodialysis, whose response must necessarily be on the basis of altered vascular reflexes rather than alterations in renal function. The effect on coronary heart disease mortality of doses as low as 0.5 g/day of ω3 polyunsaturated fatty acids has been described in both epidemiological and intervention studies [32, 33]. This presents a promising challenge for further long-term low-dose trials to reduce blood pressure and other risk factors of coronary heart disease. The unique pharmacokinetics of ω3 polyunsaturated fatty acids (e.g. time- and dose-dependent accumulaton in cell membranes) provides a sound rationale for those studies.

Studies Using Fish Diets

A comparison of the hypotensive effects of canned mackerel with that obtained with a canned herring supplement providing different amounts of ω3 polyunsaturates was performed using the same study design as in clinical trials (table 3). A significant decrease of systolic and diastolic blood pressure was reported by a mackerel diet providing 5.0 g/day of ω3 polyunsaturated fatty acids over 2 weeks in normotensive volunteers [11] as well as in hypertensive [34] and hyperlipemic [35] patients. The cross-over experiments using canned herring as control (providing only half as much ω3 polyunsaturated fatty acids as in the mackerel period) showed no hypotensive effect of the herring diet [11, 34, 35]. Although a cross-over design has certain pitfalls in the study of hypertension, such as carryover and time-related effects, these results suggest a dose-response relationship. In one study conducted in patients with mild essential hypertension, only a fall of systolic blood pressure was found [34]. The most interesting finding from the time-dependence standpoint is the significant blood pressure-lowering effect in patients with mild essential hypertension of a dietary supplement of canned mackerel providing only 3 cans/week (equivalent to 1.2 g/day of ω3 polyunsaturated fatty acids) over an 8-month period [36]. This dose might be acceptable for chronic intake by the general population,

Table 3. Effect of mackerel diet on blood pressure

First author	n	Dietary period weeks	Daily dose, g		Blood pressure mm Hg	
			EPA	DHA	systolic	diastolic
Singer, 1983 [11]	15N	2	2.2	2.8	-15**	-7**
Singer, 1985 [34]	14H	2	2.2	2.8	-12**	-4
Singer, 1986 [35]	15HLP	2	2.2	2.8	-11**	-8*
$\geq$ 160 mm Hg	6HLP	2	2.2	2.8	-17**	-5*
$<$ 160 mm Hg	9HLP	2	2.2	2.8	-7*	-3*
Singer, 1986 [36]	12H	2	2.2	2.8	-13**	-11**
		56	0.5	0.7	-9**	-7*

N = Normotensive volunteers, H = patients with mild essential hypertension, HLP = patients with hyperlipoproteinemia.
* $p < 0.05$; ** $p < 0.01$.

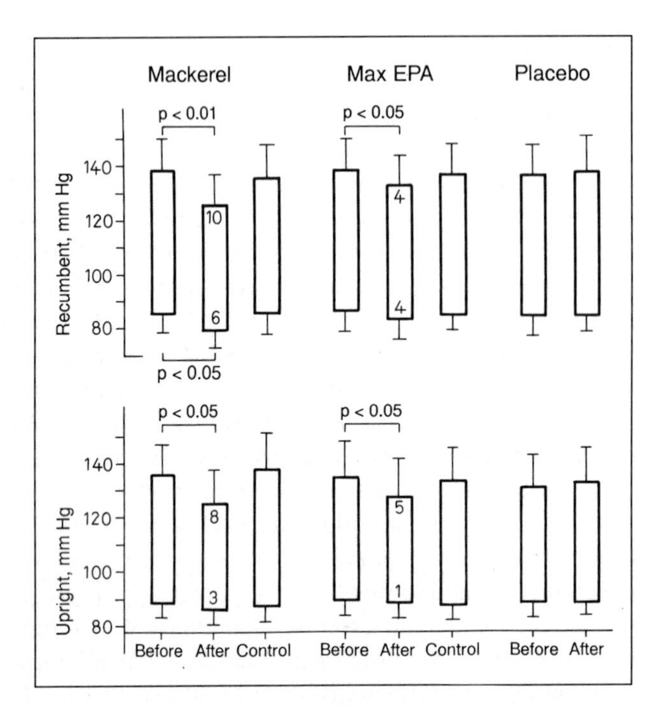

Fig. 2. Blood pressure-lowering effect of mackerel diet and fish oil supplementation, respectively, after adjustment to 2.2 g of eicosapentaenoic acid per day over 2 weeks [Own unpublished data]: Numbers within columns: change %.

being approximately the same level of intake as that which might be recommended from epidemiologic data [32, 33]. Moreover, the intake of canned fish considers the trend to fast foods in industrialized countries.

Overall, the results of the above mentioned studies indicate that a low-dose long-term ingestion of fatty fish can likewise be effective to reduce (mildly) elevated blood pressure. Unfortunately, no epidemiologic data on blood pressure and ω3 fatty acid intake are available, and because of the confounding effects of other nutrients which influence blood pressure no clear case can, so far, be made for a beneficial effect of fish consumption. The multicenter study of van Houwelingen et al. [16] using fish paste must be included in any discussion of low-dose ω3 supplementation effects. Unlike many of the studies in this area, this one included a parallel control group receiving the same amount of dietary fat supplement as meat paste. The authors observed a significant decrease of blood pressure after

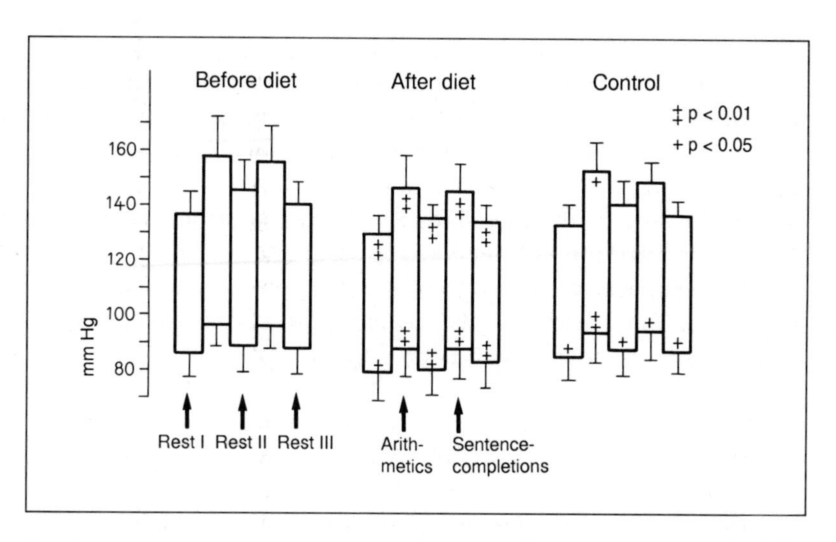

Fig. 3. Blood pressure response to acute psychological stress (arithmetics, sentence completions) before and after mackerel diet (2 cans/day over 2 weeks). Data taken from Singer et al. [37].

intake of mackerel paste, but an identical reduction in the meat paste control group. They, therefore, considered their study a negative one in regard to a hypotensive effect of fish supplementation.

From some of the available studies, it can be suggested that a fish-rich diet might be at least as effective at lowering blood pressure as fish oil supplementation. An intraindividual comparison of mackerel diet (2 cans/day) and fish oil supplementation adjusted to 2.2 g/day of eicosapentaenoic acid over 2 weeks showed a greater blood pressure-lowering effect of the former (fig. 2). It can be speculated that the bioavailability of dietary ω3 polyunsaturated fatty acids provided by canned mackerel is higher than by fish oil (some data supporting this conclusion are not demonstrated in this context). Moreover, components other than ω3 polyunsaturated fatty acids of the fish (or tomato pulp vehicle) ingested could likewise be effective to lower blood pressure.

Since excessive pressor response to stress is believed to play a role in the genesis of chronic hypertension, an important potential effect of ω3 fatty acids would be revealed if they could prevent such excessive blood pressure. An experiment addressing this point is presented in figure 3, in which an acute psychological challenge (consisting of arithmetics and sentence completion tasks) was given before and after 2 weeks of mackerel diet (2 cans/day) providing the amounts of ω3 polyunsaturated fatty acids as described [37]. The increments of blood pressure after the stressors remained unchanged, indicating that there is probably no gross change in vascular responsiveness to endogenous catecholamines during fish oil ingestion. However, the stress-induced increase of thromboxane B_2 during the stress test prior to the diet failed to occur after the fish diet [34]. More systematic studies are needed to understand the possible role of ω3 polyunsaturated fatty acids for the amelioration of the stress response in patients with essential hypertension.

Fish Oil plus Antihypertensive Drugs

Only a few reports have been published on the combination of fish oil and antihypertensive drugs. Steiner et al. [7] in a small group of 4 patients, who were given 120 or 240 mg verapamil/day in addition to 2.0 g ω3 polyunsaturated fatty acids (Promega), observed a blood pressure-lowering effect only on casual values, the self recorded values remaining unchanged. It was assumed that this was an observer error due to the knowledge of the treatment, or could also have been that the casual values in this case were actually obtained in a stressful situation of inadequate habituation to clinic measurements (compare Methodological Aspects).

In a recent study including renal allograft recipients [38], no significant changes in blood pressure were seen in a group receiving fish oil in addition to strong immunosuppressive (cyclosporin, azathioprine) and antihypertensive drugs (metoprolol, nifedipine, captopril). It was concluded that strong antihypertensive drugs could obscure a possible additional antihypertensive effect of low-dose fish oil.

In 47 patients with mild essential hypertension after a run-in period of 4 weeks a comparison of propranolol (80 mg/day) and fish oil (9 g/day) over 36 weeks alone and the combination of both (over 12 weeks) resulted in a blood pressure-lowering effect of fish oil similar to that of the β-blocker [39].

After propranolol- and fish oil-placebo, respectively, the values returned to the initial level. The combination of both propranolol plus fish oil amplified the hypotensive effect (fig. 4). During propranolol treatment there was a decrease of plasma noradrenaline, thromboxane B_2 formation and plasma renin activity that returned to the initial level after 4 weeks of

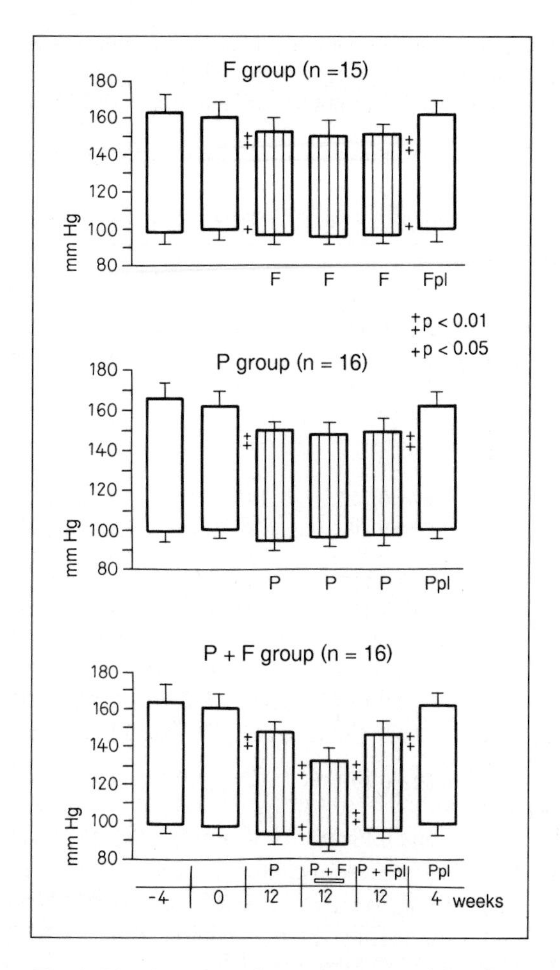

Fig. 4. Blood pressure in recumbent position from patients with mild essential hypertension before and after propranolol (80 mg/day), fish oil (9 g/day), the combination of both and placebo. Data taken from Singer et al. [39]. P = Propranolol; F = fish oil; pl = placebo.

propranolol placebo (table 4). During fish oil supplementation, there was the same decline of plasma noradrenaline and thromboxane B_2 formation, but an increase of plasma renin activity was found. The combination of propranolol plus fish oil resulted in a further decrease of plasma noradrenaline and thromboxane B_2 formation and a rise of plasma renin activity. Interestingly, total and low density lipoprotein (LDL) cholesterol, which

Table 4. Plasma noradrenaline (NA), thromboxane B_2 (TXB_2) formation and plasma renin activity (PRA) from patients with mild essential hypertension in groups receiving propranolol (80 mg/day), fish oil (9 g/day) and propranolol (P) plus fish oil (F); taken from Singer et al. [39]

Treatment period	NA nmol/l	TXB_2 ng/ml	PRA ng/ml/h
Group P (n = 16)			
At entry	1.83 ± 0.22**	138 ± 47**	1.83 ± 1.20**
On propranolol (12 weeks)	1.41 ± 0.19	79 ± 27	1.31 ± 0.54
On propranolol (24 weeks)	1.37 ± 0.20^{++}	85 ± 31^{++}	1.28 ± 0.76^{+}
On propranolol (36 weeks)	1.44 ± 0.18$^{++,**}$	74 ± 32$^{++,**}$	1.24 ± 0.87*
After propranolol placebo (4 weeks)	1.79 ± 0.17	151 ± 76	1.93 ± 0.79
Group F (n = 15)			
At entry	1.87 ± 0.25*	148 ± 53**	1.92 ± 113**
On fish oil (12 weeks)	1.49 ± 0.20	73 ± 32	3.43 ± 1.25
On fish oil (24 weeks)	1.53 ± 0.18^{+}	79 ± 32	3.27 ± 1.12^{++}
On fish oil (36 weeks)	1.62 ± 0.21	85 ± 31$^{++,**}$	2.87 ± 1.23$^{+,*}$
After fish oil placebo (4 weeks)	1.83 ± 0.17	135 ± 66	1.89 ± 1.05
Group P + F (n = 16)			
At entry	1.91 ± 0.18**	145 ± 59*	1.94 ± 1.31*
After propranolol (12 weeks)	1.45 ± 0.17**	83 ± 36**	1.32 ± 0.65**
After propranolol plus fish oil (12 weeks)	0.82 ± 0.19$^{++,**}$	41 ± 25**	3.65 ± 1.33$^{+,**}$
After propranolol plus fish oil placebo (12 weeks)	1.50 ± 0.18$^{+,**}$	92 ± 41**	1.38 ± 0.71*
After propranolol placebo (4 weeks)	1.93 ± 0.21	161 ± 87	1.76 ± 1.04

* Significance between successive treatment periods, $p < 0.05$.
** Significance between successive treatment periods, $p < 0.01$.
$^{+}$ Significance versus pretreatment level (at entry), $p < 0.05$.
$^{++}$ Significance versus pretreatment level, $p < 0.01$.

appeared to remain unchanged during propranolol therapy, was significantly lower during fish oil supplementation and during the combination of both (fig. 5). Serum triglycerides were likewise significantly lower and high density lipoprotein (HDL) cholesterol appeared significantly higher during fish oil supplementation both with and without concomitant pro-

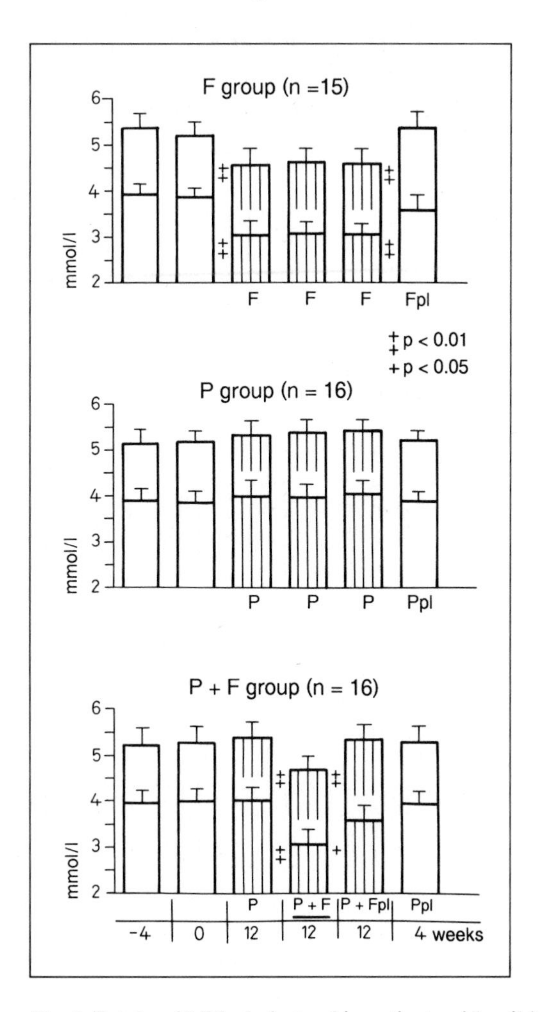

Fig. 5. Total and LDL cholesterol in patients with mild essential hypertension before and after propranolol (80 mg/day), fish oil (9 g/day), the combination of both and placebo. Data taken from Singer et al. [39]. P = Propranolol; F = fish oil; pl = placebo.

pranolol intake (fig. 6, 7). These results suggest a potentially important indication for administering fish oil in combination with antihypertensive drugs, which are known to develop adverse effects on serum lipids and lipoproteins. In addition, it could be an interesting question whether fish oil may have a drug-sparing effect.

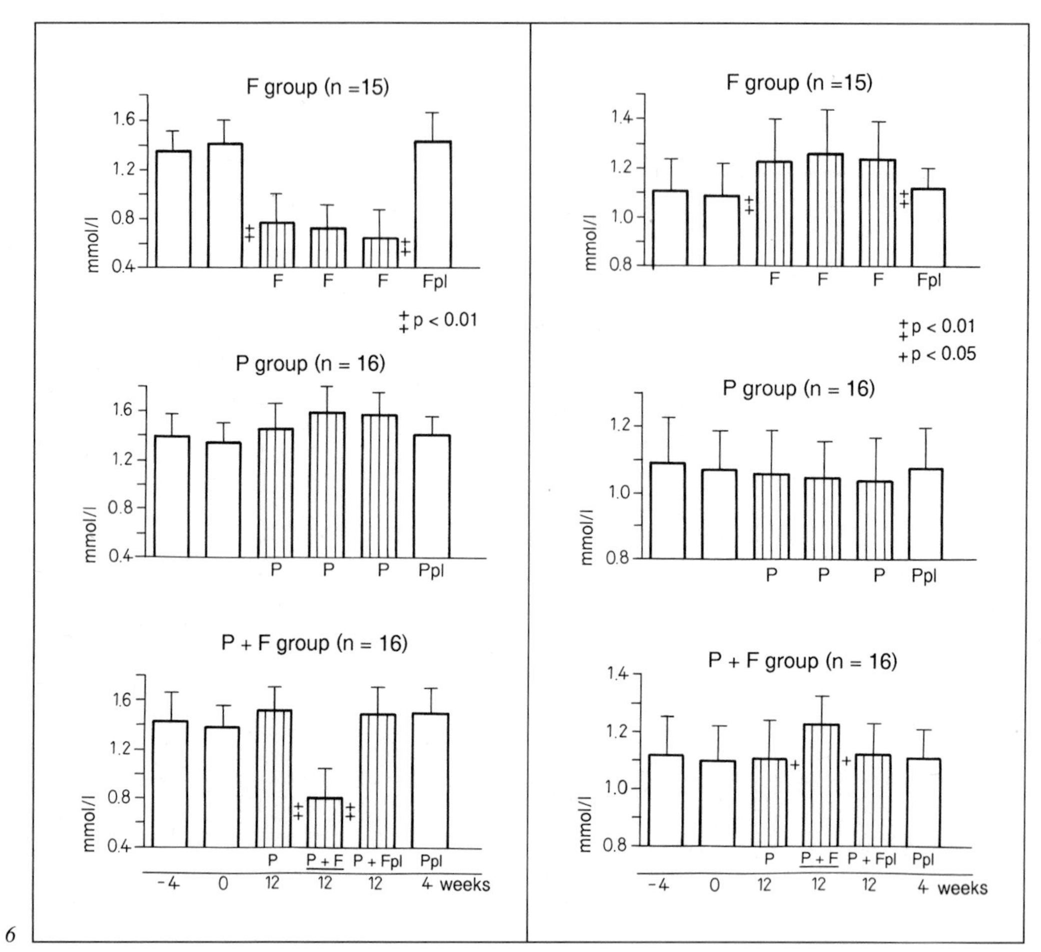

Fig. 6. Serum triglycerides in patients with mild essential hypertension before and after propranolol (80 mg/day), fish oil (9 g/day), the combination of both and placebo. Data taken from Singer et al. [39]. P = Propranolol; F = fish oil; pl = placebo.

Fig. 7. HDL cholesterol in patients with mild essential hypertension before and after propranolol (80 mg/day), fish oil (9 g/day), the combination of both and placebo. Data taken from Singer et al. [39]. P = Propranolol; F = fish oil; pl = placebo.

Suggested Mechanisms of the Blood Pressure-Lowering Effect by ω3 Polyunsaturated Fatty Acids

In clinical reports there are only occasional suggestions on the mechanisms involved in the assumed antihypertensive effects of ω3 polyunsaturated fatty acids. The most plausible explanations are a depression of thromboxane A_2 and an elevation of thromboxane A_3 and prostaglandin I_3 as indicated by excretion of their urinary metabolites [22]. Other contributing factors might be a blunted response of blood pressure to vasoconstrictor substances (noradrenaline, phenylephrine) and both intra- and extracellular electrolyte changes, such as a decrease of intracellular calcium, increased intracellular potassium and sodium excretion and a decline of serum sodium (table 5). The rise of plasma renin activity by fish

Table 5. Mechanisms of blood pressure-lowering effect by ω3 polyunsaturated fatty acids (clinical studies)

Parameters		Effect	First authors
Thromboxane A_2*		–	Knapp [22], Singer [34]
Thromboxane A_3*		+	Knapp [22]
Prostaglandin I_3*		+	Knapp [22]
Prostaglandin I_2*		(+)	Knapp [22], v. Schacky [41]
Plasma noradrenaline	Normotensives	–	Singer [11]
	Hypertensives	–	Urakaze [21]
Response to noradrenaline		–	Lorenz [12]
Response to phenylephrine		–	Knapp [42]
Intracellular calcium		–	Locher [43], Haller [20]
Intracellular potassium		+	Singer [34]
Sodium excretion		+	Lorenz [12], Urakaze [21]
Serum sodium		–	Singer [34]
Plasma renin activity		+	Singer [34], Jørgensen [40]
Microcirculation	Blood viscosity	–	Cartwright [44]
	Erythrocyte flexibility	+	Terano [45]
Renal blood flow		+	Düsing [46]
Baroreceptor reflex		+	Düsing [46]

* Measured as metabolite excretion or thromboxane B_2. + or (+) = Increased; – = decreased.

Table 6. Suggestions of future research needs

- Dose-response relationship between fish oil and blood pressure (preferably intraindividual comparisons)
- High-dose short-term and low-dose long-term studies using fish oil supplements (preferably intraindividual comparisons)
- Effect of fish oil dependent on the initial values of blood pressure
- Long-term (epidemiological) intervention trials on blood pressure control by fish-rich diet or fish oil
- Fish oil consumption and stroke rate
- Efficacy in patients with multiple risk factors, e.g. hypertension, lipid disorders, renal failure (hemodialysis) a.o.
- Effect of fish oil on life-style conditions, e.g. stress response in hypertensives
- Effect of fish oil on hemodynamics, e.g. cardiac output, peripheral resistance, baroreceptor function
- Combination of fish oil with nonpharmacological therapeutic measures, e.g. weight reduction, salt and alcohol restriction, potassium supplementation, physical activity
- Combination with antihypertensive drugs
- Studies on mechanisms involved, e.g. eicosanoids, electrolytes, vasoconstrictive substances

oil in patients with mild essential hypertension [34], as well as a high level in Eskimos [40] was interpreted as being a counterregulatory homeostatic effect. Finally, the improvement of microcirculation might contribute to the blood pressure-lowering effect of ω3 polyunsaturated fatty acids (table 5). However, no consistent results on hemodynamics are available. Thus, the literature up to the present does not allow any generalizations as to the relative importance of the many possible effects of ω3 polyunsaturated fatty acids on blood pressure regulation.

Conclusions

So far, the available data suggest that an antihypertensive effect of ω3 polyunsaturated fatty acids can usually be demonstrated in subjects with elevated blood pressure. This might be due to the phenomenon that higher initial levels are more effectively reduced (comparable to serum triglycerides according to Wilder's law of initial value) by fish oil supplementation. It must, however, be conceded that the efficacy of ω3 polyunsaturated fatty acids on blood pressure appears to be low when compared with established antihypertensive drugs, perhaps because of the prolonged time needed for relatively low doses to become adequately incorporated into membrane lipid at critical sites. On the other hand, fish oil has a unique therapeutic potential, since it might additionally benefit several disorders such as hyperlipidemia, hyperaggregability, etc., which frequently are associated with (mild) hypertension [3, 4, 47, 48]. Thus, it can be anticipated that dietary ω3 polyunsaturated fatty acids might provide an effective means of reducing the total cardiovascular risk profile via multiple effects including the reduction of blood pressure. Many questions remain to be answered from the practical point of view. Some suggestions are summarized in table 6. Apart from longitudinal studies, effects in the setting of multiple risk factors, dose-response relationships and possible potentiation of drug treatment might be attractive topics of future research.

References

1 Australian Therapeutic Trial in Mild Hypertension: Report by the Management Committee. Lancet 1980;i:1261–1267.
2 Roccella EJ: Nonpharmacological approaches to the control of high blood pressure. Final Report of the Subcommittee on Nonpharmacological Therapy of the 1984 Joint National Committee on Detection, Evaluation, and Treatment of High Blood Pressure. Hypertension 1986;8:444–467.
3 Heyden S, Schneider KA, Fodor GJ: Failure to reduce cholesterol as explanation for the limited efficacy of antihypertensive treatment in the reduction of CHD. Examination of the evidence from six hypertension intervention trials. Klin Wochenschr 1987;65:828–832.
4 Williams RR, Hunt SC, Wu LL, et al.: Concordant dyslipidemia, hypertension and early coronary disease in Utah families. Klin Wochenschr 1990;68(suppl XX):53–59.
5 Knapp HR: Omega-3 fatty acids, endogenous prostaglandins, and blood pressure regulation in humans. Nutr Rev 1989;47:301–313.
6 Herold PM, Kinsella JE: Fish oil consumption and decreased risk of cardiovascular disease: a comparison of findings from animal and human feeding trials. Am J Clin Nutr 1986;43:566–598.
7 Steiner A, Oertel R, Baettig B, et al.: Effect of fish oil on blood pressure and serum lipids in hypertension and hyperlipidemia. J Hypertens 1989;7(suppl 3):s73–s76.
8 Singer P, Wirth M, Gerike U, et al.: Age-dependent alterations of linoleic, arachidonic and eicosapentaenoic acids in renal cortex and medulla of spontaneously hypertensive rats. Prostaglandins 1984;27:375–390.
9 Singer P, Moritz V, Faulhaber HD: Some alterations of lipid metabolism in spontaneously hypertensive rats in relation to the stage of hypertension. Przeglad Lekarski 1986;43:341–349.

10 Sanders TAB, Vickers M, Haines AP: Effect on blood lipids and haemostasis of a supplement of cod-liver oil, rich in eicosapentaenoic and docosahexaenoic acids, in healthy young men. Clin Sci 1981;61:317–324.

11 Singer P, Jaeger W, Wirth M, et al.: Lipid and blood pressure-lowering effect of mackerel diet in man. Atherosclerosis 1983;49:99–108.

12 Lorenz R, Spengler U, Fischer S, et al.: Platelet function, thromboxane formation and blood pressure control during supplementation of the Western diet with cod liver oil. Circulation 1983;67:504–511.

13 Mortensen JZ, Schmidt EB, Nielsen AH, et al.: The effect of n–6 and n–3 polyunsaturated fatty acids on hemostasis, blood lipids and blood pressure. Thromb Haemostasis 1983;50:543–546.

14 Rogers S, James KG, Butland BK, et al.: Effects of a fish oil supplement on serum lipids, blood pressure, bleeding time, haemostatic and rheological variables. A double blind randomised controlled trial in healthy volunteers. Atherosclerosis 1987;63: 137–143.

15 Bruckner GG, Webb P, Greenwell L, et al.: Fish oil increases peripheral capillary blood cell velocity in humans. Atherosclerosis 1987;66:237–245.

16 van Houwelingen R, Nordøy A, van der Beek E, et al.: Effect of a moderate fish intake on blood pressure, bleeding time, hematology, and clinical chemistry in healthy males. Am J Clin Nutr 1987;46:424–436.

17 Demke DM, Peters GR, Linet OI, et al.: Effects of a fish oil concentrate in patients with hypercholesterolemia. Atherosclerosis 1988;70:73–80.

18 Bach R, Schmidt U, Jung F, et al.: Effect of fish oil capsules in two dosages on blood pressure, platelet functions, haemorheological and clinical chemistry parameters in apparently healthy subjects. Ann Nutr Metab 1989;33:359–367.

19 Norris PG, Jones CJH, Westen WJ: Effect of dietary supplementation with fish oil on systolic blood pressure in mild essential hypertension. Br Med J 1986;293:104–105.

20 Haller H, Passfall J, Bock A, et al.: Wirkung von Eicosapentaensäure (EPA) auf Blutdruck und intrazelluläres freies Calcium (Ca^{++}) bei essentieller Hypertonie. Hochdruck 1987;8:V2.

21 Urakaze M, Hanasaki N, Hamazaki T, et al.: Effects of fish oil concentrate on blood pressure of mild essential hypertensives (abstract). 8th International Symposium on Atherosclerosis, Rome, 1988, p 974.

22 Knapp HR, FitzGerald GA: The antihypertensive effects of fish oil – a controlled study of polyunsaturated fatty acid supplements in essential hypertension. N Engl J Med 1989;320:1037–1043.

23 Bønaa KH, Bjerve KS, Straume B, Gram IT, Thelle D: Effect of eicosapentaenoic and docosahexaenoic acids on blood pressure in hypertension. A population-based intervention trial from the Tromsø Study. N Engl J Med 1990;322:795–801.

24 Künzel U, Bertsch S, Beck N, et al.: Standardisierte Omega-3 Fettsäuren im Praxisalltag getestet. Therapiewoche 1989;39:1912–1917.

25 Schmidt U, Schenk N, Singer P: Blood pressure-lowering effect of fish oil in mildly hypertensive patients (abstract); in Simopoulos AP, Kifer RR, Martin RE, Barlow SM (eds): Health Effects of ω3 Polyunsaturated Fatty Acids in Seafoods. World Rev Nutr Diet. Basel, Karger, 1991, vol 66, p 556.

26 Hamazaki T, Nakazawa R, Tateno S, et al.: Effects of fish oil rich in eicosapentaenoic acid on serum lipids in hyperlipidemic hemodialysis patients. Kidney Int 1984; 26:81–84.
27 Rylance PB, Gordge MP, Saynor R, et al.: Fish oil modifies lipids and reduces platelet aggregability in haemodialysis patients. Nephron 1986;43:196–202.
28 Tenschert W, Rossodivita T, Rolf N, et al.: Langzeitwirkung einer niedrigdosierten diätetischen Gabe von Omega-3-Fettsäuren auf die Dyslipoproteinämie und das Blutdruckverhalten bei chronischen Hämodialysepatienten. Schweiz Med Rundschau 1988;77:973–977.
29 Kasim SE, Stern B, Khilnani S, et al.: Effects of Omega-3 fish oils on lipid metabolism, glycemic control, and blood pressure in type II diabetic patients. J Clin Endocrinol Metab 1988;67:1–5.
30 Olivieri O, Negri M, de Gironcoli M, et al.: Effects of dietary fish oil on malondialdehyde production and glutathione peroxidase activity in hyperlipidaemic patients. Scand J Clin Lab Invest 1988;48:659–665.
31 Jensen T, Stender S, Goldstein K, et al.: Partial normalization by dietary cod-liver oil of increased microvascular albumin leakage in patients with insulin-dependent diabetes and albuminuria. N Engl J Med 1989;321:1572–1577.
32 Kromhout D, Bosschieter EB, de Lezenne Coulander C: The inverse relation between fish consumption and 20-year mortality from coronary heart disease. N Engl J Med 1985;312:1205–1209.
33 Burr ML, Gilbert JF, Holliday RM, et al.: Effects of changes in fat, fish and fibre intake on death and myocardial reinfarction: Diet and reinfarction trial (dart). Lancet 1989;ii:757–760.
34 Singer P, Wirth M, Voigt S, et al.: Blood pressure- and lipid-lowering effect of mackerel and herring diet in patients with mild essential hypertension. Atherosclerosis 1985;56:223–235.
35 Singer P, Berger I, Wirth M, et al.: Slow desaturation and elongation of linoleic and α-linoleic acids as a rationale of eicosapentaenoic acid-rich diet to lower blood pressure and serum lipids in normal, hypertensive and hyperlipidemic subjects. Prostaglandins Leukotrienes Med 1986;24:173–193.
36 Singer P, Berger I, Lück K, et al.: Long-term effect of mackerel diet on serum lipids, blood pressure and thromboxane formation in patients with mild essential hypertension. Atherosclerosis 1986;62:259–265.
37 Singer P, Richter-Heinrich E: N-3 fatty acids reduce stress response in hypertensives. N-3 news 1987;2:2–3.
38 Urakaze M, Hamazaki T, Kashiwabara H, et al.: Favorable effect of fish oil concentrate on risk factors for thrombosis in renal allograft recipients. Nephron 1989;53: 102–109.
39 Singer P, Melzer S, Goschel M, et al.: Fish oil amplifies the effect of propranolol in mild essential hypertension. Hypertension 1990 (in press).
40 Jørgensen AK, Nielsen AH, Dyerberg J: Hemostatic factors and renin in Greenland Eskimos on a high eicosapentaenoic acid intake. Acta Med Scand 1986;219:473–479.
41 von Schacky C, Fischer S, Weber PC: Long-term effects of dietary marine omega-3 fatty acids upon plasma and cellular lipids, platelet function, and eicosanoid formation in humans. J Clin Invest 1985;76:1626–1631.

42 Knapp HR, Gregory D, Nolan S: Dietary polyunsaturates, vascular function and prostaglandins; in Galli C, Simopoulos AP (eds): Dietary ω3 and ω6 Fatty Acids. New York, Plenum, 1989, pp 283–295.
43 Locher R, Sachinidis A, Steiner A, et al.: Fischöl antagonisiert den Angiotensin II (A II)-induzierten Phosphatidylinositol (IP)-Stoffwechsel in glatten Muskelzellen. Hochdruck 1988;8:49.
44 Cartwright IJ, Pockley AG, Galloway JH, et al.: The effects of dietary omega-3 polyunsaturated fatty acids on erythrocyte membrane phospholipids, erythrocyte deformability and blood viscosity in healthy volunteers. Atherosclerosis 1985;55: 267–281.
45 Terano T, Hirai A, Hamazaki T, et al.: Effect of oral administration of highly purified eicosapentaenoic acid on platelet function, blood viscosity and red cell deformability in healthy human subjects. Atherosclerosis 1983;46:321–331.
46 Düsing R, Struck A, Scherf H, et al.: Dietary fish oil supplementation: Effects on renal hemodynamics and renal excretory function in healthy volunteers. Kidney Int 1987;31:268.
47 Hornych A, Safar M, Simon A, et al.: Thromboxane B_2 in borderline and essential hypertension; in Samuelsson B, Paoletti R, Ramwell P (eds): Advances in Prostaglandin Thromboxane and Leukotriene Research. New York, Raven Press, 1983, vol 2, pp 417–434.
48 Chen LS, Ito T, Ogawa K, et al.: Plasma concentrations of 6-keto-prostaglandin $F_{1\alpha}$, thromboxane B_2 and platelet aggregation in patients with essential hypertension. Jap Heart J 1984;25:1001–1009.

Peter Singer, MD, Heckmannufer 6, D-W–1000 Berlin 36 (FRG)

Simopoulos AP, Kifer RR, Martin RE, Barlow SM (eds): Health Effects of ω3 Polyunsaturated Fatty Acids in Seafoods. World Rev Nutr Diet. Basel, Karger, 1991, vol 66, pp 349–357

Dietary ω3 and ω6 Fatty Acids and Cardiovascular Responses to Pressor and Depressor Stimuli

David E. Mills

Department of Health Studies, University of Waterloo, Waterloo, Ontario, Canada

Introduction

The majority of studies of the effects of dietary ω3 fatty acids on blood pressure regulation in humans have focused on their effectiveness in reducing resting blood pressure in either normotensive or established hypertensive subjects [1]. These studies have failed to produce consistent results and have left unresolved the question of the potential role for dietary ω3 fatty acids in the treatment of hypertensive disorders [2]. It is possible that the physiological adaptations which occur in response to long-term elevations of blood pressure (e.g. resetting of baroreceptors, increased muscularity of resistance vessels), and which help maintain the hypertensive state, are resistant to alteration by dietary fatty acids and their metabolites. Consistent with this is the observation that blood pressure development in the genetically hypertensive spontaneously hypertensive rat is more responsive to attenuation by ω3 and ω6 fatty acids when intervention is initiated prior to the establishment of the elevated blood pressure (i.e. prior to weaning).

It was recently proposed that dietary ω6 and/or ω3 fatty acid supplementation might be more appropriate for use in the prevention of hypertension (i.e. primary intervention strategies) rather than in the reduction of blood pressure in established hypertensives (i.e. secondary and tertiary intervention strategies) [3]. In order to examine the effectiveness of dietary ω3 (and/or ω6) polyunsaturated fatty acids in a preventive context, studies of their effects on cardiovascular indices in the pre-hypertensive state, which correlate with future blood pressure development, are required. One such marker is that of cardiovascular reactivity to pressor stimuli [4, 5].

Blood Pressure Reactivity and Cardiovascular Disease

Much of the evidence for a relationship between reactivity and cardiovascular disease is based on retrospective studies. Several investigators have reported that patients with essential hypertension exhibit a hyperreactivity to physical and psychophysiological challenge [6, 7]. In addition, hyperreactivity and increased variability of pressor response to stressors have been reported in normotensive offspring of hypertensive parents and in borderline hypertensives, who have an increased risk for developing the disorder [8, 9]. To date, few prospective studies have examined the relationship between reactivity and disease. In one such study [10], reactivity to a cold pressor task was shown to be predictive of subsequent development of hypertension over a 45-year follow-up. In a second study, an exaggerated tachycardia in response to psychological challenge was shown to be a predictor of future blood pressure development in children [11]. Thus, exaggerated cardiovascular responsiveness to physical and behavioral challenge may either be a marker for, or involved in, the development of certain disorders such as ischemic heart disease and hypertension.

Animal Studies

Fatty Acids and Stress Reactivity

Several studies have investigated the effects of (1) administration of purified ω3 and ω6 fatty acids and (2) dietary supplementation with ω3 and ω6 rich oils on pressor responses to chronic psychosocial stress (isolation) in normotensive or genetically borderline hypertensive rats. While both 20:5 ω3 and 18:3 ω6 (but not 18:2 ω6 or 18:3 ω3), administered intraperitoneally in the non-esterified form via osmotic pump, attenuated pressor reactivity to stress in normotensive rats, 20:5 ω3 was accompanied by an augmentation of heart rate reactivity, whereas heart rate reactivity was suppressed in the group receiving 18:3 ω6 [12, 13]. Furthermore, on an equimolar basis, 20:5 ω3 appeared to be approximately 50% as effective as 18:3 ω6 in reducing pressor reactivity. These effects of the fatty acids did not appear to result from alterations in renal water and electrolyte excretion (unpublished) or vascular responsiveness to norepinephrine or angiotensin [14]. Subsequent studies examining the effects of dietary oil rich in various ω3 and ω6 fatty acids on reactivity to psychosocial stress demonstrated that 20:5 ω3, when administered in the triglyceride form as fish oil, had no effect on reactivity [15], even though the dosage was 100-fold that used in the previous study [13]. In contrast, 18:3 ω6 as evening primrose oil did attenuate reactivity. This observation, along with that of the suppression of circulating norepinephrine during stress by 18:3 ω6 but not 20:5 ω3 [15], suggests different sites and mechanisms of action for ω6 and ω3 fatty acids in this model, with the ω3 fatty acids possibly acting via peripheral mechanisms, such as vasodilation of resistance vessels.

Fatty Acids and Salt Sensitivity

Other animal models have been examined with respect to the effects of dietary fatty acids and reactivity to environmental pressor stimuli. In the genetically borderline hypertensive rat, a model in which blood pressure development is irreversibly affected by environmental stimuli, hypertension resulting from the replacement of drinking water with 1% saline was moderately attenuated by diets enriched with either sunflower (rich in 18:2 ω6) or fish oil (rich in 20:5 ω3 and 22:6 ω3) vs. controls receiving olive oil, and completely prevented by diets enriched with evening primrose oil (rich in 18:2 ω6 and 18:3 ω6) [16]. This effect of the various dietary fatty acids could not be explained by the observed changes in renal water and electrolyte excretion [16]. In contrast, fish oil administration augmented the pressor response to 1.5% saline in genetically spontaneously hypertensive rats vs. controls receiving a hydrogenated coconut oil/safflower oil mix [17].

Fatty Acids and Vascular Reactivity

Several investigators have examined the effects of dietary ω3 and ω6 fatty acids on vascular reactivity to pressor hormones as a possible mechanism of action for the observed effects on vascular reactivity. These studies suggest that 20:5 ω3, in either pure form or as fish oil, attenuates pressor responses to angiotensin II infusion [14, 18, 19]. Its effects on vascular responses to norepinephrine and serotonin are unclear [14, 19–21].

Human Studies

Fatty Acids and Stress Reactivity

To date, there are only a few published studies on the effects of dietary ω3 (or ω6) fatty acids on cardiovascular reactivity in man, and they are not consistent in their findings. The first examined the effects of dietary ω3 supplementation for 2-week periods, via canned mackerel (vs. herring), in a crossover design with a 3-month washout period between diets, on blood pressure and neuroendocrine responses to acute psychological stress (2 min each of mental arithmetic and sentence completion) in 14 male hypertensives [22]. In this study there were no group differences in blood pressure, circulating norepinephrine or plasma renin activity reactivity to the stress paradigm (i.e. the change from baseline to peak response), although the mackerel group (high 20:5 ω3 and 22:6 ω3) did exhibit a lower resting blood pressure after 2 weeks on the diet. Similarly, a recent study on 30 normotensive college students demonstrated that 4 weeks dietary supplementation with fish oil (18% 20:5 ω3 and 12% 22:6 ω3) had no effect on blood pressure, heart rate, or circulating catecholamine reactivity to a 2-min Stroop color-word conflict task [23] or to 2-min exposures to each of mental arithmetic, cold pressor, or favorable impression (stressful interview) tasks [24]. In contrast, supplementation with borage oil (23% 18:3 ω6) attenuated pressor and heart rate responses to these tasks vs. pretreatment [23, 24]. Interestingly, the improvement in performance on the Stroop test from pre- to post-diet was significantly greater in the borage-treated group, but not the fish oil-treated group, in comparison to an olive oil-treated control group [24].

In contrast, in a placebo-controlled (placebo unnamed), double-blind crossover trial with a placebo run-in period, fish oil supplementation reduced the systolic blood pressure and heart rate responses to a 6-min exercise test in 8 males with stable coronary heart disease, as well as reducing resting systolic blood pressure [25]. Although this study reported the statistics on the actual blood pressure and heart rate data rather than on the reactivity per se, it does appear that there was an attenuation of reactivity following fish oil supplementation.

More recently, in a double-blind study we examined the effects of dietary ω3 and ω6 supplementation on cardiovascular reactivity in unmedicated male and female borderline hypertensive subjects, all demonstrating a labile hypertension (unpublished). Following a 4-week placebo run-in period with safflower oil, subjects received 4.5 g/day of either fish oil (n = 9) or evening primrose oil (n = 8) for a 4-week period. Prior to the study and after each diet period subjects were tested for cardiovascular reactivity to a battery of acute stressors including mental arithmetic, cold pressor task, Stroop test, and cycle exercise test. Following supplementation the fish oil group exhibited lower resting diastolic blood pressure than the evening primrose oil group. Comparisons of post-supplement reactivity between the evening primrose and fish oil groups suggested that systolic blood pressure reactivity to mental arithmetic, cold pressor, and exercise and the diastolic blood pressure reactivity to exercise were lower in the evening primrose oil-supplemented group than in the fish oil-supplemented group.

Our most recently completed study involved an examination of the potential involvement of the cardiopulmonary and arterial baroreceptors in the ω6 and ω3 fatty acid effects on blood pressure regulation (unpublished). Normotensive male college students were divided into 3 treatment groups (double-blind, n = 10/group) receiving 4 weeks supplementation with 4.5 g/day of either safflower oil (79% 18:2 ω6), borage oil (23% 18:3 ω6), or fish oil (18% 20:5 ω3 and 12% 22:6 ω3). Prior to and following dietary supplementation, subjects were exposed to consecutive 5-min periods of lower body negative pressure (LBNP) at −10 and −40 mm Hg. This procedure induces a net movement of blood volume into the legs, thereby reducing thoracic volume and unloading either the low pressure cardiopulmonary baroreceptors (−10 mm Hg) or both the low and high pressure (arterial) baroreceptors (−40 mm Hg) in a simulation of hemorrhage. Blood pressure, heart rate, forearm vascular resistance, and circulating norepinephrine responses to the stimuli were measured at baseline and at each pressure level. While dietary supplements did not affect resting blood pressure, supplementation with borage oil augmented the catecholamine and forearm vascular resistance increases in response to −40 mm Hg LBNP, resulting in higher blood pressure levels during LBNP than the other diet groups. This response suggests a shift in baroreceptor sensitivity in response to 18:3 ω6. Fish oil supplementation did not appear to influence cardiovascular or neuroendocrine reactivity to the hypotensive stimulus, but decreased resting forearm vascular resistance and increased forearm blood flow.

While not being definitive, the results of human studies on ω3 and ω6 fatty acids and stress reactivity suggest that the mechanism of action of certain ω6 fatty acids include the alteration of baroreflex function in their modification of cardiovascular function, whereas that of the ω3 fatty acids appears to include peripheral mechanisms. Furthermore, it appears that dietary 18:3 ω6 is more likely to affect cardiovascular reactivity in man than the ω3 fatty acids, which exert their effects more on baseline cardiovascular function.

Fatty Acids and Salt Sensitivity

There has been no prospective work examining the relationship among dietary ω3 fatty acids, salt intake, and hypertension in man. However, several retrospective studies suggest that dietary sodium intake has a greater effect on resting blood pressure in Japanese and Okinawan populations [26, 27] and that ω3 fatty acids are unable to protect against the pressor effects of sodium in man.

Fatty Acids and Vascular Reactivity

Several investigators have studied the effects of ω3 fatty acids on vascular reactivity to pressor agents as a possible mechanism of their effects on the cardiovascular system. In one study, cod liver oil (40 ml/day) over 25 days in 8 normotensive men decreased the systolic blood pressure response to infused norepinephrine, with no observed change in pressor reactivity to angiotensin [28]. In a second report, dietary supplementation with menhaden oil (50 ml/day or 15 g ω3 fatty acids/day) for 28 days in mild hypertensives produced no change in pressor response to an infusion of phenylephrine, an α-adrenergic agonist with minimal direct cardiac effects [2]. Interestingly, in the latter study subjects taking fish oil tended to develop bradycardia, suggesting the possible involvement of central regulatory mechanisms.

In a different model, that of pregnancy-induced hypertension (preeclampsia) in which exaggerated pressor responsiveness to angiotensin is one of the earliest clinical signs, administration of evening primrose oil attenuated diastolic pressor responses to angiotensin and reduced plasma renin concentration vs. untreated controls [29]. The effects of ω3 fatty acids in this model have not been reported, possibly due to the concern that the decline in platelet count [28] and increase in serum uric acid [22] which have been reported following fish oil administration resemble those which accompany mild to severe pregnancy-induced hypertension.

Conclusions

The regulation of blood pressure under normal and pathophysiological conditions is under the complex control of many systems, ranging from local tissue autoregulatory processes to behavioral and reflex responses coordinated by the central nervous system. Furthermore, 'essential' hypertension actually consists of several subgroups (e.g. high- and low-renin types), probably representative of differing etiologies. Our limited understanding of the relationship between ω3 and ω6 fatty acids and the regulation of blood pressure by the cardiovascular system reflects the relative paucity of information on the effects of dietary polyunsaturated fatty acids on the basic physiological processes underlying the regulation of systemic blood pressure.

Several areas deserving further research include the effects of ω3 and ω6 fatty acids on (1) baroreflex regulatory processes, (2) central autonomic function, (3) renal function, (4) cardiac function, including responsiveness to autonomic stimuli, (5) local vascular responsiveness to endocrine and autonomic stimuli, (6) membrane function in excitable tissues, and (7) neuroendocrine function, including hormones produced by the heart (atrial natriuretic factor) and the central nervous system. When the relationship between ω3 and ω6 fatty acids and these control systems is understood, it will then be possible to better target their use in the prevention and treatment of the appropriate pathophysiological processes. Psychophysiological stress and other stimuli serve as tools by which the healthy and pathologic cardiovascular system can be perturbed in order to assess the function of these various control processes, as well as the effects of dietary factors in their regulation.

References

1 Knapp HR: Omega-3 fatty acids, endogenous prostaglandins, and blood pressure regulation in humans. Nutr Rev 1989;47:301–313.
2 Knapp HR, Gregory D, Nolan S: Dietary polyunsaturates, vascular function and prostaglandins; in Galli C, Simopoulos A (eds): Dietary ω3 and ω6 Fatty Acids. New York, Plenum, 1989, pp 283–295.
3 Mills DE: Essential fatty acids and hypertension: Potential for use in primary and secondary intervention strategies. Nutrition 1989;5:260–263.
4 Krantz DS, Manuck SB: Acute psychophysiologic reactivity and risk of cardiovascular disease: a review and methodologic critique. Psych Bull 1984;96:435–464.

5 Steptoe A, Melville D, Ross A: Behavioral response demands, cardiovascular reactivity and essential hypertension. Psychosom Med 1984;46:33–48.
6 Hull DH, Wolthius RA, Cortese T, et al: Borderline hypertension versus normotension: Differential response to orthostatic stress. Am Heart J 1977;94:414–420.
7 Drummond PD: Cardiovascular reactivity in borderline hypertension during behavioral and orthostatic stress. Psychophysiology 1985;22:621–629.
8 Nestel PJ: Blood pressure and catecholamine excretion after mental stress in borderline hypertension. Lancet 1969;i:692–694.
9 Harris SE, Sokolow M, Carpenter LG, et al: Response to psychologic stress in persons who are potentially hypertensive. Circulation 1953;7:874–879.
10 Wood DL, Sheps SG, Elveback LR: Cold pressor test as a predictor of hypertension. Hypertension 1984;6:301–306.
11 von Eiff AW, Gogolin E, Jacobs U, et al: Heart rate reactivity under mental stress as a predictor of blood pressure development in children. J Hypertension 1986;3:S89–S92.
12 Mills DE, Ward RP: Effects of essential fatty acid administration on cardiovascular responses to stress in the rat. Lipids 1986;21:139–142.
13 Mills DE, Ward RP: Effects of eicosapentaenoic acid on cardiovascular responses to stress. Proc Soc Exp Biol Med 1986;182:127–131.
14 Mills DE, Ward RP: Attenuation of psychosocial stress-induced hypertension by gamma linolenic acid (GLA) administration in rats. Proc Soc Exp Biol Med 1984;176:32–37.
15 Mills DE, Ward RP, Huang YS: Effects of n–6 and n–3 fatty acid supplementation on cardiovascular and endocrine responses to stress in the rat. Nutr Res 1989;9:405–414.
16 Mills DE, Ward RP, Mah M, et al: Dietary n–6 and n–3 fatty acids and salt-induced hypertension in the borderline hypertensive rat. Lipids 1989;24:17–24.
17 Codde JP, Beilin LJ, Croft KD, et al: The effect of dietary fish oil and salt on blood pressure and eicosanoid metabolism of spontaneously hypertensive rats. J Hypertens 1987;5:137–142.
18 Hui R, St-Louis J, Falardeau P: Antihypertensive properties of linoleic acid and fish oil omega-3 fatty acids independent of the prostaglandin system. Am J Hypertens 1989;2:610–617.
19 Mills DE, Summers MR, Ward RP: Gamma linolenic acid attenuates cardiovascular responses to stress in borderline hypertensive rats. Lipids 1985;20:573–577.
20 Yin K, Croft KD, Beilin LJ: Effect of pure eicosapentaenoic acid feeding on blood pressure and vascular reactivity in spontaneously hypertensive rats. Clin Exp Pharmacol Physiol 1988;15:275–280.
21 Lockette W, Webb RC, Culp BR, et al: Vascular reactivity and high dietary eicosapentaenoic acid. Prostaglandins 1982;24:631–635.
22 Singer P, Wirth M, Voigt S, et al: Blood pressure- and lipid-lowering effect of mackerel and herring diet in patients with mild essential hypertension. Atherosclerosis 1985;56:223–235.
23 Mills DE, Prkachin KM, Harvey KA, et al: Dietary fatty acid supplementation alters stress reactivity and performance in man. J Hum Hypertens 1989;3:111–116.
24 Mills DE, Ward RP: Dietary n–6 and n–3 fatty acids and stress-induced hypertension; in Horrobin DF (ed): Omega 6 Fatty Acids: Pathophysiology and Roles in Clinical Medicine. New York, Liss, 1990, pp 145–156.

25 Mehta JL, Lopez LM, Lawson D, et al: Dietary supplementation with omega-3 polyunsaturated fatty acids in patients with stable coronary heart disease. Am J Med 1988;84:45–52.
26 Prior AM: Migration, hypertension, and Pacific perspectives for prevention; in Lovenberg W, Yamori Y (eds): Nutritional Prevention of Cardiovascular Disease. New York, Academic Press, 1984, pp 137–153.
27 Tseng WP: Blood pressure and hypertension in an agricultural and a fishing population in Taiwan. Am J Epidemiol 1967;86:513–525.
28 Lorenz R, Spengler U, Fischer S, et al: Platelet function, thromboxane formation and blood pressure control during supplementation of the Western diet with cod liver oil. Circulation 1983;67:504–511.
29 Broughton Pipkin F, Morrison RA, O'Brien PMS: The effect of dietary supplementation with linoleic and γ-linolenic acids on the pressor and biochemical response to exogenous angiotensin II in human pregnancy. Prog Lipid Res 1986;25:425–429.

David E. Mills, MD, Department of Health Studies, University of Waterloo, Waterloo, Ontario N2L 3G1 (Canada)

Simopoulos AP, Kifer RR, Martin RE, Barlow SM (eds): Health Effects of ω3 Polyunsaturated Fatty Acids in Seafoods. World Rev Nutr Diet. Basel, Karger, 1991, vol 66, pp 358–366

Influence of ω3 Fatty Acids on Blood Lipids

T.A.B. Sanders

Department of Food and Nutrition Sciences, King's College London, University of London, UK

Introduction

It is well recognised that plasma lipoprotein concentrations are major risk factors for cardiovascular disease: high concentrations of cholesterol associated with very low density lipoproteins (VLDL), intermediate density lipoproteins (IDL) and low density lipoproteins (LDL) but not with high density lipoproteins (HDL) are associated with increased risk. Raised concentrations of triglycerides when associated with increased VLDL or IDL or chylomicron remnants are also predictors of risk. Recently raised concentrations of lipoprotein Lp(a) have been added to this list of risk markers for atherosclerosis and coronary heart disease (CHD). Lp(a) is believed to be an inhibitor of tissue plasminogen activator because of structural similarities between its constituent apoprotein apo(a) and plasminogen. Interestingly Lp(a) levels are increased by the HMGCoA reductase inhibitor simvastatin [1] but decreased by nicotinic acid [2].

Cross-cultural studies show an inverse relationship between the prevalence of hypertension and hyperlipidemia. For example, in Japan the prevalence of hypertension is high but the median plasma cholesterol concentration is low. However, many patients with hyperlipidemia are hypertensive. The Lipid Research Clinics program [3] found the prevalence of hypertension to be double the normal in patients with Type IIB or IV hyperlipidemia. Familial combined hyperlipidemia is often associated with hypertension and the term familial dyslipidemic hypertension [4] has been coined to describe the inherited condition. It is usually associated with hyperinsulinemia and in this respect may be similar to 'Syndrome-X'.

Individuals with 'Syndrome-X' are at greatly increased risk of CHD, these individuals typically have Type IV hyperlipidemia, impaired glucose tolerance, hypertension and hyperuricemia and are usually overweight.

The link between lipids and blood pressure does not stop there because many commonly used antihypertensive agents have influences on plasma lipid concentrations and blood glucose control [5, 6]. Diuretic drugs and Beta blockers increase VLDL levels and decrease HDL cholesterol. It has been argued that the failure of anti-hypertensive drugs in the primary prevention of coronary heart disease may be due to their adverse effects on plasma lipoprotein concentrations and glucose control. On the other hand, alpha-1 blockers do lower plasma triglycerides and increase HDL cholesterol. This may be relevant because animal studies have found that fish oil decreases alpha rather than beta adrenergic activity [7]. Commonly used lipid-lowering drugs do not affect blood pressure although some may impair blood glucose control in diabetics, for example nicotinic acid.

The ω3 polyunsaturated fatty acids have a different spectrum of effects compared to commonly used anti-hypertensive or lipid-lowering agents. An understanding of the way in which ω3 fatty acids influence the composition of plasma lipoproteins and the formed element of blood is necessary in order to explain their biological effects.

Influence on Plasma Lipids in Normal Subjects

The ω3 polyunsaturated fatty acids have different effects on plasma lipids from linoleic acid and the effects observed are strongly dose-related. Both eicosapentaenoic acid (20:5 ω3, EPA) and docosahexaenoic acid (22:6 ω3, DHA) can markedly lower plasma triglycerides and VLDL concentrations. Linoleic acid and linolenic acid do not show the same effect at comparable doses [8, 9]. These changes are also accompanied by a marked increase in the proportion of EPA and DHA in plasma and blood cell phospholipids mainly at the expense of linoleic and arachidonic acid, thus simultaneously decreasing the amount of substrate for active eicosanoid formation and increasing the proportion of inhibitor. The extent of these compositional changes is dose-dependent and is influenced by the intake of linoleic acid.

Studies in healthy volunteers do not show any significant change in total or LDL cholesterol with moderate intakes of fish oil [8, 10, 11]. Very high intakes of fish oil (24 g ω3/day) do lower the concentration of both LDL cholesterol and LDL apoB by decreasing the rate of LDL synthesis [12]. However, the reduction in LDL pool size is smaller than would be predicted from the decrease in synthesis. Moreover, despite the reduced LDL pool size there was no increase in fractional catabolic rate of LDL. This might imply down-regulation of the LDL receptors by fish oils.

Moderate intakes of EPA and DHA increase HDL2 cholesterol but high intakes decrease HDL cholesterol similar to that seen with very high intakes of linoleic acid. Dietary EPA and DHA as fish oil but not linolenic acid as linseed oil [9] decrease lecithin cholesterol acyl transferase activity and this change is accompanied by an increase in HDL2/HDL3 ratio. The increase in the average HDL particle size probably reflected reduced cholesteryl ester acceptor capacity within the smaller pool of VLDL, as well as the decline in lipid transfer activity in plasma involving transfer protein itself, LDL and HDL.

Effects on Chylomicron Clearance

Chylomicron clearance is increased in subjects following the consumption of oily fish [10]. Fish oil also tends to lead to less post-prandial hyperlipemia compared with olive oil [8]. Weintraub et al. [13] studied the effect of different types of dietary fat on postprandial lipoprotein levels. Subjects with normal apoE phenotypes received three isocaloric diets for 25-day periods. The diets provided 42% energy from fat, with 26–28% from monounsaturated fatty acid. The test diets were as follows: saturated-fat diet contained 67% saturated fat and 5% polyunsaturated fat, ω6 diet contained 30% saturated and 42% polyunsaturated (linoleic), and ω3 diet 30% saturated and 42% polyunsaturated fat (18% long-chain ω3 fatty acids, mainly EPA, and 82% linoleic acid). The subjects then received two fat tolerance tests on separate days, one with the dietary fat and the other with saturated fat. The degree of post-prandial lipemia was decreased with the polyunsaturated fats, especially with the ω3 diet, compared with the saturated-fat treatments. The saturated-fat load led to the most post-prandial lipemia regardless of previous dietary treatment. It seems likely that improved clearance of chylomicrons occurred as a result of less competition from endogenously synthesised triglycerides because neither the activity of lipoprotein lipase nor hepatic triglyceride lipase were effected.

Hyperlipidemias

Type IIa

Patients with familial hypercholesterolemia typically have low plasma triglyceride and VLDL concentrations. These are slightly lowered by fish oil supplements. However, fish oil supplements have no effect on total or LDL cholesterol concentrations [10, 14]. Patients with mild hypercholesterolemia also show no benefit in terms of cholesterol reduction [15].

Type IIb

Fish oil supplements or diets providing EPA and DHA lead to a reduction in VLDL triglycerides and cholesterol but no significant change in LDL cholesterol [10, 16, 17]. HDL cholesterol concentrations tend to increase with fish oil supplements. Increases in LDL have been noted in some but not all patients. There also appears to be sex differences with regard to the influence on HDL [17]. Significant reductions in blood pressure were also noted on treatment [17].

Type III

This lipoprotein abnormality is characterised by the accumulation of IDL which have a late pre-beta electrophoretic mobility. Patients with the disorder have the abnormal apoE2 phenotype. Fish oil supplements have been found to normalise the electrophoretic profile in most patients and this is accompanied by a reduction in plasma triglycerides and cholesterol [18]. Significant reductions in blood pressure were noted on treatment.

Type IV/V

Fish oil supplements have a marked triglyceride lowering effect in these patients [10, 14, 19]. Total cholesterol concentrations fall or remain unchanged. However, LDL cholesterol concentrations which tend to be low in these patients usually do rise. This increase in LDL is also seen with ester concentrates low in cholesterol [20]. HDL cholesterol concentrations either remain unchanged or increase. Saynor et al. [21] have shown that fat tolerance is improved in hypertriglyceridemic patients following treatment with fish oil. The response of patients with the Type V phenotype is similar to those with Type IV and the reduction in plasma triglycerides can be quite dramatic even with relatively small amounts of fish oil (10–20 ml/day).

Diabetes

Increases in LDL cholesterol have been observed in patients with insulin-dependent diabetes mellitus (IDDM) treated with fish oil [22, 23] as well as in non-insulin dependent diabetes mellitus (NIDDM) [19, 24]. Glucose control may be impaired in NIDDM.

Are the Triglyceride-Lowering Effects of Fish Oil Sustained?

As little as 6 g fish oil/day (2 g ω3 fatty acids) has a triglyceride-lowering effect in hypertriglyceridemic patients. The more commonly used dose is 3 g EPA and DHA ω3 fatty acids/day. Schectman et al. [24] have suggested on the basis of a study on a small group of patients that triglyceride-lowering effects of fish oils cannot be sustained. This suggestion is not supported by the other larger controlled trials. Miller et al. [25], in a randomised controlled trial for three months, showed that the triglyceride-lowering effect of 10 g/day MaxEPA (3.2 g ω3 fatty acids) was sustained. Moreover, Saynor et al. [21] have shown that the effect is sustained for years. The study of Schectman employed an ester concentrate as opposed to a triglyceride oil. The failure to sustain the triglyceride-lowering effect in that study could well be related to poor patient compliance.

Influence on Lp(a)

A study [26] of patients with elevated levels of Lp(a) found that fish oil led to a significant reduction in this risk factor. Similar observations have been made for nicotinic acid but not for gemfibrozil. Other workers have not confirmed these findings with fish oil [W.S. Harris, personal communication; J.P. Deslypere, personal communication]. However, there is considerable controversy regarding the standardization of Lp(a) assays. Consequently, further work in this area is needed.

Mechanisms

VLDL synthesis is decreased by moderate intakes of ω3 fatty acids [27] and at high intakes (> 20 g ω3 fatty acids/day) LDL synthesis is also decreased. There is good animal evidence that both EPA and DHA

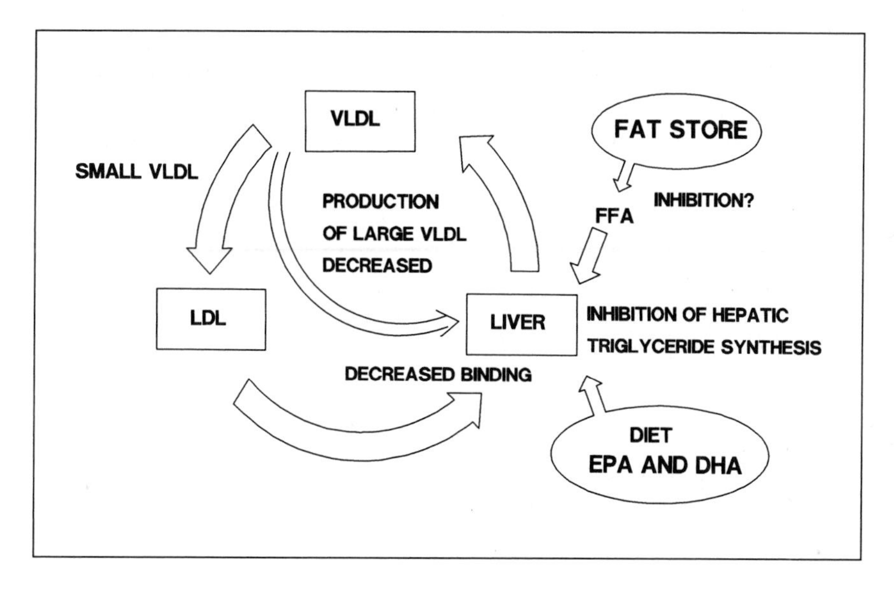

Fig. 1. Influence of EPA and DHA on lipoprotein metabolism.

decrease VLDL triglyceride synthesis. However, as with other triglyceride-lowering agents such as fibrates, LDL levels tend to increase in patients with Type IV and Type V hyperlipoproteinemias. This is a general phenomenon seen with most forms of triglyceride-lowering therapy [28] including caloric restriction as well as fibrate like drugs such as gemfibrozil. A likely explanation for this phenomenon is that a moderate intake of fish oil decreases hepatic triglyceride synthesis so that smaller than normal VLDL particles are secreted [29]. The small particles are known to be more readily converted to LDL than the larger triglyceride-rich ones (fig. 1). Schectman et al. [30] have tried to explain the heterogeneity of response to EPA and DHA on the basis of the initial LDL particle size: patients with a cholesterol:apoB ratio > 1.4 show a marked rise in LDL apoB levels.

Although EPA and DHA do act as inhibitors of hepatic triglyceride synthesis, it is possible that they also affect the activity of hormone-sensitive lipase, thus decreasing input of fatty acids for triglyceride synthesis. Lower free fatty acid (FFA) levels have been reported in some studies [Singer, personal communication] following the consumption of oily fish. Further studies are needed to investigate if EPA and DHA influence adrenergic stimulation of lipolysis.

It has recently been proposed [31] that LDL needs to undergo oxidative modification before it is taken up by macrophages. Macrophages are also capable of taking up IDL and chylomicron remnants. Foam cells which constitute an early atherosclerotic lesion are derived from lipid-laden macrophages. It has been suggested that antioxidants may protect against modification of LDL and this might explain the antiatherogenic properties of probucol. It is well known that EPA and DHA increase the requirement for antioxidant nutrients. Further studies are needed to assess the effect of EPA and DHA on the modification of LDL.

Conclusion

Fish oil is not effective in treating hypercholesterolemia but it is an effective triglyceride-lowering agent in Type IIB, III, IV and V hyperlipidemia. EPA and DHA decrease the plasma concentration of several atherogenic lipoproteins (VLDL, IDL, chylomicron remnants) and at moderate doses may increase the ratio of HDL2/HDL3 cholesterol. On the other hand, they lead to an increase in LDL levels in some patients. The potentially beneficial reduction in triglycerides and VLDL cholesterol and increase in HDL cholesterol needs to be offset by any adverse effects such as increases in LDL concentration and impaired diabetic control. It needs to be born in mind that the fatty acid composition of the lipoproteins is also altered and this might affect their atherogenic potential. It is possible that the cardioprotective effects of fish oils are independent of their effects on plasma lipids.

References

1 Kostner GM, Gavish D, Leopold B, et al: HMGCoA reductase inhibitors lower LDL cholesterol without reducing Lp(a) levels. Circulation 1989;80:1313–1319.
2 Carlson LA, Hamstem A, Apslund A: Pronounced lowering of serum levels of lipoprotein Lp(a) in hyperlipidaemic subjects treated with nicotinic acid. J Intern Med 1989;226:271–76.
3 Criqui MH, Cowan LD, Heiss G, et al: Frequency and clustering of nonlipid coronary risk factors in dylipoproteinemia: The Lipid Research Clinics Prevalence Study. Circulation 1986;73:40–41.
4 Hunt SC, Wu LL, Hopkins PN, et al: Apolipoprotein, low density lipoprotein subfraction, and insulin associations with familial combined hyperlipidemia: Study of Utah patients with Familial Dyslipidemic Hypertension. Arteriosclerosis 1989;9: 335–344.

5 Ames RP: Negative effects of diuretic drugs on metabolic risk factors for coronary heart disease. Am J Cardiol 1983;51:632–638.
6 Lowenstein J, Neusy A: Effects of prazosin and propanolol on serum lipids in patients with essential hypertension. Am J Med 1984;76(suppl 2A):79–84.
7 Reibel DK, Holahan MA, Hock CE: Effects of dietary fish oil on cardiac responsiveness to adrenoceptor stimulation. Am J Physiol 1988;254:H494–H498.
8 Sanders TA, Hinds A, Pereira CC: Influence of n–3 fatty acids on blood lipids in normal subjects. J Intern Med 1989;225(suppl 731):99–104.
9 Abbey M, Clifton P, Kestin M, et al: Effect of fish oil on lipoproteins, lecithin:cholesterol acyltransferase, and lipid transfer protein activity in humans. Arteriosclerosis 1990:10:85–94.
10 Harris WS: Fish oils and plasma lipids and lipoprotein metabolism in humans: a critical review. J Lipid Res 1989;30:785–807.
11 Rogers S, James KS, Butland BK, et al: Effects of fish oil supplement on serum lipids, blood pressure, bleeding time and rheological variables. A double blind randomised controlled trial in healthy volunteers. Atherosclerosis 1987;63:137–143.
12 Illingworth DR, Harris WS, Connor WE: Inhibition of low density lipoprotein synthesis by dietary omega-3 fatty acids in humans. Artertiosclerosis 1984;4:270–275.
13 Weintraub MS, Zechner R, Brown A, et al: Dietary polyunsaturated fats of the w6 and w3 series reduce postprandial lipoprotein levels: Chronic and acute effects of fat saturation on postprandial lipoprotein metabolism. J Clin Invest 1988;82:1884–1893.
14 Berg-Schmidt E, Varming K, Pedersen J, Dyerberg J: The effect of n–3 fatty acids on lipids and haemostatis in patients with Type IIa and Type IV hyperlipidemia. Thromb Haem 1989;62:797–801.
15 Wilt TJ, Lofgren RP, Nichol KL, et al: Fish oil supplementation does not lower plasma cholesterol in men with hypercholesterolemia. Ann Intern Med 1989;111:900–905.
16 Boberg M, Vessby B, Slinus I: Effects of dietary supplementation with n–6 and n–3 long-chain polyunsaturated fatty acids on serum lipoproteins and platelet function in hypertriglyceridaemic patients. Acta Med Scand 1986;220:153–160.
17 Dart AM, Riemersma RA, Oliver MF: Effects of Maxepa on serum lipids in hypercholesterolaemic subjects. Atherosclerosis 1989;80:119–124.
18 Molgaard J, von Schenck HV, Lassvik CL, et al: Effect of fish oil treatment on plasma lipoproteins in type III hyperlipoproteinemia. Atherosclerosis 1990;81:1–9.
19 Stacpoole PW, Alig J, Ammon L, Crockett SE: Dose-response effects on dietary marine oils on carbohydrate and lipid metabolism in normal subjects and patients with hypertriglyceridemia. Metabolism 1989;38:946–956.
20 Harris WS, Zucker ML, Dujovne CA: w3 fatty acids in hypertriglyceridemic patients: triglycerides vs methyl esters. Am J Clin Nutr 1988;48:992–997.
21 Saynor R, Verel D, Gillot T: The long-term effect of dietary supplementation with fish lipid concentrate on serum lipids, bleeding time, platelets and angina. Atherosclerosis 1984;50:3–10.
22 Haines AP, Sanders TAB, Imeson JD, et al: Effects of a fish oil supplement on platelet function, haemostatic variables and albuminuria in insulin-dependent diabetics. Thromb Res 1986;43:643–655.

23 Vandongen R, Mori TA, Codde JP, et al: Hypercholesterolaemic effect of fish oil in insulin dependent diabetic patients. Med J Austral 1988;148:141–143.
24 Schectman G, Kaul S, Cherayil GD, et al: Can the hypotriglyceridemic effect of fish oil concentrate be sustained. Ann Intern Med 1989;110:346–352.
25 Miller JP, Heath ID, Choraria SK, et al: Triglyceride lowering effect of MaxEPA fish lipid concentrate: a multicentre placebo controlled double blind study. Clin Chem 1988;178:215–260.
26 Herrmann W, Biermann J, Lindhofer HG, Kostner G: [Modification of the atherogenic risk factor Lp(a) by supplementary fish oil administration in patients with moderate physical training] Beeinflussung des atherogenen Risikofaktors Lp(a) durch supplementäre Fischölaufnahme bei Patienten mit moderatem physischem Training. Med Klin 1989;84:429–433.
27 Sanders TAB, Sullivan DR, Reeve J, Thompson GR: Triglyceride-lowering effect of marine polyunsaturates in patients with hypertriglyceridemia. Arteriosclerosis 1985; 5:459–465.
28 Kesaniemi YA, Belz WF, Grundy SM: Comparison of clofibrate and caloric restruction on kinetics of very low density lipoprotein triglycerides. Arteriosclerosis 1985;5: 153–161.
29 Sullivan DR, Sanders TAB, Trayner IM, Thompson GR: Paradoxical elevation of LDL apoprotein B levels in hypertriglyceridaemic patients and normal subjects ingesting fish oil. Atherosclerosis 1986;61:129–134.
30 Schectman G, Kaul S, Kissebah AH: Hetrogeneity of low density lipoprotein respones to fish-oil supplementation in hypertriglyceridemic subjects. Arteriosclerosis 1989;9:345–454.
31 Steinberg D, Parthasarathy S, Carew TE, Khoo JC, Witzum JL: Beyond cholesterol. Modifications of low-density lipoprotein that increase its atherogenicity. N Eng J Med 1989;320:915–923.

T.A.B. Sanders, PhD, Department of Food and Nutritional Sciences,
King's College London, University of London, Campden Hill Road,
GB–London W8 7AH (UK)

Rheumatoid Arthritis and Inflammatory Mediators

Simopoulos AP, Kifer RR, Martin RE, Barlow SM (eds): Health Effects of ω3 Polyunsaturated Fatty Acids in Seafoods. World Rev Nutr Diet. Basel, Karger, 1991, vol 66, pp 367–382

Studies of Dietary Supplementation with ω3 Fatty Acids in Patients with Rheumatoid Arthritis

Joel M. Kremer[a], *Dwight R. Robinson*[b]

[a]Department of Medicine, Division of Rheumatology, Albany Medical College, Albany, N.Y.;
[b]Massachusetts General Hospital, Harvard Medical School, Boston, Mass., USA

Introduction

Studies on the use of ω3 dietary supplementation in patients with rheumatoid arthritis first appeared in the scientific literature in the mid-1980s. Interest concerning these supplements in patients with autoimmune-induced inflammatory disease was derived from earlier investigations of their use in animal models in which dietary modifications employing ω3 fatty acids significantly improved disease manifestations and survival. The purpose of this review will be to summarize briefly the investigations employing ω3 supplements in the animal model and then more thoroughly discuss subsequent clinical investigations in patients with rheumatoid arthritis. Data on the effects of ω3 supplements on inflammatory and immune parameters will also be summarized. Lastly, we will speculate on directions of future research efforts.

Animal Studies

Certain inbred strains of mice, including the NZB, (NZB × NZW)F_1, MRL/lpr, and BxSB/Mpj strains, develop spontaneous autoimmune disease providing models for the human disease, systemic lupus erythematosus. Studies have demonstrated that dietary lipids containing ω3 fatty acids reduce the severity of diffuse proliferative glomerulonephritis in several autoimmune strains as well as reduce the severity of lymphadenopathy in the MRL/lpr strain [1, 2]. Dietary marine lipids reduce the severity of glomerulonephritis even when these lipids are withheld until after the renal disease has begun to evolve [2]. Not all forms of inflammatory autoimmune disease are suppressed by marine lipids, however. The type II collagen-induced arthritis, a model for rheumatoid arthritis, developed with the same severity in both fish oil- and beef tallow-fed rats, and the incidence of arthritis was actually increased in the fish oil group. However, the severity of adjuvant arthritis, another model for rheumatoid arthritis, was partially alleviated by fish oil [3].

In preliminary studies, two purified ω3 fatty acids, docosahexaeneoic acid (22:6; DHA) and eicosapentaenoic acid (20:5; EPA) as their ethyl esters, each were capable of reducing the severity of glomerulonephritis in (NZB × NZW)F_1 mice, based on histologic analyses, to a degree comparable to fish oils. In addition, dose-response studies demonstrate that 22:6 ω3 is more effective than 20:5 ω3, and that diets with combinations of these two ω3 fatty acids are synergistic. These studies indicate that purified ω3 fatty acids may produce optimal therapeutic effects in human clinical trials, but that mixtures of ω3 fatty acids may be more effective than either of these two major ω3 fatty acids by themselves [4].

Studies in Patients with Rheumatoid Arthritis

The possible influence of dietary alterations on the disease process in rheumatoid arthritis has generated interest for some time. A Chinese physician who developed the disease claimed that he was cured when he reverted to a diet of rice, vegetables, and fish that he consumed as a youth in his native China. He published a popular book advocating his diet as treatment for arthritis. A study published in 1983 [5] examined the effects of this diet on patients with active rheumatoid arthritis and found no discernible differences when compared to a control diet. The authors did, however, note that several patients on the rice-fish diet did improve significantly from baseline, although the improvement in the entire study cohort did not achieve statistical significance.

In a pilot study published in 1985 [6], 17 patients with rheumatoid arthritis consumed an experimental diet high in polyunsaturated fat and

Table 1. Overall assessment, pain assessment, and ARA[a] class

	Baseline	4 weeks	8 weeks	12 weeks	Follow-up	
ARA class (1–4)						
C	2.2	2.3	2.3	2.3	2.2	
E	2.0	1.9	1.9	2.1	2.0	
Patient pain[b]						
C	2.6	2.6	2.7	2.8	2.8	$p < 0.05$
E	2.8	2.5	2.6	2.5	3.6	
Patient overall[b]						
C	3.0	2.9	3.0	2.9	2.9	$p < 0.05$
E	2.8	2.6	2.9	2.7	3.6	
Physician pain[b]						
C	3.0	2.6	3.0	3.0	2.8	$p < 0.05$
E	2.9	2.6	2.7	2.6	3.1	
Physician overall[b]						
C	3.1	2.8	3.0	3.0	2.9	$p < 0.05$
E	2.9	2.7	3.0	2.6	3.1	

[a] American Rheumatism Association.
[b] 1 = None, 5 = very severe.
C = Control; E = experimental.
$p < 0.05$ for difference between follow-up and 12 week evaluations. Reprinted with permission from Kremer et al. [6].

low in saturated fat, with a daily supplement of 1.8 g EPA and 0.9 g DHA provided by 10 Max-EPA (Scherer) capsules. A control group consumed a diet with a polyunsaturated/saturated ratio of 1/4 and also took a capsule containing indigestible paraffin wax (Efamol Research). Compliance was measured by patient diaries, pill counts and gas chromatographic analysis of plasma lipids in order to document a rise in EPA in patients consuming fish oil. The study design was double-blind, controlled, randomized and of 12 weeks duration with a follow-up evaluation 1–2 months after the diets and supplements were discontinued. Standard clinical measures of arthritis activity were performed at baseline and after 4, 8 and 12 weeks and then at the time of the follow-up. The results showed a significant difference in morning stiffness between the 2 groups at the time of the 12-week evaluation, which represented a worsening in the control group while the fish oil/

Table 2. Mean values and 95% confidence intervals for the fish oil and placebo effects on rheumatoid arthritis at 7 and 14 weeks in 33 patients

Variable	Fish oil		Placebo	
	7 weeks	14 weeks	7 weeks	14 weeks
Morning stiffness, min	−2.7 (−20.9 to 5.5)	−5.9 (−22.6 to 10.8)	22.4 (−7.7 to 52.5)	49.4 (−12.8 to 111.4)
Time to fatigue, min	94.0 (8.3 to 179.7)	176.8 (83.0 to 270.6)*	11.8 (−86.3 to 109.9)	8.4 (−98.9 to 115.7)
Grip strength, mm Hg	−3.4 (−11.6 to 4.8)	9.7 (−0.1 to 19.5)	−2.0 (10.0 to 6.0)	2.9 (−7.1 to 12.9)
ARA class (I–IV)	−0.13 (−0.25 to −0.01)*	−0.18 (−0.3 to −0.06)*	−0.06 (−0.2 to 0.08)	−0.03 (−0.19 to 0.13)
Physician assessment of pain	−0.06 (−0.36 to 0.24)	−0.06 (0.36 to 0.24)	0.14 (−0.11 to 0.39)	0.06 (−0.14 to 0.26)
Patient assessment of pain	−0.14 (−0.38 to 0.10)	−0.21 (−0.52 to 0.10)	−0.06 (−0.36 to 0.24)	0.0 (−0.18 to 0.18)
Physician global assessment	−0.17 (−0.39 to 0.05)	−0.27 (−0.52 to −0.3)	0.0 (−0.14 to 0.14)	−0.9 (−0.27 to 0.09)
Patient global assessment	−0.06 (−0.36 to 0.24)	−0.11 (−0.35 to 0.13)	−0.9 (−0.27 to 0.09)	0.0 (−0.16 to 0.16)
50-ft (15 m) walking time, s	−0.01 (−0.34 to 0.32)	−0.22 (−0.69 to 0.25)	0.39 (−0.36 to 1.14)	0.39 (−0.45 to 1.53)
Tender joints, n	−2.0 (−3.4 to −0.6)*	−3.5 (−5.2 to 1.8)*	−0.8 (−2.8 to 1.2)	0.01 (−1.19 to 1.21)
Swollen joints, n	−0.4 (−4.7 to 3.9)	−2.8 (−4.3 to 1.3)*	−0.3 (−1.7 to 1.1)	−1.0 (−2.5 to 0.5)
Erythrocyte sedimentation rate, mm/h	−2.3 (−6.4 to 1.8)	−0.8 (−6.8 to 5.2)	−1.11 (−4.8 to 2.58)	−2.07 (−6.89 to 2.75)

* The univariate confidence intervals not containing 0 correspond to statistical significance at $p < 0.05$. All values for patients at 14 weeks on fish oil would remain significant if the Bonferroni correction were used, except the number of swollen joints. It should be observed that all the variables except sedimentation rate changed in a direction favorable to the fish oil supplementation. ARA = American Rheumatism Association.

Physician and patient assessments are graded on a scale of 0–4.

Reprinted with permission from Kremer et al. [7].

low-fat group remained unchanged. There were improvements in the mean number of tender joints at 12 weeks in the experimental group when compared with baseline ($p = 0.001$), which did not achieve significance vs. the control group at that time ($p = 0.16$). At the follow-up visit, a significant worsening was observed in the experimental group in patients' rating of pain and overall condition ($p = 0.02$), physicians' pain rating ($p = 0.03$), and physicians' overall evaluation ($p = 0.04$) (table 1).

A change in hemoglobin concentration between the experimental and control groups was significant at 12 weeks ($p = 0.03$), but it represented an exaggeration of a baseline difference, and the change from baseline was not significant. There was, however, a significant prolongation of the IVY bleeding time in the experimental group from 5.15 min at baseline to 6.97 min at 12 weeks. The 12-week bleeding time is still within the normal range and would not result in clinically significant bleeding. It should be noted that all patients were taking aspirin or a non-steroidal anti-inflammatory drug throughout the study.

In a subsequent clinical investigation, the effect of fish oil dietary supplementation alone was investigated in a 14-week double-blind crossover investigation with a 4-week washout period between study arms [7]. Patients with acute rheumatoid arthritis received 2.79 g EPA and 1.8 g DHA in the form of 15 Max-EPA capsules (Scherer) or olive oil. The crossover design allowed investigators to compare the effects of fish oil vs. olive oil supplementation in the same individuals while they maintained their own background diets and medications without any alterations. Thirty-three patients completed the study with satisfactory compliance documented by pill counts and gas chromatographic analysis of plasma fatty acids.

After 14 weeks on either fish oil or olive oil, patients had significantly fewer tender joints ($p = 0.007$) compared to the control period when they ingested olive oil. The time interval to the first experience of fatigue also improved in patients consuming fish oil ($p = 0.05$). No other statistically significant clinical improvements were observed in the experimental group vs. olive oil, although all of the 10 other clinical parameters measured favored fish oil. After 14 weeks significant improvements were also observed from baseline in the group consuming fish oil in American Rheumatism Association functional class, physicians global assessment of disease activity, the number of tender joints and the number of swollen joints (table 2).

Ionophore-stimulated neutrophil leukotriene B_4 (LTB_4) production also decreased by 57.8% ($p = 0.001$) when patients received the fish oil fatty acid supplement compared with when they ingested olive oil. This is of potential significance as LTB_4 is a potent pro-inflammatory substance. A significant Pearson correlation ($r = 0.393$, $p = 0.036$) was seen between decreases in neutrophil LTB_4 production observed in individual patients and decreases in the number of tender joints. A Spearman rank correlation on the same data showed a higher degree of association ($r = 0.531$, $p = 0.01$) between the variables in individual patients. Interestingly, the neutrophil LTB_4 values in the group ingesting fish oil prior to olive oil remained below the baseline value for as long as 18 weeks after these supplements were discontinued. The fact that patients were not ingesting fish oil during the follow-up period was confirmed by gas chromatographic analysis of plasma. Subjects were noted to do better clinically during the olive oil phase which occurred after (rather than before) fish oil and it was felt that the prolonged suppression of neutrophil LTB_4 production might account for the continued clinical benefits observed after the period of discontinuation of fish oil. Prolonged effects on the immune system were subsequently reported in normal volunteers ingesting fish oil and will be discussed in a subsequent section. Because of these observations, it appears that a crossover format is not appropriate to study the clinical or immune effects of fish oil in patients with inflammatory disease.

In an investigation which primarily examined potential mechanisms of action of fish oil dietary supplements in patients with rheumatoid arthritis, 12 patients with active disease consumed 20 Max-EPA (Scherer) capsules a day containing 3.6 g EPA and 2.4 g DHA daily for a period of 6 weeks [8]. The authors observed suppression of LTB_4 generation from ionophore-stimulated neutrophils of 33% from baseline after 6 weeks of fish oil ingestion ($p = 0.004$). LTB_4 and 5-HEPE were not detected at baseline but 'substantial quantities' were generated at 6 weeks. The mean sum of the quantities of neutrophil LTB_4 and LTB_5 decreased significantly after 6 weeks when compared with baseline values ($p = 0.03$). LTB_4 generation after ionophore stimulation of monocytes was also examined and decreased by 23% from baseline after 6 weeks. The differences, however, were not statistically significant. Platelet-activating factor (PAF)-acether in monocyte monolayers was examined with a standard curve using different concentrations of ionophore in a dose-dependent and time-dependent fashion. The total amount of PAF-acether produced was less at each point of the curve after 6 weeks of fish oil ingestion compared with the baseline pre-diet values (fig. 1). There was a 37% decrease in the total quantity PAF-acether produced at week 6 when compared with pre-diet values ($p =$

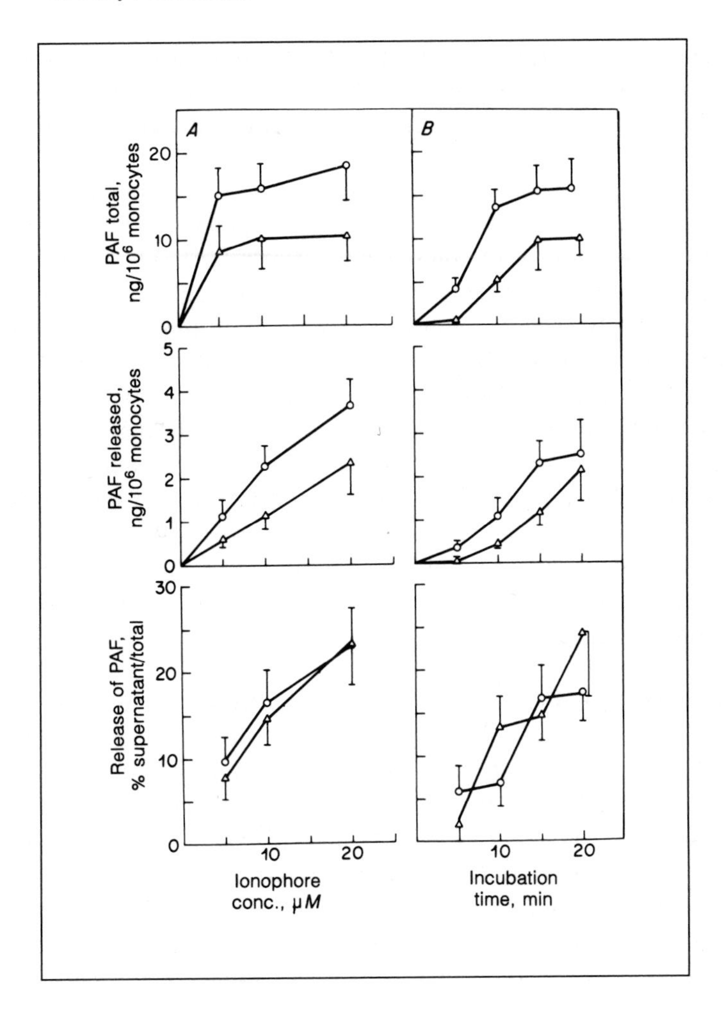

Fig. 1. Generation of platelet-activating factor (PAF)-acether from monocytes obtained from rheumatoid arthritis patients receiving dietary supplementation with fish oil. Shown are the effects of calcium ionophore activation (for details, see Patients and Methods in [8]) on the generation of total PAF-acether (supernatant + pellet), on the release of PAF-acether (supernatant), and on the percentage release of PAF-acether obtained, before fish oil supplementation (○) and after 6 weeks of supplementation (△). *A* Dose-dependent effects; *B* time-dependent effects. Values are mean ± SEM (n = 6) measured as described in Patients and Methods in [8]. Reproduced with permission from Sperling et al. [8].

0.023). This is of potential significance in rheumatoid arthritis in that PAF stimulates platelets to aggregate and produce vasoactive amines which affect endothelial permeability. Thus, PAF could contribute to the local inflammatory and immune response by facilitating the egress of pro-inflammatory cellular elements from the circulation to synovial tissue.

The investigators also examined neutrophil function, cellular membrane fluidity and fatty acid composition of cellular lipids. A standard dose-response curve of neutrophil chemotaxis to varying concentrations of LTB_4 revealed an increased response at each point of the curve at 6 weeks compared with the pre-diet period ($p = 0.016$). A similar increase in chemotaxis was observed to formylated methionine leucine phenylalanine (FMLP) ($p = 0.009$). This response seems paradoxical in that one would expect decreased chemotaxis to occur if the inflammatory response is to be dampened after fish oil ingestion. The authors believe that the observations of increased neutrophil chemotaxis post fish oil are due to a reversal of an abnormal desensitization of neutrophils to the chemotactic factor LTB_4 in patients with rheumatoid arthritis (RA), which is normalized with the addition of fish oil. The implications of this on disease activity are unclear.

Analysis of the fatty acid composition of neutrophil membranes after 6 weeks of fish oil ingestion revealed several changes. Arachidonic acid content declined significantly by 33% ($p = 0.0006$) with a simultaneous twenty-fold rise in EPA content from the pre-diet period ($p = 0.032$). DHA and docosapentaenoic acid were not detected in cellular lipids of neutrophils at any time in the investigation and were judged to be below the limit of detection. Membrane fluidity assessed by fluorescence polarization did not change although others have documented changes after fish oil ingestion. The ratio of arachidonic acid to EPA in patients' neutrophil lipids decreased from 81:1 to 27:1.

In a subsequent investigation, 46 patients with active rheumatoid arthritis completed a 12-week double-blind parallel study of fish oil vs. olive oil dietary supplementation [9]. Subjects consumed 3.2 g EPA/day and 2.0 g DHA/day given as 18 Max-EPA (Scherer) capsules. Patients continued their background medications without change. Dietary advice was given to achieve a total fat uptake of 60 g/day including supplements in both the experimental and control groups. Laboratory monitoring included ionophore-stimulated production of neutrophil LTB_4 and LTB_5 as well as superoxide production from neutrophils after stimulation with FMLP or phorbol myristate acetate (PMA).

Table 3. Effect of dietary treatments on neutrophil LTB_4 and 5-HETE production

Treatment		Hydroxy-fatty acids, ng/10^6 cells (mean ± SD)[1]			
		pre-treatment	week 4	week 12	week 16 (washout)
Fish oil (n = 19)	LTB_4	18 ± 7.5	13 ± 3.5[1]	12 ± 4.7*	15 ± 4.8
	5-HETE	53 ± 17	41 ± 15	37 ± 17**	47 ± 19
Olive oil (n = 17)	LTB_4	16 ± 5.7	14 ± 3.8	16 ± 6.4	16 ± 4.3
	5-HETE	52 ± 24	44 ± 15	55 ± 25	49 ± 12

[1] Cells stimulated with 5 μ*M* A23187 for 5 min, then extracted and analyzed by HPLC for LTB_4 and 5-HETE. Statistical analyses by paired 2-tailed t tests compared to pretreatment group.

* $p < 0.01$; ** $p < 0.02$.

Reproduced with permission from Cleland et al. [9].

The authors observed no improvements in clinical variables after 4 or 8 weeks of fish oil supplementation. At the time of the 12-week visit significant improvements were observed in the tender joint score and grip strength, while the olive oil group improved significantly in analogue pain score and grip strength. No significant improvements in the joint swelling score were observed.

Neutrophil production of LTB_4 and 5-HETE was reduced by 30% in the fish oil group after 3 months ($p = 0.01$), although the suppression of LTB_4 production achieved significance after 4 weeks. Four weeks after discontinuing fish oil, both neutrophil LTB_4 and 5-HETE production had increased towards pre-treatment values (table 3). The total production of neutrophil LTB_4 and LTB_5 was reduced compared to pretreatment LTB_4 production although this only achieved statistical significance at week 12. No significant change in total neutrophil superoxide production in response to FMLP or PMA was observed in either group.

Before reviewing the next clinical investigation of the effects of fish oil dietary supplementation in patients with rheumatoid arthritis, it is of some interest to reflect on the previously summarized investigations in order to define certain commonly observed phenomena. Clinical benefits derived from fish oil improve from baseline vs. a control group but often do not achieve statistical significance. When significant improvements are observed, they often do not occur until at least 12 weeks of dietary supplementation. This is of some interest in that Sperling et al. [8] demonstrated that significant alterations of leukotriene metabolism occur in neutrophils and monocytes after 6 weeks in RA patients consuming fish oil. Moreover, Cleland et al. [9] observed that neutrophil LTB_4 and 5-HETE production increased towards baseline as soon as 4 weeks after discontinuing the fish oil dietary supplements. This is of note in view of the seemingly paradoxical observation that clinical benefits are maintained for up to 18 weeks following the discontinuation of fish oil [7].

There is thus a discrepancy in timing between the ex vivo changes in leukotriene metabolism and the clinical improvements induced by fish oil. Why are clinical improvements delayed until 12 weeks and why are they sustained for so long when discontinuing fish oil? A study by Endres et al. [10] may help to explain these observations. Nine healthy volunteers consumed 2.75 g EPA and 1.8 g DHA daily for 6 weeks. Interleukin-1β and 1α (IL-1β, IL-1α) and tumor necrosis factor (TNF) was measured at baseline, after 6 weeks and then 10 and 20 weeks after discontinuing the supplement. IL-1β production decreased by 43% after 6 weeks of fish oil supplementation ($p = 0.048$), but a further decrease was observed 10 weeks after discontinuing the supplement when production had decreased by 61% from baseline ($p = 0.005$) (fig. 2). The production of IL-1β returned to pre-supplement values 20 weeks after discontinuing the fish oil.

IL-1 has significant potential impact on the RA disease process through a variety of mechanisms [11–20]. It is thus of relevance that fish oil dietary supplementation inhibits the production of this cytokine and that the chronology of these inhibitory effects differ from those observed with leukotriene production. It is therefore possible that fish oil-induced suppression of IL-1 contributes to the amelioration of clinical signs and symptoms of disease activity in patients with RA to a greater extent than does inhibition of leukotriene metabolism. That is, the clinical benefits observed in the studies reviewed above may be delayed relative to leukotriene inhibition because of the later inhibition of the production of IL-1.

An investigation which would examine the effects of fish oil dietary supplementation on the clinical, inflammatory and immune parameters of RA over periods of greater than 12–14 weeks was needed. In addition, it was unclear after reviewing the clinical studies summarized to this point whether dose-dependent effects of fish oil occur. The above investigations employed the same dose of fish oil dietary supplementation in all patients

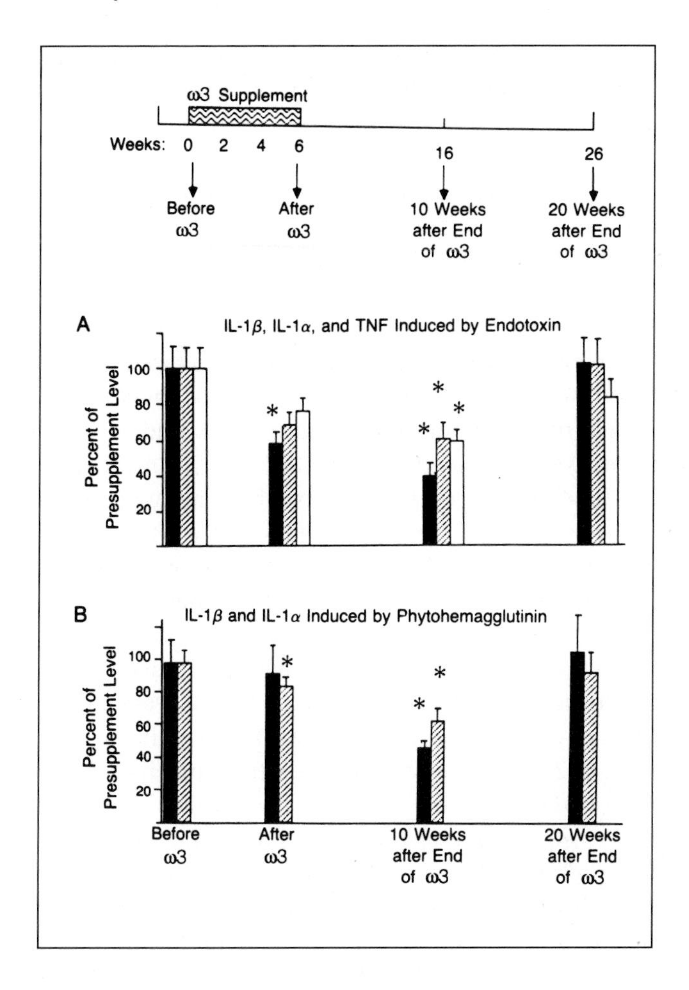

Fig. 2. Synthesis of IL-1 (solid bars), IL-1 (hatched bars), and tumor necrosis factor (TNF; open bars) by mononuclear cells stimulated with 1 ng endotoxin/ml (panel A) or 3 μg PHA/ml (panel B). The concentration of total cytokine (i.e. cell-associated plus secreted) was determined by radioimmunoassay. Concentrations are presented as percentages of presupplementation levels. With endotoxin stimulation, the concentrations were 7.4 ng IL-1β/ml, 16.0 ng IL-1α/ml, and 8.5 ng TNF/ml. With phytohemagglutinin stimulation, the concentrations were 3.0 ng IL-1β/ml and 14.0 ng IL-1α/ml. Each bar represents the mean of 27 determinations (9 donors assayed on 3 days each). The error bars denote the SEM for the 9 donors. An asterisk denotes a significant difference from the level before dietary ω3 supplementation (at $p < 0.05$). Reproduced with permission from Endres et al. [10.

regardless of body weight. A comparison of different doses of fish oil in patients with active disease over extended periods would become a logical extension of the previous work.

In a recently completed study [21], 49 patients with active rheumatoid arthritis consumed either high or low dose fish or olive oil over a period of 6 months (24 weeks). The study designed was randomized, prospective parallel and double-blind. The low-dose fish oil group ingested 27 mg/kg/day EPA and 18 mg/kg/day DHA and the high-dose group ingested twice this amount. Fish oil was supplied as an ethylester (Pharmacaps®) containing 33 mg EPA and 240 mg DHA per capsule. The olive oil group consumed 9 capsules/day. Patients maintained their background medications for arthritis without change during the study. No special dietary instructions were provided to study subjects. Clinical investigations occurred at baseline and every 6 weeks while immune parameters including ionophore-stimulated neutropil LTB_4 and LTB_5, macrophage IL-1 and IL-2 production, T and B cell response to mitogen proliferation and immunoglobulin production were measured at baseline and after 24 weeks.

Results showed that multiple clinical parameters improved from baseline in the groups consuming fish oil. The improvements were noted more commonly in the high-dose fish oil group in which 21/45 clinical measures evaluated at weeks 6–24 improved significantly from baseline, compared with 8/45 in the low-dose subjects and 5/45 in the olive oil-consuming patients. Moreover, significant changes from baseline increased with increasing duration of fish oil ingestion beyond 12 weeks. Only 5 clinical outcomes during weeks 6 and 12 of the study showed significant improvement in patients consuming fish oil, while 24 outcomes achieved significance at the time of the 18-, 24- or 30-week evaluation.

It should be noted that only the grip strength clinical outcome improved significantly when compared with the olive oil-consuming patients. Reasons for this could include common nonsignificant improvements observed in the olive oil group from baseline and the elimination of 11 patients during the study from this group for medication adjustment required by increased symptoms. (The need for increased medication during the period of investigation of dietary supplementation was considered a protocol violation). Olive oil also has potential to favorably affect the immune response in inflammatory disease [22–25].

Laboratory analysis showed that IL-1 production and release from macrophages decreased by 40.6% from baseline in the low-dose fish oil

Table 4. Fish oil effects on lipopolysaccharide-induced production of IL-1 by peripheral blood mononuclear cells in vitro

Treatment	n	Change from baseline week 0–24
Olive oil	11	-243.1 (-540.4 to 54.2)
Low-dose fish oil	18	-239.9 (-490.4 to 10.6)*
High-dose fish oil	17	-416.2 (-623.0 to 209.4)**

Values are expressed as units/ml and are means (95% confidence intervals); 1 unit is the 50% maximal proliferative point in the C3H/HeJ co-stimulation assay (see Methods in [21]).
* p = 0.059 from baseline; ** p = 0.0005.
Reproduced with permission from Kremer et al. [21].

group (p = 0.059) and by 54.7% in the high-dose group (p = 0.0005). IL-1 also decreased by 38% in the olive oil group, but the change was not statistically significant (table 4). Tritiated-thymidine incorporation after mitogen stimulation increased significantly in all 3 groups after 24 weeks compared to baseline measurements. Increased thymidine incorporation was also noted after phytohemagglutinin (PHA) stimulation but was only significant in the high-dose fish oil group. Thymidine incorporation after pokeweed mitogen (PWM) stimulation also increased significantly in all 3 groups. Production of IgG in vitro in the absence of mitogen stimulation decreased significantly in both fish oil groups from baseline, but IgG production was not affected after mitogen stimulation.

This investigation established that salutary clinical benefits after fish oil ingestion in patients with rheumatoid arthritis are more commonly observed after treatment periods of at least 18 weeks. This observation was of significance in that the previous studies had only examined potential clinical benefits over periods of 6–14 weeks. In addition, it appears that higher doses of fish oil are associated with an increased frequency of significant clinical improvements after fish oil ingestion (table 5) [21]. Moreover, fish oil ingestion is associated with significant decreases in ionophore-stimulated neutrophil LTB_4 and macrophage production of IL-1. Both of these effects could contribute to the amelioration of disease activity through a variety of mechanisms involving the inflammatory reaction [26] and the immune system [11–20]. A recent report [27] confirms the clinical benefits of fish oil ingested over a 12-week period.

Table 5. Significant improvement ($p < 0.05$) in RA patients ingesting dietary olive and fish oil [see 27]

Clinical variable	Olive oil	Fish oil	
		low dose	high dose
Number of tender joints	no	yes	yes
Number of swollen joints	no	yes	yes
Grip strength	no	yes	yes
Morning stiffness	no	no	yes
Global disease activity			
Physician evaluation	yes	yes	yes
Patient evaluation	yes	no	no
Pain			
Physician evaluation	no	no	yes
Patient evaluation	no	no	yes

Reproduced with permission from The Journal of Musculoskeletal Medicine, May, 1989.

Fish Oil in Rheumatoid Arthritis: The Next Step

Most clinicians are loathe to recommend fish oil supplements to their patients with rheumatoid arthritis because of understandable confusion about (a) the amount to prescribe; (b) the overall effectiveness; and (c) the additional expense of an 'unproven' remedy. There is therefore a need to better define the overall therapeutic value of fish oil and place these dietary supplements in their appropriate niche in relation to non-steroidal anti-inflammatory drugs (NSAIDs) and slow-acting anti-rheumatic drugs (SAARDs). Studies are therefore required to determine whether patients consuming fish oil for a long enough period of time to establish their clinical benefits (≥ 18 weeks) could discontinue NSAIDs without a deterioration in their clinical status. The investigators should monitor inflammatory and immune parameters of disease and correlate these measures with clinical outcome. This type of study would allow investigators to determine whether NSAIDs enhance or inhibit the inflammatory and immune changes induced by fish oil. The proposed hypothetical investigation will allow investigators to determine whether higher-dose fish oil supplementation for periods of at least 18 weeks could provide equivalent clinical benefits to NSAIDs (and thus allow patients to discontinue these potentially dangerous medications without clinical deterioration). It is likely that fish oil will not be widely employed by physicians treating patients with rheumatoid arthritis until its role in the therapeutic hierarchy can be established as in the hypothetical investigation described.

References

1 Prickett JD, Robinson DR, Steinberg AD: Dietary enrichment with the polyunsaturated fatty acid eicosapentaenoic acid prevents proteinuria and prolongs survival in NZB × NZW F_1 mice. J Clin Invest 1981;68:556–559.
2 Robinson DR, Prickett JD, Makoul GT, Steinberg AD, Colvin RB: Dietary fish oil reduces progression of established renal disease in (NZB × NZW)F_1 mice and delays renal disease in BXSB and MRL/1 strains. Arthritis Rheum 1986;29:539–546.
3 Prickett JD, Trentham DE, Robinson DR: Dietary fish oil augments the induction of arthritis in rats immunized with type II collagen. J Immunol 1984;132:725–729.
4 Robinson DR, Tateno S, Knoell C, Olesiak W, Xu L, Hirai A, Guo M, Colvin RB: Dietary marine lipids suppress murine autoimmune disease. J Int Med 1989; 225(suppl 1):211–216.
5 Panush RS, Carter RL, Katz P, Lowsari S, Finnie S: Diet therapy for rheumatoid arthritis. Arthritis Rheum 1983;26:462–469.
6 Kremer JM, Bigauoette J, Michalek AU, et al: Effects of manipulating dietary fatty acids on clinical manifestations of rheumatoid arthritis. Lancet 1985;i:184–187.
7 Kremer JM, Jubiz W, Michalek A, et al: Fish-oil fatty acid supplementation in active rheumatoid arthritis, a double blinded, controlled crossover study. Ann Intern Med 1987;106:497–503.
8 Sperling RI, Weinblatt M, Robin JL, et al: Effects of dietary supplementation with marine fish oil on leukocyte lipid mediator generation and function in rheumatoid arthritis. Arthritis Rheum 1987;30:988–997.
9 Cleland LG, French JK, Betts WH, et al: Clinical and biochemical effects of dietary fish-oil supplements in rheumatoid arthritis. J Rheumatol 1988;15:1471–1475.
10 Endres S, Ghorbani R, Kelley V: The effect of dietary supplementation with n–3 polyunsaturated fatty acids on the synthesis of interleukin-1 and tumor necrosis factor by mononuclear cells. N Engl J Med 1989;320:265–271.
11 Dinarello CA: Interleukin-1. Rev Infect Dis 1984;6:51–90.
12 Dinarello CA: Interleukin-1 and the pathogenesis of the acute-phase response. N Engl J Med 1984;311:1413–1418.
13 Luger TA, Charon JA, Colot M, Micksche M, Oppenheim JJ: Chemotactic properties of partially purified human epidermal cell-derived thymocyte-activating factor for polymorphonuclear and mononuclear cells. J Immunol 1983;131:816–820.
14 Schmidt JA, Mizel SB, Cohen D, Green I: Interleukin-1: a potential regulator of fibroblast proliferation. J Immunol 1982;128:2177–2182.

15 Mizel SB, Dayer JM, Krane SM, Mergenhagen SE: Stimulation of rheumatoid synovial cell collagenase and prostaglandin production by partially purified lymphocyte-activating factor (interleukin-1). Proc Natl Acad Sci USA 1981;78:2474–2477.

16 Goto M, Sasano M, Yamanaka H, et al: Spontaneous production of an interleukin-1-like factor by cloned rheumatoid synovial cells in long-term culture. J Clin Invest 1987;80:786–796.

17 Miyasaka N, Sato K, Goto M, et al: Augmented interleukin-1 production and HLA-DR expression in the synovium of rheumatoid arthritis patients. Arthritis Rheum 1988;4:480–486.

18 Di Biovine FS, Malawista SE, Nuki G, Duff GW: Interleukin-1 (IL-1) as a mediator of crystal arthritis. J Immunol 1987;138:3213–3218.

19 Gowen M, Wood DD, Ihrie EJ, McGuire MKB, Russell BGG: An IL-1-like factor stimulates bone resorption in vitro. Nature 1983;306:378–380.

20 Saklatvala J, Sarsfield SJ, Pilsworth LM: Characterization of proteins from human synovium and mononuclear leukocytes that induce resorption of cartilage proteoglycan in vitro. Biochem J 1983;209:337–344.

21 Kremer J, Lawrence DL, Jubiz W, et al: Dietary fish-oil and olive-oil supplementation in patients with rheumatoid arthritis: clinical and immunological effects. Arthritis Rheum 1990;33:810–820.

22 Traill KN, Wick G: Lipids and lymphocyte function. Immunol Today 1984;5:70–75.

23 Johnston PV: Dietary fat, eicosanoids, and immunity. Adv Lipid Research 1985;21:103–141.

24 Del Buono BJ, Williamson PL, Schlegel RA: Alterations in plasma membrane lipid organization during lymphocyte differentiation. J Cell Physiol 1986;126:379–388.

25 Erickson KL: Dietary fat modulation of immune response. Int J Immunopharmacol 1986;8:529–543.

26 Lee TH, Hoover RL, Williams JD, et al: Effect of dietary enrichment with eicosapentaenoic and monocyte leukotriene generation and neutrophil function. N Engl Med 1985;312:1217–1224.

27 van der Tempel H, Tulleken JE, Limburg PC, Muskiet FAJ, van Rijswijk MH: Effects of fish oil supplementation in rheumatoid arthritis. Ann Rheum Dis 1990;49:76–80.

Joel M. Kremer, MD, Department of Medicine, Division of Rheumatology, Albany Medical College, Albany, NY 12208 (USA)

Simopoulos AP, Kifer RR, Martin RE, Barlow SM (eds): Health Effects of ω3 Polyunsaturated Fatty Acids in Seafoods. World Rev Nutr Diet. Basel, Karger, 1991, vol 66, pp 383–390

High Affinity of Ether-Linked Lipids for ω3 Fatty Acids[1]

Fred Snyder, Merle L. Blank, Zigrida L. Smith, E.A. Cress

Medical Sciences Division, Oak Ridge Associated Universities,
Oak Ridge, Tenn., USA

Introduction

It is well established from compositional studies of phospholipid subclasses in tissues from a variety of animals that both the O-alkyl and O-alk-1-enyl ether-linked lipids are highly enriched in arachidonic acid and other polyunsaturated fatty acids [1, 2]. Examples of the chemical structures of the ether-linked phospholipids that serve as a shelter for the polyunsaturated fatty acids are illustrated in figure 1. Earlier results from our laboratory have shown the ethanolamine plasmalogens (1-alk-1′-enyl-2-acyl-*sn*-glycero-3-phosphoethanolamines) behave as a resistant storage site for arachidonic acid in the testes of rats maintained on a diet deficient in essential fatty acids [3, 4]. The ethanolamine plasmalogens could be involved in the selective release of eicosanoids under certain conditions [5–12], but a firm functional role of plasmalogens as a source of arachidonic acid in agonist-stimulated cell responses has not yet been established. Nevertheless, the high affinity of ether-linked glycerolipids for polyunsaturated fatty acids clearly emphasizes their importance in eicosanoid metabolism.

[1] This work was supported by the Office of Energy Research, U.S. Department of Energy (Contract No. DE-AC05-760R00033), the American Cancer Society (Grant BE-26U) and the National Heart, Lung and Blood Institute (Grant HL-21709-10).

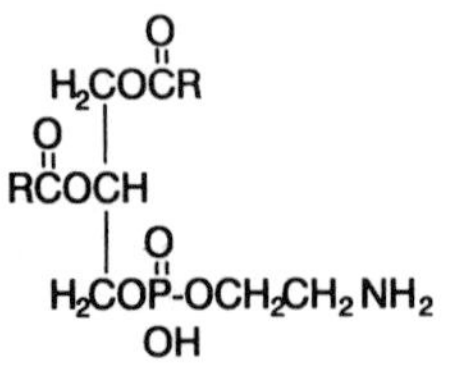

$$\begin{array}{l} \quad\ \ H_2COCR \ (\text{C=O}) \\ RCOCH \ (\text{C=O}) \\ \quad\ \ H_2COP(=O)(OH)\text{-}OCH_2CH_2NH_2 \end{array}$$

I. Phosphatidylethanolamine (1,2-diacyl-sn-glycero-3-phosphoethanolamine)

$$\begin{array}{l} \quad\ \ H_2COCR \ (\text{C=O}) \\ RCOCH \ (\text{C=O}) \\ \quad\ \ H_2COP(=O)(OH)\text{-}OCH_2CH_2\overset{+}{N}(CH_3)_3 \end{array}$$

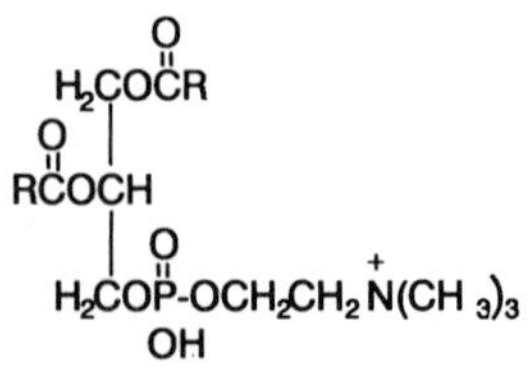

II. Phosphatidylcholine (1,2-diacyl-sn-glycero-3-phosphocholine)

$$\begin{array}{l} \quad\ \ H_2COCH_2CH_2R \\ RCOCH \ (\text{C=O}) \\ \quad\ \ H_2COP(=O)(OH)\text{-}OCH_2CH_2NH_2 \end{array}$$

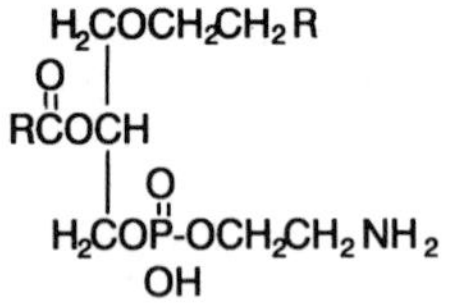

III. 1-Alkyl-2-acyl-sn-glycero-3-phosphoethanolamine

$$\begin{array}{l} \quad\ \ H_2COCH_2CH_2R \\ RCOCH \ (\text{C=O}) \\ \quad\ \ H_2COP(=O)(OH)\text{-}OCH_2CH_2\overset{+}{N}(CH_3)_3 \end{array}$$

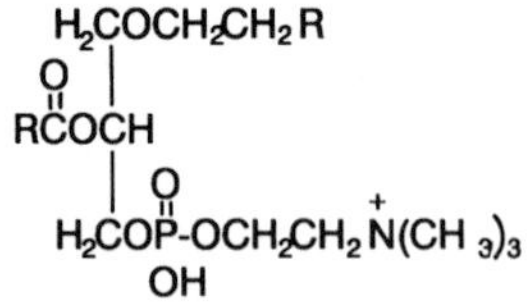

IV. 1-Alkyl-2-acyl-sn-glycero-3-phosphocholine

$$\begin{array}{l} \quad\ \ H_2COCH{=}CHR \\ RCOCH \ (\text{C=O}) \\ \quad\ \ H_2COP(=O)(OH)\text{-}OCH_2CH_2NH_2 \end{array}$$

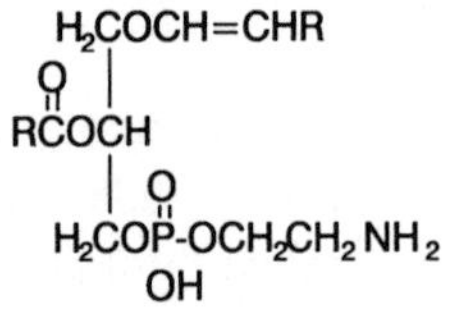

V. Ethanolamine Plasmalogen (1-alk-1′-enyl-2-acyl-sn-glycero-3-phosphoethanolamine)

$$\begin{array}{l} \quad\ \ H_2COCH{=}CHR \\ RCOCH \ (\text{C=O}) \\ \quad\ \ H_2COP(=O)(OH)\text{-}OCH_2CH_2\overset{+}{N}(CH_3)_3 \end{array}$$

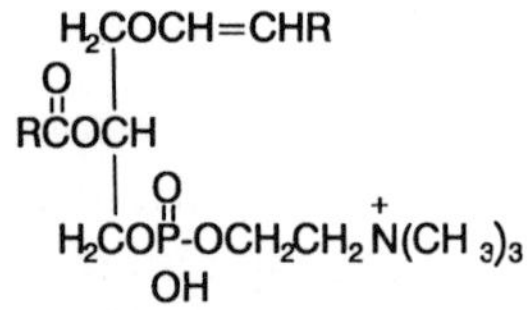

VI. Choline Plasmalogen (1-alk-1′-enyl-2-acyl-sn-glycero-3-phosphocholine)

Fig. 1. Chemical structures of various cellular phospholipid subclasses with ether linkages and their diacyl analogs.

This paper summarizes and extends some of our recent investigations of the supplementation of ω3 fatty acids (20:5 and 22:6) to a $P388D_1$ cell culture line which has macrophage-like features [13]. The $P388D_1$ cells were selected since the properties of a phospholipase A_2 that utilizes ether lipids as substrates have been well characterized in these cells [14]. In addition, the close association of ω3 fatty acids in ether-lipids found in edible fish (freshwater and ocean species) is illustrated.

Table 1. Phospholipid composition of Norwegian salmon and catfish filets (pond-raised)[1], expressed as μmol phospholipid/g wet weight

Phospholipid class/subclass	Salmon (n = 1)	Catfish[2] (n = 3)
Phosphatidylcholine	5.12	3.12±0.34
Alkylacylglycerophosphocholine	0.50	0.37±0.06
Alk-1-enylacylglycerophosphocholine (choline plasmalogen)	0.03	0.20±0.04
Phosphatidylethanolamine	1.28	0.56±0.22
Alkylacylglycerophosphoethanolamine	0.01	0.10±0.06
Alk-1-enylacylglycerophosphoethanolamine (ethanolamine plasmalogen)	0.20	0.64±0.05
Phosphatidylserine/phosphatidylinositol	0.89	0.51±0.04
Sphingomyelin	0.23	0.10±0.04

[1] Fresh fish filets were purchased at a supermarket.
[2] Values in this column are the means ± SE.

Methods

The source of the $P388D_1$ cells and the analytical procedures are identical to those previously described [13]. Fresh filets of Norwegian salmon and pond-raised catfish were purchased at a local Krogers supermarket. Cells and tissues were extracted by the Bligh and Dyer procedure (solvents containing 2% glacial acetic acid) and the classes of phospholipids from the total lipids separated by thin-layer chromatography [13]. After treating the choline- and ethanolamine-phospholipid classes with phospholipase C, benzoate derivatives of the diradylglycerols were prepared for analysis by a combination of high performance liquid chromatography and gas-liquid chromatography as described earlier [13, 15, 16].

Results and Discussion

Ether-Linked Phospholipids and Polyunsaturated Fatty Acids in Fish Filets

It is well established that fish from the sea contain significant quantities of ω3 fatty acids. However, little is known about the composition of ether-linked phospholipids or the ω3 fatty acid content of ether lipids in fresh fish consumed by humans. Table 1 illustrates the relative concentra-

Table 2. Composition of major polyunsaturated fatty acids in phospholipids of filets from Norwegian salmon and catfish[1] expressed as weight percent of total subclass[2]

Phospholipid class/subclass	Salmon				Catfish			
	20:4 (ω6)	20:5 (ω3)	22:5 (ω3)	22:6 (ω3)	20:4 (ω6)	20:5 (ω3)	22:5 (ω3)	22:6 (ω3)
Phosphatidylcholine	0.5	10.6	4.6	39.8	2.2	3.3	1.1	4.0
Alkylacylglycerophospho-choline	0.4	7.0	6.2	71.0	9.4	11.1	6.9	30.2
Phosphatidylethanolamine	1.1	7.1	5.9	46.2	8.1	6.4	2.8	13.8
Alkylacylglycerophospho-ethanolamine	–[3]	–	–	–	3.6	3.7	14.2	45.2
Alk-1-enylacylglycerophospho-ethanolamine (ethanolamine plasmalogen)	1.1	3.9	6.1	40.0	5.7	4.3	3.6	25.0
Phosphatidylserine/ phosphatidylinositol	7.7	6.8	5.1	26.4	13.6	2.7	3.1	12.5

[1] Fresh fish filets purchased at a supermarket.
[2] Distribution of acyl groups in the alkylacyl-subclasses are based on the *sn*-2 groups = 100%, while all other classes are calculated for the *sn*-1 plus *sn*-2 groups = 100%.
[3] Dashes indicate insufficient amounts of material were available for analysis.

tions of the various classes and subclasses of phospholipids in Norwegian salmon and catfish filets. The ether-linked choline and ethanolamine glycerophospholipids in salmon and catfish comprise approximately 9% and 23%, respectively, of the total phospholipids. The content of total ω3 fatty acids (20:5, 22:5, and 22:6) in the triacylglycerols was much lower in filets from freshwater catfish (1.6%) than in salmon filets (19%). However, this difference in ω3 fatty acid content between catfish and salmon, although still apparent, was not nearly as pronounced in the phospholipids from filets of the two fish (table 2).

Especially noteworthy is the very high content of ω3 fatty acids (particularly 22:6) in the ether lipids when compared to the corresponding diacyl phospholipid subclasses of the catfish, whereas the salmon had a more even distribution of ω3 fatty acids among phospholipid subclasses (table 2).

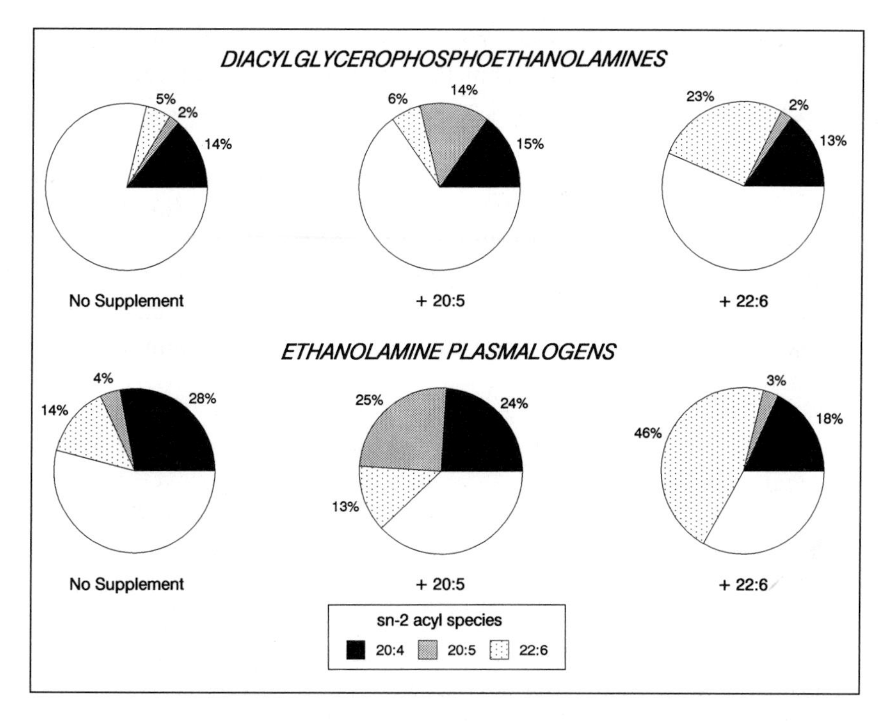

Fig. 2. Major polyunsaturated acyl moieties of phosphatidylethanolamines (1,2-diacyl-*sn*-glycero-3-phosphoethanolamines) and ethanolamine plasmalogens (1-alk-1-enyl-2-acyl-*sn*-glycero-3-phosphoethanolamines) present in $P388D_1$ cells. The unshaded portion of the circle represents the weight percentage of all other fatty acids (saturated, monoenes, dienes) present in each phospholipid subclass.

Acyl Composition of Ethanolamine-Containing Phospholipids of $P388D_1$ Cells Supplemented with ω3 Fatty Acids

Supplementation of $P388D_1$ cells with 20:5 and 22:6 fatty acids (10 μ*M* for 24 h) has no effect on their growth rate or cell viability [13]. Also the amounts of phospholipid classes or subclasses were not influenced by the ω3 supplements [13]. Although both the 20:5 and 22:6 fatty acids were incorporated into the various phospholipids of $P388D_1$ cells [13], only the ethanolamine plasmalogens (alk-1-enylacylglycerophosphoethanolamines) exhibited large increases in their content of 20:5 and 22:6 acyl moieties (fig. 2). The percentage of 22:6 in the plasmalogens was two times higher (46 vs. 23%) than that found in the corresponding diacyl species.

Both ω3 fatty acids were incorporated into phospholipids at the expense of oleic acid [13]. Supplements of the 20:5 acid caused little change in the proportion of 20:4 acyl moieties in the cellular phospholipids, whereas the P388D_1 cells supplemented with the 22:6 acid had approximately one third less arachidonate in the ethanolamine plasmalogens than the unsupplemented cells (fig. 2). Also of interest is that the enrichment of cellular phospholipids with the 20:5 acid had little influence on the distribution of the 22:6 acyl moieties and the reverse was also true [13]. Thus, effects of a specific type of ω3 fatty acid supplement on cell functions are not likely to be influenced by changes in the phospholipid content of other ω3 fatty acids.

Effects of ω3 Fatty Acids on the Metabolism of Ethanolamine Plasmalogens

In view of the close association between the ω3 fatty acids and plasmalogens, we investigated the formation of alkylacylglycerophosphoethanolamine (the immediate precursor of plasmalogens) from alkyllysoglycerophosphoethanolamine and the subsequent desaturation step that forms the O-alk-1-enyl linkage of plasmalogens. Detailed data from these experiments have been reported elsewhere [13]. Therefore, only a brief synopsis of these results are provided here.

A microsomal membrane fraction (100,000 $g \times 60$ min) isolated from the P388D_1 cells supplemented with ω3 fatty acids was capable of acylating significantly larger quantities of alkyllysoglycerophosphoethanolamine than membranes prepared from the unsupplemented cells. Approximately 46% of the acylated product was formed in the 20:5- or 22:6-treated cells versus ≈ 34% in membranes isolated from the unsupplemented cells. Moreover, the 20:5- and 22:6-supplemented cells were able to desaturate considerably more of the alkylacylglycerophosphoethanolamine into the plasmalogen form than observed in the unsupplemented cells. On the other hand, when compared to unsupplemented cells, oleic acid supplements (10 μM for 24 h) had virtually no effect either on the acylation or desaturation rates.

Molecular species analysis of labeled plasmalogens formed in experiments with intact cells using 1-[^{3}H]alkyl-2-lyso-*sn*-glycero-3-phosphoethanolamine as a precursor and plasmalogens produced by a Δ1 alkyl desaturase indicate the species containing 22:6 acyl moieties have a higher turnover rate than other molecular species [13]. These results are consistent with the acylation and desaturation rates observed with the isolated microsomal membranes described in the previous paragraph.

Conclusions

Presently it is difficult to understand the biological significance of why the ether-linked lipids contain such high amounts of polyunsaturated acids, particularly the large amounts of 22:6 (ω3) acyl groups associated with ethanolamine plasmalogens. Also possible mechanisms for the release of polyunsaturated acyl moieties from ether-linked lipids by phospholipase A_2 and the selectivity for ω3 vs. ω6 aliphatic chains are unknown. However, it is clear from our earlier results [13] that the ω3 fatty acids (especially 22:6) can have a pronounced effect on the metabolism of membrane lipids. Implication of these findings to the functional role of ω3 fatty acids and ether lipids in cellular processes require further study.

References

1 Horrocks LA: Content, composition, and metabolism of mammalian and avian lipids that contain ether groups; in Snyder F (ed): Ether Lipids: Chemistry and Biology. New York, Academic Press, 1972, pp 177–272.
2 Sugiura T, Waku K: Composition of alkyl ether-linked phospholipids in mammalian tissues; in Snyder F (ed): Platelet Activating Factor and Related Lipid Mediators. New York, Plenum Press, 1987, pp 55–85.
3 Blank ML, Wykle RL, Snyder F: The retention of arachidonic acid in ethanolamine plasmalogens of rat testes during essential fatty acid deficiency. Biochim Biophys Acta 1973;316:28–34.
4 Wykle RL, Blank ML, Snyder F: The enzymic incorporation of arachidonic acid into ether-containing choline and ethanolamine phosphoglycerides by deacylation-acylation reactions. Biochim Biophys Acta 1973;326:26–33.
5 Rittenhouse-Simmons S, Russell FA, Deykin D: Transfer of arachidonic acid to human platelet plasmalogen in response to thrombin. Biochem Biophys Res Commun 1976;70:295–301.
6 Rittenhouse-Simmons S, Russell FA, Deykin D: Mobilization of arachidonic acid in human platelets. Kinetics and Ca^{2+} dependency. Biochim Biophys Acta 1977;188: 370–380.
7 Borekman MJ, Ward JW, Marcus AJ: Phospholipid metabolism in stimulated human platelets. Changes in phosphatidylinositol, phosphatidic acid, and lysophospholipids. J Clin Invest 1980;66:275–283.
8 Daniel LW, King L, Waite M: Source of arachidonic acid for prostaglandin synthesis in Madin-Darby canine kidney cells. J Biol Chem 1981;256:12830–12835.
9 Kramer RM, Deykin D: Arachidonoyl transacylase in human platelets. Coenzyme A-independent transfer of arachidonate from phosphatidylcholine to lysoplasmenylethanolamine. J Biol Chem 1983;258:13806–13811.

10 Brown ML, Jakubowski JA, Leventis LL, et al: Ionophore-induced metabolism of phospholipids and eicosanoid production in porcine aortic endothelial cells: selective release of arachidonic acid from diacyl and ether phospholipids. Biochim Biophys Acta 1987;921:159–166.
11 Kambayashi J, Kawasaki T, Tsujinaka T, et al: Active metabolism of phosphatidylethanolamine plasmalogen in stimulated platelets, analyzed by high performance liquid chromatography. Biochem Int 1987;14:241–247.
12 Chilton FH, Connell TR: 1-Ether-linked phosphoglycerides. Major endogenous sources of arachidonate in the human neutrophil. J Biol Chem 1988;263:5260–5265.
13 Blank ML, Smith ZL, Lee YJ, et al: Effects of eicosapentaenoic and docosahexaenoic acid supplements on phospholipid composition and plasmalogen biosynthesis in $P388D_1$ cells. Arch Biochem Biophys 1989;269:603–611.
14 Lister MD, Glaser KB, Ulevitch RJ, et al: Inhibition studies on the membrane-associated phospholipase A_2 in vitro and prostaglandin E_2 production in vivo of the macrophage-like $P388D_1$ cell. J Biol Chem 1989;264:8520–8528.
15 Blank ML, Robinson M, Fitzgerald V, et al: Novel quantitative method for determination of molecular species of phospholipids and diglycerides. J Chromatogr 1984;298:473–482.
16 Blank ML, Cress EA, Snyder F: Separation and quantitation of phospholipid subclasses as their diradylglycerobenzoate derivatives by normal-phase high performance liquid chromatography. J Chromatogr 1987;392:421–425.

Fred Snyder, PhD, Medical Sciences Division, Oak Ridge Associated Universities, Post Office Box 117, Oak Ridge, TN 37831-0117 (USA)

Simopoulos AP, Kifer RR, Martin RE, Barlow SM (eds): Health Effects of ω3 Polyunsaturated Fatty Acids in Seafoods. World Rev Nutr Diet. Basel, Karger, 1991, vol 66, pp 391–400

Effects of Dietary Fish Oil on Leukocyte Leukotriene and PAF Generation and on Neutrophil Chemotaxis[1]

Richard I. Sperling

Department of Medicine, Harvard Medical School, Department of Rheumatology and Immunology, Brigham and Women's Hospital, Boston, Mass., USA

Introduction

The 5-lipoxygenase pathway enzymes, which synthesize mono- and dihydroxy fatty acids and leukotrienes from selected polyunsaturated fatty acids, differ in several ways from those of the more well-known cyclooxygenase pathway enzymes, which produce the prostaglandins and thromboxanes. The 5-lipoxygenase pathway has been identified in only a few cell types, predominantly those of bone marrow origin, and 5-lipoxygenase pathway product generation requires activation of the initial enzyme of the pathway, 5-lipoxygenase, in addition to the availability of substrate [1]. The cyclooxygenase pathway, however, is virtually ubiquitous among nucleated cells, and requires only substrate availability for product generation. Arachidonic acid (AA; 20:4 ω6), the major substrate for both the 5-lipoxygenase and cyclooxygenase pathways in humans on a typical western diet, is derived from ω6 fatty acids of land-based plants and animals. Eicosapentaenoic acid (EPA; 20:5 ω3) and docosahexaenoic acid (DHA; 22:6 ω3) are each prominent in marine organisms, primarily in the form of esterified triglycerides. Both EPA and DHA inhibit cyclooxygenase in cell-free preparations, and EPA inhibits cyclooxygenase in intact cells [2, 3]. The effects of the ω3 polyunsaturated fatty acids on the 5-lipoxygenase and

[1] This work was supported in part by a post-doctoral fellowship and a Biomedical Research Award from the Arthritis Foundation and Research Grants AI-22531 and AR-38638 from the National Institutes of Health.

PAF synthesis pathways in leukocytes from healthy volunteers and patients with defined inflammatory disorders have been evaluated.

Pathways of PAF Synthesis and 5-Lipoxygenase Pathway Metabolism of Polyunsaturated Fatty Acids

Dietary polyunsaturated fatty acids, including the ω3 polyunsaturated fatty acids, are incorporated into membrane phospholipids, generally on the 2-position. Upon activation of phospholipase A_2, 2-lyso-phospholipids are formed. The 2-lyso-phospholipids formed include 1-O-alkyl-2-lyso-phosphatidylcholine (2-lyso-PAF). Acetylation of the 2-lyso-PAF by a specific acetyl transferase yields the bioactive PAF. In addition to the formation of 2-lyso-phospholipids, unesterified fatty acids, primarily long-chain polyunsaturated fatty acids, are released simultaneously by the action of PLA_2. While AA and EPA are readily released, DHA is poorly released by PLA_2. AA is initially metabolized by 5-lipoxygenase to its 5*S* hydroperoxy derivative, 5-*S*-hydroperoxy-6-*trans*-8,11,14-*cis*-eicosatetraenoic acid (5-HPETE). Similarly, EPA and DHA are metabolized to their corresponding hydroperoxy derivatives, 5-*S*-hydroperoxy-6-*trans*-8,11,14-*cis*-eicosapentaenoic acid (5-HPEPE) and the corresponding 4- and 7-*S*-hydroperoxy-docosahexaenoic acids, respectively. DHA, however, is metabolized by 5-lipoxygenase to a much lesser extent than AA or EPA. These unstable hydroperoxy-fatty acid intermediates may, at least in part, be reduced to the corresponding alcohols known as 5-HETE, 5-HEPE, and 4- and 7-HDHA. This reduction is the final step in the metabolism of DHA by the 5-lipoxygenase pathway. 5-HPETE and 5-HPEPE may be also further metabolized by the hydroperoxy-fatty acid dehydrase activity of 5-lipoxygenase to the respective 5,6-epoxides known as leukotriene (LT) A_4(5,6-*trans*-oxido-,7,9-*trans*-11,14-*cis*-eicosatetraenoic acid), and the analogous LTA_5, derived from EPA (fig. 1) [4].

The LTA epoxide leukotrienes are unstable in aqueous environments; the non-enzymatic hydrolysis of AA-derived LTA_4 mainly yields two minimally bioactive diastereoisomeric dihydroxy-derivatives (5*S*, 12*R*- and 5*S*,12*S*-dihydroxy-6,8,10-*trans*-14-*cis*-eicosatetraenoic acids) known as the 6-trans-LTB_4 diastereoisomers. The epoxide hydrolase is expressed in the cytosol of some human cells, such as the polymorphonuclear neutrophilic leukocyte (PMN) [5], monocyte and pulmonary alveolar macrophage and is also found in erythrocytes and blood plasma [6, 7]; the action of the LTA epoxide hydrolase on LTA_4 yields the bioactive LTB_4 [1, 8–10]. Correspond-

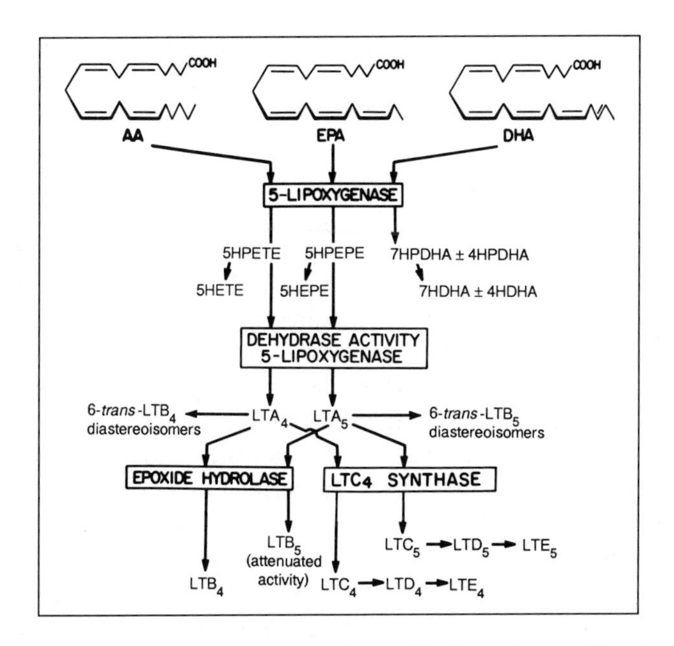

Fig. 1. Oxidative metabolism of arachidonic acid (AA, 20:4, N-6); eicosapentaenoic acid (EPA, 20:5, N-3) and docosahexaenoic acid (DHA, 22:6, N-3) by the 5-lipoxygenase pathway. 5HPETE = 5-*S*-hydroperoxyeicosatetraenoic acid; 5HETE = 5-*S*-hydroxyeicosatetraenoic acid; 5HPEPE = 5-*S*-hydroperoxyeicosapentaenoic acid; 5HEPE = 5-*S*-hydroxyeicosapentaenoic acid; 7HPDHA = 7-*S*-hydroperoxydocosahexaenoic acid; 4HPDHA = 4-*S*-hydroperoxydocosahexaenoic acid; 7HDHA = 7-*S*-hydroxydocosahexaenoic acid; 4HDHA = 4-*S*-hydroxydocosahexaenoic acid; LT = leukotriene.

ingly, the non-enzymatic hydrolysis of LTA_5 yields the analogous 6-*trans*-LTB_5 diastereoisomers, whereas the action of the LTA epoxide hydrolase on LTA_5 yields LTB_5 [11]. Nathaniel et al. [12] have demonstrated that LTA_5 covalently binds to the LTA epoxide hydrolase, thereby inactivating it.

A glutathione-S-transferase isoenzyme, known as LTC_4 synthase, present in the monocyte, eosinophil and mast cell, also further catalyzes the metabolism of the epoxide leukotrienes. This microsomal enzyme [13–15] hydrolyses LTA_4 to 5*S*-hydroxy-6*R*-S-glutathionyl-7,9-*trans*-11,14-*cis*-eicosatetraenoic acid (LTC_4) and LTA_5 to LTC_5. Sequential cleavage of the glutathione side chain yields the biologically-active 5*S*-hydroxy-6*R*-S-cysteinylglycyl-7,9-*trans*-11,14-*cis*-eicosatetraenoic acid (LTD_4) and 5*S*-hydroxy-6*R*-S-cysteinyl-7,9-*trans*-11,14-*cis*-eicosatetraenoic acid (LTE_4), and similarly LTD_5 and LTE_5 from LTC_5.

Effects of ω3 Fatty Acids on the Human 5-Lipoxygenase Pathway

While the effect of the ω3 polyunsaturated fatty acids of PAF generation and AA metabolism via the 5-lipoxygenase pathway might be evaluated best by the measurements of products generated in vivo, in the appropriate bodily fluids, until recently, limitations of the assays for LTC_4, LTD_4 and LTE_4 due to unacceptable background levels of immunoreactivity in bodily fluids [16], and for LTB_4, due to rapid, and nearly total degradation in vivo [17], precluded this approach. In the studies presented below, the effects of ω3 polyunsaturated fatty acids on product generation by ex vivo-activated isolated cells was assessed, both after in vitro incubation with exogenous fatty acids, and after dietary fish oil followed by ex vivo activation of the isolated cells. In vitro incubation with exogenous fatty acids prior to cell activation assesses the possible effects of unesterified fatty acids in bodily fluids [18, 19] whereas ex vivo activation of the isolated cells from individuals after dietary fatty acid supplementation assesses the contribution of in vivo incorporated fatty acids to the total effects of the dietary fish oil.

In the initial study of the effects of exogenous polyunsaturated fatty acids on 5-lipoxygenase pathway product generation [20], PMN were isolated from healthy donors on their usual western diet and were activated ex vivo with 10 μ*M* calcium ionophore A23187 for 5 min in the presence of 0–40 μg/ml of AA, EPA and DHA. 5-Lipoxygenase pathway product generation was quantitated by integrated optical density and/or radioimmunoassay after resolution of products by reverse-phase high-performance liquid chromatography (HPLC). In the absence of exogenous fatty acids, the only products of the 5-lipoxygenase pathway detected were the stable products derived from AA: 5-HETE, the 6-trans-LTB_4 diastereoisomers and LTB_4. The addition of concentrations of exogenous AA augmented the generation of each of these stable 5-lipoxygenase products. In the presence of quantities of exogenous EPA, these same 5-lipoxygenase metabolites of AA, derived from cellular lipids were demonstrated – as well as small quantities of the corresponding metabolites derived from exogenous EPA. The EPA-derived products differed structurally from the AA-derived products only by the presence of an additional cis-double bond in the 17-position; however, LTB_5, derived from EPA, possesses less than 10% of the bioactivity of LTB_4. In addition, in the presence of exogenous EPA, there was a significant and selective reduction in LTB_4 generation, indicating inhibition of the epoxide hydrolase. In the presence of exogenous docosahexaenoic acid the only DHA-derived 5-lipoxygenase products detected were the 7- and 4-hydroxy derivatives, and exogenous docosahexaenoic acid did not affect metabolism of endogenous AA by this pathway.

In a separate study [21], human monocyte monolayers were pre-incubated with concentrations of AA, EPA and DHA from 0–2.5 μg/ml for 4 h. After washing, the monolayers were activated with 10 μ*M* A23187; PAF was measured by bioassay and confirmed by HPLC followed by bioassay of the fractional eluates. Total PAF generation was augmented 64% after pre-incubation with the optimal concentration of 1 μg/ml of AA; pre-incubation with EPA resulted in a 28% decrease in total PAF formation and DHA pre-incubation had no effect. The changes in total PAF formation reflected parallel changes in both released and cell-retained PAF. These studies set the stage for the investigation of the effects of dietary fatty acids on the metabolism of AA via the 5-lipoxygenase pathway.

Effects of Dietary ω3 Fatty Acids in Healthy Individuals on Neutrophil Function and the 5-Lipoxygenase Pathway

Seven healthy male volunteers supplemented their usual diet for 6 weeks with 18 g daily of MaxEPA fish oil, providing 3.2 g of EPA and 2.2 g DHA daily [22]. Leukocytes were isolated and activated with the calcium ionophore in dose- and time-dependent protocols, and 5-lipoxygenase pathway product generation was quantitated by integrated optical density and/or radioimmunoassay after resolution of products by reverse-phase HPLC. After 3 weeks of dietary fish oil, 7-fold and 15-fold increases in PMN and mononuclear leukocyte cellular lipid EPA contents, respectively, were observed. These changes persisted after 6 weeks of dietary fish oil consumption. These changes in PMN cellular lipid composition after 6 weeks of the dietary fish oil supplementation were associated with a 39% decline in [^{3}H]-arachidonic acid release from PMN pre-incubated with [^{3}H]-arachidonic acid, a 60% decrease in 5-HETE and 6-trans-LTB_4 generation, and a greater than 50% decrease in LTB_4 generation. Only small quantities of the EPA-derived 5-HEPE and LTB_5 were detected after 6 weeks of dietary fish oil consumption. The suppressed [^{3}H]-arachidonic acid release after 6 weeks of dietary fish oil supplementation suggests inhibition of PLA_2 activity; the further suppression of the generation of all the 5-lipoxygenase pathway products suggests inhibition of PLA_2 and possibly 5-lipoxygenase activity as well. The changes in [^{3}H]-arachidonic acid release and arachidonate-derived 5-lipoxygenase pathway generation observed after 6 weeks of dietary fish oil consumption were not observed after 3 weeks of dietary fish oil supplementation, despite comparable neutrophil EPA contents and comparable generation of EPA-derived LTB_5 and 5-HEPE at these two time points.

The changes in mononuclear cellular lipid composition after 6 weeks of the dietary fish oil supplementation were associated with a 37% decline in [^{3}H]-arachidonic acid release in monocyte monolayers pre-incubated with [^{3}H]-arachidonic acid, an approximately 35% decrease in 5-HETE and 6-trans-LTB_4 generation, and a greater than 50% decrease in LTB_4 generation. Only small quantities of the EPA-derived 5-HEPE and LTB_5 were detected after 6 weeks of dietary fish oil consumption. The suppressed [^{3}H]-arachidonic acid release after 6 weeks of dietary fish oil supplementation suggests inhibition of PLA_2 activity; the further suppression of the generation of LTB_4 suggests inhibition of PLA_2 and possibly LTA epoxide hydrolase, as well. Again, the changes in [^{3}H]-arachidonic acid release and arachidonate-derived 5-lipoxygenase pathway product generation observed after 6 weeks of dietary fish oil consumption were not observed after 3 weeks of dietary fish oil supplementation, despite comparable monocyte EPA contents and comparable generation of EPA-derived LTB_5 and 5-HEPE at these two time points. Monocyte PAF generation was quantitated by bioassay in the same samples in which monocyte 5-lipoxygenase pathway products were measured. Total monocyte PAF generation was inhibited by more than 50% after 6 weeks of dietary fish oil, reflecting parallel inhibition of released and cell-retained PAF. The suppression of monocyte PAF generation was not observed after 3 weeks of dietary fish oil supplementation.

In this study of healthy volunteers, in addition to inhibiting 5-lipoxygenase pathway product generation, 6 weeks of dietary fish oil consumption resulted in a 71% decrease in the maximal PMN chemotactic responsiveness to LTB_4. A parallel suppression of LTB_4-mediated enhancement of PMN adherence to endothelial cell monolayers was also observed [22, 23]. These changes were not observed at the 3-week evaluation. Based on these data, dietary fish oil consumption might be expected to decrease both the margination and diapedesis of PMN into an inflammatory site in response to LTB_4. Both the alterations of PMN functional responses and the effects on fatty acid metabolism via the 5-lipoxygenase pathway returned nearly to baseline 6 weeks after discontinuing the dietary fish oil, further supporting a causal relationship to the dietary modification.

Terano et al. [24] studied the effects, individually, of 4 weeks of dietary EPA ethyl ester and DHA ethyl ester, on the function of PMN from healthy individuals. After 4 weeks of dietary EPA, small quantities of the

EPA-derived 5-lipoxygenase pathway products were observed, and LTB_4 synthesis by calcium ionophore activated PMN was inhibited 34%, without any effect observed on the synthesis of the other arachidonate-derived 5-lipoxygenase pathway products. The differing results of these two studies may reflect the differing duration and/or dose of fish oil supplementation. In addition, Terano et al. reported a significant suppression of neutrophil chemotaxis to LTB_4 and FMLP, thus confirming the findings of the previous study. DHA, however, did not effect neutrophil 5-lipoxygenation of AA and demonstrated a much less pronounced effect upon neutrophil chemotaxis; the authors attributed this effect to a rise in PMN cellular EPA content, observed after 4 weeks of DHA ethyl ester consumption [24].

Effects of Dietary ω3 Fatty Acids in Patients with Rheumatoid Arthritis on Leukocyte Function and the 5-Lipoxygenase Pathway

Twelve patients with active rheumatoid arthritis (RA) supplemented their diets daily for 6 weeks with 20 g of MaxEPA fish oil [25]. In these patients, an 18-fold rise in PMN EPA content and a concomitant 33% decline in AA content was observed after 6 weeks of fish oil supplementation. PMN and monocytes were activated ex vivo with calcium ionophore in a dose- and time-dependent protocol similar to that in the study of Lee et al. [22]. A 50% inhibition in LTB_4 generation and only small quantities of LTB_5 were observed after 6 weeks of dietary fish oil supplementation. In a clinical trial of dietary fish oil in RA, Kremer et al. [26] observed that improvement in the tender joint count correlated significantly with inhibition of LTB_4 generation, but not with the synthesis of EPA-derived LTB_5, after 12 weeks of 15 g MaxEPA consumption daily. Although, a 50% inhibition in LTB_4 generation was observed both in healthy individuals and the patients with RA, 5-HETE generation was inhibited by nearly 60% in the healthy volunteers after 6 weeks [22], but not in the RA patients [25]. In the patients with RA, a mean decrease of 15% in [^{3}H]-arachidonic acid release, which was not statistically significant, was observed. The 60% suppression of LTB_4 and 5-HETE synthesis in healthy volunteers suggests inhibition of phospholipase A_2 activity and possibly 5-lipoxygenase activity as well, whereas in the patients with RA, a selective inhibition of LTB_4 synthesis at the level of the epoxide hydrolase was observed. Nathaniel et al. [12] demonstrated that LTA_5, derived from EPA, covalently binds to the LTA epoxide hydrolase, thereby inactivating the enzyme.

Monocytes were activated ex vivo in protocols parallel to that for the PMN. Small quantities of EPA-derived LTB_5 and 5-HEPE were observed after 6 weeks of dietary fish oil. In the patients with RA, 6 weeks of dietary fish oil supplementation only minimally affected metabolism of AA by the 5-lipoxygenase pathway. The differences in the effects of 6 weeks of dietary fish oil between the healthy volunteers [22] and the patients with RA [25], may reflect the effect of the activity of the underlying disease and/or of background drug therapy. One consistent finding with regard to the effects on monocytes in the two populations was the suppression of total monocyte PAF generation, by about 50%, reflecting parallel inhibition of released and cell-retained PAF.

PMN chemotaxis to concentrations of LTB_4 and the chemotactic peptide, N-formyl-methionyl-leucyl-phenylalanine (FMLP) was evaluated in Boyden microchambers. In the patients with RA, chemotaxis to both LTB_4 and FMLP in the pre-diet period was suppressed, as compared with that of healthy volunteers [22]. After 6 weeks of dietary fish oil consumption, PMN chemotaxis to LTB_4 increased towards normal. A possible explanation of these findings is that in the pre-diet period the PMN of patients with RA were deactivated, perhaps due to in vivo LTB_4 exposure; after the fish oil diet, there was less in vivo LTB_4 exposure, and therefore less deactivation.

Summary and Conclusions

The studies of dietary fish oil supplementation in healthy volunteers demonstrate: (1) suppression of PMN LTB_4 synthesis after a minimum of 4 weeks of dietary fish oil consumption at a level of 4–6 g ω3 fatty acids daily; concomitant suppression of the other arachidonate-derived 5-lipoxygenase pathway products and decreased [^{3}H]-arachidonic acid release may be observed under certain conditions, (2) suppression of PMN chemotactic responsiveness to LTB_4 and FMLP, (3) delayed kinetics of inhibition of chemotaxis and AA metabolism relative to that of cellular lipid alteration, and (4) dietary EPA is more active than DHA in eliciting these effects. The effects of dietary EPA on monocyte function in healthy volunteers include: (1) suppression of LTB_4 synthesis concomitantly with that of the other 5-lipoxygenase pathway products and decreased [^{3}H]-arachidonic acid release, (2) suppression of PAF synthesis, and (3) delayed kinetics of inhibition of PAF generation and AA metabolism relative to that of cellular lipid alteration.

The effects of dietary fish oil in RA patients include: (1) decreased arachidonate content of cellular lipids with an augmented EPA content,

(2) decreased LTB_4 generation by PMN as an isolated effect, indicating inhibition of the epoxide hydrolase enzyme. The decrease in LTB_4 generation by PMN correlated with improvement of tender joint count in one study, (3) augmentation of depressed PMN chemotaxis to LTB_4 and FMLP, and (4) suppression of monocyte PAF generation.

From these studies one may conclude that: (1) ω3 fatty acids are incorporated into leukocyte cellular phospholipids with a concomitant loss in arachidonic acid, (2) the incorporation of ω3 fatty acids into leukocyte cellular lipids suppresses two pathways of inflammatory mediator synthesis: the 5-lipoxygenase and the PAF synthesis pathways, (3) receptor-mediated PMN functions are altered by dietary ω3 fatty acid consumption, and (4) these functional changes may be delayed vis-a-vis changes in cellular lipid composition and may vary with the underlying disease states and/or background medication.

References

1 Borgeat P, Samuelsson B: Arachidonic acid metabolism in polymorphonucleocytes: Effects of ionophore A23187. Proc Natl Acad Sci 1979;76:2148–2152.

2 Needleman P, Raz A, Minkes MS, Ferendelli JA, Sprecher H: Triene prostaglandins: prostacyclin and thromboxane biosynthesis and unique biological properties. Proc Natl Acad Sci 1979;76:944–948.

3 Corey AJ, Shih C, Cashman JR: DHA is a strong inhibitor of prostaglandin but not leukotriene biosynthesis. Proc Natl Acad Sci 1983;80:3581–3584.

4 Rouzer CA, Matsumoto T, Samuelsson B: Single protein from human leukocytes possesses 5-lipoxygenase and LTA4 synthetase activities. Proc Natl Acad Sci 1986; 83:857–861.

5 Rådmark O, Shimizu T, Jørnvall H, Samuelsson B: LTA4 hydrolase in human leukocytes. J Biol Chem 1984;259:12339–12345.

6 Fitzpatrick F, Haeggstrom J, Granstrom E, Samuelsson B: Metabolism of LTA4 by an enzyme in blood plasma: A possible leukotactic mechanism. Proc Natl Acad Sci 1983;80:5425–5429.

7 Fitzpatrick F, Liggett W, McGee J, et al: Metabolism of LTA4 by human erythrocytes. J Biol Chem 1984;259:11403–11407.

8 Fels AO, Pawlowski NA, Cramer EB, et al: Human alveolar macrophages produce LTB4. Proc Natl Acad Sci 1982;79:7866–7870.

9 Godard P, Damon M, Michel FB, et al: LTB4 production from human alveolar macrophages. Clin Res 1983;31:548A.

10 Williams JD, Czop JK, Austen KF: Release of leukotrienes by human monocytes on stimulation of their phagocytic receptor for particulate activators. J Immunol 1984; 132:3034–3040.

11 Lee TH, Mencia-Huerta J-M, Shih C, et al: Characterization and biological properties of 5,12-dihydroxy derivatives of EPA, including LTB5 and the double lipoxygenase product. J Biol Chem 1984;259:2383–2389.

12 Nathaniel DJ, Evans JF, Leblanc Y, et al: LTA5 is a substrate and an inhibitor of rat and human LTA4 hydrolase. Biochem Biophys Res Commun 1985;131:827–835.
13 Bach MK, Brashler JR, Morton DR Jr: Solubilization and characterization of the LTC4 synthetase of rat basophil leukemia cells: A novel, particulate glutathione S-transferase. Arch Biochem Biophys 1984;230:455–465.
14 Yoshimoto T, Sobermann RJ, Lewis RA, Austen KF: Isolation and characterization of LTC4 synthetase of rat basophil leukemia cells. Proc Natl Acad Sci 1985;82: 8399–8403.
15 Owen WJ Jr, Soberman RJ, Yoshimoto T, et al: Synthesis and release of LTC4 by human eosinophils. J Immunol 1987;138:532–538.
16 Heavey DJ, Soberman RJ, Lewis RA, et al: Critical considerations in the development of an assay for sulfidopeptide leukotrienes in plasma. Prostaglandins 1987;33: 693–708.
17 Serafin WE, Oates JA, Hubbard WC: Metabolism of LTB4 in the monkey: Identification of the principal nonvolatile metabolite in the urine. Prostaglandins 1984;27: 899–911.
18 Dole VP: A relation between non-esterified fatty acids in plasma and the metabolism of glucose. J Clin Invest 1956:35:150–154.
19 Waku K, Lands WEM: Control of lecithin biosynthesis in erythrocyte membranes. J Lipid Res 1968;9:12–18.
20 Lee TH, Mencia-Huerta J-M, Shih C, et al: Effects of exogenous arachidonic acid, eicosapentaenoic acid and docosahexaenoic acids on the generation of 5-lipoxygenase pathway products by ionophore activated human neutrophils. J Clin Invest 1984;74:1922–1933.
21 Sperling RI, Robin J-L, Kylander KA, et al: The effects of N-3 polyunsaturated fatty acids on the generation of PAF-acether by human monocytes. J Immunol 1987;139: 4186–4191.
22 Lee TH, Hoover RL, Williams JD, Sperling RI, et al: Effect of dietary supplementation with eicosapentaenoic acid and docosahexaenoic acid on in vitro neutrophil and monocyte leukotriene generation and neutrophil function. N Engl J Med 1985; 312:1217–1223.
23 Hoover RL, Karnovsky MJ, Corey EJ, et al: LTB4 action on endothelium mediates augmented neutrophil/endothelial adhesion. Proc Natl Acad Sci 1984;81:2191–2193.
24 Terano T, Seya A, Harai A, et al: Effect of oral administration of highly purified EPA and DHA on eicosanoid formation and neutrophil function in healthy subjects; in Lands WEM (ed): Proceedings of the AOCS Short Course on Polyunsaturated Fatty Acids and Eicosanoids. Champaign IL, American Oil Chemists' Society, 1987, pp 133–138.
25 Sperling RI, Weinblatt M, Robin J-L, et al: Effects of dietary supplementation with marine fish oil on leukocyte lipid mediator generation and function in rheumatoid arthritis. Arth Rheum 1987;30:988–997.
26 Kremer JM, Jubiz W, Michalek A, et al: Fish oil fatty acid supplementation in active rheumatoid arthritis. Ann Intern Med 1987;106:497–503.

Richard I. Sperling, AM, MD, Department of Medicine, Harvard Medical School, Department of Rheumatology and Immunology, Brigham and Women's Hospital, Boston, MA 02115 (USA)

Simopoulos AP, Kifer RR, Martin RE, Barlow SM (eds): Health Effects of ω3 Polyunsaturated Fatty Acids in Seafoods. World Rev Nutr Diet. Basel, Karger, 1991, vol 66, pp 401–406

Effects of ω3 Fatty Acid Supplements on ex vivo Synthesis of Cytokines in Human Volunteers

Comparison with Oral Aspirin and Ibuprofen[1]

Stefan Endres[a], *Simin N. Meydani*[b], *Charles A. Dinarello*[c]

[a] Medizinische Klinik, Klinikum Innenstadt der Universität München, FRG; [b] USDA Human Nutrition Research Center on Aging, Tufts University, Boston, Mass., and [c] Dept. of Medicine, Tufts University – New England Medical Center, Boston, Mass., USA

Biological Activities of Interleukin-1

Aside from its pyrogenic effects interleukin-1 (IL-1) influences a wide array of biological functions [reviewed in 1]. Originally described in purified material from monocyte supernatants, most effects have been confirmed using recombinant IL-1 [2].

IL-1 is a cofactor in activating T-lymphocytes inducing synthesis of interleukin-2 and interleukin-2 receptors. It increases antibody synthesis from B cells and acts in synergy with interferons in the enhancement of natural killer cell activity. IL-1 has been found to be identical to hemopoietin-1, which was recognized by its ability to synergize with colony stimulating factors in affecting hemotopoietic stem cell proliferation [3], and it induces the release of granulocytes from the bone marrow.

IL-1 triggers several changes in metabolism. Hepatic synthesis of the acute phase proteins (serum amyloid A, C-reactive protein, fibrinogen, α_2-macroglobulin), participating in different stages of host defense, is increased. Concurrently, synthetic capacity is economized by a drop in

[1] These studies are supported by NIH Grant AI 15614. S.E. is supported by grant EN-169/2 of the Deutsche Forschungsgemeinschaft.

albumin synthesis and more amino acids are liberated by induction of skeletal muscle proteolysis. Fibroblast collagen production is enhanced, possibly contributing to the process of wound healing. IL-1 has also been identified as an osteoclast activating factor.

IL-1 has profound effects on vascular endothelial cell function. It activates human endothelial cells in vitro to synthesize and release prostaglandin I_2 and E_2 [4]. While these two arachidonic acid metabolites increase blood flow, IL-1 also orchestrates a cascade of cellular and biochemical events that lead to vascular congestion, clot formation and cellular infiltration.

IL-1 induces other lymphokines (interleukin-2, interleukin-3, interleukin-6; granulocyte macrophage colony stimulating factor [GM-CSF], interferon-β, -γ; tumor necrosis factor [TNF], lymphotoxin) and can serve as an autocrine signal promoting the expression of its own gene [5].

Modulation of Cytokine Synthesis by ω3 Fatty Acids

To date the only known pharmacologic agents to reduce cytokine synthesis are corticosteroids and ciclosporin A. We have completed a study to investigate whether dietary supplementation with ω3 fatty acids contained in fish-oils, affects the synthesis of IL-1 and tumor necrosis factor (TNF) [6]. The main forms of ω3 fatty acids are eicosapentaenoic acid and docosahexaenoic acid.

Since IL-1 and TNF are principal mediators of inflammation, reduced production of these cytokines may contribute to the amelioration of inflammatory symptoms in patients taking ω3 supplementation. Therefore, we decided to investigate the effects of ω3 fatty acids on the synthesis of the cytokines IL-1 and TNF.

Nine healthy volunteers added 18 g of fish-oil concentrate (Max EPA®) per day to their normal diet. In vitro production of IL-1 and TNF was determined by incubating peripheral blood mononuclear cells for 24 h with different stimuli. At the end of the incubation, the cells were lysed by freeze-thawing to obtain total, that is cell-associated plus secreted cytokine. IL-1β [7], IL-1α [8] and TNF [9] were measured by specific radioimmunoassays.

Figure 1 illustrates the production of IL-1β during the course of the study. Our findings demonstrate that ω3 fatty acid supplementation reduced the ability of IL-1β production induced by endotoxin. The effect

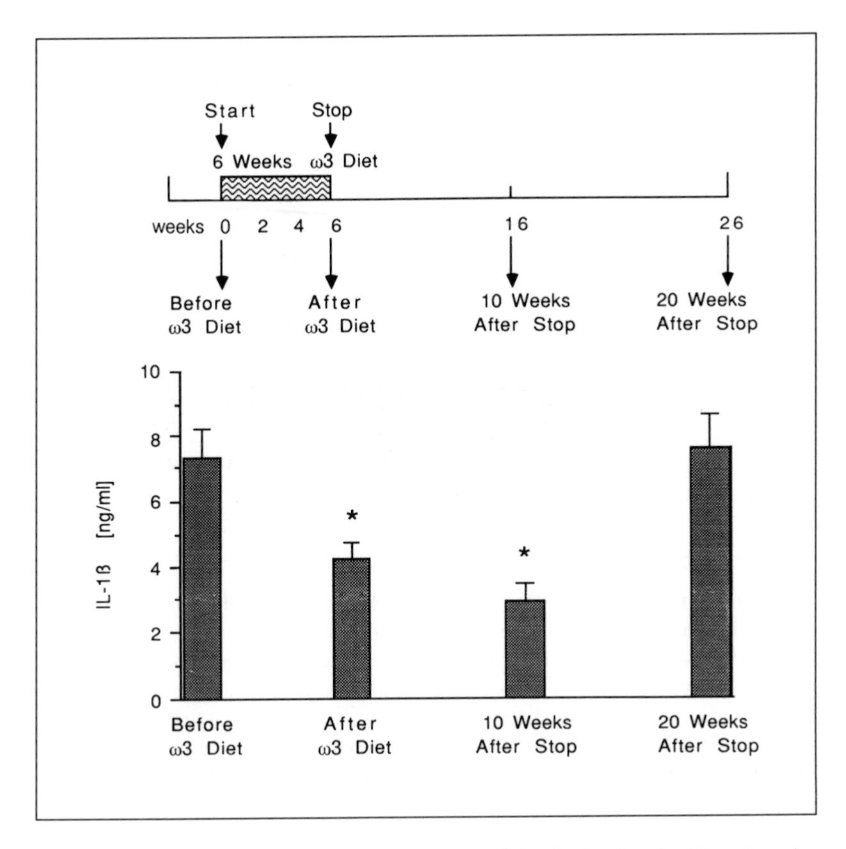

Fig. 1. Influence of ω3 fatty acids on production of IL-1β stimulated with endotoxin. Mononuclear cells were incubated for 24 h with 1 ng/nl of endotoxin. IL-1β was determined by RIA. The bars represent the mean values for 9 volunteers with error bars as the standard error of the mean. * indicates significant difference from the baseline (before ω3 diet) at $p < 0.05$.

was most pronounced 10 weeks after stopping the supplementation and suggests prolonged incorporation of ω3 fatty acids into a pool of circulation MNC. The capacity of the MNC from these donors to synthesize IL-1β returned to the presupplement level 20 weeks after ending the supplementation. Similar results were observed when we measured IL-1α and TNF.

There are several clinical implications of the present findings. There is a pathophysiologic rationale for therapeutic trials with ω3 fatty acids in

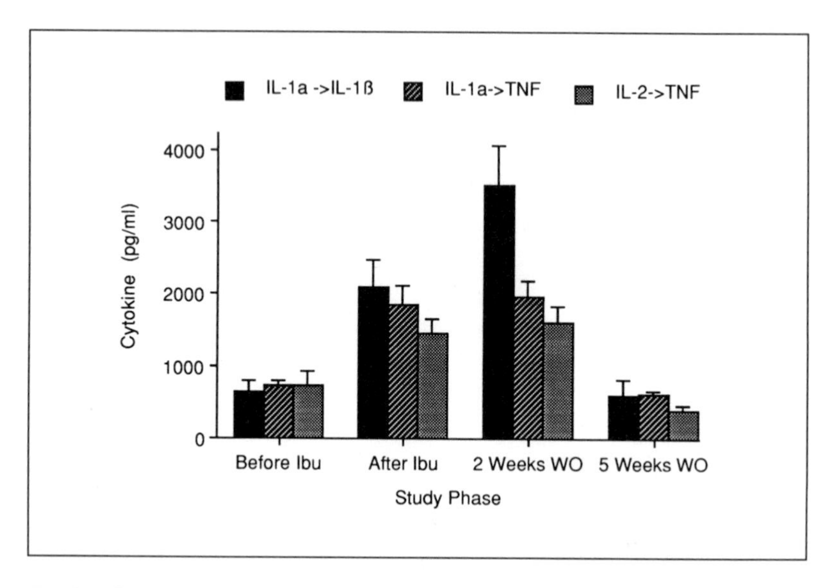

Fig. 2. Influence of oral ibuprofen on IL-1α-induced production of IL-1β and TNF and on IL-2-induced production of TNF. Mononuclear cells were incubated for 24 h with IL-1α (100 ng/ml) or IL-2 (1,000 U/ml) as stimulants. Induced TNF and IL-1β determined by RIA. Determinations were performed during four study phases: before starting ibuprofen medication, immediately after 12 days of ingestion of 200 mg ibuprofen/day, and during the washout (WO) phase, 2 and 5 weeks after discontinuation of ibuprofen. The bars represent the mean values for 7 volunteers with error bars as the standard error of the mean.

certain diseases with documented involvement of inflammatory cytokines such as IL-1 and TNF. For example, rheumatoid arthritis [10], psoriasis and type I diabetes mellitus are likely candidates.

Our findings may also have relevance for atherogenesis. As described above, IL-1 and TNF induce the expression of intercellular adhesion molecule-1 on the surface of vascular endothelial cells, stimulate the production of platelet activating factor, plasminogen activator inhibitor and procoagulant activity in endothelial cells, and induce smooth muscle cell proliferation [11]. Furthermore IL-1 and TNF are generated in the vessel wall itself, by endothelial and smooth muscle cells [12]. Thus, since the vessel wall forms both a cardinal target tissue and a source of cytokines, a suppressive effect on their generation by a ω3 supplemented diet may contribute to its protective effect against atherosclerosis.

Modulation of Cytokine Synthesis by Oral Cyclooxygenase Inhibitors

In subsequent studies we have investigated the effect of oral aspirin [13] and of oral ibuprofen [14] on the production of IL-1 and TNF. Aspirin and ibuprofen lead to an *increase* of the ex vivo production of IL-1β by mononuclear cells (fig. 2). This is particularly remarkable since in most other biological systems, e.g. the inhibition of platelet function, ω3 fatty acids and cyclooxygenase inhibitors like aspirin, induce a parallel effect.

References

1 Dinarello CA: Interleukin-1. Rev Infect Dis 1988;10:168–189.

2 Dinarello CA, Cannon JG, Mier JW, Bernheim HA, LoPreste G, Lynn DL, Love RN, Webb AC, Auron PE, Reuben RC, Rich A, Wolff SM, Putney SD: Multiple biological activities of human recombinant interleukin-1. J Clin Invest 1986;77: 1734–1739.

3 Moore MA, Warren DJ: Synergy of interleukin-1 and granulocyte colony stimulating factor: In vivo stimulation of stem-cell recovery and hemopoietic regeneration following 5-fluorouracil treatment of mice. Proc Natl Acad Sci USA 1987;84:7134–7138.

4 Rossi V, Breviario F, Ghezzi P, Dejana E, Mantovani A: Interleukin-1 induces prostacyclin in vascular cells. Science 1985;229:1174–1176.

5 Dinarello CA, Ikejima T, Warner SJC, Orencole SF, Lonnemann G, Cannon JG, Libby PL: Interleukin-1 induces interleukin-1. I. Induction of circulating interleukin-1 in rabbits in vivo and in human mononuclear cells in vitro. J Immunol 1987; 316:379–385.

6 Endres S, Ghorbani R, Kelley VE, Georgilis K, Lonnemann G, van der Meer JWM, Cannon JG, Rogers TS, Klempner MS, Weber PC, Schaefer EJ, Wolff SM, Dinarello CA: The effect of dietary supplementation with ω3 polyunsaturated fatty acids on the synthesis of interleukin-1 and tumor necrosis factor by mononuclear cells. N Engl J Med 1989;320:265–271.

7 Endres S, Ghorbani R, Lonnemann G, van der Meer JWM, Dinarello CA: Measurement of immunoreactive interleukin-1β from human mononuclear cells: Optimization of recovery, intrasubject consistency and comparison with interleukin-1a and tumor necrosis factor. Clin Immunol Immunopath 1988;49:424–438.

8 Lonnemann G, Endres S, van der Meer JWM, Cannon JG, Dinarello CA: A radioimmunoassay for human interleukin-1 alpha: Measurement of IL-1 alpha produced in vitro by human blood mononuclear cells stimulated with endotoxin. Lymphokine Res 1988;7:75–85.

9 van der Meer JWM, Endres S, Lonnemann G, Cannon JG, Ikejima T, Okusawa S, Gelfand JA, Dinarello CA: Concentrations of immunoreactive human tumor necrosis factor alpha produced by human mononuclear cells in vitro. J Leukocyte Biol 1988;43:216–223.

10 Krane SM, Dayer JM, Simon LS, Byrne S: Mononuclear cell conditioned medium containing mononuclear cell factor (MCF), homologous with interleukin-1, stimulates collagen and fibronectin synthesis by adherent rheumatoid synovial cells: effects of prostaglandin E_2 and indomethacin. Collagen Relat Res 1985;5:99–117.
11 Libby P, Ordovas JM, Auger KR, Robbins AH, Birinyi LK, Dinarello CA: Endotoxin and tumor necrosis factor induce interleukin-1 gene expression in adult human vascular endothelial cells. Am J Path 1986;124:179–186.
12 Warner SJC, Auger KR, Libby P: Human interleukin-1 induces interleukin-1 gene expression in human vascular smooth muscle cells. J Exp Med 1987;165:1316–1321.
13 Endres S, Cannon JG, Ghorbani R, Dempsey RA, Sisson SS, Lonnemann G, van der Meer JWM, Wolff SM, Dinarello CA: In vitro production of IL-1β, IL-1α, TNF and IL-2 in healthy subjects: Distribution, effect of cyclooxygenase inhibition and evidence of independent gene regulation. Eur J Immunol 1989;19:2327–2333.
14 Endres S, Whitaker RED, Ghorbani R, Meydani SN, Dinarello CA: Oral aspirin or ibuprofen increase the ex vivo synthesis of inflammatory cytokines (submitted for publication).

Charles A. Dinarello, MD, Division of Geographic Medicine and Infectious Diseases, Department of Medicine, Tufts University – New England Medical Center, 750 Washington Street, Boston, MA 02111 (USA)

Simopoulos AP, Kifer RR, Martin RE, Barlow SM (eds): Health Effects of ω3 Polyunsaturated Fatty Acids in Seafoods. World Rev Nutr Diet. Basel, Karger, 1991, vol 66, pp 407–416

Effects of ω3 Fatty Acids on Glucose and Lipid Metabolism in Non-Insulin-Dependent Diabetes mellitus

Bengt Vessby

Department of Geriatrics, Uppsala University, Uppsala, Sweden

Introduction

Non-insulin-dependent diabetes mellitus (NIDDM) is characterised by an impaired glucose homeostasis with a relative lack of insulin, usually associated with increased concentrations of peripheral insulin, with a decrease of the peripheral insulin sensitivity. Abnormalities of the lipoprotein lipid composition are common, especially increased triglyceride and low high-density lipoprotein (HDL) cholesterol concentrations, and many patients have an elevated blood pressure. Also, NIDDM is associated with an increased platelet aggregability and tendency to thrombus formation with an impaired fibrinolytic capacity. All these factors may contribute to the high incidence of atherosclerotic cardiovascular disease. Dietary advice is the basis of all treatment of diabetes mellitus. Diabetic patients are today prescribed a diet with a restricted fat content but with a rather high content of carbohydrates, especially fibre-rich complex carbohydrates. Up to the last years, however, there has been limited interest in the question of fat quality in the diabetic diet. This is somewhat surprising considering that nearly all the metabolic disturbances which are suggested to increase the risk for premature development of cardiovascular disease may be modified by changes of the dietary fat quality.

The Greenland Eskimoes, consuming a high-fat diet, have been reported to have a very low incidence of diabetes mellitus [1] and also a remarkably good glucose tolerance. It has been suggested that the low incidence of diabetes mellitus may be due to the high intake of long-chain, polyunsaturated ω3 fatty acids of marine origin. ω3 fatty acids have been shown to reduce increased triglyceride levels in hypertriglyceridaemic patients [2]. Also, substitution of ω3 fatty acids for ω6 fatty acids in the diet of rats fed high-fat diets have been reported to prevent the development of insulin resistance [3]. Based on these observations, and others, it has been suggested that ω3 fatty acids may be of value in the treatment of NIDDM.

During the last few years there have been several reports concerning the effects of supplementation of fish oils rich in ω3 fatty acids in the diet of patients with non-insulin-dependent diabetes mellitus. Mostly the studies have contained few subjects and there are as yet only a limited number of controlled studies. The duration and the design of the trials have varied. All these factors may have contributed to the partly conflicting results of the studies.

Effects of ω3 Fatty Acids on the Glucose Homeostasis in NIDDM

There are today many indications that addition of ω3 fatty acids, especially when given in high concentrations, may impair the glucose homeostasis in NIDDM patients, but suggestions of beneficial effects have also been forwarded. Popp-Snijders et al. [4] studied six NIDDM patients for eight weeks who were given 3 g ω3 fatty acids in addition to their ordinary diet. There was a tendency to increased blood glucose and reduced serum insulin concentrations although none of these changes were significant. The fasting-insulin/blood-glucose ratio was, however, significantly decreased. The same authors reported an increased metabolic clearance rate of glucose during a simultaneous infusion of glucose and insulin which suggests an improved in vivo insulin sensitivity.

In contrast to this, Glauber et al. [5] found increased fasting blood glucose concentrations when supplying six obese men with NIDDM with 5.5 g ω3 fatty acids during four weeks. They also reported a decreased glucose tolerance during a mixed meal with a significant increase of the mean glucose area under the curve. The glycosylated haemoglobin rose significantly. At the same time, the fasting insulin levels fell slightly (n.s.) and the mean insulin level during the course of a meal, administered as a liquid formula, was significantly reduced by 30%. The insulin response to intravenous glucose was further reduced by fish oil supplementation as was the glucagon-stimulated insulin response. There was, however, no change of the glucose disposal rate (at supraphysiological insulin levels) while the mean basal hepatic glucose output was significantly elevated. There was a significant correlation between the fasting blood glucose concentrations and the basal rate of hepatic glucose output. The concentration of free fatty acids in plasma remained unchanged. After ω3 fatty acid withdrawal the fasting glucose returned to baseline.

In another study Friday et al. [6] gave eight men with NIDDM 8g ω3 fatty acids during eight weeks, in addition to their ordinary diet. However, the subjects were instructed to decrease their fat intake to compensate for the extra calories in the marine lipid concentrate. Also in this study the fasting plasma glucose levels and the meal-stimulated glucose response increased significantly after ω3 fatty acid supplementation. No significant changes were seen in fasting or meal-stimulated plasma insulin, glucose disposal after an intravenous glucose injection or insulin-to-glucagon ratios. In response to the plasma glucose alterations higher insulin levels would have been expected suggesting an impairment in insulin secretion in response to ω3 fatty acids.

Kasim et al. [7] supplied 22 NIDDM patients without overt hyperlipoproteinaemia with 2.7 g ω3 fatty acids per day during eight weeks. A transient increase of fasting blood glucose and HbA_1 was seen. The lack of a persistent effect on this rather limited dose of ω3 fatty acids may, in connection with the data by Popp-Snijders et al. [4], suggest that the effects on the blood glucose concentrations may be more pronounced when higher doses are given.

Schectman et al. [8] performed a controlled cross-over study in 13 NIDDM patients in a single-blind study comparing the effects of 12 g MaxEPA, containing 4 g ω3 fatty acids, with those of 12 g safflower oil. At the end of the cross-over study the patients continued on a higher dose of ω3 fatty acids corresponding to 7.5 g per day. The fasting blood glucose was significantly increased only on the high dose, but the HbA_1 increased during both treatment periods compared with baseline. The postprandial blood glucose values were assessed only after the low dose of ω3 fatty acids showed increased blood glucose concentrations at 10 pm in the evening. There was no change of fasting or postprandial C-peptide concentrations. In a follow-up study by the same group [9] hypertriglyceridaemic patients, including six NIDDM patients, were given a high dose of ω3 fatty acids (9.8 g/day) during three months followed by a low dose during the next three months (3.9 g/day). This was done in order to investigate if the initial effects of the high doses of fish oil concentrates could be sustained when

Table 1. Effects of ω3 fatty acids on glucose homeostasis in NIDDM

Fasting blood glucose	± or ↑
Meal-stimulated glucose response	↑
Glycosylated haemoglobin	± or ↑
Fasting serum insulin concentrations	± or ↓
Stimulated insulin response	± or ↓
Insulin/glucose ratio	↓
Basal hepatic glucose output	↑
Peripheral glucose disposal	±

↑ = Increased; ± = unchanged; ↓ = decreased.

followed by a low dose maintenance therapy. The mean blood glucose concentrations, as well as the HbA_1 concentrations, increased significantly among the diabetic patients and remained virtually unchanged throughout the study independent of the reduction of the initial dose of ω3 fatty acids.

In a controlled, double-blind, crossover study of two periods of 8 weeks each, Vessby and Boberg [10] treated 14 NIDDM patients with either 10 g MaxEPA (3 g ω3 fatty acids) or 10 g of a placebo preparation (olive oil). The changes of fasting blood glucose concentrations were significantly different between the two treatment periods as was the fractional removal rate of glucose during the intravenous insulin tolerance test which was lower after addition of ω3 fatty acids than after placebo. The insulin and C-peptide concentrations during fasting and after i.v. glucagon injection were unchanged. In a subsequent study with a similar design it was shown that the increased blood glucose concentration was not due to an impaired glucose disposal as studied with the euglycaemic clamp technique (unpublished observations). On the other hand, the blood glucose concentrations after intravenous glucose were higher after ω3 fatty acids, than on admission or after placebo, while the serum insulin levels did not increase which is compatible with an impairment in the capacity to release insulin in response to a glucose challenge.

Taken together (table 1), it is obvious that ω3 fatty acids, at least when given in higher doses and during limited periods of time, may impair the glucose control in patients with NIDDM. Although the reason for this is not clear, data indicate that this may be due to an impaired pancreatic insulin response to glucose as well as to an elevated basal hepatic glucose production due to increased availability of gluconeogenic precursors or increased hepatic resistance to fasting levels of insulin. Most studies have failed to show any change in the peripheral glucose elimination [5, 11] although diverging results have been reported [4]. Increased postprandial blood glucose concentrations were also seen in two patients with lipotrophic diabetes [12] treated with high doses of ω3 fatty acids, while the C-peptide concentrations in fasting and after a test meal were reduced. Lardinois et al. [13] have reported that normal subjects given mixed test meals of varying fatty acid composition show greater serum insulin response to meals enriched with polyunsaturated fats as compared to those containing more saturated fatty acids. However, in NIDDM patients no such differences in serum insulin or C-peptide responses were seen, indicating an inability in these subjects to evoke a greater insulin response to polyunsaturated than to saturated fatty acids.

Interestingly, it has also been shown that NIDDM patients given an ordinary diet rich in polyunsaturated fatty acids, especially fat fish, during a controlled crossover study on a metabolic ward, showed higher blood glucose concentrations than when they were given a diet with an identical nutrient composition containing mainly saturated fat [14]. The content of ω3 fatty acids in the fish-enriched diet corresponded to approximately 5 g per day with an ω6/ω3 ratio of about 4. Blood glucose concentrations after a standard test meal were higher during the polyunsaturated fat diet while the insulin concentrations were lower than on the diet rich in saturated fat, changes which were similar to those recorded after addition of high doses of fish oil concentrates rich in ω3 fatty acids. Also, the urinary glucose excretion was higher on the polyunsaturated fat-rich diet while insulin concentrations and C-peptide concentrations after glucagon, as well as the k value during the intravenous insulin tolerance test, were unchanged.

Effects of ω3 Fatty Acids on Lipoprotein Composition in NIDDM

Generally, most studies including hypertriglyceridaemic NIDDM patients have shown reductions of the serum triglycerides and very low-density lipoprotein (VLDL) concentrations after administration of ω3 fatty acids [6–10]. In contrast, the study by Kasim et al. [7], including type 2 patients without overt hyperlipoproteinaemia, did not show any changes of either the serum triglyceride or serum total cholesterol concentrations. In the few patients with slightly increased triglycerides there was, however,

Table 2. Effects of ω3 fatty acids on lipoprotein composition in NIDDM

Serum triglycerides	↓ (or ±)
Serum cholesterol	↓ or ±
Serum apo B	± or ↑
VLDL triglycerides	↓
LDL cholesterol	↑ or ±
LDL apo B	↑
LDL cholesterol/LDL apo B	↓
HDL cholesterol	± or ↑
Serum apo A-I	±
Serum apo A-II	(↓)

↑ = Increased; ± = unchanged; ↓ = decreased.

a transient decrease after four weeks together with a similarly transient increase of the lipoprotein-lipase activity.

In contrast to the supposedly beneficial effects of ω3 fatty acids on the VLDL concentrations, the effects on the LDL cholesterol concentration and the LDL composition are more questionable. NIDDM patients typically have fairly normal LDL cholesterol concentrations. Generally, there have been no further reductions of LDL cholesterol after supplementation with fish oil [5–7, 9, 10]. Rather, there have been increases of LDL cholesterol and/or the LDL-apo B concentrations [8], effects which may be dose-dependent. Schectman and coworkers have in two studies [8, 9] reported significantly reduced ratios between the LDL cholesterol and the LDL-apo B concentrations indicating a changed LDL composition. The increase in this ratio may represent an increase in a dense LDL subspecies which may be particularly atherogenic [15].

The high-density lipoprotein (HDL) cholesterol concentrations are low in NIDDM patients. They have generally been unchanged [5, 7, 8, 10] or slightly increased [9] in response to supplementation with ω3 fatty acids. The apo A-I concentrations have in most cases been unaffected [7–10] while the apo A-II concentrations have been reported to be reduced in one study [10].

Taken together (table 2), the effects of ω3 fatty acids on the lipoprotein composition in NIDDM are hard to evaluate with regard to coronary heart disease risk. While the reduced triglyceride concentrations should be beneficial, the unchanged or even increased LDL concentrations, with a change of LDL composition, may be less desirable. It is worth noting that the HDL changes, which have been reported in some studies, do not seem to be directly coupled to the VLDL reductions [9] as is usually seen in non-diabetic patients, indicating a possible direct effect of ω3 fatty acids on the HDL levels.

The reduction of VLDL concentrations by ω3 fatty acids is probably due to a decreased production of triglyceride-rich lipoproteins, rather than to an increased removal. Enhanced insulin production, elevated free fatty acid levels and poor glycaemic control have all been linked to an elevated VLDL production in NIDDM [16]. During treatment with ω3 fatty acids, reduced VLDL levels are achieved in spite of increased plasma glucose concentrations and unchanged free fatty acid and insulin concentrations indicating that other mechanisms may contribute to the reduced hepatic lipoprotein production.

The long-term effects of ω3 fatty acids on lipoprotein-lipid levels in NIDDM are as yet not elucidated. In the study by Schectman et al. during six months [9] the efficacy was reduced during long-term treatment and there was no significant remaining reduction of the triglyceride concentrations after five months of treatment, in spite of a continuous daily supplementation with 3.9 g ω3 fatty acids. On the whole, it can not be excluded that the effects on lipoprotein concentrations and composition may be different in NIDDM patients and in non-diabetic subjects.

Other Effects of ω3 Fatty Acids on Cardiovascular Risk Factors in NIDDM

Although the effects of ω3 fatty acids on blood glucose concentrations and lipoprotein-lipid concentrations are somewhat questionable, other effects may counteract these effects and yet contribute to a reduced cardiovascular risk in NIDDM patients. Thus, reductions of blood pressure levels have been reported also in NIDDM patients after addition of ω3 fatty acids [7]. A reduced tendency for platelet aggregation has been described in insulin-dependent diabetics [17]. Whether this is true also in NIDDM is at present not known.

In non-diabetic patients ω3 fatty acids have been suggested to increase the fibrinolytic capacity [18]. In contrast, in a recent controlled and blinded randomised study in NIDDM patients, there was a significant increase of the plasminogen activator inhibitor type 1 (PAI-1) activity after ω3 fatty acids compared with when the patients were given the placebo preparation containing olive oil [B. Vessby, unpublished observation]. A suggested coupling between reduction of serum triglycerides and improvement of the fibrinolytic capacity was not present in these patients. This is in accordance with other recent reports showing elevated PAI-1 levels in healthy [19, 20] and hyperlipidaemic [21] subjects after supplementation with fish oil.

Areas of Further Research

The data concerning the effects of ω3 fatty acids in NIDDM patients are still scarce and incomplete. Most studies have been performed in small groups of patients during limited periods of time. Very few studies have been controlled and blinded. The studies have included heterogeneous groups of subjects with different degrees of impairment of the insulin secretion capacity. Some have been treated with diet only, many with antidiabetic drugs and some with insulin. Thus, there is a great need for further, controlled studies in well-defined groups of subjects. There is also need for studies of long-term effects and further studies on dose-response relationship.

Mechanistic studies are needed to determine the regulation of fatty acid synthesis and the fatty acid composition in different tissues after administration of ω3 fatty acids. ω3 fatty acids are precursors of prostaglandins and leukotrienes. It is known that prostaglandins are capable of augmenting as well as diminishing the pancreatic insulin release. Other possible regulating effects of eicosanoids on the pancreatic islet function include modulation of islet blood flow and increased responsiveness as reviewed recently [22]. Thus eicosanoids may modulate the pancreatic islet function in many ways. Also, altered membrane fluidity, due to incorporation of ω3 fatty acids, may influence the function of membrane-associated proteins in other tissues in the body.

It is today far too early to definitely conclude anything about the benefits or drawbacks of ω3 fatty acid treatment in NIDDM patients. The effects of ω3 fatty acids may be related to the dose. Today evidence is lacking regarding long-term effects of treatment with low doses. Possibly, negative effects on glucose homeostasis and lipid metabolism may be considered at high doses only, while other beneficial effects, e.g. on platelet reactivity, may be retained also at lower doses. Subjects with NIDDM are advised to increase their fish consumption including a higher intake of fatty fish. However, general use of different fish oil preparations with high concentrations of ω3 fatty acids in NIDDM patients should be discouraged before further studies have been performed. This may be in contrast to the situation in insulin-dependent diabetic patients, where ω3 fatty acids may have more beneficial effects.

References

1 Sagild U, Jorgen Littauer C, Jespersen S, et al: Epidemiological studies in Greenland 1962–1964: I. Diabetes mellitus in Eskimos. Acta Med Scand 1966;179:29–39.
2 Illingworth DR, Connor WE, Hatcher LF, et al: Hypolipidaemic effects of n–3 fatty acids in primary hyperlipoproteinaemia. J Int Med 1990;225 (suppl 731):91–97.
3 Storlien LH, Kraegen EW, Chisholm DJ, et al: Fish oil prevents insulin resistance induced by high-fat feeding in rats. Science 1987;237:885–888.
4 Popp-Snijders C, Schouten JA, Heine RJ, et al: Dietary supplementation of omega-3 polyunsaturated fatty acids improves insulin sensitivity in non-insulin-dependent diabetes. Diabetes Research 1987;4:141–147.
5 Glauber H, Wallis P, Griver K, et al: Adverse metabolic effects of omega-3 fatty acids in non-insulin-dependent diabetes mellitus. Ann Int Med 1988;108:663–668.
6 Friday KE, Childs MT, Tsunehara CH, et al: Elevated plasma glucose and lowered triglyceride levels from omega-3 fatty acid supplementation in type 2 diabetes. Diabetes Care 189;12:276–281.
7 Kasim SE, Stern B, Khilnani S, et al: Effects of omega-3 fish oils on lipid metabolism, glycaemic control, and blood pressure in type 2 diabetic patients. J Clin Endocrinol Metab 1987;67:1–5.
8 Schectman G, Kaul S, Kissebah AH: Effect of fish oil concentrate on lipoprotein composition in NIDDM. Diabetes 1988;37:1567–1573.
9 Schectman G, Kaul S, Cherayil GD, et al: Can the hypotriglyceridaemic effect on fish oil concentrate be sustained? Ann Int Med 1989;110:346–352.
10 Vessby B, Boberg M: Dietary supplementation with n–3 fatty acids may impair the glucose homeostasis in patients with non-insulin-dependent diabetes mellitus. J Int Med 1990 (in press).
11 Annuzzi G, Riccardi G, Capaldo G, et al: Omega-3 fatty acid dietary supplementation in type 2 (non-insulin-dependent) diabetic patients: Effects on glucose and lipid metabolism. Abstr. 25th Annu Meet Eur Assoc Study of Diabetes, Lisbon, 1989.
12 Stacpoole PW, Alig J, Kilgore LL, et al: Lipodystrophic diabetes mellitus. Investigations of lipoprotein metabolism and the effects of omega-3 fatty acid administration in two patients. Metabolism 1988;37:944–951.
13 Lardinois CK, Starich GH, Mazzaferri EL, et al: Effect of source of dietary fats on serum glucose, insulin, and gastric inhibitory polypeptide response to mixed test meals in subjects with non-insulin-dependent diabetes mellitus. J Am Coll Nutr 1988;7:129–136.
14 Vessby B, Karlström B, Boberg M, et al: An increased proportion of polyunsaturated fatty acids in the diet influences the blood glucose control in NIDDM patients. Abstr Fifth Int Symp Diabetes Nutr, Sorrent, 1987.

15 Sniderman AD, Wolfson C, Teng B, et al: Association of hyperapobetalipoproteinemia with endogenous hypertriglyceridemia and atherosclerosis. Ann Intern Med 1982;97:833–839.
16 Greenfield M, Kolterman D, Olefsky J, et al: Mechanisms of hypertriglyceridaemia in diabetic patients with fasting hyperglycaemia. Diabetologia 1980;18:441–446.
17 Haines AP, Sanders TAB, Imerson JD, et al: Effects of a fish oil supplementation on platelet function, haemostatic variables and albuminuria in insulin dependent diabetics. Thromb Res 1986;43:643–655.
18 Barcolli U, Glas-Greenwalt P, Pollak VE: Enhancing effect of dietary supplementation with omega-3 fatty acids on plasma fibrinolysis in normal subjects. Thromb Res 1985;30:307–312.
19 Emeis JJ, Houwelingen AC v, Hoogen CM v d, Hornstra G: A moderate fish intake increases plasminogen activator inhibitor type-I in human volunteers; in Houwelingen AC v (ed): Fish against thrombosis? Thesis, Leiden 1988.
20 Berg Schmidt E, Varming K, Ernst E, et al: Dose-response studies on the effect of n–3 polyunsaturated fatty acids on lipids and haemostasis. Thromb Haemostasis 1990;63:1–5.
21 Berg Schmidt E, Ernst E, Varming K, et al: The effect of n–3 fatty acids on lipids and haemostasis in patients with type IIa and IV hyperlipidaemia. Thromb Haemostasis 1989;62:797–801.
22 Robertson RP: Eicosanoids as pluripotential modulators of pancreatic islet function. Diabetes 1988;37:367–370.

Bengt Vessby, MD, Department of Geriatrics, Uppsala University,
P.O. Box 12042, S–75012 Uppsala (Sweden)

Simopoulos AP, Kifer RR, Martin RE, Barlow SM (eds): Health Effects of ω3 Polyunsaturated Fatty Acids in Seafoods. World Rev Nutr Diet. Basel, Karger, 1991, vol 66, pp 417–424

Dietary Supplementation with ω3 Fatty Acids in Insulin-Dependent Diabetes mellitus

Tonny Jensen

Steno Memorial Hospital, Gentofte, Denmark

Introduction

In the late 1970s Dyerberg et al. [1] observed that Greenland Eskimos with a traditionally high consumption of seal and other seafood had a low prevalence of atherosclerosis and a low mortality from myocardial infarction. They suggested that dietary fish had some properties that could potentially prevent coronary artery disease, and they claimed that the long-chain ω3 polyunsaturated fatty acids eicosapentaenoic acid (C 20:5 ω3) and docosahexaenoic acid (C 22:6 ω3) were the active components of fish. Their findings stimulated a number of experimental and clinical studies revealing that ω3 fatty acids have potentially antiatheromatous effects [2, 3], and recently, epidemiological and intervention studies have shown an inverse dose-response relation between fish consumption and death from coronary heart disease [4, 5].

Since atherosclerotic vascular disease is a major cause of morbidity and mortality in patients with diabetes mellitus [6–8], the effects of fish oil on cardiovascular risk factors have recently been evaluated in these patients [9, 10]. Although it seems clear that some of the beneficial effects of fish oil are the same in diabetic patients as in nondiabetic subjects, the close relationship between lipid and glucose metabolism and the possible different pathophysiological mechanisms involved in atherogenesis in diabetic patients compared with nondiabetic subjects preclude extrapolation of all findings in the nondiabetic population to patients with diabetes mellitus [11, 12]. Moreover, the effects of ω3 fatty acids apparently are different in patients with insulin-dependent or non-insulin-dependent diabetes mellitus and in patients with or without microvascular complications.

The purpose of this paper is to discuss the perspectives of ω3 fatty acids in the prevention and treatment of atherosclerotic vascular disease in insulin-dependent diabetic patients. The present knowledge of the beneficial and potentially deleterious effects of ω3 fatty acids will be described, and attempts to identify subgroups of patients in whom dietary fish most safely may be recommended will be proposed.

Effects of ω3 Polyunsaturated Fatty Acids

Mechanisms of Action

A number of established and hypothetical cardiovascular risk factors are affected by ω3 fatty acids. In most studies they reduce plasma triglyceride and very low-density lipoprotein (VLDL) cholesterol [3, 13]. In contrast, the effects of fish oil consumption on low-density lipoprotein (LDL) cholesterol and high-density lipoprotein (HDL) cholesterol have been inconsistent, possibly reflecting differences in the characteristics of the patients studied, and in the various brands and doses of fish oil used. However, in at least three separate studies a paradoxical elevation of LDL cholesterol levels was noted [9, 14, 15], and Thiery and Seidel [16] recently reported that supplementation with fish oil resulted in enhancement of atherosclerosis in rabbits. In all these studies a whole-fish-oil concentrate with a relatively high proportion of saturated fatty acids was used.

Eicosapentaenoic acid is both substrate and an inhibitor of the cyclooxygenase pathway [17] which metabolizes arachidonic acid to prostaglandins and thromboxane A_2. It thereby reduces the production of the platelet-aggregating thromboxane A_2 [18] and increases the production of the vasodilatory prostaglandin I_3 [19]. The net result is a change in hemodynamic balance towards a more vasodilatory state, with less platelet aggregation. These effects together with reduced blood viscosity [20], reduced vasospastic response to catecholamines and angiotensin [21], and increased activity of endothelium-derived relaxation factor [22] may all explain the antihypertensive action of fish oil observed in most studies [23, 24]. The ω3 fatty acids are also incorporated in cell membrane phospholipids and may modify receptors, transport pathways and enzymes, related to cell surfaces [25].

Possible Prevention of Atherosclerosis

A number of animal studies have provided considerable evidence that fish oil prevents atherosclerosis. Weiner et al. [2] found that dietary supplementation with 30 ml cod liver oil daily for 8 months in a hyperlipidemic swine model markedly retarded progression of coronary atherosclerosis. The protective effect of cod liver oil was unrelated to changes in plasma concentrations of lipoproteins, but associated with platelet arachidonic acid content and serum thromboxane B_2. They hypothesised that ω3 fatty acids mainly inhibit platelet-endothelial cell interaction or have a direct protective effect on the vascular wall. Similar prevention by fish oil of atherosclerosis progression has been described in dogs [26] and monkeys [27]. With the increasing evidence that atherosclerosis starts with damage to the endothelium with subsequent platelet activation and release of chemoattractants and growth factors, efforts to intervene against these early steps in atherosclerosis seem attractive [28].

In a retrospective epidemiologic study, Kromhout et al. [4] found a strong inverse relation between fish consumption during a 20-year period and coronary artery disease. Other studies have found similar inverse relations between intake of fish and cardiovascular mortality [29, 30]. Interestingly, in the study by Kromhout et al. the relation between intake of fish and coronary artery disease was apparent even with the consumption of small amounts of ω3 fatty acids, suggesting that intake of 100–150 g fish per week over long periods has beneficial biological effects. Recently Burr et al. [5] in a 2-year prospective intervention study, found that intake of 200–400 g fish per week significantly reduced the overall mortality in men who had recovered from myocardial infarction.

Insulin-Dependent Diabetes mellitus and ω3 Fatty Acids

Beneficial and Potential Deleterious Effects

In 1986 Haines et al. [9] published the first detailed study on the effects of fish oil supplementation in 41 insulin-dependent diabetic patients. They gave 19 patients 15 g fish oil (MaxEpa) daily and 22 patients 0.6 g olive oil daily for six weeks and found a significant reduction in platelet thromboxane production and a significant increase in LDL cholesterol during fish oil supplementation compared with olive oil supplementation. No changes were noted in blood pressure or glycosylated hemoglobin in that study. The increase in LDL cholesterol was unexpected and the authors proposed that fish oil supplementation might hasten the conversion of VLDL cholesterol to LDL cholesterol [31], or alternatively it could be the saturated fat (approximately 30%) in MaxEpa that was responsible for the rise in LDL cholesterol. Also Tilvis et al. [32] found in insulin-dependent diabetic women reduced platelet thromboxane production during fish oil supplementation. In a smaller study of insulin-dependent diabetic patients, Miller et al. [33] found no changes in LDL cholesterol, fasting blood glucose or glycosylated hemoglobin during dietary supplementation with 20 g MaxEpa daily for eight weeks. Recently, in a double-blind crossover study from the Steno Memorial Hospital [34], it was found that dietary supplementation with cod liver oil significantly reduced the blood pressure from 146/90 to 139/85 mm Hg in albuminuric insulin-dependent diabetic patients. During cod liver oil supplementation the HDL cholesterol increased and the VLDL cholesterol and triglyceride decreased. No changes were observed in glomerular filtration rate, degree of albuminuria or glycosylated hemoglobin during supplementation with cod liver oil.

Thus, ω3 fatty acids seem not to interfere with glucose metabolism in insulin-dependent diabetic patients. A blood pressure lowering effect is seen in hypertensive patients, but not in patients without hypertension. As to the effects on lipid metabolism conflicting results have been reported, probably related to different brands and doses of fish oils used.

Future Perspectives and Recommendations of ω3 Fatty Acids in Diabetes mellitus

Not all diabetic patients are at same risk for cardiovascular disease. The increased mortality of cardiovascular disease is for the major part explained by an extremely high mortality in the subgroup of patients developing clinical nephropathy [35, 36]. These patients have a number of established cardiovascular risk factors as atherogenic changes in plasma lipoproteins [37] and elevated blood pressure [38]. They also have signs of vascular vulnerability. The permeability of the whole vascular bed is clearly elevated in patients with incipient and clinical nephropathy [39], and recently, elevated plasma levels of von Willebrand factor and impaired fibrinolytic response to exercise have been demonstrated in these patients, suggesting that they have endothelial cell dysfunction or damage [40]. Apart from beneficial effects on blood pressure and plasma concentrations of lipoproteins, cod liver oil intake also resulted in partial normalization of the elevated vascular permeability in albuminuric patients [34]. In support of a direct protective effect of dietary fish on the vascular wall, Wahlquist et al. [41] recently reported that diabetic fish eaters have higher arterial compliance compared with diabetic non-fish eaters. Thus, although the many diverse effects of ω3 fatty acids are far from being elucidated in insulin-dependent diabetic patients with albuminuria, the intervention on several clustered risk factors in these high-risk patients is very attractive. However, only future studies can settle whether long-term intake of ω3 fatty acids will be effective in prevention of atherosclerosis.

In the future, more studies of the effects of ω3 fatty acids on the development of microvascular complications are awaited. Recent animal studies have suggested that fish oils may have protective effects on the development of diabetic cardiomyopathy [42] and on the progression of renal disease [43]. However, it is still preliminary to say whether such effects can be extrapolated to humans.

In conclusion, the effects of dietary supplementation with fish oils are different in diabetic patients compared with nondiabetic subjects. Most beneficial effects are seen in insulin-dependent diabetic patients with albuminuria. These patients have a number of established and hypothetical cardiovascular risk factors and constitute a high risk group for early cardiovascular death. Therefore, long-term fish oil studies, intervening on several clustered risk factors, seem attractive in such patients.

References

1 Dyerberg J, Bang HO: Hemostatic function and platelet polyunsaturated fatty acids in Eskimos. Lancet 1979;ii:433–435.

2 Weiner BH, Ockene IS, Levine PH, Cuenoed HF, Fisher M, Johnson BS, Daoud AS, Jarmolych J, Hosmer D, Johnson MH, Natale A, Vandreuil C, Hoogasian JJ: Inhibition of atherosclerosis by cod liver oil in a hyperlipidemic swine model. N Engl J Med 1986;315:841–846.

3 Phillipson BE, Rotherock DW, Connor WE, Harris WS, Illingworth DR: Reduction of plasma lipids, lipoproteins, and apoproteins by dietary fish oils in patients with hypertriglyceridemia. N Engl J Med 1985;312:1210–1216.

4 Kromhout D, Bosschieter ED, de Lezenne Coulander C: The inverse relation between fish consumption and 20-year mortality from coronary heart disease. N Engl J Med 1985;312:1205–1209.

5 Burr ML, Fehily AM, Gilbert JF, Rogers S, Holliday RM, Sweetnam PM, Elwood

PC, Deadman NM: Effects of changes in fat, fish, and fibre intakes on death and myocardial reinfarction: Diet and Reinfarction Trial (DART). Lancet 1989;ii:757–761.

6 Kannel WB, McGee DL: Diabetes and vascular disease. The Framingham Study. J Am Med Ass 1979;241:2035–2038.

7 Jarrett RJ, McCartney P, Keen H: The Bedford Survey: Ten years mortality rates in newly diagnosed diabetics, borderline diabetics and normoglycaemic controls and risk indices for coronary heart disease in borderline diabetics. Diabetologia 1982;22: 79–84.

8 Fuller JH, Shipley MJ, Rose G, Jarrett RJ, Keen H: Mortality from coronary heart disease and stroke in relation to degree of glycaemia: The Whitehall Study. Br Med J 1983;287:867–870.

9 Haines AP, Sanders TAB, Imeson JD, Mahler RF, Martin J, Mistry M, Bickers M, Wallace PG: Effects of a fish oil supplement on platelet function, haemostatic variables and albuminuria in insulin-dependent diabetics. Thromb Res 1986;43:643–655.

10 Kasim SE, Stern B, Khilnani S, McLin P, Baciorowski S, Jen K-L: Effects of omega-3 fish oils on lipid metabolism, glycaemic control, and blood pressure in type II diabetic patients. J Clin Endocrinol Metab 1988;67:1–5.

11 Sorisky A, Robbins DC: Fish oil and diabetes. The net effect. Diabetes Care 1989; 12:302–304.

12 Axelrod L: Omega-3 fatty acids in diabetes mellitus. Gift from the sea? Diabetes 1989;38:539–543.

13 Mehta JL, Lopez LM, Lawson D, Wargovich TJ, Williams LL: Dietary supplementation with omega-3 polyunsaturated fatty acids in patients with stable coronary heart disease: effects on indices of platelet and neutrophil function and exercise performance. Am J Med 1988;84:45–52.

14 Demke DM, Peters GR, Linet OI, Metzler CM, Klott KA: Effects of a fish oil concentrate in patients with hypercholesterolemia. Atherosclerosis 1988;70:73–80.

15 Sullivan DR, Sanders TA, Trayner IM, Thompsom GR: Paradoxical eevation of LDL apoprotein B levels in hypertriglyceridemic patients and normal subjects ingesting fish oil. Atherosclerosis 1986;61:129–134.

16 Thiery J, Seidel D: Fish oil feeding results in an enhancement of cholesterol-induced atherosclerosis in rabbits. Atherosclerosis 1987;63:53–56.

17 Needleman P, Raz A, Minkes MS, Ferrendelli JA, et al: Triene prostaglandins: prostaglandin and thromboxane biosynthesis and unique biological properties. Proc Natl Acad Sci 1979;76:944–948.

18 Siess W, Roth P, Scherer B, Kurzmann I, Bøhlig B, Weber PC: Platelet-membrane fatty acids, platelet aggregation, and thromboxane formation during a mackerel diet. Lancet 1980;i:441–444.

19 Fischer S, Weber PC: Prostaglandin I_3 is formed in vivo in man after dietary eicosapentaenoic acid. Nature 1984;307:165–168.

20 Popp-Sniders C, Schouten JA, van der Meer J, van der Veen EA: Fatty fish-induced changes in membrane lipid composition and viscosity of human erythrocyte suspensions. Scand J Clin Lab Invest 1986;46:253–258.

21 Lorenz R, Spengler U, Fischer S, Duhm J, Weber PC: Platelet function, thrombox-

ane formation and blood pressure control during supplementation of the Western diet with cod liver oil. Circulation 1983;67:504–511.

22 Shimokawa H, Lam JY, Chesebro JH, Bowie EJ, Vanhoutte PM: Effects of dietary supplementation with cod-liver oil on endothelium-dependent responses in porcine coronary arteries. Circulation 1987;76:898–905.

23 Mortensen JZ, Schmidt EB, Nielsen AH, Dyerberg J: The effects of n–6 and n–3 polyunsaturated fatty acids on hemostasis, blood lipids and blood pressure. Thromb Haemostasis 1983;50:543–546.

24 Norris PG, Jones CH, Weston MJ: Effect of dietary supplementation with fish oil on systolic blood pressure in mild essential hypertension. Br Med J 1986;293:104–105.

25 Leaf A, Weber PC: Cardiovascular effects of n–3 fatty acids. N Engl J Med 1988; 318:549–557.

26 Landymore RW, MacAulay M, Sheridan B, Cameron C: Comparison of cod-liver oil and aspirin-dipyridamole for the prevention of intimal hyperplasia in autologous vein grafts. Ann Thorac Surg 1986;41:54–57.

27 Davis HR, Bridenstine RT, Vesselinovitch D, Wissler RW: Fish oil inhibits development of atherosclerosis in rhesus monkeys. Atherosclerosis 1987;7:441–449.

28 Ross R: The pathogenesis of atherosclerosis: an update. N Engl J Med 1986;314: 488–500.

29 Shekelle RB, Missell LV, Paul O, Shryock AM, Stamler J: Fish consumption and mortality from coronary heart disease. N Engl J Med 1985;313:820.

30 Norell SE, Ahlbom A, Feychting M, Pedersen NL: Fish consumption and mortality from coronary heart disease. Br Med J 1986;293:426.

31 Rose HG, Haft GK, Juliano J: Clofibrate-induced low density lipoprotein elevation. Therapeutic implications and treatment by cholesterol resin. Atherosclerosis 1976; 23:413–427.

32 Telvis RS, Rasi V, Viinikka L, Ylikorkala O, Miettinen TA: Effects of purified fish oil on platelet lipids and function in diabetic women. Clin Chim Acta 1987;164: 315–322.

33 Miller ME, Anagnostou AA, Ley B, Marshall P, Steiner M: Effects of fish oil concentrates on hemorheological and hemostatic aspects of diabetes mellitus: A preliminary study. Thromb Res 1987;47:201–214.

34 Jensen T, Stender S, Goldstein K, Hølmer G, Deckert T: Partial normalization by dietary cod-liver oil of increased microvascular albumin leakage in patients with insulin-dependent diabetes and albuminuria. N Engl J Med 1989;321:1572–1577.

35 Borch-Johnson K, Kreiner S: Proteinuria: value as predictor of cardiovascular mortality in insulin-dependent diabetes mellitus. Br Med J 1987;294:1651–1654.

36 Jensen T, Borch-Johnsen K, Kofoed-Enevoldsen A, Deckert T: Coronary heart disease in young type 1 (insulin-dependent) diabetic aptients with and without diabetic nephropathy: Incidence and risk factors. Diabetologia 1987;30:144–148.

37 Jensen T, Stender S, Deckert T: Abnormalities in plasma concentrations of lipoproteins and fibrinogen in type 1 (insulin-dependent) diabetic patients with increased urinary albumin excretion. Diabetologia 1988;31:142–145.

38 Feldt-Rasmussen B, Borch-Johnsen K, Mathiesen ER: Hypertension in diabetes as related to nephropathy. Early blood pressure changes. Hypertension 1985;7(suppl II):18–20.

39 Feldt-Rasmussen B: Increased transcapillary escape rate of albumin in type 1 (insulin-dependent) diabetic patients with microalbuminuria. Diabetologia 1986;29:282–286.
40 Jensen T, Bjerre-Knudsen J, Feldt-Rasmussen B, Deckert T: Features of endothelial dysfunction in early diabetic nephropathy. Lancet 1989;i:461–463.
41 Wahlqvist ML, Lo CS, Myers KA: Fish intake and arterial wall characteristics in healthy people and diabetic patients. Lancet 1989;i:944–946.
42 Black SC, Katz S, McNeill JH: Cardiac performance and plasma lipids of omega-3 fatty acid-treated streptozocin-induced diabetic rats. Diabetes 19889;38:969–974.
43 Ito Y, Barcelli U, Yamashita W, Weiss M, Glas-Greenwalt P, Pollak VE: Fish oil has beneficial effects on lipids and renal disease in nephrotic rats. Metabolism 1988;37:352–357.

Tonny Jensen, MD, Steno Memorial Hospital, DK–2820 Gentofte (Denmark)

Simopoulos AP, Kifer RR, Martin RE, Barlow SM (eds): Health Effects of ω3 Polyunsaturated Fatty Acids in Seafoods. World Rev Nutr Diet. Basel, Karger, 1991, vol 66, pp 425–435

ω3 Polyunsaturated Fatty Acid Constituents of Fish Oil and the Management of Skin Inflammatory and Scaly Disorders

Vincent A. Ziboh

Department of Dermatology T.B. 192, University of California, Davis, Calif., USA

Introduction

Three major families of polyunsaturated fatty acids (PUFAs) are characteristic of the mammalian species: the ω9, ω6, and the ω3 families. The ω6 and ω3 PUFAs are defined by the position of the double bond closest to the terminal methyl group of the fatty acid molecule[1]. For instance, in the ω6 family, the first double bond occurs between the sixth and seventh carbons from the methyl group end of the molecule whereas in the ω3 family the first double bond occurs between the third and fourth carbons. These 18-carbon structural forms are shown in figure 1. Desaturation and elongation of these 18-carbon shorter PUFAs into the longer-chain PUFAs occur in vivo within each series without altering the methyl end of the molecule. These basic structures cannot be synthesized de novo by vertebrate animals nor are the ω3 and ω6 families of PUFAs interconvertible. Thus, these 18-carbon PUFAs must be obtained from diet.

The longer-chain ω3 PUFAs: eicosapentaenoic acid (20:5 ω3, EPA) and docosahexaenoic acid (22:6 ω3, DHA) are especially rich in fish and

[1] Fatty acids and acyl groups are denoted 18:2 ω6, 18:3 ω3 and so on, with the first number representing the number of carbons in the acyl chain, the number following the colon indicating the number of methylene interrupted cis-double bonds and the number of 'ω' (or 'n') indicating the number of carbon atoms from the methyl end of the acyl chain to the nearest double bond.

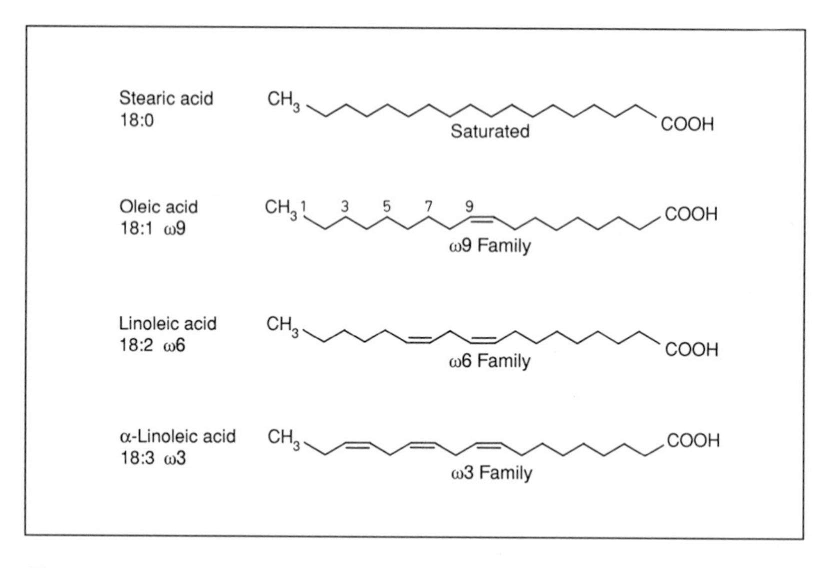

Fig. 1. Representative structures of saturated, monounsaturated (ω9), and polyunsaturated (ω6 and ω3) fatty acids.

marine animals. Although the ω3 PUFAs constitute only a few percent of the total fatty acids in most mammalian tissues, DHA in particular is present at high concentrations in the retina, cerebral cortex, testis and sperm [1]. These levels are relatively high in these tissues despite the disparity of dietary intakes of these ω3 PUFAs [2, 3]. The consistent presence of the ω3 PUFAs in these tissues provided strong evidence for the possible biological role of DHA in these tissues. The biological role of the ω3 PUFAs in other tissues are less clear, however there is an increasing body of evidence which suggest that the ω3 PUFAs may function to regulate the metabolism and function of the ω6 PUFAs.

Eicosanoids in Cutaneous Disorders

The possibility that altered arachidonic acid (AA) metabolism may play a role in the pathogenesis of cutaneous scaly disorders was first described by Hammarstrom et al. [4]. These investigators reported abnormally high levels of precursor AA and its lipoxygenase product 12-

hydroxy-eicosatetraenoic acid (12-HETE) in the lesions (plaques) of psoriatic patients. Psoriasis is a human chronic scaly disorder, which is characterized by sharply defined, thickened and erythematous plaques. Also observed in these lesions are polymorphonuclear (PMN) cells [5] known to elaborate inflammatory mediators. These observations for the first time suggest that the increased AA and its metabolites in the plaques of the psoriatic patients may contribute at least in part to the pathogenesis of the disease.

The reports [6, 7] that polymorphonuclear leukocytes from rabbit peritoneal cavity can transform AA into a major novel lipoxygenase product leukotriene B_4 (LTB_4) (5S, 12R-dihydroxy-6,8,10,14-eicosatetraenoic acid) as well as into other minor products, 5S, 12R-dihydroxy-6,8,10-trans-14 cis-eicosatetraenoic acids, excited a flurry of investigations to determine the role of these substances in the pathology of a variety of clinical disorders. A determination for the level of these novel lipoxygenase products of AA in skin was accomplished by Brain et al. [8] who reported that fluid derived from abraded psoriatic plaques contained elevated LTB_4-like activity. In another study, the in vitro incubations of enzyme preparations from clinically involved psoriatic epidermis with AA revealed the transformation of AA into products with chromatographic mobilities similar to LTB_4 and 12-HETE [9]. Taken together, these reports indicated the presence of pro-inflammatory as well as the total generation of LTB_4 in the lesions.

These early biochemical studies were followed by a flurry of direct challenges of normal human skin with a variety of proinflammatory mediators of the lipoxygenase pathway to determine whether or not psoriasis-like lesions can be induced. For instance, the intracutaneous injections of products of the 5-lipoxygenase pathway (LTB_4, LTC_4, LTD_4 and LTE_4) into normal human skin elicited inflammation, characterized by erythema, wheal formation and neutrophil infiltration [10]. Similarly, the topical application of a single dose of LTB_4 to skin of normal individuals produced erythema and swelling which were characterized histologically by intraepidermal neutrophil microabscesses [11]. Taken together, these reported findings suggest that products of the 5- and 12-lipoxygenase pathways may contribute at least in part to the inflammatory aspect of the disease. Although LTB_4, LTC_4 and LTD_4 have been reported to stimulate the in vitro DNA synthesis of cultured human keratinocytes [12], it is premature at this time to conclude that they are the sole contributors of the marked hyperproliferation of psoriasis.

Management of Psoriasis

Management of Disease by Antiproliferative Agents

The proliferative nature of psoriasis had prompted the therapeutic use of a variety of systemic and topical antiproliferative agents. Some of these agents include: methotrexate, 5-fluorouracil, anthralin, coal tar plus ultra-violet light (UVB) and psoralens combined with long-wave ultraviolet light (PUVA). Retinoids, both systemic and topical, have also been used to arrest the proliferative aspects of the disease. Similarly, a variety of steroids have been used to arrest both the inflammatory and proliferative aspects of the disease. Although these therapeutic managements have proved beneficial to varying degrees, the accompanying side effects have been very undesirable.

Pharmacological Management by Inhibitors of Arachidonic Acid Metabolism

The recognition that AA metabolism is altered in psoriasis prompted attempts to inhibit the generation of proinflammatory lipoxygenase products (particularly LTB_4 and 12-HETE) which are markedly elevated in the psoriatic lesions. The first such effort was the systemic administration to psoriatic patients of benoxaprofen (2-[4-chlorophenyl]-α-methyl-5-benzoxazole acetic acid), a non-steroidal antiinflammatory drug and a weak inhibitor of both the lipoxygenase and cyclooxygenase pathways [13] by Kragballe and Herlin [14], and Allen and Littlewood [15]. These studies resulted in the impressive clinical clearance of the lesions and strongly suggested that inhibition of the generation of products of the AA metabolism (particularly the 5-lipoxygenase pathway) may be critical in the management of at least the inflammatory aspects of psoriasis. These excitements were tempered by the toxicity that resulted from this therapeutic management which resulted in its ban from further clinical use. It's success, however, enkindled great interest in attempts to seek substances that can suppress the generation of inflammatory eicosanoids with negligible side effects as a possible approach to alleviating the disease activity of psoriasis.

Dietary Management of Psoriasis by ω3 Polyunsaturated Fatty Acids

Reports that Greenland Eskimos whose major dietary oil is fish or marine oil [16] and who exhibit a low incidence of psoriasis [17] prompted us to test the efficacy of dietary fish oil (Max-EPA) which is rich in ω3 series of PUFAs: eicosapentaenoic acid (20:5 ω3) and docosahexaenoic

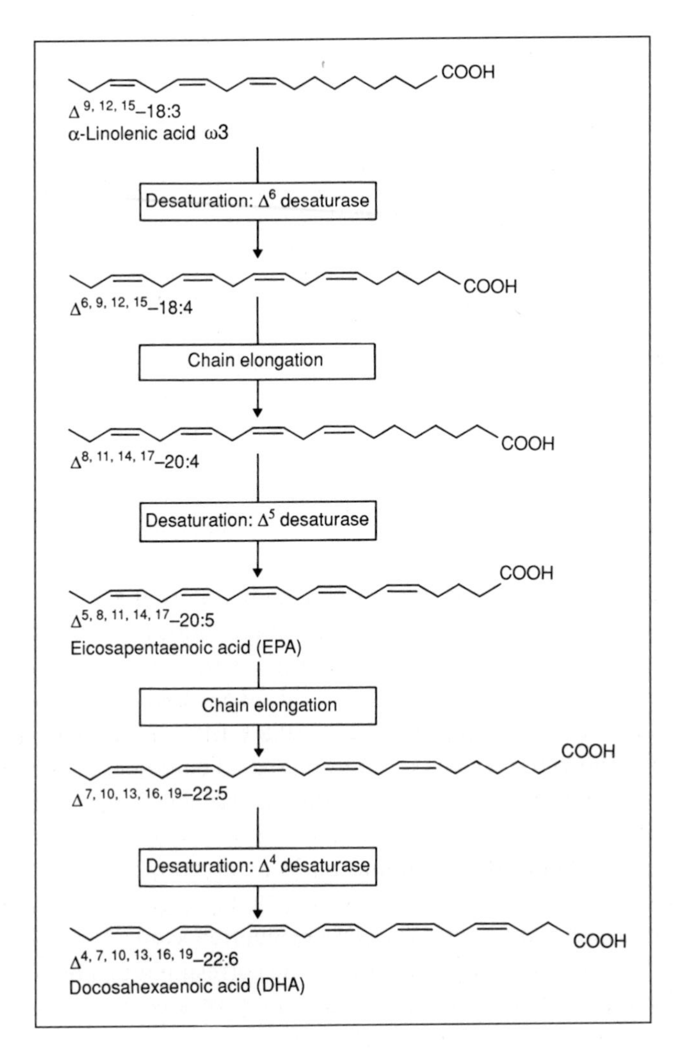

Fig. 2. Oxidative desaturation and elongation of ω3 polyunsaturated fatty acids (18:3 ω3).

acid (22:6 ω3) in a group of psoriatic patients [18]. The oxidative desaturation and elongation of ω3 polyunsaturated fatty acids is shown in figure 2. Our rationale was to test whether the supplementation of psoriatic diets with ω3 PUFAs could suppress the in vivo neutrophil generation of proinflammatory ω6 eicosanoids, particularly LTB_4, known to accumulate

Table 1. Diet in psoriasis study

Constituents of administered diet	Excluded foods
Fish	Fats and oils
Poultry (white meat)	Red meat
Fruits	Whole milk dairy products
Vegetables	Baked goods
Grains	Casseroles
Skim milk dairy products	Nuts
Alcohol	Egg yolk
Carbonated beverages	Salad dressings
Coffee and tea	Gravies

in the lesion, and whether such a suppression could alleviate the symptoms of psoriatic lesions.

Each of the thirteen patients in our open study followed a diet low in fat, particularly low in AA, linoleic acid, and saturated fats (table 1). Total daily caloric intake was calculated for each patient in order to maintain pre-study body weight. The total Max-EPA oil taken daily was approximately 60 g which contained approximately 10.8 g EPA, 7.2 g DHA and 0.6 g AA per day, respectively. All patients applied an emollient (Unibase) twice daily to their entire body to stabilize the disease activity. Blood was taken at 2-week intervals to obtain serum and for the isolation of neutrophils. Furthermore, epidermal keratome biopsies of both involved and uninvolved skin were taken prior to the Max-EPA supplementation, then at 2-week intervals for eight weeks. Baseline evaluation scores for scaling, erythema, and thickness were compared with the corresponding post-therapy scores for all the patients completing the full 8-week course of study. No placebo group was followed in this study.

Clinical and Biochemical Evaluations of Participating Patients

Clinical findings from the above 8-week study revealed that approximately 60% of the participating patients demonstrated mild to moderate improvement of their psoriatic lesions. Improved clinical response correlated with a high EPA/DHA ratio attained in the epidermal tissue as shown in figure 3. Interestingly, two patients in the study who had moderately

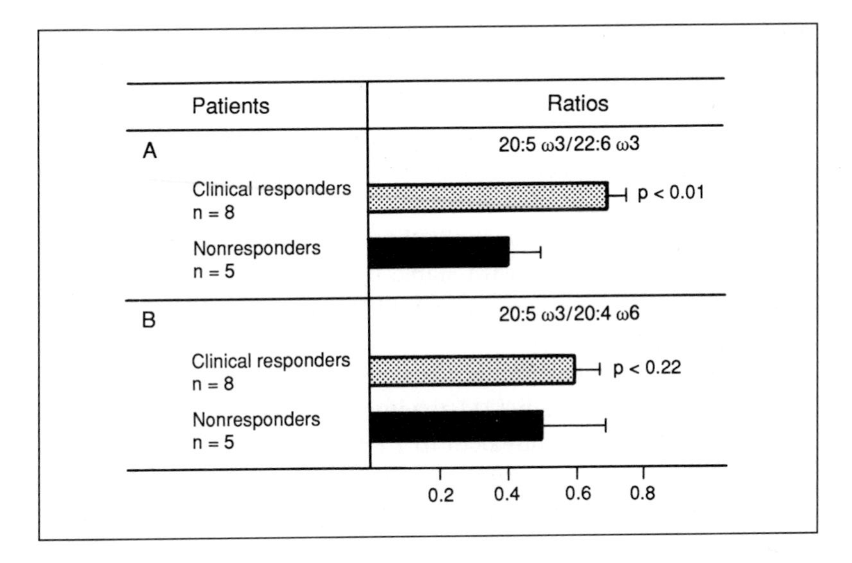

Fig. 3. Polyunsaturated ratios in psoriatic lesions in clinical responders and nonresponders. *A* EPA (20:3 ω5)/DHA (22:6 ω3) ratio. *B* EPA (20:5 ω3)/AA (20:4 ω6) ratio.

severe psoriatic arthritis prior to the study reported significant alleviation of their symptoms. Five of the patients also reported a marked decrease in the pruritus associated with their psoriatic lesions. This later finding was also observed in another dietary study of psoriatic patients [19] with fish oil.

Biochemically, the ingested ω3 PUFAs contained in the Max-EPA were rapidly incorporated up into the serum lipids, neutrophil lipids and epidermal lipids. However, it is the epidermal lipid profiles that revealed significant correlations of the ω3 PUFAs with activity of the disease. The serum fatty acids were mainly useful for monitoring the dietary compliance of the participants whereas the neutrophil profiles were similar to those in the epidermis.

Comments

Report of the preceding study suggests that dietary fish oil containing ω3 PUFAs can exert minimum to moderate clearance of the lesions of psoriasis. The above beneficial effects have also been confirmed in two other dietary studies [19, 20]. In another open trial study [20], the investigators provided fish oil containing approximately 12 g EPA/day to 10 psoriatic patients for 6 weeks and reported moderate beneficial effects in 8. Additionally, in this study the authors reported a marked suppression of LTB_4 generation by the Ca^{2+} ionophore-induced peripheral blood polymorphonuclear leukocytes in vitro. In two later double-blind studies the authors supplemented diets of psoriatic patients with lower levels of ω3 PUFAs. For instance in one double-blind, randomized, placebo-controlled trial study of 28 patients with stable psoriasis [19] the patients received Max-EPA capsules equivalent to 1.8 g EPA/day for 8–12 weeks. The investigators after 8 weeks reported a lessening of itching, erythema and scaling in the active treatment group when compared to the placebo group. In another double-blind study [21], the investigators reported that the administration of Max-EPA equivalent to 1.8 g EPA/day for 8 weeks resulted in no statistical difference in clinical manifestations of the psoriasis between the active treatment group and the placebo group. The reason for the discrepancy in these two low-level ω3 PUFA double-blind studies is not immediately clear. It should be noted, however, that in these two latter studies the content of administered EPA (1.8 g) is several-fold less than those previously used in the two open studies. Furthermore, these discrepancies raise the critical issue of establishing the true contents of EPA/DHA in fish oil being administered to patients, constant monitoring for authenticity, as well as a close monitoring of compliance by the patients. Although in three of the four referenced studies beneficial effects of the ω3 PUFAs were indicated, the mechanism(s) for the observed effects are unclear. To evolve a possible mechanism for the observed beneficial effects from dietary fish oils, we investigated the metabolic fate of EPA (20:3 ω6) and DHA (22:6 ω3) in a guinea pig epidermal model. Data from these studies revealed that both 20:5 ω3 and 22:6 ω3 are readily transformed by epidermal 15-lipoxygenase into 15-hydroxyeicosapentaenoic acid (15-HEPE) and 17-hydroxydocosahexaenoic acid (17-HDHE), respectively [22]. To ascertain whether or not these metabolites exert any effect on leukocyte ability to generate proinflammatory LTB_4, we incubated Ca^{2+} ionophore-stimulated rat basophilic leukemia (RBL) cells with 15-HEPE and 17-HDHE, respectively. Interestingly, both 15-HEPE and 17-HDHE inhibited the ability of these cells to generate LTB_4 in vitro [22]. Data from such in vitro study is shown in figure 4. Thus, the conversion of the tissue-incorporated ω3 PUFAs into endogenous 15-lipoxygenase metabolites with the ability to suppress local generation of LTB_4 provides an attractive mechanism by which the fish oil may exert its beneficial effects on inflam-

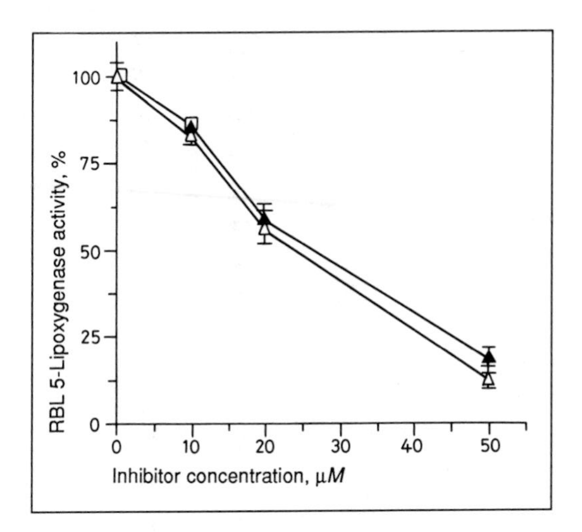

Fig. 4. Inhibitory effects of 15-HEPE (▲) and 17-HDHE (△) on the activity of 5-lipoxygenase from RBL-cell homogenates. Each point represents the mean ± SEM of three experiments. Approximate IC_{50} values in μ*M* for 15-HEPE and 17-HDHE are 28 and 25, respectively.

matory/hyperproliferative skin diseases. A speculative scenario of the possible modulatory effect of EPA contained in the fish oil on the generation of 5- and 12-lipoxygenase products from arachidonic acid is shown in figure 5. Pathway A illustrates the dietary ingestion of vegetable oil (safflower or corn oil), its desaturation and elongation into arachidonic acid. The terminal AA is metabolized by the PMNs via the 5-lipoxygenase into proinflammatory leukotrienes, particularly LTB_4. Pathway B, on the other hand, illustrates metabolism of 20:5 ω3 via the cyclooxygenase and lipoxygenase pathways. Notable in the pathways is the metabolism of 20:5 ω3 via the 15-lipoxygenase pathway, which in epidermis generates 15-HEPE, a potent inhibitor of 5- and 12-lipoxygenase pathways. These metabolic transformations underscore the possibility that dietary intake of fish oil does result in epidermal uptake of the ω3 PUFAs, the local transformations of these ω3 PUFAs by epidermal 15-lipoxygenase into hydroxy acids. These in turn could inhibit the generation of inflammatory mediators by infiltrating neutrophils. It seems reasonable to speculate that the efficacy of fish oil must depend on the quantity of ω3 PUFAs ingested, amount incorporated into epidermis, their release and metabolism into antiinflammatory metabolites. Furthermore, the local tissue generation of antiin-

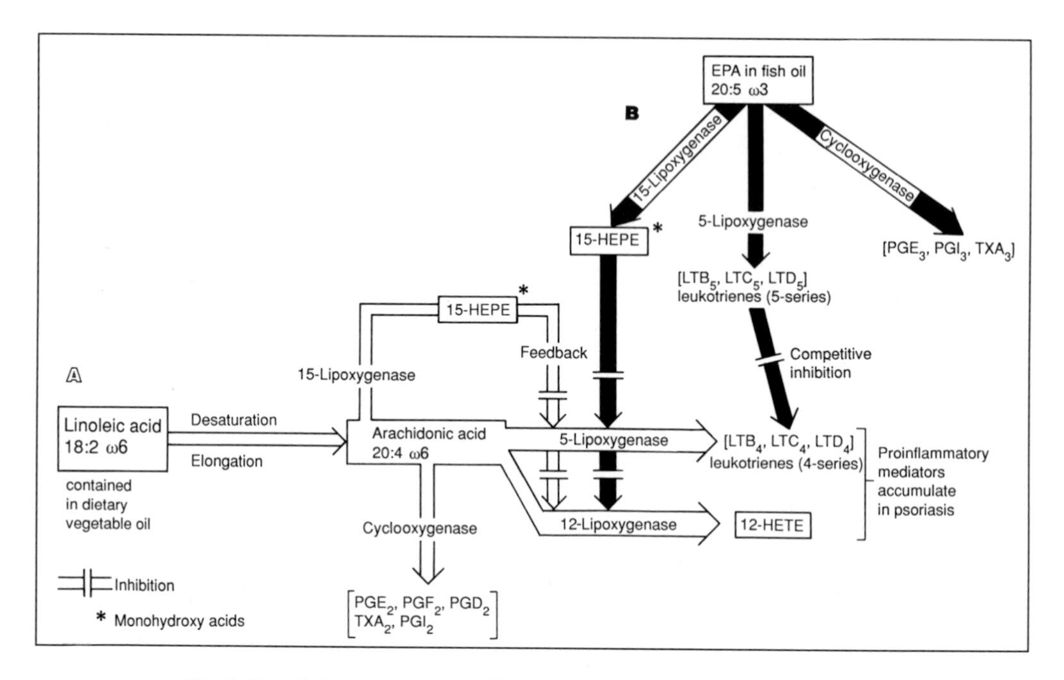

Fig. 5. Speculative modulatory effects of fish oil on the 5-lipoxygenation of arachidonic acid.

flammatory metabolites underscore the need for further investigations into the role of ω3-derived 15-lipoxygenase products in normal and inflammatory/hyperproliferative skin disorders.

References

1 Tinoco J: Dietary requirements and functions of alpha-linolenic acid in animals. Prog Lipid Res 1982;21:1–45.

2 Crawford MA, Casperd NM, Sinclair AJ: The long chain metabolites of linoleic and linolenic acids in liver and brain in herbivores and carnivores. Comp Biochem Physiol 1976;54B:395–401.

3 Crawford MA, Sinclair AJ: Nutritional influences in the evolution of mammalian brain; in Ciba Foundation Symposium: Lipids, Malnutrition and the Developing Brain. Amsterdam, Elsevier, 1972, pp 267–287.

4 Hammarstrom A, Hamberg M, Samuelsson B, Duell EA, Strawiski M, Voorhees JJ: Increased concentration of non-esterified arachidonic acid. 12*L*-hydroxy-5,8,10,14-eicosatetraenoic acid, prostaglandin E_2 and prostaglandin $F_{2\alpha}$ in epidermis of psoriasis. Proc Natl Acad Sci 1975;72:5130–5134.

5 Pinkus H, Mehregan AM: The primary histologic lesion of seborrhoeic dermatitis and psoriasis. J Invest Dermatol 1966;46:109–116.
6 Borgeat P, Hamber M, Samuelsson B: Transformation of arachidonic and homo-gamma-linolenic acid by rabbit polymorphonuclear leukocytes. J Biol Chem 1976; 251:7816–7824.
7 Borgeat P, Samuelsson B: Transformation of arachidonic acid by rabbit polymorphonuclear leukocytes: Formation of a novel dihydroxyeicosatetraenoic acid. J Biol Chem 1979;254:2643–2646.
8 Brain S, Camp R, Dowd P, et al: Psoriasis and leukotriene B_4. Lancet 1983;2:762–763.
9 Ziboh VA, Casebolt TL, Marcelo CL, et al: Biosynthesis of lipoxygenase products by enzyme preparations from normal and psoriatic skin. J Invest Dermatol 1984;83: 426–430.
10 Soter NA, Lewis RA, Corey EJ, Austin FK: Local effects of leukotrienes (LTC_4, LTD_4, LTE_4 and LTB_4) in human skin. J Invest Dermatol 1983;80:115–119.
11 Camp R, Jones RR, Brain S, et al: Production of intraepidermal microabscesses by topical application of leukotriene B_4. J Invest Dermatol 1984;82:202–204.
12 Kragballe K, Desjarlais L, Voorhees JJ: Leukotrienes B_4, C_4 and D_4 stimulate DNA synthesis in cultured human epidermal keratinocytes. Br J Dermatol 1985;113:43–52.
13 Dawson W, Boot JR, Harvey J, et al: The pharmacology of benoxaprofen with particular reference to effects on lipoxygenase pathway. Europ J Rheumatol Inflamm 1982;5:61–68.
14 Kragballe K, Herlin T: Benoxyprofen improves psoriasis: A double-blind study. Arch Dermatol 1983;119:548–552.
15 Allen BR, Littlewood SM: Benoxaprofen: Effect on cutaneous lesions of psoriasis. Br Med J 1982;285:1241–1244.
16 Dyerberg J, Bang HO: Haemostatic function and platelet polyunsaturated fatty acids in Eskimos. Lancet 1979;ii:433–435.
17 Kromann N, Green A: Epidemiological studies in the Uppernavik District, Greenland. Acta Med Scand 1980;200:401–406.
18 Ziboh VA, Miller C, Kragballe K, et al: Effects of dietary supplementation of fish oil on neutrophil and epidermal fatty acids: Modulation of the clinical course of psoriatic subjects. Arch Dermatol 1986;122:1277–1282.
19 Bittinger SB, Cartwright I, Tucker WFG, et al: A double-blind, randomized, placebo-controlled trial of fish oil in psoriasis. Lancet 1988;i:378–380.
20 Maurice PDL, Allen BR, Barkeley ASJ, et al: The effects of dietary supplementation with fish oil in patients with psoriasis. Br J Dermatol 1987;117:599–606.
21 Bjornboe A, Smith AK, Bjornboe G-EAA, et al: Effect of dietary supplementation with n–3 fatty acids on clinical manifestations of psoriasis. Br J Dermatol 1988;118: 77–83.
22 Miller CC, Yamaguchi RY, Ziboh VA: Guinea pig epidermis generates putative anti-inflammatory metabolites from fish oil polyunsaturated fatty acids. Lipids 1989;24:998–1003.

Vincent A. Ziboh, MD, Department of Dermatology, T.B. 192,
University of California, Davis, CA 95616 (USA)

Simopoulos AP, Kifer RR, Martin RE, Barlow SM (eds): Health Effects of ω3 Polyunsaturated Fatty Acids in Seafoods. World Rev Nutr Diet. Basel, Karger, 1991, vol 66, pp 436–445

Fish Oil in Combination with Other Therapies in the Treatment of Psoriasis

B.R. Allen

Department of Dermatology, University Hospital, Queen's Medical Centre, Nottingham, UK

Introduction

Psoriasis is an extremely common inflammatory disorder of the skin characterised, in its most common form, by the development of persistent, scaling red plaques particularly over the extensor aspects of joints.

The pathological changes seen in skin lesions include accentuation of the dermal papillae and alterations in the capillaries they contain; keratinocyte hyperproliferation with parakeratosis; and infiltration of the epidermis with neutrophils, T lymphoctes and macrophages. On a quantitative basis neutrophils predominate in the infiltrate, and in the rare pustular form of the disease clinically apparent pustules may cover the skin. It is believed that the local production of chemoattractants, growth factors and other biologically active compounds may play a role in causing the lesions.

In a Western European Caucasian population the prevalence has been assessed at about 2%, but there are significant racial differences and the disease is much less common in the negro and mongoloid races. It is said to be of negligible frequency in the American Indians and is rare amongst Eskimo (Inuit) people. The family incidence is high as is the pair-wise concordance rate in monozygotic twins. The risk that the identical twin of an affected person will get the disease is 70% [1].

Although the exact mode of inheritance is not known it is believed to be multifactorial with environmental factors playing a part in its expression. This is seen most clearly in the Köbner phenomenon where non-specific local injury to the skin induces a psoriatic lesion and in guttate psoriasis where a preceding streptococcal sore throat may occur in 80% of cases.

Theoretical Benefits of Fish Oil

There seems little doubt that eicosanoid metabolism is disturbed in psoriasis with high levels of LTB_4, LTC_4, 12-HETE and free AA all being described in the skin. For a review see Camp [2]. Similar changes are not seen in other inflammatory skin diseases such as eczema. It is not clear whether the disturbances are primary or the consequence of other preceding phenomena. For example levels of IL-1 are very high in normal stratum corneum and drop in lesional psoriatic skin, suggesting activation which might, as a secondary phenomenon, stimulate eicosanoid metabolism.

It is not clear either whether the abnormality lies in the skin itself or elsewhere, for example in the neutrophils. We [3] were unable to detect higher levels of LTB_4 formation in stimulated peripheral blood PMNLs but in mixed PMNLs and platelets increased amounts of 12-20-diHETE were found [4] suggesting that omega oxidation of eicosanoids might be increased. This concept was supported by subsequent work showing evidence of increased production of omega oxidation products from psoriatic cells [5] which might indicate increased enzyme kinetics.

Research into the possibility that dietary supplements of fish oil might be beneficial in psoriasis was stimulated by the observation that the incidence of psoriasis was low in Eskimos [6] suggesting the possibility that the preliminary studies describing benefit in rheumatoid arthritis [7] and ulcerative colitis [8] might extend to psoriasis, although the low prevelance in Mongoloid races has to be taken into account.

At the theoretical level there is good reason to believe that increasing the amount of EPA in the diet might be beneficial. In view of the potent pro-inflammatory properties of the 5-lipoxygenase products of AA metabolism, especially LTB_4, reducing their level or activity might be expected to be beneficial. There are two possible mechanisms whereby partial substitution of AA by EPA in the diet can influence psoriasis.

Firstly EPA is known to suppress LTB_4 synthesis in stimulated PMNLs and this is something which our own study confirmed [9]. The fact that this was decreased synthesis rather than increased breakdown was demonstrated by the finding that breakdown products decreased as well. This accords with studies on rats given EPA supplements in which, at the site of sponge implants, the level of PGE_2 was considerably greater than LTB_4 [10], implying active inhibition of synthesis rather than competition for substrate.

The scond effect of EPA supplementation is the synthesis of LTB_5 and its isomers [11, 12]. LTB_5 has approximately 10% of the activity of LTB_4 in causing aggregation, degranulation and chemokinesis of PMNLs [13–15] and is only weakly agonistic in potentiating bradykinin-induced plasma exudation in rabbit skin [10].

The Effects of Fish Oil as Sole Therapy

There have now been a number of trials of dietary supplements with fish oil in the treatment of psoriasis [16]. The improvement correlated with high epidermal EPA/DCHA ratios attained in the epidermal tissue specimens. Our own findings have been similar. We studied 10 patients with active psoriasis resistant to conventional treatment. They were put on a low fat diet supplemented with up to 50 ml fish oil (MaxEPA®), providing 9 g of EPA per day, for at least 6 weeks. Measurements were made of clinical status, plasma and platelet fatty acid levels and, in 8 patients,of the eicosanoid profile of the PMNs stimulated with calcium ionophore A23187 before and after dietary supplementation. There was a marked increase in the mean EPA/AA ratios in the plasma (0.083 to 1.099) and platelets (0.036 to 0.281) reflecting the dietary change. Tests of platelet behaviour did not alter. There was a significant drop in LTB_4 production by the PMNLs (means 228 and 121 ng/1.5 $\times$ 10^7 cells before and after treatment) and smaller non-significant decreases in the non-enzymatic isomers of LTB_4 (68 to 44 and 70 to 45 ng/1.5 $\times$ 10^7 cells, respectively). In 6 out of 8 patients a small HPLC peak with the retention time of LTB_5 appeared after dietary supplementation but this was not quantified. In 7 out of 8 patients there was a substantial diminution of Ω-oxidation products of LTB, also not quantified, reflecting the decrease in LTB_4. It was concluded that dietary supplementation with EPA is an effective way of lowering LTB_4 production. Eight patients showed an improvement in their psoriasis but the effect was only a modest one. This might indicate that LTB_4 does not play a major role in the pathogenesis of psoriasis, or, alternatively, a greater improvement might have been achieved with a further reduction in LTB_4 production, but this could be difficult to achieve without the availability of preparations containing a higher concentration of EPA.

Bittiner et al. [17] carried out a double-blind trial on 28 patients with stable chronic plaque psoriasis comparing a daily supplement of 10 capsules of MaxEPA with 10 capsules of olive oil as a control. After 8 weeks treatment there was a significant reduction in the itching, erythema and scaling of the active group with a trend towards a reduction in the body surface area involved. No change occurred in the placebo group. Bjørneboe et al. [18] were, however, unable to detect any improvement in their patients.

Many consider topical treatment in psoriasis preferable to systemic therapy, and a small single-blind within-patient pilot study in 11 patients [19] comparing 10% MaxEPA in Unguentum Merck® with the base alone showed, after 7 weeks, that, on objective criteria, 8 patients had improved and all but 2 regarded themselves as better subjectively.

Combination Therapies

In view of the suggested role of LTB_4 in the psoriatic process a major question arising from these studies is why significant suppression of this PMNL-derived inflammatory mediator is accompanied by such a relatively modest clinical improvement. It is possible that in vivo cutaneous levels do not decrease to the same extent as the levels generated in the stimulated PMNLs ex vivo. Also dietary EPA supplementation in normal subjects causes a reduced biological responsiveness of PMNLs with respect to both their adherence to endothelial cell monolayers and their chemotactic response to LTB_4 which may be independent of the diminution of LTB_4 synthesis. This suggests another explanation for the anti-inflammatory effect seen in psoriasis where the capillaries are histologically abnormal. Another alternative lies with the sulphidopeptido-leukotrienes. LTC_4 which is vasodilatory in human skin [20] is found in increased quantities in psoriatic plaques [21] and could contribute to the pathogenesis of the lesion. Lee et al. [12] have demonstrated that leukotrienes C_4 and C_5 are equipotent and unlike leukotrienes B their synthesis is not inhibited by dietary EPA administration. Since the sulphidopeptide-leukotrienes are also products of 5-lipoxygenase activity, it can also be concluded that increased EPA intake does not have an inhibitory effect on this enzyme. Dietary fish oil intake therefore cannot be equated with the administration of a 5-lipoxygenase inhibitor.

Ultraviolet B

Some of the commonly used therapies for psoriasis are given in table 1. All are empirical and often the response is incomplete. It has therefore been the practice of dermatologists over many years to combine dif-

Table 1. Therapy for psoriasis

Topical	Physical	Systemic	Obsolete (but effective)
Anthralin	PUVA	Antimitotics	Arsenic
Coal tar	UVB	Azathioprine	Benoxaprofen
Salicylic acid		Hydroxyurea	Mercury
Corticosteroids		Methotrexate	Razoxane
		Cyclosporin	
		Retinoids	

ferent modalities in an attempt to produce an additive effect. For example the Goeckerman regime dates back to 1925 and combines the use of UVB and coal tar, and the Ingram regime, commonly used in the United Kingdom, combines tar baths followed by UVB irradiation with the use of an anthralin (dithranol) paste. Not all combinations are desirable, however, and the use of UVB or Psoralens and Ultraviolet A (PUVA) together with methotrexate should be avoided in view of a greatly increased risk of inducing skin cancer.

The mild anti-inflammatory effect achieved by the administration of fish oil would appear to offer the opportunity to provide a modest background benefit to supplement other therapies. In a double-blind, placebo-controlled trial using fish oil and low dose UVB, Gupta et al. [22] have indeed shown benefit. Using a double-blind, randomized, parallel design they treated 18 patients suffering from stable plaque psoriasis for 15 weeks using 10 capsules twice daily of either MaxEPA (giving a daily dose of EPA of 3.6 g and DCHA of 2.4 g) or olive oil as a placebo. During weeks 3–11 they also recieved phototherapy with whole-body UVB. The UVB used was in suberythema doses to maximise the relative effect of the fish oil. Both at the end of therapy and 4 weeks later the patients treated with fish oil showed an improvement that was statistically significant for all parameters compared to the control group.

In so-called PUVA therapy the effect of long-wave ultraviolet radiation (320–400 nm) which by itself has little effect on psoriasis is enhanced by the administration of photosensitising psoralens either topically or systemically. As damage to the skin in the form of rapid aging and an increased risk of squamous cell carcinoma accompany the long-term use of PUVA, it would be highly beneficial if a similar response from fish oil to that achieved with UVB could be demonstrated.

Retinoid Therapy

As hyperkeratosis is a major feature of psoriasis the investigation of vitamin A derivatives for their potential as specific manipulators of the keratinisation process has been logical. Vitamin A itself is too toxic for use in pharmacologically active doses but in the derivatives isotretinoin and etretinate the ratio between efficacy and toxicity has been greatly improved. Both of these synthetic retinoids are in widespread use today for the treatment of acne and psoriasis, respectively. They are not without side effects, however, and muco-cutaneous symptoms, hair loss and teratogenicity remain a problem. In psoriasis the effect of etretinate is often only partial and so combination regimes using, for example, etretinate plus anthralin or etretinate plus PUVA are popular. Both of the synthetic retinoids cause hyperlipidaemia which is dose-related and reverses when treatment is stopped. The cause of this hyperlipidaemia is not yet known but the evidence seems to point to increased production of VLDL. The mechanisms behind the elevation of serum lipids have recently been reviewed by Marsden [23].

Whilst working in our department, Marsden [24, 25] demonstrated that the administration of fish oil would decrease retinoid-induced hyperlipidaemia. He performed two studies. In the first, 16 males aged 17–35 years suffering from acne were treated with 1 mg/kg/day of isotretinoin for a total of 12 weeks. After 8 weeks they were randomly allocated into two groups, 8 in each, and received in addition either 15 ml MaxEPA or 15 ml olive oil daily for 2 weeks. In the second study 3 males and 4 females aged 31–69 years with psoriasis were treated with 50 or 75 mg/day etretinate for 16 weeks, and 15 ml MaxEPA oil daily was added for weeks 8–12. On isotretinoin, triglyceride levels increased by 0.7 ± 0.1 mmol/l (mean ± SEM) at 8 weeks ($p < 0.01$) and were reduced by 0.6 ± 0.1 mmol/l after 2 weeks MaxEPA ($p < 0.05$). Two weeks after MaxEPA was stopped the levels rose again by 0.06 ± 0.1 mmol/l; levels did not alter with olive oil. The changes in cholesterol were less marked. There was an increase of 0.7 ± 0.1 mmol/l at 8 weeks and a reduction of 0.3 ± 0.05 mmol/l 2 weeks after MaxEPA was stopped; here again levels did not alter on placebo. With etretinate triglyceride levels increased by 0.9 ± 0.1 mmol/l at 8 weeks ($p < 0.01$), were decreased by 0.4 ± 0.07 mmol/l after 4 weeks MaxEPA (NS) and increased by 0.4 ± 0.05 mmol/l 2 weeks after MaxEPA was stopped. These findings, which have been supported by the work of Lowe et al. [26], suggest that MaxEPA reverses hypertriglyceridaemia due to isotretinoin and etretinate therapy and can reduce cholesterol levels, too. Large retinoid-induced increases in triglyceride are probably confined to subjects with pre-existing hypertriglyceridaemia but smaller increases are almost universal. If treatment is continued for long periods the changes in serum lipids and lipoproteins are liable to be associated with increased atherogenesis and an increased risk of ischaemic heart disease. Hence the addition of EPA supplements may have a valuable part to play in the management of patients on long-term retinoid therapy.

Cyclosporin

One of the most interesting recent developments in the therapy of inflammatory skin disease has been the discovery of the beneficial effects of cyclosporin, particularly in the treatment of atopic dermatitis and psoriasis. The results of studies currently being completed on psoriasis seem to indicate a greater than 50% reduction in PASI (Psoriasis Area and Severity Index) in 97% of patients taking 5 mg/kg/day with complete clearance in 76%. The onset of improvement is rapid but unfortunately so is relapse when treatment is stopped and studies are in hand to look at low-dose maintenance regimes. The mechanism of action of cyclosporin is believed to be through suppression of T lymphocyte function although this is not the only action of the drug [27]. A major drawback to the use of cyclosporin in non-life-threatening conditions is its nephrotoxicity which is believed to reflect a general vasculo-toxicity rather than a specific effect on the kidney itself. Thus the reduction in GFR which can be demonstrated within 90 min of giving a single dose is thought to be due to spasm of the afferent arteriole of the renal glomerulus. The mechanism of this vascular action is the subject of debate but it is possible that it results from a disturbance of eicosanoid metabolism. In cultured human umbilical endothelial cells cyclosporin has been shown to suppress PGI_2 formation [28], and the administration of iloprost, a synthetic analogue of prostacyclin, will enhance the effect of cyclosporin on cardiac allograft survival [29]. Looking at the other side of the equation, Petric et al. [30] found that, in a rat model, urinary TxB_2 excretion rose from 30.6 to 60.8 ng/24 h ($p < 0.001$) following 48 h on cyclosporin 50 mg/kg/day. This is a high dose, above the therapeutic range in humans, but the observation does raise the question as to whether dietary supplements of $\omega 3$ polyunsaturated fatty acids, by increasing the formation of inactive TxA_3 without interfering with the vasodilatory action of PGI_2 and PGI_3, might be helpful.

Two recent studies on rats [31, 32] suggest that this might be the case. After 14 days pre-treatment with olive oil or fish oil the rats received 50 mg/kg/day of cyclosporin for an additional 14 days. Glomerular filtration was significantly higher in the group treated with fish oil. Histologically there was a reduction in the vacuolar changes of the proximal tubule usually seen with high-dose cyclosporin treatment. A recent pilot study in Holland [33] investigating otherwise healthy patients suffering from chronic plaque psoriasis and being treated with cyclosporin showed less impairment of renal function in those treated with EPA than in the controls. The total renal vascular resistance, which usually rises with cyclosporin therapy was unchanged in those on EPA. These results are in agreement with another study in patients on cyclosporin therapy following renal transplantation in whom 12-week fish oil therapy resulted in improvement in glomerular filtration rate [ref. 13 in 34].

A multi-centre double-blind trial is currently in progress in Holland with the aim of clarifying the situation.

References

1 Brandrup F, Holm N, Grunnet N, et al: Psoriasis in monozygotic twins: Variations in expression in individuals with identical genetic constitution. Acta Derm Venereol Stockh 1982;62:229–236.

2 Camp RDR: Role of arachidonic acid metabolites in psoriasis and other skin diseases; in Lewis A, Ackerman N, Otterness I (eds): New Perspectives in Anti-Inflammatory Therapies. Advances in Inflammation Research. New York, Raven Press, 1988, vol 12, pp 163–172.

3 Maurice PD, Bather PC, Allen BR: Arachidonic acid metabolism by polymorphonuclear leukocytes in psoriasis. Br J Dermatol 1986;114:57–64.

4 Maurice PDL, Allen BR, Heptinstall S, et al: Arachidonic acid metabolism by peripheral blood cells in psoriasis. Br J Dermatol 1986;114:553–566.

5 Maurice PDL, Camp RDR, Allen BR: The metabolism of leukotriene B4 by peripheral blood polymorphonuclear leukocytes in psoriasis. Prostaglandins 1987;33:807–818.

6 Hellgren L: Psoriasis: The prevelance in sex age and occupational groups in the total population in Sweden. Stockholm, Almquist and Wiksell, 1967.

7 Kremer JM, Bigaouette J, Michalek AV, et al: Effects of manipulation of dietary fatty acids on clinical manifestations of rheumatoid arthritis. Lancet 1985;i:184.

8 McCall T, O'Leary D, Bloomfield J, et al: The effect of eicosapentaenoic acid in treatment and neutrophil function of patients with ulcerative colitis. 6th International Conference on Prostaglandins and Related Compounds. Abstract, p. 42, 1986.

9 Maurice PDL, Allen BR, Barkley ASJ, et al: The effects of dietary supplementation with fish oil in psoriasis. Br J Dermatol 1987;117:599–606.

10 Terano T, Salmon JA, Higgs GA, et al: Eicosapentaenoic acid as a modulator of inflammation. Biochem Pharmacol 1986;35:779–785.

11 Prescott SM, Zimmerman GA, Morrison AR: The effects of a diet rich in fish oil on human neutrophils: Identifaction of leukotriene B5 as a metabolite. Prostaglandins 1985;30:209–227.

12 Lee TH, Mencia-Huerta J-M, Shih C, et al: Characterisation and biologic properties of 5,12-dihydroxy derivatives of eicosapentaenoic acid, including leukotriene B5 and the double lipoxygenase products. J Biol Chem 1984;259:2383–2389.

13 Prescott SM: The effect of eicosapentaenoic acid on leukotriene generation and neutrophil function. J Biol Chem 1984;258:7615–7621.

14 Terano T, Salmon JA, Moncada S: Biosynthesis and biological activity of leukotriene B5. Prostaglandins 1984;27:217–232.

15 Lee TH, Hoover RL, Williams JD, et al: Effect of dietary enrichment with eicosapentaenoic and docosaenoic acids on in vitro neutrophil and monocyte leukotriene generation and neutrophil function. N Engl J Med 1985;312:1217.

16 Ziboh VA, Cohen KA, Ellis CN, et al: Effects of dietary supplementation with fish oil on neutrophil and epidermal fatty acids. Arch Dermatol 1986;122:1277–1282.

17 Bittiner SB, Tucker WFG, Cartwright I, et al: A double-blind, randomised, placebo-controlled trial of fish oil in psoriasis. Lancet 1988;i:378–380.

18 Bjørneboe A, Kleymeyer-Smith A, Bjørneboe G-E: Effect of dietary supplementation with n–3 fatty acids on clinical manifestations of psoriasis. Br J Dermatol 1988; 118:77–83.

19 Dewsbury CE, Graham P, Darley CR: Topical eicosapentaenoic acid (EPA) in the treatment of psoriasis. Br J Dermatol 1989;120:581.

20 Bisgaard H, Kristensen J , Sondergaard J: The effect of Leukotriene C4 and D4 on cutaneous blood flow in humans. Prostaglandins 1982;23:797–801.

21 Brain SD, Camp RDR, Kobza Black AK, et al: Leukotrienes C4 and D4 in psoriatic skin lesions. Prostaglandins 1984;29:611–619.

22 Gupta AK, Ellis CN, Tellner DC, et al: Double-blind, placebo-controlled study to evaluate the efficacy of fish oil and low-dose UVB in the treatment of psoriasis. Br J Dermatol 1989;120:801–807.

23 Marsden JR: Lipid metabolism and retinoid therapy. Pharmacol Ther 1989;40:55–65.

24 Marsden JR: Effect of dietary fish oil on hyperlipidaemia due to isotretinoin and etretinate. Hum Toxicol 1987;6:219–222.

25 Marsden JR: Reduction of retinoid hyperlipidaemia with MaxEPA. Br J Dermatol 1987;116:450.

26 Lowe NJ, Borok ME, Ashley JM, et al: Fish oil reduces hypertriglyceridaemia in psoriatic patients receiving etretinate therapy. Arch Dermatol 1988;124:177.

27 Borel JF: Basic science summary. Transplant Proc 1988;22:(suppl 2):722-730.

28 Voss BL, Hamilton KK, Samara ENS, McKee PA: Cyclosporine suppression of endothelial prostacyclin generation. Transplantation 1988;45:793–796.

29 Rowles JR, Foegh ML: The synergistic effect of cyclosporine and iloprost on survival of cardiac allografts. Transplantation 1986;42:94–96.

30 Petric R, Freeman D, Wallace C, et al: Effect of cyclosporine on urinary prostanoid excretion, renal blood flow and glomerulotubular Function. Transplantation 1988; 45:883–889.

31 Elzinga L, Kelley VE, Houghton DC, et al: Fish oil vehicle for cyclosporin lowers renal thromboxanes and reduces experimental toxicity. Transplant Proc 1987;19: 1403–1406.
32 Elzinga L, Kelley VE, Houghton DC, et al: Modification of experimental nephrotoxicity with fish oil as the vehicle for cyclosporin. Transplantation 1987;43:271–274.
33 Stoof TJ, Korstanje MJ, Bilo HJG, et al: Does fish oil protect renal function in cyclosporin treated psoriasis patients? Sandoz in House Data 1989.
34 Homan van der Heide JJ, Bilo HJG, Tegzess AM, et al: Omega-3 polyunsaturated fatty acids improve renal function in renal transplant recipients treated with cyclosporine A: Annu Meet Am Soc Nephrol 1988, Abstr 21.
35 Miller SJ, Raza A, Shinefeld HR, Elias PM: In vitro and in vivo antistaphylococcal activity of human stratum corneum lipids. Arch Dermatol 1988;124:209–215.

B.R. Allen, MD, Department of Dermatology, University Hospital,
Queen's Medical Centre, Nottingham, NG7 2UH (UK)

Cancer

Simopoulos AP, Kifer RR, Martin RE, Barlow SM (eds): Health Effects of ω3 Polyunsaturated Fatty Acids in Seafoods. World Rev Nutr Diet. Basel, Karger, 1991, vol 66, pp 446–461

Dietary ω3 Fatty Acids and Cancer: An Overview

Claudio Galli, Ritva Butrum

Institute of Pharmacological Sciencs, University of Milan, Milan, Italy; National Cancer Institute, EPN-212 A, National Institutes of Health, Bethesda, Md., USA

Introduction

The concept that dietary fat affects the incidence of cancer and influences tumor development is widely accepted, based on both epidemiological and experimental evidence.

There are, however, several important questions which need a clear answer, and the two major ones are: (a) do specific types or families of fatty acids (the saturates, the monoenes, the ω6 and ω3 series) exert differential effects (e.g. stimulation or inhibition)?, (b) which are the mechanisms responsible for these effects? This is important in order to focus the attention on processes which are relevant for cancer development and, at the same time, can be controlled by dietary fatty acids (FA), and is also important in order to identify parameters which can be assessed and measured in clinical studies.

This introduction is aimed to present an overview of: (a) the evidence for the impact of FA on cancer, based on epidemiological and experimental data, with special reference to the ω3 FA; (b) the mechanisms proposed for the effects of dietary FA, as a basis for further studies on the effects of FA on specific processes; (c) some consideration will also be given to the alterations of lipid and fatty acid metabolism in cancer patients. This type of information is important in order to appreciate how cancer may affect fatty acid metabolism and utilization in order to sustain growth and dissemination, and is useful in order to define dietary strategies for the control of these alterations and to improve the nutritional status of the patients.

Epidemiological Studies

Breast and colon cancers, the major malignancies in Western Societies, are both associated with modern affluence and may share etiological factors. Epidemiological evidence has implicated that the diet is potentially important and that fat is one of the major components correlating with increased cancer. Experiments carried out in animal models of carcinogenesis support these obervations, but they also indicate that the mechanisms responsible for the influences of fat on tumor development are rather complicated. Understanding of the mechanisms is, thus, a prerequisite for an application of experimental data to prevention of human cancer.

Breast and Colon Cancer

Most of the information provided by epidemiological investigations is derived mainly from case-control studies, based on comparisons of dietary practices between cancer patients and controls. More dependable data are provided by prospective cohort studies, in which dietary practices are measured among a large number of people and are then related to subsequent risk of cancer. These data, which are only beginning to be available due to the long duration of these studies, strongly suggest that, among environmental factors, the diet has a primary role for the risk of breast and especially of colon cancers.

Some similarity is present between the epidemiology of breast and colon cancers, but dietary factors appear to play a more significant role in the incidence of colon cancer, suggesting that this is likely to be more responsive to prevention through changes of the diet. Although a substantial genetic contribution to the risk of breast and colon cancers has been identified [1], both cancers certainly involve a strong interaction of genetics and the environment and, among nongenetic factors, the diet appears to be the major candidate [2]. Each type of cancer, however, has distinct risk factors.

For breast cancer several risk factors, mostly related to reproductive behavior, have been identified. Earlier age at first birth, number of children [3], early menarche and menopause, all support a role of sex hormones in the etiology of the disease. Reported changes of blood or urinary levels of sex hormones in breast cancer may, however, be a consequence of the disease. Differences in rates between countries have been attributed to differences in the intake of fat and animal fat in particular. However, a recent survey in rural counties in China has suggested that factors related to industrialization may seriously confound international comparisons of cancer rates, and case control studies in various countries [4–6] have shown that for breast cancer there is only a weak association with fat intake. Prospective studies [6, 7] have shown also weak associations, after controlling for total energy intake, with incidence of breast cancer for total, saturated and polyunsaturated fat. These data, thus, do not support very strongly the dietary fat hypothesis and they are compatible with only a very weak association with fat intake.

More clear relationships between fat intake and the occurrence of mammary tumors have been observed in animal experiments, and this will be discussed later on.

In contrast to breast cancer, there is stronger epidemiological evidence that intakes of animal fat or meat are associated with risk of colon cancer [8–10]. Although total energy intake may also be a factor [11], in studies carried out in France [12] and Belgium [13] intake of vegetable fat was inversely related to risk of colon cancer.

The suggested mechanisms for the influence of dietary fat on the incidence of colon cancer are: (a) higher fat intake enhances the excretion of bile acids, and (b) colonic bacteria convert bile acids to carcinogenic substances [14, 15].

Other dietary factors appear to influence the risk for colon cancer, such as certain types of fibers, which are thought to reduce the transit time for fecal matter in the gut and to dilute potential carcinogens in the feces [16]. Especially intake of fiber from fruit and vegetables, containing also active agents such as vitamins, protease inhibitors and β-carotenes, appears to be protective. Adjustment for total energy intake is, however, essential for a correct evaluation of these correlations. In addition, alcohol intake has been advocated as a risk factor for both breast and colon cancer.

In comparing breast and colon cancers, important differences in the etiology, or the period of life during which the factors act are indicated by the different relationships between changes in diet and mortality rates for the two tumors in Japan. In this country a striking rise in mortality from colon cancer, but only a small change in breast cancer mortality, has occurred between 1955 and 1975, when fat intake increased from 10 to about 25% of total energy intake [17]. In general, although dietary fat appears to be correlated with incidence rates of tumors, especially with colon cancer, data on the correlations with specific FA are still limited.

Epidemiological data on the impact of the ω3 FA on cancer are rather scarce and do not present a clearcut evidence of health benefit as that described for the effects on thrombotic, atherosclerotic, immune and inflammatory processes. A study on the effects of fish intake on risk for breast cancer in a large number of women showed that moderate alterations of fat and fish intake did not influence significantly the incidence of this type of tumor [18]. Another recent study [19] compared breast cancer incidence and mortality rates with approximate fish consumption in humans and reported an inverse association between percent calories consumed from fish and the incidence of breast cancer. In the report on the MRFIT study [20], it has been indicated that the 18:3 ω3/18:2 ω6 ratio and the ratio of total ω3/total ω6 in the diet are negatively correlated with the risk for cancer mortality, whereas there is no correlation between the levels of 18:2 ω6 and cancer. Apparently, the levels of linoleic acid (LA) intake in the population at study were quite above the values which are critical in order to show any effect, whereas the low levels of the ω3 were in the range where changes may more effectively influence processes involved in cancer development.

A more mechanistic approach to investigate the relationships between dietary fat and tumors is obtained in experiments on animal models, in which steps interposed between intake of specific FA, FA metabolism and factors involved in tumor development can be studied in detail.

Dietary FA, with Special Consideration to the ω3, and Experimental Carcinogenesis

The process of carcinogenesis is interpreted on the basis of a two-stage model consisting of an *initiation phase* followed by *tumor promotion.* During initiation a normal cell is altered becoming a latent cancer cell, as a consequence of the interactions, followed by alterations, between the carcinogen and cell genoma. During promotion, the altered genes are expressed and autonomous cellular growth, that is not responsive to normal physiologic regulatory signals, takes place. The possible effects of dietary fats have been tested in this two-stage model in order to evaluate the steps affected by exogenous and dietary agents.

Initiation may be affected by dietary FA through various mechanisms (e.g. carcinogen metabolism, uptake, absorption or dilution). Tumor promotion is less clearly understood than initiation, but there is evidence that dietary fat may enhance or inhibit this stage of carcinogenesis. Although it has been postulated that dietary fat might initiate carcinogenesis in man and other mammals, e.g. through DNA damage induced by oxidized fats or products of lipid oxidation [21], rodent feeding studies with oxidized fats have not supported this hypothesis. Certain fats, notably animal fats, have been reported to enhance initiation, but this might be due to contaminants present in fat rather than to the fats per se [22].

More convincing evidence is available on the effects of dietary fat on tumor promotion using various experimental models. Rodent mammary tumors (spontaneous, carcinogen-induced, transplantable, benign and carcinomatous) are well studied tumor models, but various others (e.g. colon, lung, prostatic, pancreatic, fibrosarcomas) have been used. Several experiments have shown that a high intake of dietary fats enhance tumorigenesis either spontaneous or induced by carcinogens. LA (18:2 ω6) has been shown to favor tumor promotion, and maximal stimulation of tumor growth is obtained with about 4% of total calories, i.e. at a level in the diet somewhat higher than that required for optimal EFA intake. When this amount of LA is present in the diet, however, additional dietary fat from other sources will have similar enhancing effects.

Mechanisms of the Effects of FA on Cancer

The mechanisms proposed for the tumor promoting effects of LA and of the ω6 FA in general are worth being considered in detail. In fact, the rationale for studies with the use of ω3 FA, aimed to inhibit tumor development, is based on the assumption that these FA will counteract the influence of the ω6 FA on specific processes in tumor develoment. A number of experimental studies in various cancer models, such as mammary, colon, lung, prostatic, pancreatic, fibrosarcoma, which have been reviewed by Karmali [23], and others described in this volume, have indeed shown that oils rich in ω3 FA reduce tumor incidence, growth and metastatic spread in various ways: they may reduce growth, or enhance latency of induction, or stimulate tumor loss, or they may not show any tumor-promoting activity in comparison to ω6 FA. It should be pointed out, however, that some studies have used levels of dietary

LA which are inadequate to meet EFA requirements, in protocols used for ω3 FA, and that the preparation of diets rich in FA which are prone to oxidative losses, requires great care and monitoring of the presence of lipid peroxides.

The following mechanisms have been proposed for the opposite effects of ω6 and ω3 FA.

Immune System Activity

Although evidence (pro and con) that dietary fats can influence immune system activity has been provided by various reports, relatively limited data are available on the relationships between this effect and experimental carcinogenesis. Various studies, however, have shown that diets rich in unsaturated FA of the ω6 series increase tumor growth through processes mediated by depression of cell-mediated immune responses. Decreased tumor lysis [24], or enhanced growth associated with reduced response of peripheral blood lymphocytes [25], or reduced responsiveness of individual cytotoxic lymphocytes [26] have been reported after high dietary LA.

Although effects of ω3 FA in immunologically mediated processes in experimental cancer are limited, several reports indicate that they may attenuate multiple defects in the immune response in various experimental pathologies. Fish oil was shown, e.g. to lower adrenal weight and serum C_3 levels, indicating improved immune responses in postburn treatment to animals [27]; to prevent metabolic alterations in endotoxic shock models [28], and to modulate the immunosuppressive effects of ciclosporin and thus reduce nephrotoxicity [29]. In addition to the effects on tumor cells which may be mediated by the immune system, various types of FA have been shown to affect growth of normal and tumor cells in vitro when added to the culture media. These effects, however, cannot be directly compared with those obtained after oral intake of FA and are void of immunological mechanisms. Clearly the hypothesis of immune system modulation as a mechanism responsible for the effects of dietary FA on tumors is very appealing, although other mechanisms appear to be also involved. Further research is obviously needed.

Prostaglandin Synthesis

Prostaglandins, which have an active role in cell proliferative processes in a variety of tissues, may also play a role in tumor growth.

Although prostaglandins of the 2 series appear to inhibit proliferation of tumor cells in vitro at concentrations several orders of magnitude greater than the physiological ones [30], various studies have shown that PGE_2 has inhibitory effects on the immune system. It inhibits mitogenesis of both T and B cells in vitro at concentrations of the nanomolar range and decreases IL-1 and 2 production and the response of T cells to IL-2 [31]. In addition, PGE_2 inhibits various other aspects of the immune response, such as lymphokine secretion, macrophage collagenase synthesis, natural killer cell activity, and the tumoricidal activity of activated macrophages [32, 33]. Finally, this prostaglandin may enhance immunoglobulin production and inhibit T cell suppression of B cell functions [34].

On the other side, products of the arachidonic acid (AA) lipoxygenase, such as leukotrienes and possibly others, also modify lymphocyte function in vitro [35] and may stimulate IL-1 production and γ-interferon synthesis [36]. Due to the general immunosuppressive effects of the prostaglandins of the 2 series, diets favoring the formation of these AA products may promote tumorigenesis. It is therefore postulated that administration of ω3 FA, which results in reduced formation of the prostaglandins of the 2 series, should also counteract the inhibition of the immune system. It has been reported, however, that also ω3 FA reduce the formation of products such as IL-1, IL-2 and TNF, which are part of the immune response, by mononuclear cells [37]. The inhibition of cytokine production by ω3 FA may be mediated by reduced cytokine gene expression as a consequence of the direct inhibiting activity of the newly formed PGE_3 and of the indirect inhibiting activity of LTB_5, counteracting the stimulating activity of LTB_4 [38].

It is clear from the reported experiments that various eicosanoids have different effects on immune cells, but the role of each compound in vivo and the influence of FA of the ω6 and ω3 series on these parameters in vivo in cancer are still unclear. Support for the hypothesis that tumorigenesis induced by high dietary levels of ω6 unsaturated FA is mediated by enhanced prostaglandin synthesis, comes also from the observation that prostaglandin inhibitors suppress tumor growth, under these dietary conditions [39]. It has been reported, however, that PG-inhibitors suppress tumorigenesis also in animals fed low levels of unsaturated fat [40]. In conclusion it appears that the relationships between dietary fats, prostaglandin activity and tumorigenesis must be investigated by directly evaluating prostaglandin levels in tumor cells of animals, after administration of varying amounts and types of fats.

Lipid Peroxy Radicals

The growth promoting effects of unsaturated fats on tumor cells have been attributed to the formation of lipid oxidation products, such as epoxides or peroxides, rather than to the parent compounds. These products can in fact be activators of cell proliferation, as shown in various experimental models [41, 42]. In addition, various antioxidants (selenium plus vitamin E, BHT, propyl-gallate), which are effective scavengers of lipid peroxy radicals and oxygen radicals, inhibit chemically induced tumorigenesis in animals fed diets rich in unsaturated FA [43, 44]. The radical hypothesis is questioned, however, by results suggesting an effect of the antioxidant on carcinogen metabolism rather than on formation of lipid oxidative products [45] and by studies with negative effects of certain antioxidants such as BHA [46]. The tumor inhibiting activity of ω3 FA, which should increase lipid peroxidation, is also in contrast with the lipid peroxy- and oxygen-radical hypothesis of tumorigenesis induced by unsaturated FA.

Membrane Fluidity

Fluidity of cell membranes is dependent upon the lipid composition of the membrane (e.g. cholesterol, cholesterol/phospholipid ratio, phospholipid proportions, type of FA). Unsaturated FA are considered to enhance membrane fluidity, in contrast to saturates. In vitro studies with cells in culture have shown that supplementation of FA modifies membrane fluidity. These results are, however, misleading, since intracellular lipid inclusion bodies containing triacylglycerols are formed, under these conditions. In addition, studies with biological membranes have shown that the influence on membrane fluidity, measured through polarization techniques, is dependent more upon other parameters, e.g. cholesterol content, rather than upon the unsaturation of the FA [47]. In addition, feeding studies in mammals have shown that the proportion of total unsaturated FA and the total level of unsaturation of membrane lipids remain constant, in spite of gross changes in the amount of unsaturates in the diet [48]. Thus, although dietary FA might modulate various membrane parameters (e.g. membrane-associated enzymes, membrane receptors, tumor antigens, prostaglandin synthesis) and possibly membrane fluidity, this latter does not appear to be directly correlated with the total level of FA unsaturation in the membrane. The observation that tumor growth is reduced in animals fed ω3 FA, in spite of the accumulation of these highly unsaturated compounds in the tumor, is also in contrast with the hypothesis of a tumor-promoting activity of FA supposedly inducing higher membrane fluidity, at least in certain specific lipid domains in membranes.

Intercellular Communication

Intercellular communication, which modulates cell growth and differentiation, may also have a role in tumor promotion. One type of intercellular communication, e.g. metabolic cooperation based on passage of low molecular weight (growth-regulatory?) molecules, through *gap junctions* in membrane structures, is blocked by tumor promoters [49]. Unsaturated FA inhibit metabolic cooperation, whereas saturates do not [50], but it is not known whether or not intercellular communication (metabolic cooperation) in normal or neoplastic cells can be altered by fat. It is possible that inhibition of metabolic cooperation by unsaturated fats is a result of enhanced cellular proliferation induced by these FA, since proliferating cells communicate less than nonproliferating cells. This point should, however, be specifically addressed.

Hormone Secretion

Several growth factors (estrogen, androgen, progesterone, glucocorticoids, insulin, insulin GF, PDGF, epidermal GF, prolactin, thyroid hormones) are important for the regulation of tumor breast cancer in vitro [51]. Early studies have suggested that high levels of dietary fat may stimulate tumorigenesis of hormone-sensitive (e.g. mammary) tumors by stimulating hormone secretion (prolactin and ovarian steroids). Subsequent studies did not confirm this hypothesis and no definite relationship has been provided between dietary fats and other endocrine secretory activities (e.g. growth hormone, glucocorticoids, insulin), except for an increased secretion of thyroid hormones induced by high fat diets. Increased thyroid secretion may be associated with increased development of neoplastic tissue (e.g. tumors of the mammary gland).

Pilot studies reported in this volume [52] suggest, however, that dietary fats alter sex steroid and pituitary-derived hormones, and that ω3 FA lower the production of estrogen and prolactin as well as that of other growth factors (EGF, TGF, IGF, etc.) secreted both externally or internally by tumor cells.

On the other side, hyperalimentation of fat, especially unsaturated, stimulates development and growth of tumors which are not dependent upon hormone secretion, such as tumors of colon, pancreas, kidney, lung, lymphatic system, stomach, liver, adrenals, thyroid and salivary glands. The hypothesis that enhanced endocrine secretion may be involved in the (mammary) tumorigenesis induced by hyper-alimentation with fat enriched in ω6 FA should, thus, be explored in greater detail. However, dietary FA may modify various growth factors in different cell lines and this may result in altered oncogene expression and/or altered growth factor receptors in plasma membranes.

Hormone Responsiveness

Experimental evidence has been provided showing that low fat consumption results in reduced levels of hormone receptors (prolactin, estrogen) in the mammary gland. On the other side, elevation of fat consumption above physiological levels does not modify hormone receptor levels [52]. However, it has been shown that elevation of dietary fat from 0 to 5% stimulates hormone-induced growth responsiveness of mammary glands [53]. In spite of a lack of direct data of the effect of dietary FA on receptor-mediated responses to various hormones, various data indicate that the generation of intracellular mediators of signal transduction, in various types of cells, such as formation of inositol phosphates in platelets [54], or Ca^{2+} mobilization in cultured rat myocardial cells [55], are depressed by ω3 FA.

These are some of the mechanisms which have been proposed on the basis of data obtained in a number of studies, and additional ones, such as modulation of oncogene expression, will be described in this session. These are not alternative mechanisms, but in specific conditions may contribute to the overall effects of dietary FA on cancer. It is not surprising that so many mechanisms have been proposed, since so many diversified effects of FA on a number of processes have been described. It will be important to assess the relevance of each mechanism in specific conditions in order to transfer the information to clinical studies.

Lipid Alterations in Cancer

Serum lipid profiles are frequently altered in cancer patients and reduction of total cholesterol, especially of the LDL-cholesterol subfraction, has been described as a consequence of the presence of the tumor [56]. There may be also a relation between low serum cholesterol levels and increased cancer morbidity and mortality [57]. Reduced LDL levels may result from increased expression of LDL receptors in malignant cells [58] leading to enhanced clearance of this lipoprotein [59]. Changes of FA profiles in plasma lipids have been reported in few studies. The major observation is an elevation of the oleic/stearic acid ratio in red blood cell lipids [60], suggesting enhanced conversion of stearic to oleic acid, and, more recently a marked reduction of LA has been reported in cholesterol esters (CE). The decrease of LA in CE is significantly correlated with body weight loss [61].

These data clearly indicate that impairments of lipid and FA metabolism and transport through lipoproteins are present in cancer. These alterations may also result from weight loss and malnutrition which are associated with the disease. No information is available on possible deficiencies of ω3 FA in cancer patients.

General Conclusions and Research Needs

Dietary FA have been shown by epidemiological investigations to affect tumor incidence rates in population groups, although it is not clear whether specific types of FA are responsible for the dietary fat-cancer incidence correlations. Selection of FA subclasses in the evaluation of dietary composition is an essential prerequisite in prospective studies of dietary fat and cancer, in order to separately assess the effects of ω3 vs. ω6 FA.

The limited epidemiological data relating ω3 fatty acid intake to cancer risk forces a greater dependence on studying mechanisms in animal models. Extrapolation from animal studies to humans is always fraught with difficulty, but without corresponding human data, it becomes particularly perilous. Most work in animals has been conducted under extreme conditions, that is, high total fat diets with either very high or very low ratios of ω6 to ω3 fatty acids. It is difficult to speculate about the practical implications of this work. Dietary studies that more closely mimic human consumption pattern are needed.

In addition, several specific questions appear relevant in studying the effects of selected FA, such as the ω3, in the diet, on cancer development: e.g. to dissect the direct effects on cell tumors from those mediated by modifications of general processes, such as immune function, and hormone secretion and activity. Information obtained in studies on experimental tumor models could then be compared with clinical data. Also the effects of ω3 FA on specific aspects of tumor diffusion, such as metastatic processes, should be investigated more specifically at both the experimental and clinical levels.

Important aspects of studies with tumor models are the mechanism(s) responsible for the modulation of tumor growth by unsaturated FA of the two series. Since FA accumulation in tissues, and metabolism are appreciably affected by other components in the diet, such as antioxidants, microelements, vitamins, etc., these dietary components should also be carefully considered. This may help in defining the contribution of various processes, such as peroxidation and eicosanoid formation, in the effects of fats on tumors.

In general, although potential mechanisms of cancer inhibition by ω3 fatty acids in animals have been studied, too little attention has been paid to their likely relevance to the prevention of cancer in humans. For example, consider that although much interest in the immune modulating effects of ω3 fatty acids (at least stronger than that exerted by the ω6 fatty acids) has been expressed, chemically induced tumors are relatively nonantigenic and the role of the immune system in human cancer has not been effectively demonstrated. Perhaps, it is best to consider those mechanisms that are thought to be most related to human cancer, and then to study how ω3 fatty acids modulate these mechanisms. Certainly, in all cases it is also necessary to distinguish between prevention and treatment. The role of ω3 fatty acids in cancer prevention may be limited by the difficulty in significantly increasing intake of foods rich in these fatty acids for a sustained period of time.

Finally, more detailed studies on the alterations of lipid and FA metabolism in cancer patients may help in appreciating some of the mechanisms through which cancer cells may maintain their growth rate, in evaluating the status of the unsaturated FA in the patient, and the requirements for fatty acid supplementation.

References

1 Anderson DE: Genetic study of breast-cancer: Identification of a high risk group. Cancer 1974:34:1090–1097.

2 Doll R, Peto RJ: The causes of cancer. Quantitative estimates of avoidable risks of cancer in the United States today. J Natl Canc 1981;66:1191–1308.

3 Kampert JB, Whittermore AS, Paffenbarger RS, et al: Combined effect of childbearing, menstrual events, and body size on age-specific breast cancer. Am J Epidemiol 1988;128:962–979.

4 Rohan TE, McMichael AJ, Baghurts PA: A population-based case control study of diet and breast-cancer in Australia. Am J Epidemiol 1988;128:478–489.
5 Hirohata T, et al: Occurrence of breast cancer in relation to diet and reproductive history: A case-control study in Fukuoka, Japan. Natl Cancer Inst Monogr 1985;69: 187–190.
6 Willet WC, Stampfer MJ, Colditz GA, et al: Dietary-fat and the risk of breast cancer. N Engl J Med 1987;316:22–28.
7 Jones DY, Schatzki A, Green SB, et al: Dietary-fat and breast cancer in the national health and nutrition examination survey. I. Epidemiologic follow-up-study. J Natl Canc 1987;79:465–471.
8 Potter JD, McMichael AJ: Diet and cancer of the colon and rectum. A case control study. J Nat Canc 1986;76:557–569.
9 Bristol JB, Emmett PM, Heaton KW, et al: Sugar, fat, and the risk of colorectal cancer. Br Med J 1985;291:1467–1470.
10 Graham S, Marshall J, Haughey B, et al: Dietary epidemiology of cancer of the colon in Western New York. Am J Epidemiol 1988;128:490–503.
11 Willett W, Stampfer MJ: Total energy-intake implications for epidemiologic analysis. Am J Epidemiol 1986;124:17–27.
12 Macquart-Moulin G, Riboli E, Cornee J, et al: Case-control study on colorectal cancer and diet in Marseilles. Int J Canc 1987;38:183–191.
13 Tuyns AJ, Haelterman M, Kaaks R: Colorectal cancer and the intake of nutrients: oligosaccharides are a risk factor, fats are not. A case-control study in Belgium. Nutr Cancer 1987;10:181–196.
14 Reddy BS: Diet and excretion of bile-acids. Cancer Res 1981;41:3766–3768.
15 Goldin BR, Swenson L, Dwyer J, et al: Effect of diet and *Lactobacillus acidophilus* supplements on human fecal bacterial enzymes. J Natl Canc Res 1980;64:255–261.
16 Reddy BS: Dietary Fiber and Colon Carcinogenesis. A Critial Review; in Vahouny GV, Kutchevsky D (eds): Dietary Fiber in Health and Disease. Plenum Press, 1982, p. 265–286.
17 Mishina T, Watanabe H, Araki H, et al: Epidemiological study of prostatic cancer by matched pair analysis. Prostate 1985;6:423–436.
18 Stampfer MJ, Willett WC, Colditz GA, et al: Intake of cholesterol, fish and specific types of fat in relation to risk of breast cancer; in Lands WEM (ed): Polyunsaturated Fatty Acids and Eicosanoids. Champaign. AOCS, 1987, pp 248–252.
19 Kaizer L, Boyd NF, Kriukov V, Tritcher D: Fish consumption and breast cancer risk: An ecological study. Nutr Cancer 1989;12:61–68.
20 Dolecek TA, Grandits G: Dietary polyunsaturated fatty acids and mortality in the multiple risk factor intervention trial (MRFIT); in Simopoulos AP, Kifer RR, Martin RE, Barlow SM (eds): Health Effects of ω3 Polyunsaturated Fatty Acids in Seafoods. World Rev Nutr Diet. Basel, Karger, 1991, vol 66, pp 205–217.
21 Albanes D: Total calories, body weight, and tumor incidence in mice. Cancer Res 1987;47:1987–1992.
22 Sylvester PW, Russell M, Ip MM, et al: Comparative effects of different animal and vegetable fats fed before and during carcinogen administration on mammary tumorigenesis, sexual maturation and endocrine function. Cancer Res 1986;46:757–762.

23 Karmali RA: Omega-3 FA and cancer: A review; in Lands WEM (ed): Polyunsaturated fatty Acids and Eicosanoids. Champaign, AOCS, 1987, pp 222–233.
24 Gabor H, Hillyard LA, Abraham S: Effect of dietary fat on growth kinetics of transplantable mammary adenocarcinoma in Balb/c mice. J Natl Cancer Inst 1985;74:1299–1305.
25 Wagner DA, Naylor PH, Kim U, et al: Interaction of dietary fat and the thymus in the induction of mammary tumors by 7,12-dimethylbenz(α)anthracene-induced rat mammary carcinoma. Cancer Res 1982;42:1266–1273.
26 Thomas IK, Erickson KL: Lipid modulation of mammary tumor cell cytolysis: Direct influence of dietary fats on the effector component of cell-mediated cytotoxicity. J Natl Cancer Inst 1985;74:675–680.
27 Trocki O, Heyd TJ, Weymack JP, et al: Effects of fish oil on postburn metabolism and immunity. JPEN 1987;2:521.
28 Pomposelli JJ, Flores EA, Bistrian BR, et al: Long-term fish oil enriched diets attenuate the metabolic effects of endotoxin in guinea pigs. JPEN 1988;12(suppl):40.
29 Kirkman RL, Barrett LV, et al: A fish oil vehicle increases the therapeutic index of cyclosporine. Surg Forum 1987;38:394.
30 Balazsovits, J, Mills G, Falk J, et al: Prostaglandins inhibit proliferation of the murine P815 mastocytoma by decreasing cytoplasmic free calcium levels (Ca^{2+}). Prostaglandins 1988;36:191–204.
31 Goodwin JS, Geuppens J: Regulation of the immune response by prostaglandins. J Clin Immunol 1983;3:295–310.
32 Herman J, Rabison AR: Prostaglandin E_2 depresses natural cytotoxicity by inhibiting interleukin-1 production by large granular lymphocyts. Clin Exp Immunol 1984;57:380–384.
33 Marshall LA, Johnston PV: The influence of dietary essential fatty acids on a rat immunocompetent cell, prostaglandin synthesis and mitogen-induced blastogenesis. J Nutr 1985;115:1572–1580.
34 Staite ND, Panayi GS: Prostaglandin regulation of B lymphocyte function. Immunol Today 1986;5:175–178.
35 Rola-Pleszczynsnki M: Immuno regulation by leukotrienes and other lipooxygenase metabolites. Immunol Today 1985;6:302–307.
36 Johnson HM, Torres BA: Leukotrienes: positive signals for regulation of γ-interferon production. J Immunol 1984;132:413–416.
37 Endres S, Ghorbani R, Kelley VE, et al: the effect of dietary supplementation with n–3 polyunsaturated fatty acids on the synthesis of interleukin-1 and tumor necrosis factor by mononuclear cells. N Engl J Med 1989;320:265–271.
38 Endres S, Meydani SN, Dinarello CA: Effects of ω3 fatty acid supplements on ex-vivo synthesis of cytokines in human volunteers. Comparison with oral aspirin and ibuprofen; in Simopoulos AP, Kifer RR, Martin RE, Barlow SM (eds): Health Effects of ω3 Polyunsaturated Fatty Acids in Seafoods. World Rev Nutr Diet. Basel, Karger, 1991, vol 66, pp 401–406.
39 Carter CA, Milholland RJ, Shea W, et al: Effect of the prostaglandin synthetase inhibitor indomethacin on 7,12-dimethylbenz(α)anthracene-induced mammary tumorigenesis in rats fed different levels of fat. Cancer Res 1983;43:3559–3562.
40 Kollmorgen GM, King MM, Kosanke SD, et al: Influence of dietary fat and indomethacin on the growth of transplantable mammary tumors in rats. Cancer Res 1983;43:4714–4719.

41 Slaga TJ, Klein-Szanto AJP, Triplett LL, et al: Skin tumor-promoting activity of benzoyl peroxide, a widely used free radical-generating compound. Science 1981; 213:1023–1025.
42 Bull AW, Nigro ND, Golembieski WA, et al: In vivo stimulation of DNA synthesis and induction of ornithine decarboxylase in rat colon by fatty acid hydroperoxides, autoxidation products of unsaturated fatty acids. Cancer Res 1984;44:4924–4928.
43 Horvath PN, Ip C: Synergistic effect of vitamin E and selenium in the chemoprevention of mammary carcinogenesis in rats. Cancer Res 1983;43:5335–5341.
44 King MM, McCay PB: Modulation of tumor incidence and possible mechanisms of inhibition of mammary carcinogenesis by dietary antioxidants. Cancer Res 1983; 43(suppl):2485s–2490s.
45 King MM, McCay PB, Kosanke SD: Comparison of the effect of butylated hydroxytoluene on N-nirosomethylurea- and 7,12-dimethylbenz(α)-anthracene-induced mammary tumors. Cancer Lett 1981;14:219–226.
46 Ip C: Dietary vitamin E intake and mammary carcinogenesis in rats. Carcinogenesis 1982;3:1453–1456.
47 Owen JS, Bruckdorfer KR, Ray RC, et al: Decreased erythrocyte membrane fluidity and altered lipid composition in human liver disease. J Lipid Res 1982;23:124–132.
48 Galli C, White HB Jr, Paoletti R: Brain lipid modifications induced by essential fatty acid deficiency in growing male and female rats. J Neurochem 1970;17:347–355.
49 Murray AW, Fitzgerald DJ: Tumor promoters inhibit metabolic cooperation in cocultures of epidermal and 3T3 cells. Biochem Biophys Res Commun 1979;91: 395–401.
50 Aylsworth CF: Effects of lipids on gaP junctionally mediated intercelluar communication: Possible role in the promotion of tumorigenesis by dietary fat; in Rogers A, Birt D, Mettlin C, Ip C (eds): Dietary Fat and Cancer. New York, Alan Liss, 1986, pp. 607–622.
51 Welsch CW: Interrelationship between dietary fat and endocrine processes in mammary gland tumorigenesis; in Rogers A, Birt D, Mettlin E and Ip C (eds): Dietary Fat and Cancer. New York, Alan Liss, 1986, pp 623–654.
52 Fernandes G, Venkatraman JT: Modulation of breast cancer growth in nude mice by ω3 lipids; in Simopoulos AP, Kifer RR, Martin RE, Barlow SM (eds): Health Effects of ω3 Polyunsaturated Fatty Acids in Seafoods. World Rev Nutr Diet. Basel, Karger, 1991, vol 66, p 488.
53 Welsch CW, DeHoog JV, O'Conor DH, et al: Influence of dietary fat levels on development and hormone responsiveness of the mouse mammary gland. Cancer Res 1985;45:6147–6154.
54 Medini L, Colli S, Mosconi C, et al: Diets rich in n–9, n–6 and n–3 fatty acids differentially affect the generation of inositol phosphates and of thromboxane by stimulated platelets, in the rabbit. Biochem Pharmacol 1990,39:129–133.
55 Hallaq H, Leaf A: Effects of ω3 and ω6 polyunsaturated fatty acids on the action of cardiac glycosides on cultured rat myocardial cells; in Simopoulos AP, Kifer RR, Martin RE, Barlow SM (eds): Health Effects of ω3 Polyunsaturated Fatty Acids in Seafoods. World Rev Nutr Diet. Basel, Karger, 1991, vol 66, pp 250–254.
56 Miller SR, Tartter PI, Papatestas AE, et al: Serum cholesterol and human cancer. J Natl Cancer Inst 1981;67:297–300.

57 Sherwin RW, Wentworth DN, Cutter JA, et al: Serum cholesterol levels and cancer mortality in 361,662 men screaned for the Multiple Risk Factor Intervention Trial. J Am Med Ass 1987;257:943–948.
58 Vitols S, Gahrton G, Bjorkholm M, et al: Hypocholesterolaemia in malignancy due to elevated low-density-lipoprotein-receptor activity in tumor cells: Evidence from studies in leukemic patients. Lancet 1985,ii:1150–1154.
59 Henriksson P, Eriksson M, Ericsson S, et al: Hypocholesterolaemia and increased elimination of low density lipoproteins in metastatic cancer of the prostate. Lancet 1989;ii:1178–1180.
60 Wood CB, Habib NA, Thompson A, et al: Increase of oleic acid in erythrocytes associated with malignancies. Br Med J 1985;291:163–165.
61 Mosconi C, Agradi E, Gambetta A, et al: Decrease of polyunsaturated fatty acids and elevation of the oleic/stearic acid ratio in plasma and red blood cell lipids of malnourished cancer patients. JPEN 1989;13:501–504.

Claudio Galli, MD, Institute of Pharmacological Sciences, University of Milan,
Via Balzaretti 9, I–20133 Milan (Italy)

Simopoulos AP, Kifer RR, Martin RE, Barlow SM (eds): Health Effects of ω3 Polyunsaturated Fatty Acids in Seafoods. World Rev Nutr Diet. Basel, Karger, 1991, vol 66, pp 462–476

ω3 Fatty Acid Diet Effects on Tumorigenesis in Experimental Animals

William T. Cave, Jr.[1]

University of Rochester School of Medicine and Dentistry Endocrine Unit, St. Mary's Hospital, Rochester, N.Y., USA

Introduction

Both human epidemiological surveys and animal studies have established that high fat diets are associated with a high incidence, and accelerated development, of certain tumors. Recently, experimental studies have been undertaken to evaluate the different tumor promoting capabilities of qualitatively different types of fat. As part of this comprehensive evaluation, investigators have studied the effects of diets containing high proportions of ω3 polyunsaturated fatty acids (PUFA) on the development of a variety of transplanted and carcinogen-induced animal tumors. In many of these studies, they have directly compared the effects of the ω3 PUFA-supplemented diets with those of equivalent ω6 PUFA-supplemented diets. Using such animal models, it has been demonstrated that tumors of at least four organs are responsive to dietary lipid interventions. These include the breast, colon, pancreas, and prostate. This review summarizes much of the currently available information on these ω3 PUFA dietary effects on tumorigenesis, and discusses several of the biochemical mechanisms of action proposed to explain them.

[1] The author wishes to acknowledge support from NIH grant CA-30629.

Mammary Neoplasms

The ability of high levels of dietary fat to enhance mammary tumorigenesis in animal models has been known for almost half a century [1]. Numerous laboratories have documented the validity of this relationship in a variety of rodent mammary tumor models: spontaneous tumors, transplanted tumors, carcinogen-induced tumors, diethylstilbestrol-induced tumors, and X-ray-induced tumors [2]. Collectively, these results have led most investigators to conclude that dietary fat has a unique mammary tumor-promoting activity. Subsequent experiments evaluating the relative tumor-promoting capabilities of the different classes of fat have led to the conclusion that some important differences do exist. For example, Carroll and Hopkins [3], using the 7,12-dimethylbenz(a)anthracene (DMBA)-induced tumor model, have shown that, when rats were fed diets containing high levels of ω6 PUFA, such as 20% (w/w) cottonseed oil, sunflower oil, or corn oil, they had considerably more tumors than rats fed equivalent dietary levels of saturated fats such as coconut oil, butter, or tallow. Subsequent in vivo and in vitro studies have confirmed these findings and have supported the general conclusion that high ω6 PUFA diets are more effective in promoting mammary tumorigenesis than high saturated fat diets [4, 5]. Once this difference in tumor promoting ability between saturated and ω6 polyunsaturated fat was appreciated, it was quickly realized that the structural molecular differences among the different types of PUFA might affect their tumor-promoting potential. As a result, a number of experiments have been undertaken recently to evaluate the tumorigenic capabilities of the ω3 PUFAs. These investigations have involved both carcinogen-induced mammary tumors, and transplanted mammary tumors.

Carcinogen-Induced Tumorigenesis

Some of the earliest reports on the effects of dietary ω3 PUFA on carcinogen-induced mammary tumor development were those of Jurkowski and Cave [6]. Using the N-methyl-N-nitrosourea (NMU) [CAS: 684-93-5]-induced mammary tumor model of Gullino et al. [7], they presented data from two experiments. In the first one, they divided 6-week-old female Buffalo rats into three diet groups: 0.5% menhaden oil (MO), 3% MO, and 20% MO. At 50, 78, and 106 days of age the rats were given NMU intravenously through their tail veins. When their tumors reached approximately 2 cm in diameter, the animals were killed, and selected tissues were stored for later fatty acid analysis. In the second experiment, following the same tumor induction protocol, they placed the rats on five diets. 0.5% MO, 20% MO, 0.5% corn oil (CO), 20% CO, and Formulab Chow 5008 (Ralston Purina). The results of both experiments showed that there was a progressive lengthening of the tumor latent period, as well as a reduction in tumor incidence and tumor burden, as the percentage of MO increased in the diets. In contrast, the rats on the 20% CO diet had a high tumor incidence and the shortest latent period. The fatty acid profiles of the tumor and liver membrane lipids showed that the different types of PUFA in the diet did significantly alter the cellular lipid composition. The cell membrane lipids of the MO diet groups had higher proportions of eicosapentaenoic acid (EPA) than did those of the CO groups; and this proportion was highest in the membrane lipids of the animals on the 20% MO diet. These findings were interpreted as evidence that diets with high levels of ω3 PUFA could affect mammary tumor development, and alter membrane lipid composition, in a distinctly different manner than diets composed of equivalent amounts of ω6 PUFA.

In further studies [8], these investigators have also examined the effects of dietary mixtures of ω3 and ω6 PUFA on tumor development. Using the same mammary tumor model, they divided the rats into five diet groups: 20% CO, 15% CO + 5% MO, 5% CO + 15% MO, 20% MO, and rat chow. Their results showed that, while the tumor latent periods of the animals on the 15% CO + 5% MO diet were equivalent to those of the 20% CO group, those of the 15% MO + 5% CO group were intermediate between the extreme values of the 20% CO and 20% MO groups. The fatty acid profiles of the tumor membrane lipids from each diet group revealed that there was a positive correlation between the amount of CO in the diet and the proportion of linoleic acid (LA; C18:2), and between the amount of menhaden oil in the diet and the proportion of EPA (C20:5). It was concluded that, if the level of an ω6 PUFA in the diet was sufficiently elevated (i.e. CO > 15%), small amounts of ω3 PUFA had no significant inhibitory effect on tumor promotion. However, if the amount of ω6 PUFA in the diet was reduced, the non-tumor promoting effects of the ω3 PUFA were expressed. These results seemed to differ somewhat from those reported for mixtures of saturated fats and ω6 PUFA; because in those latter studies a small amount of an ω6 PUFA (3% CO) added to a large amount (17%) of a saturated fat, such as tallow, enhanced mammary tumor development in a manner equivalent to 20% CO [9].

Using the DMBA mammary tumor model, Braden and Carroll [10] were also able to demonstrate that different polyunsaturated fats had different tumor-enhancing capabilities. They administered 5 mg DMBA to 50-day-old female Sprague-Dawley rats and then put them into one of the following diets 1 week later: 20% CO, 10% CO, 3% CO, 20% MO, 10% MO, and 3% MO. As expected, they found that the tumor yield increased when the levels of CO increased. In contrast, however, they noted that increasing dietary levels of MO seemed to progressively reduce the tumor incidence and prolong the tumor latent period.

In other studies using DMBA-treated rats, Karmali et al. [11] compared the effects of the following diets: 23.5% CO, 18.5% CO + 5% fish oil (FO), 8% CO + 15.5% FO, and 3% CO + 20.5% FO. They also reported that the rats receiving the diets with the highest content of ω3 PUFA had a reduced incidence of mammary tumors, but noted that these tumors had shortened latent periods. These observations led them to conclude that, when the dietary ratio of ω3 to ω6 PUFA was 1 and 4, the ω3 PUFA appeared to provide a moderate protective effect against the development of DMBA-induced mammary tumors.

Most recently, Abou-El-Ela et al. [12] have reported the results of their experiments in which they treated 50-day-old Sprague-Dawley female rats with a single intragastric injection of 10 mg DMBA and then, 21 days later, divided them into three diet groups: 20% MO, 20% CO, and 20% primrose oil (PO). The rats were killed 16 weeks following DMBA administration and tissue samples saved for eicosanoid analysis. They observed a similar tumor incidence in all groups, but noted that the tumor number was reduced in both the PO and MO groups. The tumor tissues of the MO-fed rats produced less PGE, 6-keto-F_{1a}, and LTB_4 than those of either of the other groups. It was proposed that these differences in eicosanoid production might account for the distinctive mammary tumorigenic responses associated with each of the different types of PUFA.

In a subsequent study [13], using the same experimental model, these investigators evaluated the tumorigenic effects of the following six diet groups: 20% CO, 20% CO + 0.004% indomethacin (I), 20% CO + 0.5% difluoromethylornithine (D), 20% CO + I + D, 15% MO + 5% CO, 15% MO + 5% CO + D. At 112 days after DMBA administration, the animals were killed, and selected tumor tissues assayed for PGE and LTB_4 synthesis. Postmicrosomal fractions of the selected tumors were used for ornithine decarboxylase (ODC) assay. The results indicated that the rats fed 20% CO and 20% CO + I had significantly greater mammary tumor incidence than rats fed either MO or D. PGE production rates were highest in the 20% CO group, and reduced in the I and D groups, and the inhibition was additive when both I and D were combined. PGE production was lower in the tissues of the MO diet groups than in those of either the I or D groups, and this production was even further reduced in the tissues of the MO groups also receiving D. LTB_4 production appeared enhanced in the tissues from the 20% CO + I rats, but significantly depressed in the D and MO groups. ODC activity was higher in the tumors of the 20% CO group than in the tumors of any of the other diet groups. It was concluded that diets containing an ω3:ω6 PUFA ratio of 1.2 reduced tumorigenesis, PGE and LTB_4 production, and ODC activity. Moreover, it appeared that the mammary tumor promotion induced by 20% CO diet was not reduced by inhibiting cylooxygenase or ODC activities unless it was accompanied by an inhibition of the lipoxygenase pathway. These investigators have thus proposed that LTB_4 may be importantly involved in tumor promotion in this model, and that the tumor-inhibiting effects of dietary MO may be more directly related to an inhibition of the 5-lipoxygenase pathway than to an inhibition of the cyclooxygenase pathway.

Transplant Tumorigenesis

One of the first reports on the effects of ω3 fatty acids on the development of transplanted mammary tumors was published by Karmali et al. [14]. They subcutaneously transplanted pieces of the R3230AC mammary adenocarcinoma into female F344 rats fed either normal rat chow or rat chow supplemented with a fish oil (MaxEPA) at three different dosages: 100, 200, and 400 μl/rat/day. They began their dietary intervention 1 week before transplantation and continued it until the animals were killed 3 weeks after transplantation. Their results indicated that there was reduced transplant growth in the fish oil groups relative to the controls. Furthermore, the tumors that grew in the fish oil groups had diminished contents of several different PGs and TxB_2. It was hypothesized that the tumor-inhibitory effects of the fish oil supplement were linked to the inhibitory effect of the ω3 PUFAs, EPA and docosahexaenoic acid (DHA) on arachidonate metabolism.

Gabor and Abraham [15] studied the growth of a transplantable BALB/c mammary adenocarcinoma in 6- to 8-week-old female BALB/c mice maintained on diets containing variable (0–10%) amounts of either CO, MO, or hydrogenated cottonseed oil (HCTO). They reported that the tumors in the 10% MO and 10% HCTO groups were significantly smaller than those in the 10% CO group. They also noted that the normally observed tumor-enhancing effects of the CO disappeared when the dietary ratio of MO to CO reached 9:1. Their fatty acid analyses revealed that the lipid compositions of the tumor cells were significantly affected by the type of fat in the diet, and their labelled-thymidine uptake studies indicated that the rate of tumor cell loss was lower in the mice on the 10% CO diet than in those on either the 10% MO or 10% HCTO diet. They concluded that diets high in ω3 PUFA could inhibit the development of this transplantable tumor, and they suggested that this effect might be mediated by a mechanism that accelerated tumor cell loss.

Kort et al. [16] have published data on the comparative effects of diets containing either 25% cacao butter (CB) or 25% FO on tumor development using the BN472 transplantable mammary adenocarcinoma in female BN/Bi mice. They started their mice on their respective diets at 3 weeks of age and inoculated them with their transplants at 12 weeks of age. Following transplantation, one half of the rats on the CB diet were switched to the FO diet. They removed the tumors 18 days after inoculation and then killed the animals 13 days later. They found that tumor growth was significantly lesser in the animals maintained on the FO diet than in those of either the CB group or the CB-FO group. They concluded from these observations that the tumor-inhibiting effect of the fish oil diet was greater when the diet was initiated before transplantation than when it was given only after transplantation.

Borgeson et al. [17] have recently compared the effects of 10% FO diet with those of a 10% CO diet on human mammary carcinoma (MX-1) transplant growth in female heterozygous BALB/c nu/+ athymic (nude) mice. They observed that the growth rates of the transplants in the animals on FO diets were significantly depressed relative to those in the CO diet group. Moreover, the transplants in the FO group appeared more responsive to chemotherapeutic drugs mitomycin C and doxyrubicin. They discovered an increase in the activity of the mitochondrial enzyme carnitine acyltransferase in the tissues of the FO-treated animals and they have hypothesized that it is this type of dietary ω3 PUFA induced alteration in enzyme activity that allows mammary tumors to become more susceptible to chemotherapeutic agents.

Colonic Neoplasms

Over the past decade, a number of experiments have indicated that, while ω6 PUFA diets enhance chemically induced colonic tumorigenesis, diets containing equivalent levels of coconut oil, olive oil, or trans-fat do not [17–21]. These results suggest that, like the mammary tumor models, colonic neoplasms are also sensitive to qualitative differences in the fatty acid composition of dietary fat. Accordingly, most recent studies, with few exceptions [22, 23], indicate that high ω3 PUFA diets similarly inhibit colonic tumorigenesis.

Carcinogen-Induced Tumorigenesis

In one of the first studies published on this subject, Reddy et al. [24] compared the effects of differing amounts of dietary ω3 and ω6 PUFA on colon tumor development initiated in 5-week-old male F344 rats treated with the carcinogen azoxymethane (AOM) [CAS: 25843-45-2]. They administered the AOM subcutaneously, weekly for 2 weeks, and then, 4 days following the last dose, started the animals on one of the following diets: 4% MO + 1% CO; 22.5% MO + 1% CO; 23.5% CO, and 5% CO. Thirty-four weeks later, they terminated the experiment. Their results showed that the large intestines of the mice receiving either one of the MO-containing diets or the 5% CO diet had a significantly lower tumor incidence and a smaller tumor number than those from mice fed the 20% CO Diet. They concluded from this that high ω3 PUFA diets did not have the tumor-enhancing potential exhibited by high ω6 PUFA diets. No definitive biochemical explanation was given to account for this difference, but they postulated two potential mechanisms might exist: (1) that ω3 PUFA may have an inhibitory effect on the colonic concentration of secondary bile acids, or (2) that the ω3 PUFA in the diet may be altering colonic eicosanoid metabolism.

In subsequent studies [25], using the same tumor model, they investigated the efficacy of various mixtures of dietary ω3 and ω6 fatty acids on colon carcinogenesis. They divided their rats into six dietary groups: 4% MO + 1% CO; 17.6% CO + 5.9% MO; 11.8% CO + 11.8% MO; 5.9% CO + 17.6% MO; 5% CO; and 23.5% CO; and after the rats had been on their diets 38 weeks, they killed them. The results showed no statistically significant differences in the incidence and multiplicity of colon tumors between the animals fed the 5% CO diet and the 1% CO + 4% MO diet. The rats fed high fat diets containing 17.6% CO + 5.9% MO; 11.8% CO + 11.8% MO; or 5.9% CO + 17.6% MO also showed no enhancement in tumor incidence. The animals fed the diet containing 23.5% CO, however, had an increased incidence of tumors relative to those fed the 5% CO diet. The total number of colonic tumors was significantly lower in animals fed diets containing 5.9% CO + 17.6% MO, 5% CO, and 1% CO + 4% MO than in those fed the 23.5% CO diet; and the number of adenocarcinomas was slightly but not significantly lower in animals fed the 17.6% CO + 5.9% MO diet, and 11.8% CO + 11.8% MO diet than in the animals fed the 23.5% CO diet. On the basis of these results, it was concluded that high fat intake is a necessary but not sufficient condition for colon tumor promotion, and that this process is significantly influenced by the relative proportions of ω3 and ω6 PUFA in the diet.

Minoura et al. [26] have confirmed these results using a similar 'in vivo' animal model. They placed weanling male Donryu rats on either an ω3 PUFA diet (made by adding 4.7% EPA ethyl ester + 0.3% LA ethyl ester to a basal semipurified, fat-free diet), or an ω6 PUFA diet (made by adding 5% LA ethyl ester to the basal diet). They treated 40 rats from each diet group with weekly injections of AOM subcutaneously for 11 weeks. Fifteen weeks after the last injection they killed 30 rats in each group and examined their intestines. They killed the remaining rats 20 weeks after the last injection of AOM and used their tumors and normal colon tissue for prostaglandin and lipid analysis. They found that the animals on the ω3 PUFA diet had a significantly lower tumor incidence and tumor number than those fed the ω6 PUFA diet. The fatty acid analyses of the lipids from the tumors and normal colonic mucosa indicated that there were clearly defined diet-related differences, and the eicosanoid measurements demonstrated that the levels of PGE_2 in the colonic tumors of the ω6 PUFA diet group were markedly higher than those of the tumors in the ω3 PUFA diet group. They concluded that, compared to diets with a high LA content, those with a high EPA content significantly inhibit colon carcinogenesis. They suggested that EPA-induced alterations in PGE_2 metabolism may either directly affect tumor development by reducing the rate of neoplastic cell proliferation or indirectly affect tumorigenesis by modifying cellular and humoral immune responses. They did not, however, exclude the possibility that EPA may also influence tumor development by altering bile acid metabolism.

Nelson et al. [27] have also studied this ω3 PUFA effect on colon tumorigenesis using the 1,2-dimethylhydrazine (DMH)-induced tumor model. They divided 50 5-week-old male Sprague-Dawley rats into three diet treatment groups: 22 rats received a diet containing 17% MO added to a powder Purina rat chow base; a similar number received a diet containing 17% CO added to the same rat chow base; and 6 received the rat chow with no additional lipid. After being on these diets for 1 week, they gave 19 rats in each of the high rat groups, and 3 animals from the rat chow group, 6 weekly injections of DMH subcutaneously, leaving the remaining animals as controls. They killed 3 DMH-treated animals in each of the diet groups, together with 3 rat chow controls, 4 weeks after the last injection of the carcinogen and collected their plasma for peroxide analysis. They kept all of the remaining animals on their diets until they terminated the experiment 17 weeks after the last DMH injection. They observed that the rats on the MO-supplemented diets had fewer colorectal tumors than those on the CO-supplemented diet and reported that there were no significant differences in the plasma peroxide concentrations between the dietary groups, whether or not they had been treated with carcinogen. They concluded that high ω3 PUFA diets were less supportive of tumor development than were high ω6 PUFA diets, and that, if diet-induced changes in colonic peroxides were, in any way, related to this difference in tumor development, it could not be determined by alterations in plasma peroxide levels.

Transplant Tumorigenesis

Most recently, Cannizzo and Broitman [28] have reported their results from studies on the development of transplanted colon carcinoma cells in mice maintained on diets containing varying amounts of either ω6 PUFA or ω3 PUFA. They divided weanling BALB/c ByJ mice into four diet groups: a 5% safflower oil (SO) group; a 24.7% SO group; a 5% MO group; and a 24.7% MO group. After 30 days, they divided the mice within each diet group into 4 subgroups: a tumor growth assay group; a colonization assay group; a time-to-death group, and a control group. In the tumor growth assay group they performed laparotomies on each of the animals and injected 1×10^6 CT-26 cells into the subserosa of their colons. They killed the mice at 7, 14, 21, and 28 days after inoculation. In the time-to-death groups, they inoculated the animals in an identical manner and observed their length of survival. They gave the animals in the colonization assay groups a tail vein injection of 1×10^5 tumor cells and, after 21 days, killed them and removed their lungs for quantification of tumor deposits. Their results indicated that the 24.7% SO group had the largest tumors, followed by the 5% SO group. No statistical differences in tumor growth were observed between the two MO groups, but both of the MO groups had greatly reduced tumor sizes, relative to the SO groups, at all time points. The mortality rate of the 24.7% SO group was greater than that of either the 5% SO group or the MO groups. The colonization assay revealed that the mice in the 24.7% SO group had significantly more tumors than those in any of the other dietary groups. Collectively, these results indicated that SO enhanced pulmonary colonization and colonic tumor growth in a dose-dependent manner. Conversely, they demonstrated that MO at both concentrations reduced colonic tumor growth and at high concentrations inhibited pulmonary colonization. On the basis of these findings it was hypothesized that dietary ω3 PUFA can substitute for the native ω6 PUFA in actively metabolizing cells and consequently interfere with the processes necessary for the accelerated growth and survival of tumor cells.

Pancreatic Neoplasms

The information currently available on the effects of dietary lipid on pancreatic tumorigenesis has come primarily from the work of O'Connor et al. [29, 30]. By administering the carcinogen azaserine (AZA) [diazoacetate serine ester; CAS: 115-02-6] intraperitoneally to 21-day-old male Wistar rats; they have been able to induce the formation of preneoplastic atypical acinar cell nodules (AACN) and adenocarcinomas. With this model, they have been able to demonstrate that high ω6 PUFA diets (e.g. 20% CO or 20% SO) fed during the post-initiation period significantly increase the incidence of pancreatic neoplasms beyond that normally found in animals fed either a normal rat chow diet or an 18% hydrogenated coconut oil + 2% CO diet [31].

Carcinogen-Induced Tumorigenesis

In their studies designed to evaluate the effects of high ω3 PUFA diets [29], O'Connor et al. gave rats a single dose of AZA (30 mg/kg body weight) intraperitoneally at 14 days of age and divided them into three diet groups at 21 days of age. These diet groups consisted of a 20% MO and a 20% CO group, each using dried cod as their major protein source, and a 20% CO group using casein as the protein source. They observed that the number and size of the preneoplastic lesions were reduced in the 20% MO diet group relative to the 20% CO groups. They concluded that this reduction in tumor growth was most likely due to the high content of ω3 PUFA in the MO, but acknowledged the possibility that a potential deficiency of LA in the fish oil may have been a contributing factor.

Therefore, they undertook another series of experiments in order to define this ω3 PUFA effect more completely [30]. In the first experiment, they divided their rats into nine diet treatment groups which each had the same quantitative amount of PUFA, but an ω3:ω6 ratio that varied from 0.01 to 7.0 (20% CO, 19% CO + 1% MO, 17% CO + 3% MO, 15% CO + 5% MO, 10% CO + 10% MO, 5% CO + 15% MO, 3% CO + 17% MO, 1% CO + 19% MO, 20% MO). They killed the animals after 4 months and discovered that, as the dietary ratio of ω3 to ω6 PUFA increased, there was a significant decrease in the extent of preneoplastic development, whether expressed as AACN/cm^3, or as percent volume of pancreas occupied by AACN. Furthermore, using regression analysis, they noted that there was also a statisically significant, but somewhat unstable, decrease in AACN diameter. Their fatty acid analyses of red cell membrane lipids confirmed that, as the percentage of MO in the diet increased, the amount of ω3 PUFA increased and the ω6 PUFA decreased. Serum TxB_2, PGE_2, and 6-keto-PGF_{1a} levels were also noted to be reduced in the rats fed diets with high ω3:ω6 PUFA ratios.

In a second experiment, they performed a crossover study, in which the animals after tumor induction were maintained on either 20% MO or 20% CO for 2 months and then switched to the opposite diet for 2 months. They also included a 20% MO group and a 20% CO group in order to establish the effects of a non-switched regimen. When they killed the animals at the end of 4 months, they observed that, as before, the 20% CO diet group again had the greatest number of AACN/cm^3 pancreas and the 20% MO group the least. The rats switched from 20% CO diet to the 20% MO diet, however, had a smaller number of AACN/cm^3 and lesser volume percentage of pancreas occupied by foci than the rats in the 20% CO diet group while the rats switched from the 20% MO diet to 20% CO diet had more tumor development than the 20% MO diet group. They concluded that the dietary ratio of ω3 to ω6 PUFA does significantly influence AACN development and that dietary alterations in this ratio, even several months after tumor induction, can influence neoplastic development.

Prostatic Neoplasms

The role of dietary fat in prostate cancer has been difficult to evaluate in the laboratory because of the lack of sensitive animal models. Some progress, however, has been made in this area over the last 5 years. Pollard and Luckert [32], using 3-month-old male Lobund-Wistar rats with subcutaneous implants of testosterone propionate, have been able to demonstrate that animals fed a 20% CO diet have a greater incidence of spontaneous prostate adenocarcinomas with a shorter latent period than animals fed a 5% CO diet. These results have been interpreted as indicating that a high ω6 PUFA diet can additively enhance the known promotional effect of testosterone in this model system.

Recently, efforts to better define this lipid-tumor interrelationship have led several investigators to examine the effects of dietary intervention on the development of transplanted human prostatic carcinoma cells in the nude mouse. At least two of these have evaluated the effects of high ω3 PUFA diets.

Transplant Tumorigenesis

In one of the first experiments of this type, Karmali et al. [33] divided 20 4-week-old athymic male mice (Swiss nu/nu) into two diet groups: 23.52% CO, and 20.52% FO (MaxEPA) + 3% CO. They inoculated the mice subcutaneously with 1×10^6 cells from a cultured line of DU-145 human prostatic tumor cell line at 6 weeks of age and killed them 8 weeks later. They observed that the mean tumor volumes and weights of the FO-fed mice were significantly lower than those of the CO-fed mice. They also noted that the tumor cells from the FO diet-treated animals had histological evidence of a reduced content of acid phosphatase, as well as a reduced content of immunoassayable PGE_2.

Rose and Cohen [34], in a similar experiment, compared the development of cultured DU-145 tumor cells in male nude mice (Balb/c CD-1) maintained on a 23% CO diet with ones fed a 17% MO + 6% CO diet. They started the mice on their diets at 6 weeks of age and then, three weeks later, redivided the animals in each group into two subgroups. They gave one subgroup a subcutaneous transplant of 1×10^6 cells and the other subgroup a transplant of 5×10^6 cells. Six weeks after the inoculation of the tumor cells they killed the animals. They noted that there was no difference in tumor latency among the mice given 5×10^6 cells regardless of their diet, but in the mice receiving 1×10^6 cells the slope of the regression line for the animals fed the fish oil diet was significantly less than that of those fed the CO diet. They concluded from this observation that the MO diet could retard the progression of the transplanted cells, but that specific quantification of the cell inocula was needed to appropriately demonstrate the effects of initial tumor burden on this response.

Conclusions

Collectively, the data derived from these lipid-sensitive animal models of tumorigenesis demonstrate that high ω3 PUFA diets consistently diminish tumor development, while high ω6 PUFA diets enhance it. The biochemical mechanisms responsible for these effects remain unclear. The fatty acid profiles of the membrane lipids of these tumors have repeatedly shown that they are very sensitive to qualitative alterations in dietary lipid composition. Such membrane lipid changes may affect the responsiveness of tumor cells to endocrine, paracrine, or autocrine growth factors. Alterations in eicosanoid metabolism have also been frequently associated with these membrane lipid changes. Typically, ω3 PUFA-fed animals have been reported to have a reduced production of prostaglandins of the 2-series and LTB_4. Such eicosanoid alterations may have direct effects on tumor cell development or they may modulate immune mechanisms which have an influence on tumor cell survival. In colon tumorigenesis it is possible that ω3 PUFA may play an additional role through their effects on bile acid metabolism. Finally, the recently reported effects of ω3 PUFA on the induction of specific mitochondrial enzymes may also eventually prove important. Regardless of these uncertainties concerning mechanisms of action, however, the experimental data from these animal tumor models clearly substantiates the conclusion that qualitative as well as quantitative differences in dietary lipids can modify tumor lipid metabolism. The fact that some of these diet-induced modifications are associated with significant differences in tumor development strengthens the argument that a better understanding of lipid-tumor interrelationships may have important preventive and therapeutic medical implications.

References

1 Cave WT: Differential effects of specific types of dietary lipid on mammary tumor development; in Abraham S (ed): Carcinogenesis and Dietary Fat. Boston, Kluwer, 1989, pp 85–99.

2 Welsch CW, Aylsworth CF: The interrelationship between dietary lipids, endocrine activity, and the development of mammary tumors in experimental animals; in Perkins EG, Visek WJ (eds): Dietary Fats and Health. Champaign, American Oil Chemists Society, 1983, pp 760–816.

3 Carroll KK, Hopkins GJ: Dietary polyunsaturated fat versus saturated fat in relation to mammary carcinogenesis. Lipids 1979;14:155–158.

4 Kidwell WR, Monaco ME, Wicha MS, et al: Unsaturated fatty acid requirements for

growth and survival of a rat mammary tumor cell line. Cancer Res 1978;38:4091–4100.

5 Wicha MS, Liotta LA, Kidwell WR: Effects of free fatty acids on the growth of normal and neoplastic rat mammary epithelial cells. Cancer Res 1979;39:426–435.

6 Jurkowski JJ, Cave WT: Dietary effects of menhaden oil on the growth and membrane lipid composition of rat mammary tumors. J Natl Cancer Inst 1985;74:1145–1150.

7 Gullino PM, Pettigrew HM, Grantham FH: N-nitrosomethylurea as mammary gland carcinogen in rats. J Natl Cancer Inst 1975;54:401–414.

8 Cave WT, Jurkowski JJ: Comparative effects of omega-3 and omega-6 dietary lipids on rat mammary tumor development; in Lands WEM (ed): Proceedings of the AOCS Short Course on Polyunsaturated Fatty Acids and Eicosanoids. Champaign, American Oil Chemists' Society, 1987, pp 261–266.

9 Cave WT, Jurkowski JJ: Dietary lipid effects on the growth, membrane composition and prolactin-binding capacity of rat mammary tumors. J Natl Cancer Inst 1984;73: 185–191.

10 Braden LM, Carroll KK: Dietary polyunsaturated fat in relation to mammary carcinogenesis in rats. Lipids 1986;21:285–288.

11 Karmali RA, Doshi RU, Adams L, et al: Effect of n–3 fatty acids on mammary tumorigenesis, in Samuelsson B, Paoletti R, Ramwell PW (eds): Advances in Prostaglandin, Thromboxane, and Leukotriene Research. New York, Raven Press, 1987, vol 17, pp 886–889.

12 Abou-El-Ela S, Prasse KW, Carroll R, et al: Eicosanoid synthesis in 7.12-dimethylbenz(a)anthracene-induced mammary carcinomas in Sprague-Dawley rats fed primrose oil, menhaden oil, or corn oil diet. Lipids 1988;23:948–954.

13 Abou-El-Ela S, Prasse KW, Carroll R, et al: Effects of d,1-2-difluoromethyl-ornithine and indomethacin on mammary tumor promotion in rats fed high n–3 and/or n–6 fat diets. Cancer Res 1989;49:1434–1440.

14 Karmali RA, Marsh J, Fuchs C: Effect of omega-3 fatty acids on growth of a rat mammary tumor. J Natl Cancer Inst 1984;73:457–461.

15 Gabor H, Abraham S: Effect of dietary menhaden oil on tumor cell loss and the accumulation of mass of a transplantable adenocarcinoma in BALB/c mice. J Natl Cancer Inst 1986;76:1223–1229.

16 Kort WJ, Weijma IM, Vergroesen AJ, et al: Conversion of diets at tumor induction shows the pattern of tumor growth and metastasis of the first given diet. Carcinogenesis 1987;8:611–614.

17 Borgeson CE, Pardini L, Pardini RS, et al: Effects of dietary fish oil on human mammary carcinoma and on lipid-metabolizing enzymes. Lipids 1989;24:290–295.

18 Broitman SA, Vitale JJ, Vavrousek-Jakuba E, et al: Polyunsaturated fat, cholesterol, and large bowel tumorigenesis. Cancer 1977;40:2455–2463.

19 Reddy BS, Maeura Y: Tumor promotion by dietary fat in azoxymethane-induced colon carcinogenesis in female F344 rats: influence of amount and source of dietary fat. J Natl Cancer Inst 1984;72:745–750.

20 Reddy BS, Tanaka T, Simi B: Effect of different levels of dietary trans fat or corn oil on azoxymethane-induced colon carcinogenesis in F344 rats. J Natl Cancer Inst 1985;75:791–798.

21 Sakaguchi M, Minoura T, Hiramatsu Y, et al: Effects of dietary saturated and unsaturated fatty acids on fecal bile acids and colon carcinogenesis induced by azoxymethane in rats. Cancer Res 1986;46:61–65.
22 Nigro ND, Bull AW, Boyd ME: Inhibition of intestinal carcinogenesis in rats: effect of difluoromethylornithine with piroxicam or fish oil. J Natl Cancer Inst 1986;77:1309–1313.
23 Fady C, Reisser D, Lagadec P, et al: In vivo and in vitro effects of fish-containing diets on colon tumor cells in rats. Anticancer Res 1988;8:225–228.
24 Reddy B, Maruyama H: Effect of dietary fish oil on azoxymethane-induced colon carcinogenesis in male F344 rats. Cancer Res 1986;46:3367–3370.
25 Reddy B, Sugie S: Effect of different levels of omega-3 and omega-6 fatty acids on azoxymethane-induced colon carcinogenesis in F344 rats. Cancer Res 1988;48:6642–6647.
26 Minoura T, Takata T, Sakaguchi M, et al: Effect of dietary eicosapentaenoic acid on azoxymethane-induced colon carcinogenesis in rats. Cancer Res 1988;48:4790–4794.
27 Nelson RL, Tanure JC, Andrianopoulos G, et al: A comparison of dietary fish oil and corn oil in experimental colorectal carcinogenesis. Nutr Cancer 1988;11:215–220.
28 Cannizzo F, Broitman SA: Postpromotional effects of dietary marine or safflower oils on large bowel or pulmonary implants of CT-26 in mice. Cancer Res 1989;49:4289–4294.
29 O'Connor TP, Roebuck BD, Peterson F, et al: Effect of dietary intake of fish oil and fish protein on the development of 1-azaserine-induced preneoplastic lesions in the rat pancreas. J Natl Cancer Inst 1985;75:959–962.
30 O'Connor TP, Roebuck BD, Peterson F, et al: Effect of dietary omega-6 fatty acids on development of azaserine-induced preneoplastic lesions in rat pancreas. J Natl Cancer Inst 1989;81:858–863.
31 Roebuck BD, Yager JD, Longnecker DS, et al: Promotion by unsaturated fat of azaserine-induced pancreatic carcinogenesis in the rat. Cancer Res 1981;41:3961–3966.
32 Pollard M, Luckert PH: Promotional effects of testosterone and dietary fat on prostate carcinogenesis in genetically susceptible rats. Prostate 1985;6:1–5.
33 Karmali RA, Reichel P, Cohen LA et al: The effects of dietary omega-3 fatty acids on the DU-145 transplantable human prostatic tumor. Anticancer Res 1987;7:1173–1180.
34 Rose DP, Cohen LA: Effects of dietary menhaden oil and retinyl acetate on the growth of DU 145 human prostatic adenocarcinoma cells transplanted into athymic nude mice. Carcinogenesis 1988;9:603–605.

William T. Cave, Jr., MD, University of Rochester School of Medicine and Dentistry, Endocrine Unit, St. Mary's Hospital, Rochester, NY 14611 (USA)

Simopoulos AP, Kifer RR, Martin RE, Barlow SM (eds): Health Effects of ω3 Polyunsaturated Fatty Acids in Seafoods. World Rev Nutr Diet. Basel, Karger, 1991, vol 66, pp 477–487

ω3 Fatty Acids and Cancer Metastasis in Humans

Jennifer Man-Fan Wan, Beatrice S. Kanders, Marilyn Kowalchuk, Howard Knapp, Debra J. Szeluga, John Bagley, George L. Blackburn[1]

Nutrition/Metabolism Laboratory, Cancer Research Institute, New England Deaconess Hospital, Harvard Medical School, Boston, Mass., USA

Introduction

In recent years, the nutritional aspects of cancer causation have become one of the most intriguing and extensively studied environmental factors in cancer research. The priority to improve anti-neoplastic therapy in such cancers as the breast, colon, and prostate by means of dietary manipulation, has increased as the response to adjuvant chemotherapy, and immunotherapy has remained limited [1].

Epidemiologic evidence for a nutritional influence has been substantiated by widespread reports that high levels of dietary fat stimulate tumor development in many experimental cancer systems [2–4]. Animal model studies are highly consistent with a fat hypothesis, and evidence has indicated that the primary effect of fat is exerted on the promotion stage of mammary carcinogenesis.

At present, the mechanism of action of lipid in cancer promotion remains unclear. Manipulation of dietary lipid may alter both the physical and biologic properties of cell membranes, including membrane fluidity [5], receptor binding sites, modulation of intracellular hormone action [6], and immunoresponsiveness to mitogens [7]. Any of these alterations could affect signal transduction by growth factors, hormones, and nutrient transport to the tumor from the host. Furthermore, alterations in membrane structure and phospholipid turnover by dietary lipids can promote the activity of a membrane-bound enzyme, protein kinase C. Activation of this enzyme has been implicated in tumor promotion [8], probably through its

[1] The authors wish to acknowledge Michelle Kienholz for her assistance and Tracey Long for typing the manuscript.

role in transmembrane signalling and through its association with other hormones and a number of growth factors and oncogenes.

The strongest evidence indicates that prostanoids, e.g., prostaglandin (PG), thromboxane (TX), and prostacyclin (PGI) of the 2 series produced from the oxygenation of arachidonic acid (AA; 20:4ω6), an ω6 polyunsaturated fatty acid (ω6 PUFA) may indirectly promote the growth of cancer. Both animal and human cancers of the breast have been reported to produce large quantities of prostaglandins [9, 10]. Prostaglandin E_2 has been associated with the inhibition of various aspects of immune response, including the suppression of cytokines such as interleukin 1 (IL-1) [11]; attenuation of T-cell proliferation and function [12]; and inhibition of macrophage- and natural killer cell-mediated cytotoxicity toward malignant cells [13]. These reports indicate that excess production of PGE_2 from dietary lipid rich in ω6 PUFAs may amplify the level of immunosuppression, and indirectly promote tumor growth.

Fish Oil and Cancer

Evidence that ω6 PUFAs and the prostaglandins derived from them can affect immune function and promote cancer has stimulated interest in studying the effect of fish oil on tumor growth. Fish oil contains mainly ω3 PUFAs, such as eicosapentaenoic acid (EPA; 20:5ω3) and docosahexaenoic acid (DHA; 22:6ω3) that can partially antagonize the overproduction of eicosanoids derived from ω6 PUFA by competing as a substrate for cyclooxygenase [14, 15]. At the same time, ω3 PUFA produces the alternative 3-series and 5-series eicosanoids which are less inflammatory [15].

Several laboratory reports have confirmed the potential role of ω3 PUFAs in reducing tumor promotion. Animal models such as the R3230 AC mammary tumor in rats [16], mammary adenocarcinomas in mice [17], and colon cancer in rats [18] all appear to support the hypothesis that a menhaden fish oil-supplemented diet significantly reduces tumor incidence and growth. Reduction in tumor promotion has been associated with a marked inhibition of the prostaglandin E_2 production together with a reduction in the amount of AA in the cell membrane.

Undoubtably, the modulatory action of ω3 PUFAs on macrophage function suggests promise in their use during cancer treatment. Macrophage activation in response to tumor cells often results in the production of degradative enzymes and secretion of a number of biologically active

Table 1. Effects of long-term ω3 PUFAs in breast cancer

	Fish	Beef	Veg	Low fat
Weight at inoculation, g	103.5	102.9	106.1	108.8
Median survival, days (95% confidence interval)	52 (0–54)	34* (0–44)	42* (0–44)	43 (0–50)
Primary tumor weight at death, g (% body weight)	21.3 (16.8)	33.3 (27.9)	26.1 (20.2)	35.9 (24.7)
Metastasis weight at death, g (% body weight)	1.8 (1.4)	1.6 (1.3)	2.6 (2.0)	0.9 (0.6)
Non-tumor body weight at death, g	106.7	84.8	99.0	106.6

* $p < 0.03$, generalized Wilcoxon, Log-Rank test compared with Fish.

products, such as interleukin-1, tumor necrosis factor, interferons, and other growth factors. These biologic products are in turn responsible for amplifying the host immune response. In a dose-dependent fashion, diets high in ω3 PUFAs were able to affect the fatty acid composition in the phospholipids of immune cell populations including splenocytes, thymocytes, and lymphocytes [19, 20]; and to reduce macrophage production of PGE_2 in animals with mammary [16] and colon cancer [18]. Modulation of immune cell function by alteration of eicosanoid metabolism (i.e by reducing PGE_2 production) with ω3 fatty acids may indirectly inhibit tumor promotion via immunomodulation.

To determine the effects of ω3 PUFAs on breast cancer promotion, we have chosen the 13762 MAT (mammary ascites tumor) tumor model in rat, and have evaluated the long-term and short-term effects of three high-fat (20% by weight) diets: ω3 PUFAs (fish oil); ω6 PUFAs (vegetable oil); saturated fat (beef tallow), and one low-fat (5% by weight corn oil) control diet on tumor growth. This tumor model bears characteristics similar to those of human cancer. It is a transplantable, metastatic adenocarcinoma that is hormone-sensitive. Our preliminary results support the hypothesis that ω3 PUFAs can prolong survival and slow tumor growth in animals with transplantable, metastatic breast cancer. Long-term treatment (started prior to tumor implantation) (table 1) improves survival and slows tumor growth better than short-term treatment (started at the time of tumor implantation) (table 2).

Table 2. Effects of long-term ω3 PUFAs on tumor growth after excision

	Fish	Beef	Veg	Low fat
Weight at inoculation, g	108.8	107.1	108.6	106.1
Primary tumor weight at excision, g	4.1	6.9	6.5	4.8
(% body weight)*	(3.2)	(5.5)	(5.0)	(3.3)
Metastasis weight at death, g	0.1	0.8	0.1	0.0
(% body weight)	(0.1)	(0.8)	(0.1)	(0.0)
Non-tumor body weight at death, g*	139.4	118.6	127.8	168.1

* $p < 0.05$, 1-way ANOVA.

By using a ^{14}C-leucine tracer, we have further characterized the mechanisms of ω3 PUFAs action on the growth of the 13762 mammary tumor by measuring the rates of protein synthesis and breakdown. Estimates from leucine kinetics indicate that chronic feeding of a diet enriched with ω3 PUFAs can slow tumor growth through modulation of both tumor protein synthesis and breakdown [21]. Our findings suggest that modifications of the ω3/ω6 PUFA ratio by diets high in ω3 PUFAs may have a beneficial role as adjunctive anti-neoplastic therapy for breast cancer.

Clinical Trial:
Effect of High-Fat Diets on ω3 Fatty Acids on Blood Lipids in Human

Studies in humans have attempted to determine the effect of fish oil therapy on changes in phospholipid fatty acid composition. However, only a few studies have controlled for the amount of dietary fat during fish oil supplementation in free-living subjects. To date, no study has addressed the possible inhibitory effects of high-fat diets on the incorporation rate of ω3 fatty acids into cell membranes. Since competition exists between ω3 PUFAs and ω6 PUFAs, a reduction in dietary fat and specifically in ω6 fatty acids may allow for increased incorporation of ω6 PUFAs into membrane. In most human studies, fish oil supplements have been given as part of the usual American diet, which is high in fat (35–40% of total calories). This high dietary fat consumption may alter the competition between AA and EPA for cyclooxygenase, thus decreasing the ratio of the bioactive prostacyclins to thromboxanes.

Table 3. Baseline demographics, mean ± SD

Diet	n	Age	% IBW	Weight, kg	BMI, kg/m^2
High fat	12	33±6	102±9	74±9	24±2
Low fat	12	32±4	105±6	82±7*	24±1

* $p < 0.05$.

To test the effect of different levels of dietary fat on the incorporation of ω3 fatty acid into cell membranes, we have compared a diet low in fat (20% of total calories) supplemented with 8 g/day of EPA to the typical American diet (35% fat) also supplemented with 8 g/day of EPA. The special low-fat diet has been formulated to allow for an ω3/ω6 ratio of 1.5:2.0.

Twenty-four young (20–50 years) healthy, normal-weight (<111% IBW) males were selected for study. Following a two-week screening, which included a history and physical and biochemical measurements (chemistry profile, lipid panel, bleeding time, serum thromboxane, urinary prostaglandins, and three-day food record), subjects were placed on a run-in diet that provided 35% calories from fat. Food diaries were validated against semi-quantitative concentrations of serum fatty acid fractions.

After the two-week run-in phase, subjects were randomly assigned to either the low-fat (20% of total calories from fat) or high-fat (35% of total calories from fat) dietary regime for a four-week active intervention period. Supplementation consisted of 10–12 EPA capsules per day to provide approximately 7.5 g of ω3 fatty acids per day. Subjects were seen on weeks 1, 3, and 4 to document dietary and supplement compliance and to dispense the supplement. Data were collected at week 4 and compared to run-in values. The data from the biochemical measurements and nutrient intake were analyzed by use of non-parametric t test.

Baseline demographics (table 3) shows that all groups were well balanced at baseline with respect to age, percent of ideal body weight (IBW), and body mass index (BMI); baseline weight was significantly higher in the low-fat group.

Preliminary results of the dietary intake data, lipoprotein analysis, and bleeding time are presented in tables 4–6. Dietary intake was comparable at baseline between the two groups (table 4). Subjects assigned to the high-fat diet had a run-in fat intake of 111 g (39% of total calories), which

Table 4. Daily macronutrient intake, mean ± SD

	High fat		Low fat	
	run-in	week 4	run-in	week 4
Calories	2641±471	2289±604	2870±656	1934±587**
Protein, %	17±3	17±3	16±3	20±5*
Carbohydrate, %	44±7	43±5	47±7	60±9**
Fat, %	39±6	40±5	36±7	20±5**
Fat, g	111±15	112±20	114±25	42±12*
Polyunsaturated, g	25±7	22±7	24±10	10±2**
Monounsaturated, g	40±5	41±7	40±8	14±4**
Saturated, g	37±6	41±11	40±12	14±6**

* $p < 0.05$ vs. own run-in. ** $p < 0.0001$ vs. own run-in.

Table 5. Change in lipoproteins, mean ± SD

	High fat		Low fat	
	run-in	week 4	run-in	week 4
Cholesterol mg/dl	188±44	183±38	187±48	169±39*
LDL, mg/dl	122±42	123±37	120±47	108±34
HDL, mg/dl	50±10	50±13	48±6	46±8
Triglycerides, mg/dl	86±38	61±33*	83±23	77±50
Ratio, total:HDL	3.9±1	40±2	4.0±1	3.8±1

* $p < 0.05$ vs. own run-in.

Table 6. Change in bleeding time, mean ± SD

	High fat		Low fat	
	run-in	week 4	run-in	week 4
Bleeding time, s	5±1	4.5±1	4±2	5.5±2

remained at 112 g (40% of total calories) at week 4. Subjects assigned to the low-fat group had a run-in fat intake of 114 g (36% of total calories), which was reduced to 42 g (20% of total calories) at week 4. The low-fat group had a significant reduction in total fat intake.

Compliance to fish oil supplement (as measured by pill count) was 100% and 98% for the high- and low-fat groups, respectively. There were no significant differences between the two treatment groups at run-in for any of the lipoprotein measures presented (table 5). In the high-fat group, only triglycerides were significantly decreased during the intervention; all lipoprotein measures remained unchanged throughout the intervention. The low-fat group, however, experienced significant reductions in cholesterol but not low-density lipoproteins (LDL), triglycerides, and the total cholesterol to high-density lipoprotein (HDL) ratio. There was no significant change in bleeding time by group assignment (table 6). Still pending are the results from the analysis of fatty acids, serum thromboxanes, and urinary prostaglandins.

The results of our preliminary study show that subjects demonstrated good compliance to the dietary interventions and the supplement regimen. As expected, subjects randomized to low-fat group experienced significant reductions in dietary fat intake and in serum cholesterol.

Can Fish Oil Attenuate Metastasis?

Tumor growth and spread presuppose a sequence of events involving unique tumor-host interactions. For example, continued growth of neoplastic tissue requires a milieu in which immune mechanisms, which would normally be extended to inhibit tumor growth or spread, be suppressed. Subtle alterations in host-tumor balances may affect those processes that are responsible for recognition and destruction of tumor by host immune cells, as well as the survival and growth of shed tumor cells.

Once a tumor cell is in circulation, it can undergo a variety of cell-cell interactions. Some of these interactions may be deleterious to the tumor cell (i.e., interactions with macrophages). One type of cell-cell interaction that facilitates metastasis is the interaction between the circulating tumor cell and the host platelet. For the tumor cell to metastasize, it must penetrate the basement membrane of the host organ. The interaction of tumor cells with host platelets directly helps them to adhere on endothelial basement membrane. The adherence of platelets and tumor cells cause second local proteolysis associated with breakdown of the basement membrane components. This allows migration and locomotion of the tumor cell through the defect in the extracellular matrix into other organs [22] and continue to proliferate.

The ability of tumor cells to activate platelets is well described [23, 24]. Both human and animal tumor cells have been shown to induce platelet aggregation [25, 26]. Agents known to modify platelet aggregation such as aspirin have been tested as anti-metastatic agents [27]. Honn [28] and Honn et al. [29] have proposed that tumor cells can alter the critical thromboxane A_2/prostacyclin I_2 (TXA_2/PGI_2) balance to induce platelet activation, which in turn favors metastasis. Honn et al. [30, 31] have demonstrated that intravenous injection of PGI_2 into mice reduces lung colony formation of B16a tumor cells by greater than 70%, and the use of PGI_2 in combination with a phosphodiesterase inhibitor (theophylline) reduces metastasis effectively. TXA_2 produced in the platelets is a vasoconstrictor and promotes platelet aggregation, whereas PGI_2 produced in the endothelial cells of arteries and veins retards and reverses platelet aggregation. This evidence points to the significance of maintaining a well-adjusted ratio of TAX_2/PGI_2 to prevent tumor spreading.

In the clinical setting, prostacyclin has been considered as a desirable endogenous platelet inhibitor in which platelet-vascular interactions are increased. The accumulation of platelets on the surface of the injured vessel may well be prevented by a continuous operative infusion of prostacyclin analogs. Unfortunately, use of prostacyclin analogs in this way has been associated with dose-dependent side effects (restlessness, jaw pain, flushing, nausea, headaches, and hypotension) [32]. Furthermore, prostacyclin can only be administered parenterally and therefore is not suited for long-term anti-metastatic therapy.

The therapeutic effects of systemic infusion of prostacyclin analogs may be achieved through low doses of these drugs when given during a diet high in ω3 PUFAs, which lower platelet activity, thereby reducing drug-related toxicity. The beneficial anti-thrombotic effect of fish oil comes from the EPA component that is converted to a PGI_2-like compound (PGI_3), which has anti-platelet aggregatory activity, whereas the corresponding TXA_2-like compound (TXA_3) produced from EPA is inactive [33–35]. The net result should be strongly anti-aggregatory and vasodilatory and may attenuate tumor metastasis.

PGE_2 derived from ω6 PUFAs has been implicated as a component that promotes tumor growth activity. Its production by both host tissues and tumor may in turn amplify the level of immunosuppression by mediating the activation and function of suppressor T-lymphocytes, and hence limiting the activation of macrophages by feedback inhibition [19, 20]. In the presence of vast amounts of prostaglandins, the immune system would become 'exhausted' and eventually lose its capability to control tumor spread (metastasis). This hypothesis also provides the rationale for adjuvant therapies directed at modulating host prostaglandin synthesis by diets high in ω3 PUFAs so as to prevent or to delay metastases.

Conclusion

The current observations provide the rationale for adjuvant therapies directed at modulating host prostaglandins synthesis to prevent tumor promotion and metastasis. Further research should focus on the competition between ω3 PUFAs (fish oil) and ω6 PUFAs (vegetable oil) to favor the synthesis of TXA_3 over that of the potent platelet agonist, TXA_2. Modulation of prostaglandin metabolism by dietary manipulation is possible and has been demonstrated in free-living subjects with supplementation of fish oil capsules.

Knapp et al. [36] showed that supplementing the usual diet of the free-living subjects with 50 ml MaxEPA fish oil (10 g/day) for 1 week was capable of prolonging bleeding time by 25–30% and of reducing plasma beta-thromboglobulin by 50%. These workers demonstrated that 4-week supplementation of fish oil capsules was able to reduce serum TXB_2, the excretion of TXB_2 metabolite, and of PGI_2 metabolite. While the excretion of PGI_3 metabolite was increased dramatically, the TXB_3 metabolite was increased only slightly.

Since competition exists between ω3 and ω6 fatty acids, reduction in dietary fat may allow for an increase in the incorporation of ω3 fatty acids with low-dose supplementations. The proposed low-fat, high-ω3 fatty acid diet by us may modulate the precursor for thromboxanes and prostaglandins. As it relates to tumor promotion, the diet may alter the rate of tumor cell attachment in lymph nodes and in lung and liver tissue via immunomodulation and antiplatelet aggregation. These effects of EPA as part of a practical, low-fat diet in free-living individuals may prove beneficial for altering several causes of mortality, such as metastasis, atherosclerosis, chronic inflammatory diseases, and other diet-related (immunosuppressive) diseases.

References

1 Bonadona G, Valagussa P: Adjuvant system therapy for resectable breast cancer. J Clin Oncol 1985;3:259–275.

2 Carroll KK: Lipid and carcinogenesis. J Environ Pathol Toxicol 1980;3:253–271.

3 Chan PC, Dao TL: Enhancement of mammary carcinogenesis by a high-fat diet in Fischer, Long-Evans, and Sprague-Dawley Rats. Cancer Res 1981;41:164–167.

4 Hopkins GJ, West CE, Hard GC: Effect of dietary fats on the incidence of 7,12-dimethylbenz(a)anthracene-induced tumors in rats. Lipids 1976;11:328–333.

5 McMurchie E, Raison JK: Membrane lipid fluidity and its effect on the activation energy of membrane-associated enzymes. Biochim Biophys Acta 1979;554:364–374.

6 Rillema JA: Action and metabolism of intracellular mediators in neoplastic mammary cells; in McGuire WL (ed): Breast Cancer. New York, Plenum Press, 1979, vol 3, pp 117–147.

7 Kollmorgen GM, Sansing WA, Lehman AA, et al: Inhibition of lymphocyte function in rats fed high fat diet. Cancer Res 1979;39:3458–3462.

8 Gullem JG, O'Brian CA, Fitzer CJ, et al: Altered levels of protein kinase C and Ca^{2+}-dependent protein kinases in human colon carcinomas. Cancer Res 1987;47: 2036–2039.

9 Karmali RA, Welts A, Thaler HT, et al: Prostaglandins in breast cancer: Relationship to disease stage and hormone status. Br J Cancer 1983;48:649–696.

10 Kollmorgen GM, King MM, Kosanke SD: Influence of dietary fat and indomethacin on the growth of transplantable mammary tumors in rats. Cancer Res 1983;43: 4714–4719.

11 Knudsen PJ, Dinarello CA, Strom TB: Prostaglandins post-transcriptionally inhibit monocyte expression of interleukin-1 activity by increaseing intracellular cyclic adenosine monophosphate. J Immunol 1986;137:3189–3194.

12 Goodwin JS, Webb DR: Regulation of the immune response by prostaglandins. Clin Immunol Immunopathol 1980;15:106–122.

13 Plescia OJ, Pontieri GM, Brown J, et al: Amplification by macrophages of prostaglandin-mediated immunosuppression in mice bearing synthetic tumors. Prostaglandins Leukotrienes Med 1984;16:205–223.

14 Lee TH, Hoover RL, Williams JD, et al: Effect of dietary enrichment with eicosapentaenoic acids on in vitro neutrophil acid monocyte leukotriene generation and neutrophil function. N Eng J Med 1985;312:1217–1224.

15 Goldman DW, Pickett WK, Goetzl EJ: Human neutrophil chemotactic and degranulating activities of leukotriene B_5 derived from eicosapentaenoic acid. Biochem Biophys Res Commun 1983;197:282–288.

16 Karmali RA, Marsh J, Fushs C: Effect of omega-3 fatty acids on growth of a rat mammary tumor. J Natl Cancer Inst 1984;73:457–461.

17 Abraham S, Hillyard C, Hillyard LA: Effect of dietary 18-carbon fatty acids on growth of transplantable mammary adenocarcinomas in mice. J Natl Cancer Inst 1983;71:601–605.

18 Jurkowski JJ, Cave WJ Jr: Dietary effects of menhaden oil on the growth and membrane lipid composition of rat mammary tumors. J Natl Cancer Inst 1985;74:1145–1150.

19 Marshall LA, Johnston PV: Effect of dietary alpha-linoleic acid in the rat on fatty acid profiles of immunocompetent cell populations. Lipids 1983;18:737–742.
20 Marshall LA, Johnston PV: The influence of dietary essential acids on rat immunocompetent cell prostaglandin synthesis and mitogen-induced blastogenesis. J Nutr 1985;115:1572–1580.
21 Wan JMF, Chu CC, Istfan NW, et al: Effect of omega-3 PUFAs on protein kinetics of a rat mammary tumor (Abstract). FASEB 1988;A350:21.
22 Liotta LA: Tumor invasion: Role of the extracellular matrix. Cancer Res 1986;46: 1–7.
23 Gasic GJ, Boettiger D, Catalfamo JL, et al: Aggregation of platelets and cell membrane vesiculation by rat cells transformed in vitro by Rous sarcoma virus. Cancer Res 1978;38:2950–2955.
24 Cavanaugh PG, Sloane BF, Bajkowski AS, et al: Purification and characterization of platelet aggregating induced by tumor cells and their shed membrane vesicles. Clin Exp Metastasis 1985;1:297–307.
25 Laghi F, DiRoberto PF, Panici PB, et al: Coagulation disorders in patients with tumors of the uterus. Tumor 1983;69:349–353.
26 Zacharski LR, Rickles FR, Henderson WG, et al: Platelets and malignancy. Rationale and experimental design for the VA Cooperative Study of RA-233 in the treatment of cancer. Am J Clin Oncol 1982;5:593–609.
27 Honn KV, Onoda JM, Menter JD, et al: Prostacyclin/thromboxanes and tumor cell metastasis; in Honn KV, Sloane BF (eds) Hemostatic Mechanisms and Metastasis. The Hague, Martinus Nijhoff, 1985, pp 207–231.
28 Honn KV: Inhibition of tumor cell metastasis by modulation of the vascular prostacyclin/thromboxane A_2 system. Clin Exp Metastasis 1983;1:103–114.
29 Honn KV, Busse WD, Slone BF: Prostacyclin and thromboxanes. Implications for their role in tumor cell metastasis. Biochem Pharmacol 1983;32:1–11.
30 Honn KV, Cicone B, Skoff A: Prostacyclin: A potent antimetastatic agent. Science 1981;212:1270–1272.
31 Honn KV, Cicone B, Skoff A, et al: Thromboxane synthetase inhibitors and prostaglandin can control tumor cell metastasis. J Cell Biol 1980;87:64–68.
32 Block PC: Percutaneous transluminal coronary angioplasty; in Connor WE, Bristow JP (eds): Coronary Heart Disease: Prevention, Complications, and Treatment. Philadelphia, Lippincott, 1985, pp 405–418.
33 Ahmed AA, Holub BJ: Alteration and recovery of bleeding times, platelet aggregation and fatty acid composition of individual phospholipids in platelets of human subjects receiving a supplement of cod-liver oil. Lipids 1984;19:617–624.
34 Dyerberg J, Bang HO, Staffersen E, et al: Eicosapentaenoic acid and prevention of thrombosis and atherosclerosis? Lancet 1978;ii:117–119.
35 Fischer S, Weber PC: Prostaglandin I_3 is formed in vivo in man after dietary eicosapentaenoic acid. Nature 1984;307:165–168.
36 Knapp HR, Reilly IAG, Alessandrini P, et al: In vivo indices of platelet and vascular function during fish-oil administration in patients with atherosclerosis. N Engl J Med 1986;314:937–942.

Jennifer Man-Fan Wan, PhD, Cancer Research Institute,
New England Deaconess Hospital, 194 Pilgrim Road, Boston, MA 02215 (USA)

Simopoulos AP, Kifer RR, Martin RE, Barlow SM (eds): Health Effects of ω3 Polyunsaturated Fatty Acids in Seafoods. World Rev Nutr Diet. Basel, Karger, 1991, vol 66, pp 488–503

Modulation of Breast Cancer Growth in Nude Mice by ω3 Lipids[1]

G. Fernandes, J.T. Venkatraman

Department of Medicine, The University of Texas Health Science Center at San Antonio, San Antonio, Tex., USA

Introduction

In recent years, diet and nutritional factors are considered to be of major importance in the etiology of human cancer [1–3]. Epidemiological studies have suggested a close association between the intake of fat and incidence of breast cancer [4–6]. Excessive fat in the diet has been reported to enhance the growth of chemically induced colon and mammary tumors, as well as transplantable carcinomas [7–9]. The effect of various types of dietary fat on mammary carcinogenesis is an area of intense interest. Several studies have shown that high dietary fat intake, particularly vegetable fat (polyunsaturated), is related to the increased incidence of breast, prostate and colon cancer [10, 11]. On the other hand, diets based on marine oils containing high levels of ω3 fatty acids appear to have beneficial effects against several types of malignant tumors when compared to diets containing vegetable oils [12, 13]. The ω3 fatty acids of marine origin primarily differ from arachidonic acid (ω6) levels and inhibit both cyclooxygenase and lipoxygenase pathways which may be playing a crucial role in modulating the induction and proliferation of tumor cells by altering both immune function and oncogene expression as well as have a

[1] This research was supported in part by grants from NIH AG-83417, The Kleberg Foundation and the National Dairy Board. We would like to thank Dr. Kent Osborne for providing tumor cell lines and Mr. Vikram Tomar for his technical assistance.

direct action on autocrine growth factors. Several earlier beneficial effects of ω3 fatty acids mostly noted in cardiovascular disease have been closely linked to the changes in plasma lipids, blood pressure and platelet functions. However, since ω3 fatty acids are found to act effectively on inflammatory cells, several experimental studies are now in progress to compare ω3 lipid action on induction and proliferation of malignant cells sensitive to eicosanoid metabolism.

ω3 Fatty Acids and Cancer Protection

Although for many years quality and quantity of dietary lipids were known to modulate tumor incidence in rodents [5, 7], only recent experimental evidence has suggested that ω3 lipid diet protects against the development of mammary tumors and autoimmune disease in animals [14, 15]. Further, even though several suggestive reports alluded to the possible effects of fish or fish oil intake on lowering cancer incidence in selected North American populations, yet there has been minimal epidemiological investigation to link any possible protective role of fish consumption or fish oil intake in the development of breast cancer incidence in general population. However, Kaizer et al. [16] recently compared breast cancer incidence and mortality rates with approximate estimates of fish consumption in humans and reported an inverse association between percent calories consumed from fish and the incidence of breast cancer which was found to correlate with a protective effect of fish consumption. Their results were consistent with recent experimental animal data suggesting that ω3 fatty acids obtained from fish may play a protective role against breast cancer. In experimental animals, the effects of fish oils have been investigated for mammary, colon, lung, pancreatic, prostatic and fibrosarcoma tumor growth in animals. In 1984, Karmali et al. [17] were the first to report a significant reduction in the size of transplantable R3230 AC mammary tumor in female Fischer-344 rats when they supplemented diet with high levels of ω3 fatty acids and their subsequent experiments further suggested the importance of optimal ω3/ω6 ratio in reduction of tumor growth. Braden and Carroll [18] have reported high levels of ω3 (10% or 20%) to markedly inhibit tumorigenesis. Other studies [19] have indicated moderate inhibitory effect of fish oil on DMBA-induced mammary tumors. Fish oil (20%) has been reported to lengthen tumor latent period and reduce tumor incidence in NMU-induced mammary tumors, azoxyme-

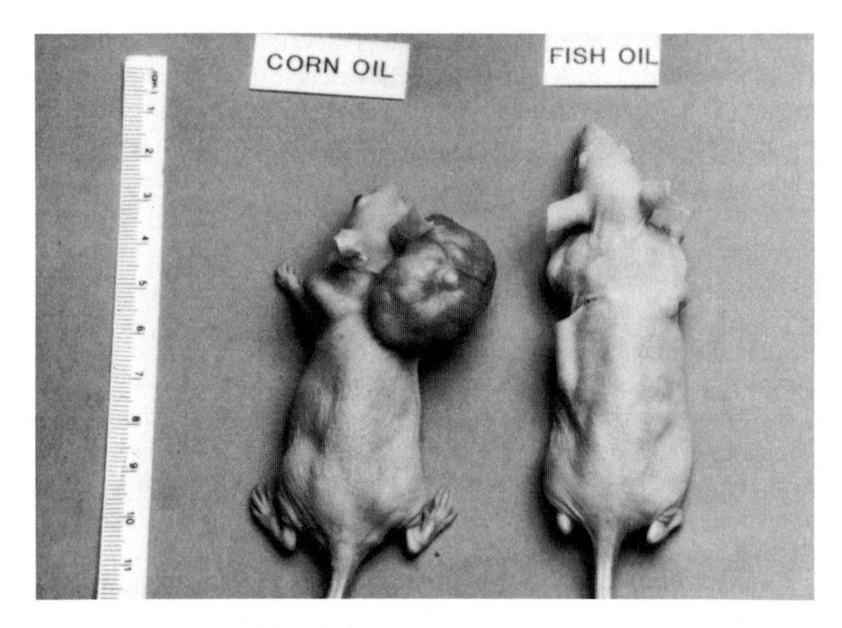

Fig. 1. Effect of dietary lipids on growth of human breast cancer cells in nude mice. Nude mouse on corn oil diet exhibited large tumor compared to the fish oil fed mouse.

thane-induced colon cancers, pancreatic preneoplastic lesions as well as transplantable mammary adenocarcinoma compared to other lipid sources [20, 21]. In addition, antimetastatic effects of fish oil has also been observed by us [22] and other investigators indicating the need to study the role of ω3 for possible eventual therapeutic usage in humans.

ω3 Lipid and MCF-7 Cells

We have recently begun to study the effects of various dietary lipids on the growth and metastasis of well-characterized estrogen receptor-positive (MCF-7) human breast cancer cells in nude mouse model. The immune-deficient nude mouse model has been well accepted as a useful animal model system by several investigators to grow and for examining cell kinetics of human-origin tumor cells, particularly breast carcinoma cells [23, 24]. Estrogen-dependent (MCF-7) as well as estrogen-independent (MDA-MB 231) cell lines have been studied extensively to determine the role of

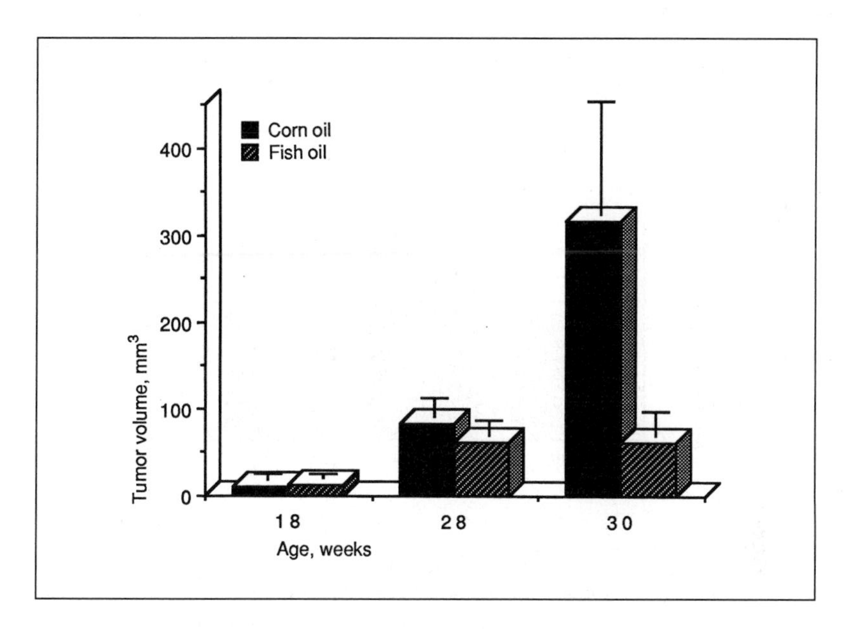

Fig. 2. Influence of dietary lipids on the volume of breast tumor in nude mice. Tumor growth was faster in mice fed corn oil compared to the fish oil-fed group.

growth factors and estrogen and anti-estrogen therapy on growth rate of tumor cells [25, 26].

To examine the role of dietary lipids, we fed 4-week-old female athymic nude mice (Balb/c-nu/nu) semipurified diet containing 20% corn oil or 20% fish oil (menhaden) for 4 weeks (other lipids, such as butterfat and Crisco, were also studied but not included in this paper). MCF-7 (estrogen-dependent) cells, cultured and maintained in vitro, were injected subcutaneously (5×10^6 cells) to each mouse. Both body weight and tumor growth were monitored weekly until the termination of the study.

The MCF-7 tumor take in nude mice fed diets containing corn oil was 17/24 (71%) compared to fish oil-fed group which had much less tumor take – 5/22 (23%). In these studies, we supplemented equal levels of antioxidants (vitamin E and TBHQ) both for corn oil and fish oil diets. The rate of tumor growth and/or volume was higher in animals fed the corn oil diet compared to those fed the fish oil diet (fig. 1, 2). Visible pulmonary metastasis nodules were higher only in corn oil-fed animals injected with MDA-MB231 cells and less in mice fed fish oil diet (fig. 3).

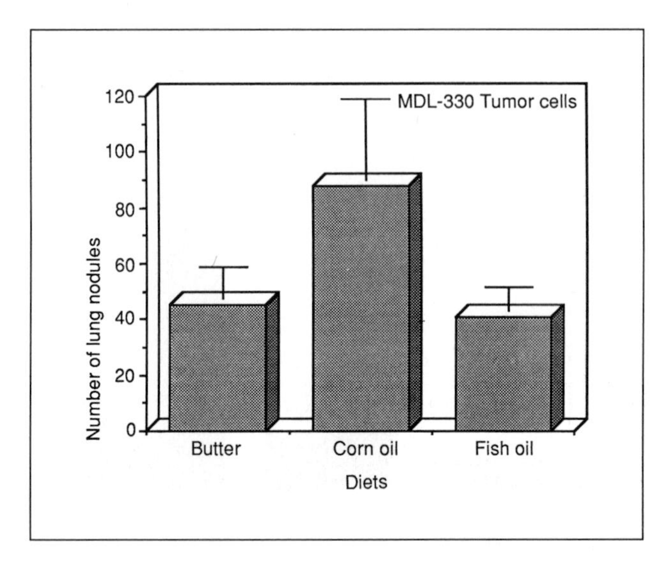

Fig. 3. Effect of dietary lipids on metastasis of MDL tumor to lungs in nude mice. Tumor metastasis was higher in mice fed corn oil compared to butterfat or fish oil.

Table 1. Serum hormones and tissue PGE_2 levels

	Corn oil	Fish oil
Serum hormones		
Estrogen, pg/ml	378.00 ± 32.0	245.00 ± 17.00
Prolactin, ng/ml	11.60 ± 1.8	7.30 ± 2.10
Tissue PGE_2 levels, ng/mg/30 min at 37 °C		
Spleen	1.92 ± 0.23	0.58 ± 0.14
Tumor	0.22 ± 0.07	0.14 ± 0.0

Effect of Dietary Fat on Hormonal Levels

To determine the changes in hormonal levels, we measured serum estrogen, prolactin levels, and PGE_2 levels in the tissue by using radioimmunoassays. Results are shown in table 1.

The mice fed corn oil had produced higher levels of estrogen and prolactin in the serum compared to the group fed dietary fish oil suggesting that changes in serum hormone levels may have an important role in reg-

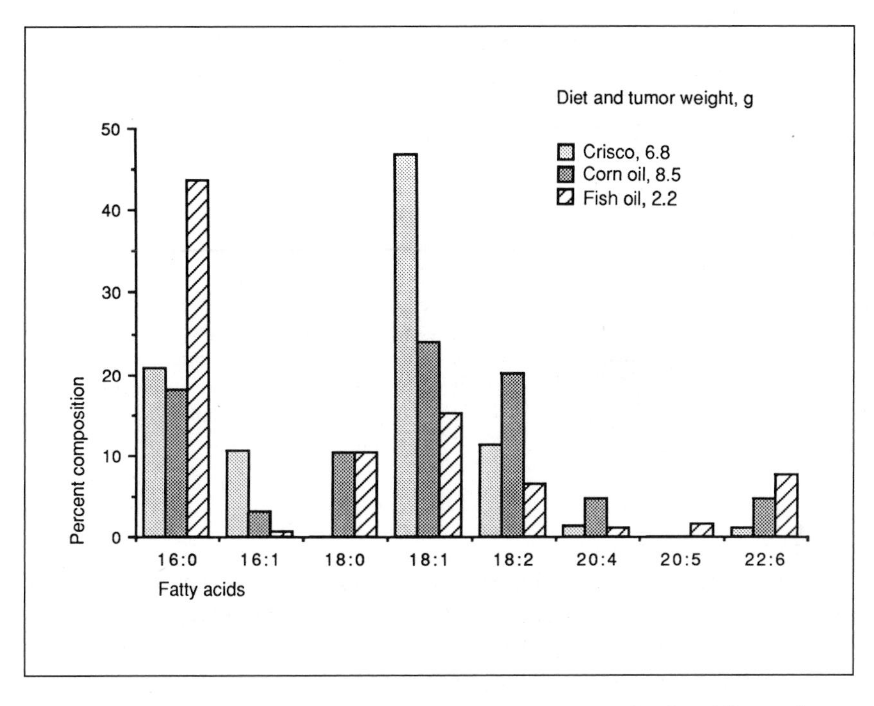

Fig. 4. Fatty acid composition of MCF-7 tumors in nude mice fed different dietary lipids. The fatty acid composition of the dietary lipid was reflected in the tumor tissue.

ulating the growth rate of transplanted tumor cells. Dietary lipids also had a striking effect on the tissue prostaglandin E_2 (PGE_2) levels in nude mice. For instance, the level of PGE_2 production was found considerably higher in corn oil-fed animals as compared to the group fed fish oil diet both in tumor and spleen tissues.

Fatty Acids

As dietary fats are known to alter tissue fatty acid composition and thereby alter PGE_2 production, we decided to analyze the fatty acid composition to relate changes in PGE_2 levels in the tumor tissue. The fatty acid composition of the dietary lipids was closely related to changes in the fatty acid composition of tumor and spleen tissue of each dietary group. As expected, tumors from animals fed corn oil had higher incorporation of 18:2 and 20:4 while fish oil-fed animals had higher levels of ω3 fatty acids (20:5 and 22:6) in their tumors (fig. 4).

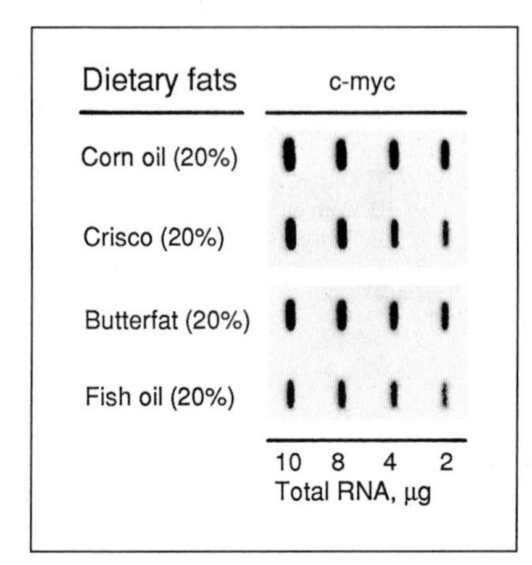

Fig. 5. Effect of dietary lipids on the expression of c-myc oncogene in MCF-7 tumors grown in nude mice. c-myc expression in the tumor tissue was higher in animals fed corn oil compared to those fed fish oil.

Oncogene Expression

Experiments were carried out to measure for changes in oncogene expression in tumor tissues obtained from animals fed various fats. c-myc mRNA expression was found higher in the corn oil-fed group (also compared to the group fed Crisco and butterfat) and was low in tumors from fish oil-fed mice (fig. 5).

When the SDS-PAGE-separated polypeptides were electrophoretically transferred on to immobilon-P and immunoblotted with P21 antibodies, the level of P21 proteins was higher in tumor lysate from corn oil-fed animals compared to the fish oil group (fig. 6) (human melanoma tumors grown in nude mice were also compared to fish oil-fed groups), indicating that increased Ras-P21 protein level is closely linked to the rapid growth rate of tumor cells which can be modulated by feeding a diet containing ω3 lipids.

Dietary Lipids and Breast Cancer Growth: Possible Mechanisms

For a number of years, inhibition of breast tumor growth was achieved by restricting the calorie intake whereas diets containing higher levels of dietary lipids consistently developed tumors earlier. The incidence was always higher in genetically susceptible strains of female mice. A number of mechanisms have been proposed to explain the modulation of mammary tumorigenesis in experimental animals by increased dietary fats [27]. For example, immune suppression [28], prostaglandin production [29, 30],

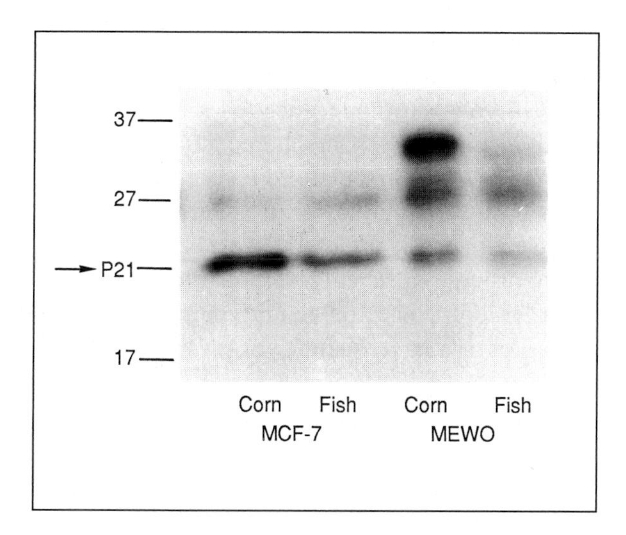

Fig. 6. Western blot showing the effects of dietary lipids on the ras-P21 protein levels in tumor tissues of nude mice. P21 protein expression was higher in mice fed corn oil compared to those fed fish oil.

free radical formation [31], membrane fluidity changes [32, 33], intracellular transport system modulation [34], increased caloric utilization [35], increased mammotrophic hormone secretion [36], overexpression of oncogenes and certain growth factors are some of the possible mechanisms through which mammary tumorigenesis may be modulated. Recently we briefly reviewed the literature on micronutrients and cancer and have emphasized the role of various dietary micronutrients, including the role of vitamins as antioxidants in preventing lipid peroxidation as well as rapid growth of mouse melanoma and its metastasis in the lungs of C57BL/6 mice [37] maintained on high polyunsaturated dietary fat. We have also previously reported the beneficial effects of calorie/fat restriction on maintenance of suppressor T-cell functions and mammary tumor incidence [39]. Further, caloric restriction was also found to decrease prolactin and anti-MMTV antibody levels and decrease mammary tumor virus particles in C3H/Bi mice [38, 39].

Recently, however, several investigators have also found that diets based on fish oil containing high levels of ω3 fatty acids, such as eicosapentanoic acid (EPA) and docosahexaenoic acid (DHA), do show beneficial effects against breast cancer compared to corn oil diets. These ω3 fatty acids differ from arachidonic acid (ω6) and inhibit both the cyclooxygenase and lipoxygenase enzyme systems. We, too, have observed decreased MCF-7 human tumor cell growth in nude mice fed a fish oil-containing diet indicating that ω3 lipids may have some therapeutic value in controlling tumor growth. In spite of several recognized favorable health benefits of fish oil, skepticism still prevails due to inconsistent clinical or experimental data resulting mostly from the use of undefined or poor-quality ω3 lipids, as well as failure to provide sufficient levels of antioxidant to fish oil and other test diets in order to prevent lipid oxidation in storage as well as of host tissue after ingestion of a large quantity of ω3 lipids.

Antioxidants, Free Radicals and Cancer

The relationship of antioxidants to tumorigenesis may be rationalized by noting that free radicals can initiate swift changes in DNA [40]. Inhibitory effects of antioxidants on tumorigenesis in animals has been extensively reported [41–44]. Vitamin E, butylated hydroxytoluene (BHT), vitamin C, and selenium are reported to have antitumorigenic effect [37, 41]. The incorporation of ω6 and ω3 fatty acids containing double bonds may alter membrane bilayer fluidity and could increase membrane lipid peroxidizability [33]. Therefore, antioxidant requirements for maintaining membrane integrity in immune cells may increase dramatically when fish oil diet is consumed, particularly to maintain adequate blood complement and clotting mechanisms, including prevention of excess free radical formation. We have noticed that when high levels of fish oil are fed in the absence of additional vitamin E supplementation, a significant decrease of vitamin E levels occurs in membranes and serum resulting in increased liver mitochondrial and microsomal lipid peroxidation in mice [45]. When vitamin E supplementation was inadequate, it lead to decreased natural killer cell activity and increased frequency of myocarditis lesions in mice injected with coxsackie virus [46] and caused early deaths, indicating a careful enrichment of vitamin E is essential to prevent any adverse effects to the host animals. Proper experimental and clinical trials are still required to study the effect of long-term intake of ω3 lipids on membrane-associated functions as well as immune functions.

Phospholipid Fatty Acid Composition of Membrane Lipids and Cancer

Cellular functions, including cytotoxic cell-mediated immune parameters, may be directly affected by modifications in membrane lipid composition. The modifications may alter physicochemical characteristics of membranes and thereby alter cellular functions, including carrier-mediated transport, properties of certain membrane-bound glycoproteins and/or enzymes, binding of hormones to receptors, phagocytosis, endocytosis, cytotoxicity of NK and Mφ, prostaglandin production, and finally cell growth and differentiation [32, 47]. Many functional responses are caused directly by membrane lipid structural changes which affect bulk lipid fluidity or specific lipid domains. Transporters, receptors and enzymes may be highly sensitive to changes in the structure of their lipid microenvironment, leading to changes in their activity and specifically production of various cytokines and PGs, which are known to act adversely on immune cells [48, 49]. Further, the changes in lipids and vitamin E levels may have direct effects on protein kinase C (PK-C) [50–52] and thereby may alter tumor cell proliferation and/or immune cell response to growth factors.

Endocrine System and Breast Cancer

Growth of human breast cancer cells in vitro is regulated by several growth factors, including estrogen, androgen, progesterone, glucocorticoids, insulin, insulin growth factors (IGF I and II), cathepsin-D, platelet-derived growth factor (PDGF), epidermal growth factors (EGF-1), prolactin and thyroid hormone [53–56]. About one third of human breast cancers are hormone-dependent and require estrogen for maximal growth. Therapy designed to reduce the estrogen concentration or to inhibit the effects of estrogen by competitive blockade of the estrogen receptor results in regression of these tumors [57, 58]. The mechanisms by which endocrine therapy inhibits tumor growth have not been completely defined. Others have proposed that either inhibition of pathways leading to increased cell proliferation and/or stimulation of pathways leading to cell loss could be involved [59].

Shafie and Grantham [60] found hormonal environment to be very important in the rate of solid MCF-7 tumor growth in nude mice. 17β-estradiol, insulin, and prolactin all have positive effects on breast tumor proliferation in nude mice. The multiplicity of growth stimulating hormones for breast cancer cell culture systems in vitro suggests the possibility that indirect mediators of estrogen effects may play important regulatory roles in vivo. Estrogen regulation of MCF-7 cells is associated with induction of TGF-α and IGF-1 and repression of TGF-β. Although the role of different lipids in modulating endocrine hormones is well recognized, well-defined dietary studies are still required to understand the effects of various lipids on modulating the interactions of hormones and growth factors with tumor cells of human origin. Our pilot studies do suggest that dietary lipids may alter sex steroid and pituitary-derived hormones. However, it is not yet clear how any one of the above or other possible mechanism(s) that is involved by feeding ω3 lipids in lowering the production of estrogen and prolactin levels as well as possibly other growth factors (EGF, TGF, IGF etc) secreted both externally or internally by tumor cells. New studies are required to measure the changes induced by ω3 fatty acids for several growth factors in different tumor cell lines which may indeed also alter the expression of oncogenes and/or growth factor receptor mRNA levels on plasma membranes.

Oncogenes and Breast Cancer

Human breast cancer cell lines have amplified or activated ras oncogene which appears to modify estrogen dependency of tumors for proliferation [61, 62]. In MCF-7 cells, the c-ras^H gene is amplified but not mutated [63]. Increased expression of c-*ras* protein [64], loss of a normal c-ras^H allele [65], and acquisition of abnormal c-*ras* alleles [66] appear to correlate with progression of breast cancer to more aggressive forms. A whole series of other cellular proto-oncogenes are expressed in tumor specimens [67]. The products of these oncogenes (members of the ras family as well as myc, myb, fms, fos, fes) include those localized in plasma membrane, nucleus, and cytoplasm. Two other oncogenes, c-erb B and neu, are both closely related to the EGF receptor and have also been detected in breast cancer cell lines and tumor biopsies [68, 69]. In our pilot studies we noted that both c-myc expression as well as c-*ras* protein level are reduced in MCF-7 tumors in fish oil-fed mice. Others have also noted less P21-ras protein in vitro in tumor cells cultured in the presence of ω3 fatty acids [70]. Extending these studies on more precise molecular aspects, particularly on plasma membrane growth factor receptor functions, may contribute to understanding the influences of various dietary lipids and hormonal action on tumor cell differentiation and proliferation both in in vivo and in vitro systems (fig. 7). The mechanism by which excess quantity and quality of dietary fat intake (both saturated and polyunsaturated fats) exerts its effects on tumor growth is still a highly complex issue and remains very much a controversial area of study. However, results obtained by dietary lipid therapy both in normal and in nude mice appear to indicate that ω3 lipid should be studied more extensively to evaluate its future possible role as a dietary supplement to reduce distant metastasis as

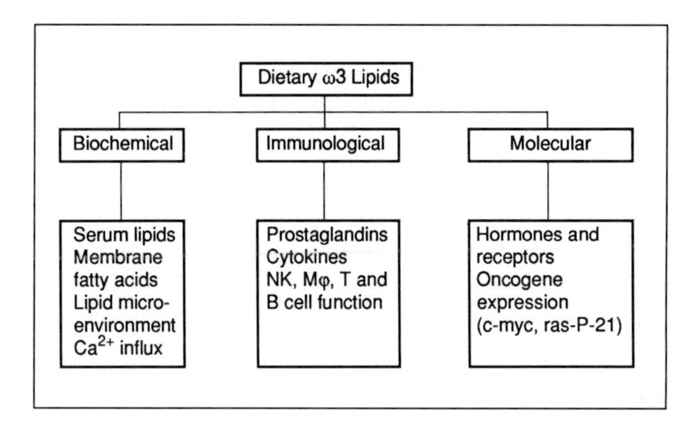

Fig. 7. Possible mechanisms of ω3 lipid action on modulating tumor growth.

well as to evaluate whether ω3 could also have a protective role in reducing drug toxicity as well. However, more detailed experimental studies are still required to study the effect of various commonly consumed dietary lipids at different levels in the diets against both estrogen receptor-negative and -positive tumor cell growth and metastasis before considering its usage as an adjunct therapeutic value for breast cancer patients in the coming years.

In summary, several recent observations have described the beneficial effects of dietary ω3 fatty acids in reducing the incidence of carcinogen-induced tumors in animal models. We in the past have observed reduced breast tumors in mice fed restricted amount of food. The present study was designed to compare the growth of MCF-7 estrogen receptor-positive human breast cancer-origin tumor cells in nude mice maintained on semi-purified diets containing 20% polyunsaturated (corn oil), and ω3 fatty acids (menhaden oil) levels. The results revealed that the rate of MCF-7 tumor cell growth and the tumor volume was found significantly higher in corn oil-fed mice and was significantly lower in fish oil-fed mice. Moreover, pulmonary metastasis of MDA-MB231 cells was higher in corn oil-fed mice when compared to the fish oil-fed group. Serum and tumor tissue analysis indicated decreased serum estrogen and prolactin levels as well as less tissue content of PGE_2. Reduced c-myc oncogene mRNA levels and lesser levels of P21 protein were found in tumor tissues of fish oil-fed animals than either corn oil- or butterfat-fed mice. Both reduction of PGE_2

and the endocrine hormones by fish oil appear to act on reducing the expression of oncogene mRNA levels which may be one of the mechanisms involved in reducing human breast cancer cell growth in nude mice.

References

1 Cancer Facts and Figures. Atlanta, American Cancer Society, 1988.
2 Carroll KK: Lipids and carcinogenesis. J Environ Pathol Toxicol Oncol 1980;3: 253–271.
3 Wynder EL, MacCornack FA, Stellman SD: The epidemiology of breast cancer in 785 United States Caucasian women. Cancer 1978;41:2341–2354.
4 National Academy of Sciences Committee on Diet Nutrition and Cancer: Diet, nutrition, and cancer. Washington, National Academic Press, 1982.
5 Irai A, Terano T, Saito H, Tamura Y, Yoshida S: Clinical and epidemiological studies of eicosapentaenoic acid in Japan; in Lands WEM (ed): Polyunsaturated Fatty Acids and Eicosanoids. Champaign, American Oil Chemists' Society, 1987, pp 9–24.
6 Carroll KK, Chandra RK (eds): Fish oils and cancer; in Health Effects of Fish and Fish Oils. St. John's, ARTS Biomedical Publishers and Distributors, 1989, pp 395–408.
7 Wynder EL, Hirayama T: Comparative epidemiology of cancers in United States and Japan. Rev Med 1977;6:567–594.
8 Reddy BS, Maeura Y: Tumor promotion by dietary fat in azoxymethane-induced colon carcinogenesis in female Fischer 344 rats: Influence of amount and source of dietary fat. J Natl Cancer Inst 1984;72:745–750.
9 Kollmorgen GM, King MM, Kosanke SD, Do C: Influence of dietary fat and indomethacin on the growth of transplantable mammary tumors in rats. Cancer Res 1983;43:4714–4719.
10 Carroll KK, Braden LM, Bell JA, Kalamegham R: Fat and cancer. Cancer 1986; 58(suppl):1818–1825.
11 Reddy BS: Diet and colon cancer: Evidence from human and animal models; in Reddy BS, Cohens LA (eds): Diet, Nutrition, and Cancer; a Critical Evaluation. Boca Raton, CRC Press, 1986, vol 1, pp 47–66.
12 Carroll KK, Braden LM: Dietary fat and mammary carcinogenesis. Nutr Cancer 1985;6:254–259.
13 Reddy BS, Maryama H: Effect of dietary fish oil (menhaden oil) on colon carcinogenesis. Fed Proc 1986;45:(Abst. 5360).
14 Carroll KK: Fish Oils and Cancer; in Chandra RK (ed): Health Effects of Fish and Fish Oils. St. John's, ARTS Biomedical Publishers and Distributors, 1989, pp 395–408.
15 Fernandes G: Effect of dietary fish oil supplement on autoimmune disease: changes in lymphoid cell subsets, oncogene mRNA expression and neuroendocrine hormones; in Chandra RK (ed): Health Effects of Fish and Fish Oils. St. John's, ARTS Biomedical Publishers and Distributors, 1989, pp 409–433.

16 Kaizer L, Boyd NF, Kriukov V, Tritchler D: Fish consumption and breast cancer risk: An ecological study. Nutr Cancer 1989;12:61–68.
17 Karmali RA: Do tissue culture and animal model studies relate to human diet and cancer? Prog Lipid Res 1986;25:533–538.
18 Braden LM, Carroll KK: Dietary polyunsaturated fat in relation to mammary carcinogenesis in rats. Lipids 1986;21:285–288.
19 Cave WT Jr, Jurkowski JJ: Comparative effects of omega-3 and omega-6 dietary lipids on rat mammary tumor development; in Lands WEM (ed): Proc AOAC Short Course on Polyunsaturated Fatty Acids and Eicosanoids. Champaign, American Oil Chemists' Society, 1987, pp 261–266.
20 Karmali RA, Marsh J, Fuchs C: Effect of omega-3 fatty acids on growth of rat mammary tumor. J Natl Cancer Inst 1984;73:457–461.
21 Karmali RA: Eicosanoids in neoplasia. Prev Med 1987;16:493–502.
22 Fernandes G: Inhibition of MCF-7 estrogen-dependent human breast cancer cell growth in nude mice by omega-3 fatty acid diet. Clin Res 1989;37:466A.
23 Fogh J, Giovanella BC (eds): The Nude Mouse in Experimental and Clinical Research. New York, Academic Press, 1982, vol 2.
24 Seibert K, Lippman ME: Influence of tamoxifen treatment on heterotransplanted tumors in nude mice. Clin Res 1984;32:422–428.
25 Taylor IW, Hodson PJ, Green MD, Sutherland RL: Effects of tamoxifen on cell cycle progression of synchronous MCF-7 human mammary carcinoma cells. Cancer Res 1983;43:4007–4010.
26 Pawson T: Growth factors, oncogenes and breast cancer; in Paterson AHG, Lees AW (eds): Fundamental Problems in Breast Cancer. Boston, Martinus Nijhoff, 1987, pp 155–173.
27 Chan PG, Ferguson KA, Dao TL: Effects of different dietary fats on mammary carcinogenesis. Cancer Res 1983;43:1079–1083.
28 Hillyard LA, Abraham S: Effect of dietary polyunsaturated fatty acids on growth of mammary adenocarcinomas in mice and rats. Cancer Res 1979;39:4430–4437.
29 Karmali RA, Welt S, Thaler HT, Lefevre F: Prostaglandins in breast cancer: Relationship to disease stage and hormone. Brit J Cancer 1983;48:689–696.
30 Dupont J: Essential fatty acids and prostaglandins. Prev Med 1987;16:485–492.
31 King MM, Bailey DM, Gibson DD, Pitha JV, McCay PB: Incidence and growth of mammary tumors evidenced by 7,12-dimethylbenzanthracene as related to dietary content of fat and anti-oxidant. J Natl Cancer Inst 1979;63:657–663.
32 Spector AA, Yorek MA: Membrane lipid composition and cellular function. J Lipid Res 1985;26:1015–1035.
33 Stubbs CD, Smith AD: The modification of mammalian membrane polyunsaturated fatty acids composition in relation to membrane fluidity and function. Biochim Biophys Acta 1984;779:89–137.
34 Aylsworth CF, Jones C, Trosko JE, Meites J, Welsch CW: Promotion of 7,12-dimethylbenzanthracene induced mammary tumorigenesis by high dietary fat: Possible role of intracellular communication. J Natl Cancer Inst 1984;72:637–645.
35 Boissonneault GA, Elson CE, Pariza MW: Net energy effects of dietary fat on chemically-induced mammary epithelial cells in vivo and in vitro. J Natl Cancer Inst 1986;76:335–338.

36 Chan PC, Didoto F, Cohen L: High dietary fat, elevation of rat serum prolactin and mammary cancer. Proc Soc Exp Biol Med 1975;149:133–135.
37 Fernandes G, Venkatraman JT: Micronutrient and lipid interactions in cancer. Ann NY Acad Sci 1990;587:78–91.
38 Fernandes G, Yunis EJ, Good PA: Suppression of adenocarcinoma by the immunological consequences of caloric restriction. Nature 1976;263:504–507.
39 Sarkar N, Fernandes G, Telang NT, Kourides FA, Good RA: Low calorie diet prevents the development of mammary tumors of C3H mice and reduces circulating prolactin level, murine mammary tumor virus expression, and proliferation of mammary alveolar cells. Proc Natl Acad Sci 1982;79:7758–7762.
40 Cadena E: Biochemistry of oxygen toxicity. Ann Rev Biochem 1989;58:79–110.
41 Micozzi MS: Foods, micronutrients and reduction of human cancer; in Moon TE, Micozzi MS (eds): Nutrition and Cancer Prevention – Investigating the Role of Micronutrients. New York, Dekker, 1989, pp 213–241.
42 Kinsella JE, Lokesh B, Broughton S, Whelan J: Dietary polyunsaturated fatty acid and eicosanoids: Potential effects on the modulation of inflammatory and immune cells; an overview. Nutrition 1990;6:24–43.
43 Das UN, Begin ME, Ells G, Huang YS, Horrobin DF: Polyunsaturated fatty acids augment free radical generation in tumor cells in vitro. BBRC 1987;145:15–24.
44 Lauds WEM (ed): Polyunsaturated fatty acids and eicosanoids. Proc AOCS 1987.
45 Laganiere S, Yu BP, Fernandes G: Increased membrane lipid peroxidation caused by a high omega-3 fatty acid diet is reversed by dietary vitamin E; in Bendich A, Tengerdy R, Phillips M (eds): Antioxidants, Nutrients and Immune Response. New York, Plenum Press, 1989.
46 Jauntt D, Godney EK, Lutton W, Fernandes G: Role of natural killer cells in experimental murine myocarditis. Semin Immunopathol 1989;11:51–59.
47 Thomas IK, Erickson KL: Lipid modulation of mammary tumor cell cytolysis: Direct influence of dietary fat on the effector component of cell-mediated cytotoxicity. J Natl Cancer Inst 1985;74:675–680.
48 Lewis MG, Kaduce TL, Spector AA: Effect of essential polyunsaturated fatty acid modifications on prostaglandin production by MDCK canine kidney cells. Prostaglandins 1981;22:747–760.
49 Lokesh BR, German B, Kinsella JE: Differential effects of docosahexaenoic acid and eicosapentaenoic acid on suppression of lipoxygenase pathway in peritoneal macrophages. Biochim Biophys Acta 1988;958:99–107.
50 Nishizuka Y: The role of protein kinase C in cell surface signal transduction and tumor promotion. Nature 1986;308:693–698.
51 Welsch CW: Enhancement of mammary tumorigenesis by dietary fat: review of potential mechanisms. Am J Clin Nutr 1987;45:192–202.
52 Mahoney CW, Azzi A: Vitamin E inhibits protein kinase C activity. Biochem Biophys Res Commun 1988;154:694–697.
53 Shafie SM: Estrogen and the growth of breast cancer: New evidence suggests indirect action. Science 1980;209:701.
54 McGuire WL, Garza DL, Chamness GC: Evaluation of estrogen receptor assays in human breast cancer tissue. Cancer Res 1977;37:637.
55 Osborne CK, Coronado EB, Robinson JP: Human breast cancer in the athymic nude mouse: Cytostatic effect of long-term anti-estrogen therapy. Eur J Cancer Clin Oncol 1987;23:1189.

56 Osborne CK, Hobbs K, Clark GM: Effect of estrogen and anti-estrogens on growth of human breast cancer cells: role of the estrogen receptor. Cancer Res 1985;45: 584–590.
57 Osborne CK, Hobbs K, Trent JM: Biological differences among MCF-7 human breast cancer cell lines from different laboratories. Breast Cancer Res Treat 1987;9: 111–121.
58 Ozzello L, Sordat M: Behavior of tumors produced by transplantation of human mammary cell lines in athymic nude mice. Eur J Cancer 1980;16:553.
59 Soule HD, McGrath CM: Estrogen responsive proliferation of clonal human breast carcinoma cells in athymic nude mice. Cancer Lit 1980;10:177–189.
60 Shafie SM, Grantham FH: Role of hormones in the growth and progression of human breast cancer cells (MCF-7) transplanted into athymic nude mice. J Natl Cancer Inst 1981;67:51–56.
61 Pawson T: Growth factors, oncogenes and breast cancer; in Paterson AHG, Lees AW (eds): Fundamental Problems in Breast Cancer. Boston, Martinus Nijhoff, 1987, pp 155–171.
62 Fasano O, Birnbaum D, Edlund L, Fogh J, Wigler M: New human transforming genes by tumorigenicity assay. Mol Cell Biol 1984;4:1695–1705.
63 Kasid A, Knabbe C, Lippman ME: Effect of v-ras^{H} oncogene transfection on estrogen-independent tumorigenicity of estrogen-dependent human breast cancer cells. Cancer Res 1987;47:5733–5738.
64 Alitalo K, Schwab M: Oncogene amplification in tumor cells. Adv Cancer Res 1986; 74:235–281.
65 Theillet C, Lidereau R, Escot C, Hutzell P, Brunet M, Gest J, Schlom J, Callahan R: Loss of a C-H-ras-1 allele and aggressive human primary breast carcinomas. Cancer Res 1986;46:4776–4782.
66 Lidereau R, Escot MC, Theillet C, Champene M, Brunet M, Gest J, Callahan R: High frequency of rare alleles of the human C-Ha-ras-1 proto-oncogene in breast cancer patients. J Natl Cancer Inst 1986;77:699–703.
67 Weinstein BI: The origins of human cancer: Molecular mechanisms of carcinogenesis and their implications for cancer prevention and treatment. Cancer Res 1988;48: 4135–4143.
68 Downward J, Yarden Y, Mayes E, Scrace G, Totly N, Stockwell P, Ullrich A, Schlessinger J, Waterfield MD: Close similarity of epidermal growth factor receptor and v-erb B oncogene protein sequences. Nature 1984;307:521–527.
69 Bargmann CL, Hung MC, Weinberg RA: The new oncogene encodes an epidermal growth factor receptor-related protein. Nature 1986;319:226–234.
70 Telang NT, Bockman RS, Modak MJ, Osborne MP: The role of fatty acids in murine and human mammary carcinogenesis: An in vitro approach; in Abraham S (ed): Carcinogenesis and Dietary Fat. Boston, Kluwer Academic Publishers, 1989, pp 427–451.

G. Fernandes, PhD, Department of Medicine, The University of Texas Health Science Center at San Antonio, 7703 Floyd Curl Drive, San Antonio, TX 78284-7872 (USA)

Appendix

Simopoulos AP, Kifer RR, Martin RE, Barlow SM (eds): Health Effects of ω3 Polyunsaturated Fatty Acids in Seafoods. World Rev Nutr Diet. Basel, Karger, 1991, vol 66, pp 506–537

Poster Session I

(Abstracts)

Effect of Dietary ω3 Fatty Acids (FA) on Eye and Brain Development in Very Low Birth Weight Neonates (VLBWN)

Ricardo Uauy, David Birch, Eileen Birch, Dennis R. Hoffman
Department of Pediatrics and Human Nutrition Center, UT Southwestern Medical Center and Retina Foundation of the Southwest, Dallas, Tex., USA

To evaluate whether ω3 FA are essential for normal eye and brain development in the human, we obtained rod and cone full field electroretinograms (ERG) and assessed acuity by visual evoked potentials (VEP). Healthy VLBWN (< 1,500 g) born at (mean ± SD) 30.5 ± 1.3 weeks were enrolled by day 10 of life to receive human milk (HM; n = 10) containing ω6 and ω3 essential FA (EFA), or were randomized (n = 40) to formulas: A, containing mainly linoleate (L) as EFA with ω6/ω3 = 80; B, containing L and α-linolenate (LN) as EFA with ω6/ω3 = 9; C, providing L, LN, plus long-chain ω3 polyunsaturated fatty acids (LCPUFA) from marine sources [C20:5 ω3 (EPA) and C22:6 ω3 (DHA)]. ERGs done at 35.7 ± 1.2 and 57 ± 1.2 weeks post conception were used to derive 2 μV threshold (T) from Naka-Rushton plots for rods: T 2 μV and critical flicker frequency (CFF) 20–60 Hz served as indices of cone function. Pattern reversal VEPs at both ages assessed the visual cortex. Upon entry and at times of visual testing plasma, red blood cell (RBC) and cheek cell FA composition was measured by gas chromatography focusing on LCPUFA. Results (mean ± SD) are shown in the table. Diet effects were tested using ANOVA.

	RBC ω3 LCPUFA % total FA		Rod ERG log T[1]		Cone ERG log T[1]		VEP Acuity log MAR[2]	
	36 weeks	57 weeks	36 weeks	57 weeks	36 weeks	57 weeks	36 weeks	57 weeks
HM	6.85 ± 2.3	4.68 ± 1.3	0.41 ± 0.6	−1.39 ± 0.2	−0.11 ± 0.2	−0.18 ± 0.2	0.89 ± 0.09	0.41 ± 0.25
A	3.20 ± 1.8	0.99 ± 0.6	1.08 ± 0.4	−1.38 ± 0.3	0.05 ± 0.2	−0.20 ± 0.4	1.06 ± 0.17	0.65 ± 0.11
B	5.26 ± 1.9	1.92 ± 0.4	0.65 ± 0.6	−1.36 ± 0.2	0.08 ± 0.4	−0.19 ± 0.2	1.05 ± 0.14	0.64 ± 0.22
C	6.01 ± 2.5	5.49 ± 1.5	0.48 ± 0.6	−1.37 ± 0.2	−0.04 ± 0.3	−0.21 ± 0.2	1.03 ± 0.08	0.54 ± 0.15
p<	0.008	0.0001	0.01	0.9	0.4	0.9	0.027	0.036

[1] [scotopic troland-seconds (td-sec)]; [2] minutes of arc.

Diet-induced changes in plasma and tissue lipids were highly significant. EPA was highest in plasma in group C, tissue EPA content was minimally increased. Rod function was significantly correlated to DHA and total ω3 FA content of RBC membrane lipids at 36 weeks. Cone function including CFF was not modified by diet. VEP acuity was best for HM-fed at 36 weeks, formula groups were not different; but at 57 weeks both groups receiving DHA (HM and C) had significantly lower log MAR (higher acuity) than groups A and B. Our results support an essential role for ω3 fatty acids in the functional development of the retina and of the visual cortex in the human. Furthermore LN may be insufficient to fully support the functional development of the eye and brain, thus preformed LCPUFAs may also be considered essential for small premature infants.

Supported by NIH HD 22380, EY 05236, EY 05235 and CP Foundation.

Arachidonic Acid (AA) in Plasma and Red Blood Cell (RBC) Phospholipids (PL) During Follow-Up of Preterm Infants: Occurrence, Dietary Determinants and Functional Relationships

S.E. Carlson[a], *J.M. Peeples*[a], *S.H. Werkman*[a], *R.J. Cooke*[a], *W.M. Wilson, III*[b]
[a]Newborn Center, Departments of Pediatrics and Obstetrics and Gynecology;
[b]Boling Center for Developmental Disabilities, University of Tennessee, Memphis, Tenn., USA

Plasma PL AA (g/100 g total fatty acid, μg/ml) and RBC AA (g/100 g total fatty acid) in preterm infants declined as much as 33% between 33 weeks postconception (PCA) when they were receiving $>$ 110 kcal/kg of a low linoleic acid-containing preterm formula and 38 weeks (after they had been fed a term formula with a higher linoleic acid concentration). Mean AA did not return to values found at enrollment until infants had been weaned from formula and begun on a mixed diet including sources of AA (meat and eggs). Stepwise regression was used to determine if ω3 supplementation of formula with eicosaepentaenoic and docosahexaenoic acid, or history of meat and egg consumption, was related to the AA content of plasma and RBC PLs, and if PL AA was related to growth and development in infancy. At all times during infancy (38, 48, 57, 69, 79 and 93 weeks PCA), AA was significantly lower in infants fed an ω3-supplemented term formula compared to controls. The type of formula predicted up to 30% of the variance in plasma PL AA. Meat and egg intake were recorded as servings/week (1 egg yolk = 1 serving) from diet histories obtained at each visit. Mean weekly combined egg and meat intake increased from 1.9 servings/week at 69 weeks to 8.4 and 15.0 servings at 79 and 93 weeks. Egg and meat intake significantly predicted higher plasma and RBC PL AA at 79 and 93 weeks (contribution to r^2 = 0.10–0.15). Plasma PL AA (μg/ml) or the ratio of plasma PL AA (μg/ml) to EPA (μg/ml) significantly predicted growth (length and weight at 69, 79 and 93 weeks, r^2 = 0.15–0.24) and development (Fagan Infantest at 79 and 93 weeks, Bayley Mental and Psychomotor Developmental Indices at 93 weeks, r^2 = 0.10–0.22). DHA supplementation has been shown to prevent declines in RBC PL DHA and improve visual acuity in preterm infants. These data suggest that there is also a need to prevent declines in PL AA of preterm infants, especially when feeding ω3 supplemented diets.

Essential Fatty Acid (EFA) Status in Normal and Complicated Pregnancies

A.C.v. Houwelingen, G. Hornstra, M.D.M. Al
Department of Human Biology, Limburg University, Maastricht, The Netherlands

To assess the EFA status of neonates born after uncomplicated pregnancies, fatty acid compositions were determined, isolated from the walls of umbilical veins and arteries, reflecting fetal EFA supply and backflow, respectively. In each cord, the arteries (backflow) contained significantly less fatty acids of the linoleic (ω6) and linolenic (ω3) families, and more mead acid (20:3 ω9) and other fatty acids of the oleic (ω9) family than the veins (supply). These results may indicate that the EFA requirement of peripheral fetal tissue is not adequately covered. The observed marginal EFA status of normal neonates is based on biochemical evidence only, and therefore it does not necessarily have functional consequences. However, we found a significant negative correlation between the relative amount of mead acid in serum phospholipids (the presence of which might indicate a poor EFA status), and the Apgar score one minute after birth in normal neonates. Moreover, in cases attended by a less optimal placenta transfer, such as pregnancy-induced hypertension (PIH), the situation may become worse.

In mothers with PIH the serum linoleic acid (18:2 ω6, LA) content was significantly lower compared to mothers with normal pregnancies. Most other ω6 fatty acids tended to be higher under hypertensive conditions. Total serum ω3 polyenes were also higher in PIH. This indicates that PIH is associated with an increased unsaturation of maternal serum phospholipids.

The lower LA status of mothers with PIH could indicate that they have a higher demand for more desaturated and/or more elongated fatty acids of the ω6 family, as a result of which more of the available LA is used for conversion. Consequently, the lower LA status in mothers with PIH may be of metabolic rather than dietary origin.

The higher maternal demand for fatty acids of the ω6 family in PIH may be secondary to an increased fetal requirement for polyunsaturated fatty acids under this condition. The increased unsaturation of maternal serum phospholipids could facilitate the placental transfer of long-chain polyunsaturated fatty acids. As a result, the neonatal EFA status after PIH differs only slightly from babies born after normal pregnancies.

Influence of Membrane Fatty Acid Manipulation on Endothelial Cell Functional Properties and Eicosanoid Production

R.C.R.M Vossen[a], *M.C.E. van Dam-Mieras*[b], *G. Hornstra*[a], *R.F.A. Zwaal*[a]
[a]Department of Biochemistry, Limburg University, Maastricht, The Netherlands;
[b]Open University, Heerlen, The Netherlands

Endothelial cells are a contributing factor in the pathogenesis of atherosclerosis. As the endothelial cell membrane fatty acid composition reflects that of dietary lipids, dietary-induced changes in membrane-related endothelial parameters could – at least partly – explain the correlation between dietary lipids and cardiovascular disease. For example, dietary-induced changes in ω3 polyunsaturated fatty acid content of endothelial cell membranes may influence cellular eicosanoid metabolism and other functions relevant to atherogenesis.

In this study changes in endothelial cell fatty acid composition were induced in vitro by culturing the cells in fatty acid supplemented media. For example, an increase in eicosapentaenoic acid content of endothelial cell phospholipids was induced by supplementation of the culture medium with this fatty acid. This was accompanied by a decrease in arachidonic acid content. Comparable results were obtained with docosahexaenoic acid. An increase in arachidonic acid was obtained by supplementation with saturated fatty acids or with arachidonic acid itself. Furthermore, these lipid-modified endothelial cells showed a high elongase activity, but a low desaturase activity.

Considerable alterations induced in membrane fatty acid composition in vitro did not greatly influence such membrane-related parameters as polymorphonuclear leukocyte adhesion and endothelial cell procoagulant activity. However, a positive correlation was found between the endothelial cell arachidonic acid content and the production of eicosanoids when the change in arachidonic acid was induced by supplementation of other fatty acids to the culture medium. For example, endothelial cells manipulated with eicosapentaenoic acid showed a very low eicosanoid production, while palmitic acid-manipulated cells showed a high eicosanoid production. In contrast, when an increase in arachidonic acid content was produced by addition of arachidonic acid itself to the culture medium, the eicosanoid production did not increase. The results indicate that endothelial cells can keep a functional homeostasis notwithstanding considerable changes in membrane fatty acid composition. However, eicosanoid production was influenced by membrane fatty acid manipulations. The availability of arachidonic acid for the formation of local mediators probably has important consequences for cellular communication during hemostatic and inflammatory processes and therefore merits further study.

Fish- and Vegetable Polyenes Increase Platelet Prostanoid Sensitivity: Implications for Platelet Aggregation

J.W.M. Heemskerk[a, b], *G. Hornstra*[a], *M.A.H. Feijge*[b]
Departments of [a]Human Biology and [b]Biochemistry, University of Limburg, Maastricht, The Netherlands

Rats were given diets containing high amounts of either ω3 or ω6 fatty acids, provided by fish oil or sunflowerseed oil, respectively. Washed platelets obtained from these animals aggregated more actively in response to collagen or thrombin than platelets of rats fed a diet containing hydrogenated coconut oil, rich in saturated fatty acids. This difference in platelet aggregation was not related to differences in platelet membrane fluidity (measured by steady-state fluorescence anisotropy of DPH- or TMA-DPH-labeled platelets), to formation of TxB_2 by activated platelets, or to the platelet arachidonate content in various phospholipid classes.

The difference in platelet response to collagen was no longer observed in platelets pretreated with indomethacin. Co-activation of indomethacin-treated platelets with the thromboxane A_2 mimetic U46619, however, restored the diet-related differences in platelet aggregation induced by collagen. These results strongly suggest that dietary fats rich in ω3 and ω6 polyenes significantly increase platelet TxA_2 sensitivity, which seems a major determinant of the diet-induced differences in collagen-induced platelet aggregability. Comparable results have been obtained for fish oil-fed rabbits, the platelets of which appeared more sensitive to the aggregation-inhibiting effect of prostacyclin than platelets of animals fed various vegetable oils.

Thrombin-induced platelet aggregation is also increased upon feeding fish oil to rats and rabbits. This effect is independent from the prostanoid pathway, since pretreatment of the platelets with indomethacin did not obliterate the diet-induced differences. Studies with essential fatty acid-deficient rats demonstrated that the increased thrombin sensitivity of platelets obtained from these animals is normalized by columbinic acid (18:3 $\omega 6_c,9_c,13_t$), which is not a prostanoid precursor but possesses several structural functions of essential fatty acids. Therefore, differences in thrombin-induced aggregation as observed after feeding essential polyunsaturated fatty acids could also result from structural changes in the platelet membrane.

Production of Biomedical Test Materials from Menhaden Oil

J.D. Joseph, S.B. Galloway, R.R. Kifer
National Marine Fisheries Service, Charleston Laboratory, Charleston, S.C., USA

Test materials currently in production and available for distribution to investigators include vacuum deodorized menhaden oil, concentrates of ω3 polyunsaturated fatty acid ethyl esters ($\geq$ 85% ω3 esters), and limited quantities of purified 20:5 ω3 and 22:6 ω3 that are produced by physical and chemical separation techniques. Partially refined menhaden oil, winterized, alkali-refined, and bleached by the supplier, is deodorized in a two-stage wiped film molecular still to reduce cholesterol, organic contaminants, and fishy odors and flavors to very low or undetectable levels. During production, antioxidants are added to the oil unless a special request for antioxidant-free oil is being filled. Until shipped to the researchers or used for production of ω3 concentrates, the oil is stored at −10 °C in stainless steel vessels under a vacuum.

In the production of concentrates, refined oil is first transesterified to produce ethyl esters which are then concentrated by reacting the esters with urea dissolved in hot ethyl alcohol under a N_2 blanket. Upon cooling, the straight-chained saturated and monounsaturated esters form adducts with urea that precipitate from the alcoholic solution, thereby concentrating the non-adducted polyunsaturates. Following removal of the alcohol in a film evaporator, the concentrate is washed with dilute HCl and deionized water to remove residual urea and alcohol. Finally, the neat esters are distilled in a two-stage glass wiped film molecular still to reduce the percentage of 16-carbon polyunsaturates and to eliminate oxidation products, polymers, color bodies, and any remaining cholesterol. Prior to shipment or soft-gel encapsulation, the concentrates are stored at −40 °C in stainless vessels under N_2.

To produce purified 20:5 ω3 and 22:6 ω3 ethyl esters, each $\geq$ 95% pure, a portion of the distilled concentrate is fractionated by supercritical-fluid CO_2 technology to yield fractions of 75% pure EPA and 85% pure DHA. These products are then further purified in a process-scale high performance liquid chromatograph equipped with a 10 cm × 60 cm C18 reverse phase column, using 80% ethyl alcohol as the eluent. The esters are recovered from the alcoholic solution by evaporation of the solvent in a single-stage glass/Teflon wiped film still. Purified esters are stored at $-40\,^\circ C$ in brown glass bottles under N_2.

The test materials are packaged to best meet the needs of the individual investigators. For oils, options include soft-gel encapsulation or bulk shipment in suitably sized containers, and with or without antioxidants. Concentrates are available in capsules or in bulk volumes, but purified esters, only in small amber glass vials with Teflon-lined caps.

FOODBASE: International Food Consumption Database

J.S. Douglass, K.H. Fleming, B.J. Petersen, R.R. Butrum
Technical Assessment Systems, Inc., Washington, D.C., USA

Technical Assessment Systems, Inc. (TAS) in conjunction with the U.S. National Cancer Institute (NCI), is working on a three-year project to collect international food consumption data. A second goal of the project is to construct a database describing the information that has been collected. The database system, FOODBASE, will contain descriptive information on food consumption surveys conducted worldwide since 1940. The database will also contain evaluations, according to TAS-defined criteria, of the quality and usefulness of individual surveys for investigating diet-disease relationships. Food consumption and nutrient content source data from three selected surveys will be included in the system. In addition, the descriptive database will be linked to an extensive bibliography of international food consumption literature.

FOODBASE will be a powerful tool for nutritionists, epidemiologists, medical researchers, and other health professionals to use in assessing the intake of foods and food constituents throughout the world. It will be equally useful to researchers investigating diet-disease relationships, and to nutrition educators, food science and industry researchers, and nutritional anthropologists.

The system is designed for users with little or no computer experience and will operate on IBM-compatible computers. It will allow sorting and analysis of data, and all database fields will be searchable.

TAS is now evaluating, coding and entering information on food consumption surveys. They are continuing to contact food consumption survey experts throughout the world to obtain additional data.

History of Fat in the Human Diet

A.J. Sinclair, K.O. O'Dea
Department of Applied Biology, RMIT, Melbourne, Vic.;
Department of Human Nutrition, Deakin University, Geelong, Vic., Australia

There is considerable debate at the present time about the composition of an ideal diet from the point of view of its fat content and fatty acid composition. Concern is expressed about the level of fat in the diet and arguments are presented for lowering the level to 20 or 25% fat calories or even lower; alternative views include greater consumption of oleic acid-rich oils without lowering the fat content of the diet. In addition, much discussion has centered on the ideal or appropriate balance of fatty acids and particularly the ideal ω6 to ω3 ratio. Evidence has come from the diets of a wide range of populations throughout the world at the present time and from previous eras.

We have had an opportunity to examine the composition of a variety of traditional foods used in the diet of Australian aborigines, and in addition to examine the effect of such food on the plasma lipid profile and plasma fatty acid composition in people of both aboriginal and european origin.

We have concentrated on foods/foodstuffs of animal rather than vegetable origin and the types of animals (terrestrial and marine) analysed were based on actual foods eaten during the course of several expeditions where aborigines reverted back to their traditional hunter/gatherer lifestyle. Foods were collected immediately after the animals were killed and the samples were usually frozen within 2 h. Fasting plasma samples were taken from the subjects before and after the period(s) of up to seven weeks on the traditional diets.

There were striking consistencies in the lipid content and fatty acid composition of the animal foods independent of geographical location. The muscle samples were uniformly low in fat (<2.6% wet weight), with a high proportion of PUFA which is consistent with the data of Crawford et al. for large wild African herbivores. In terms of the actual PUFA composition, these meats could be divided into three main groups: rich in ω3 PUFA which were sea and estuarine in origin; rich in ω6 PUFA which were land-based, including marsupial mammals, reptiles and birds; and rich in both these PUFA types which included land-based, fresh water and coastal animals. Since muscle is the most important component of the animal carcass in terms of its contribution to dietary energy, its low fat content and high proportion of PUFA has a major influence on the hunter-gatherer diet overall.

The effects of traditional hunter-gatherer diets, which were low in fat and with a relatively high proportion and content of 20- and 22-carbon PUFA of both the ω6 and ω3 series, on plasma lipids and plasma fatty acid patterns were quite consistent in all studies: there were reductions in plasma triglyceride levels and a decrease in the proportion of linoleic acid in the major plasma lipid fractions and an increase in the proportion of the long-chain PUFA such as arachidonic, eicosapentaenoic and docosahexaenoic acids. The rise in the proportion of arachidonate and other long-chain PUFA reflected closely their presence in the diets.

These findings are not necessarily consistent with the current trends in dietary manipulation which seek to depress the level of arachidonate and raise that of the ω3 PUFA. The physiological significance of these findings will be discussed.

Flavonoids as Stabilizers of Fish Oil: An Alternative to the Use of Synthetic Antioxidants

A. Valenzuela[a], *J. Sanhueza*[a], *L.A. Loyola*[b], *G. Morales*[b], *A. Garrido*[a], *C. Skorin*[c], *F. Solis de Ovando*[c], *C. Necochea*[c], *F. Leighton*[c]
[a]INTA, Universidad de Chile; [b]Universidad de Antofagasta;
[c]Universidad Católica de Chile, Santiago, Chile

Flavonoids constitute a large group of natural compounds, ubiquitous in photosynthesizing cells, characterized in general by two benzene rings joined by a C3 structure condensed as a six-member ring which changes with the nature of the flavonoid. Some carbons, at any of the three rings, are often hydroxylated. Several hundred species of flavonoids have been identified, among the millions of theoretically possible members of the flavonoid class. Flavonoids, among other properties, exhibit high binding affinity to biological polymers and heavy metal ions; they catalyze electron transport and scavenge free radicals. Because of these properties, their liposolubility, and after some preliminary observations on their hydroperoxide reducing properties, we decided to try them as antioxidants for polyunsaturated oils, specifically fish oil.

The present study involves some well known flavonoids, often employed in biological studies, as well as species isolated from native Chilean plants, among which novel species were found.

For stabilization studies, fish oil containing approx. 30% ω3 fatty acids, after winterization was submitted to high-vacuum distillation (10^{-3} mm Hg, 140 °C, in 750-ml batches) and was kept in the dark under N_2 at 4 °C. Peroxidation was studied for 48 h under air at 60 °C, sampling every 12 h. Peroxide content (AOAC) and malondialdehyde formation were monitored. Measurements were made in 10-ml aliquots, placed in Petri dishes, into which 1 mg/g of *DL*-α-tocopherol (Roche) or flavonoid was added.

Among commercially available flavonoids tried, quercetin was found to exhibit a marked antiperoxidation effect. Among the species isolated from native plants, three species were found with potent antiperoxidation activity, comparable or better than *DL*-α-tocopherol. Their structure was determined (Loyola et al., in press): compound III, a trihydroxyflavone previously isolated from other sources; compound IV, a novel dihydroxyflavone; and compound Pt2, a novel trihydroxyflavanone. In addition to the antiperoxidative capacity of each of these compounds, we have started to evaluate mixtures. Marked synergism was found among *DL*-α-tocopherol and some flavonoids, and also among different flavonoids. For example, 0.3 mg Pt2 + 0.7 mg quercetin per gram oil are 4 times more effective than *DL*-α-tocopherol.

In conclusion, flavonoids, added as single species, or in mixtures of flavonoids or with other antioxidants, have been evaluated as stabilizers of polyunsaturated oil, employing fish oil. The results indicate that these compounds could be used as natural antioxidants, replacing the synthetic compounds used today. Further physicochemical and toxicological evaluations are required to assess the effectiveness of these antioxidants.

Supported by projects UNDP CHI/88/017 and FONDECYT 250/88.

ω3 Fatty Acids in 40 Species of Warm Water Finfish

Julia S. Lytle, Thomas F. Lytle
Gulf Coast Research Laboratory, Ocean Springs, Miss., USA

Certain dietary polyunsaturated fatty acids (PUFA), especially the ω3 PUFA series of fish oils, are demonstrated to be efficient in lowering plasma cholesterol and preventing atherosclerosis. Recent nutritional studies have reported a wide variety of health benefits by adding ω3 fatty acids to the diet. Both because health benefits have been studied on populations consuming cold water fish and because most fishes analysed have been fishes from cold waters, the public as well as the scientific community has the misconception that warm water fishes are not good ω3 sources. Our studies report individual, seasonal and geographical fatty acid variations in muscle tissue of 40 species of finfish from the warm waters of the Gulf of Mexico. For some species, data exists for variations due to size, age, sex and sexual maturity.

Results indicate that warm water fishes can be a valuable source of dietary ω3 fatty acids. Relative and absolute concentration values were compared to published values of cold water fishes. The fatty acid ranges, by season, were: total ω3 fatty acids (g/100 g wet tissue), winter 0.17–0.30, spring 0.06–0.96, summer 0.16–1.0, fall 0.16–0.70. These values are slightly lower than those found in some cold water fishes, and as happens in cold water fishes, higher quantities of ω3 acids generally occur in fattier fishes. The ω3 fatty acid profiles are unique for each species with greater variations shown between species than within a species. For certain fishes, there were distinct differences between males and females; these differences were most pronounced in advanced stages of gonadal maturation. There was an overall enrichment of ω3 fatty acids during summer and spring months compared to winter and fall months.

If the antagonistic action of dietary ω3 acids in the production of potent prostaglandins from a predominance of ω6 fatty acids is an important factor to consider when choosing a diet including seafood, then a particular balance of ω3 and ω6 dietary acids may be more important than the absolute ω3 amounts in the diet. Expressed as ratios of ω3/ω6 fatty acids, the warm water fishes equal or exceed ratios found in most cold water fishes. Ratios of ω3/ω6 in the 40 species of Gulf fish range from 1.4–9.7, whereas some cold water fishes, including those high in absolute amounts of ω3 such as Sockeye salmon and Albacore tuna, range from 3.5–4.6.

ω3 Fatty Acids in the U.S. Food Supply

N.R. Raper, J. Exler
Human Nutrition Information Service, U.S. Department of Agriculture,
Hyattsville, Md., USA

Research on ω3 fatty acids has focused primarily on their mechanisms of action and optimum dietary levels needed to achieve certain beneficial health effects. Little research has been conducted on levels in diets. This study estimates levels and sources of the ω3 fatty acids eicosapentaenoic acid (EPA) and docosahexaenoic acid (DHA) in the U.S. food supply, dating from 1935. These ω3 fatty acids are found primarily in fish.

Food supply data are one of the five components of the National Nutrition Monitoring System, a system of federal activities which monitor the dietary and nutritional status of the U.S. population. The U.S. food supply series, dating from 1909, provides data on per capita quantities of foods available for consumption in the national food supply and quantities of nutrients provided by these foods. These data are useful as indicators of trends in food and nutrient levels over time rather than as measures of ingestion.

Estimates of the ω3 fatty acid content of the food supply are based on annual estimates of per capita food use and data on the ω3 fatty acid content of foods. The method used to calculate the nutrient content of the food supply is to multiply the pounds per capita per year of each food by the nutrient value of the edible portion per pound of food.

Despite a 6.5 pound per capita increase in the use of finfish and shellfish between 1935–1939 and 1985, the level of EPA was the same in 1985 as in 1935–1939, approximately 50 mg per capita per day. The level was unchanged because use of fatty types of fish, which are generally better sources of EPA and DHA, has declined over the years. Consequently, fish contributed less EPA in 1985 than in 1935–1939. However, poultry, a source of EPA and DHA since the 1960's due to the inclusion of fish meal in their diets, offset this decline.

The level of DHA increased from approximately 60 to 80 mg per capita per day between 1935–1939 and 1985. Despite a decrease in the use of fatty fish, the amount of DHA from fish was the same in 1985 as in 1935–1939. Greater use of canned white tuna, which is a more concentrated source of DHA than EPA, offset the decline from less use of other fatty fish. Poultry, which is also a more concentrated source of DHA than EPA, contributed to the overall increase in DHA.

These food supply estimates of ω3 fatty acids are currently the only source of data on ω3 fatty acids in the American diet.

Trans ω3 Eicosapentaenoic and Docosahexaenoic Acid Isomers Exhibit Different Inhibitory Effects on Arachidonic Acid Metabolism in Human Platelets Compared to the Respective Cis Fatty Acids

S.F. O'Keefe[a], *M. Lagarde*[b], *A. Grandgirard*[a], *J.L. Sebedio*[a]
[a]INRA, Station de Recherches sur la Qualité des Aliments de L'Homme, Unité de Nutrition Lipidique, Dijon;
[b]INSERM U205, Chimie Biologique INSA, Villeurbanne, France

ω3 trans geometrical isomers of 20:5 ω3 and 22:6 ω3 were isolated from rats fed heated linseed oil. The ability of these acids to inhibit 20:4 ω6 metabolism by human platelets was examined.

The concentrations required to inhibit 50% of platelet aggregation (IC50) induced by 2.5 μM 20:4 ω6 were higher for the 20:5 Δ17t isomer compared to all cis 20:5 ω3; means 29.2 and 7.6 μM, respectively ($p < 0.05$). There were no significant differences in IC50 between 22:6 Δ19t and all cis 22:6 ω3; means 4.3 and 5.6 μM, respectively ($p < 0.05$). Inhibition of 20:4 ω6 metabolism by cyclooxygenase was similar for 20:5 Δ17t and 20:5 ω3 when examined at their IC50 values but comparison at equal concentrations indicated that 20:5 ω3 was a significantly better inhibitor ($p < 0.05$). The ability to inhibit platelet aggregation was paralleled by cyclooxygenase inhibition as determined by thromboxane B2 and 12-hydroxyheptadecatrienoic acid formation. 22:6 Δ19t appeared to inhibit cyclooxygenase more completely than 22:6 ω3, examined at their IC50 values or at similar concentrations ($p < 0.05$).

20:5 ω3 and 22:6 ω3 isomers having an ω3 cis or trans bond appear to have similar modes of action although levels required for effectiveness are different for the C20 acids.

Biological Effects of Hydroxylated ω3 Fatty Acids Produced by Platelets

J.W. Karanian, H.Y. Kim, A.M. Yoffe, N. Salem, Jr.
Laboratory of Clinical Studies, DICBR, National Institute of Alcohol Abuse and Alcoholism, Bethesda, Md., USA

Substrate requirements for lipoxygenase enzymes are not strictly prescribed and ω3 fatty acids such as eicosapentaenoic acid (20:5) and docosahexaenoic acid (22:6), may be subject to lipoxygenation. In the human platelet production of 12-HEPE from exogenous 20:5 ω3 and 14-HDHE from exogenous 22:6 ω3 have been reported. Platelets from all of the mammalian species thus far tested produce hydroxylated docosahexaenoate (HDHE). About 50% of the HDHE biosynthesized by the human platelet was found in the platelet supernatant. HDHE may therefore be synthesized intracellularly and released or may be produced by a plasma membrane complex on the cell surface. The inhibition of HDHE production by specific lipoxygenase inhibitors such as baicalein (12-LO inhibitor) and the stereochemical purity of the metabolites indicate enzymatic production. The calcium-ionophore A23187 stimulated the production of HDHE from human platelets in a dose-dependent manner with significant increases at concentrations as low as 0.2 μ*M*.

The ω3 metabolites formed by human platelets are biologically active as they are capable of inducing smooth muscle contractions. For example, HDHE (10 μ*M*) induced a mild contraction of airway smooth muscle and this is associated with an increase in leukotriene production. Both NDGA, a lipoxygenase inhibitor, and FPL55712, a leukotriene receptor antagonist, were capable of a partial blockade of this response. This was a relatively specific effect since alpha-, beta-, histaminergic and serontonergic receptor antagonists did not alter the HDHE-induced contraction. Perhaps more significantly HDHE at concentrations as low as 250 n*M* antagonized the contractile effect of a thromboxane-agonist but not that of other eicosanoids or catecholamine agonists tested on vascular smooth muscle. In comparison, preliminary studies with HEPE indicate a less potent action on thromboxane-induced vascular contraction. A similar antagonism of thromboxane-induced platelet aggregation by HDHE and HEPE has been reported. It is hypothesized that platelet HDHE is capable of directly inhibiting agonist interaction with the thromboxane receptor.

When rats were fed a corn oil- or fish oil-based diet for 21 days, both the 20:5 ω3 and 22:6 ω3 content rose in platelet from trace levels in the corn oil group to at least 11% and 6%, respectively, in the fish oil group. The ω3 PUFAs appear to have replaced the ω6 PUFAs since the levels of the latter dropped significantly. An increase in both 12-HEPE from non-detectable levels to 2.86 ± 0.1 nmol/ml and 14-HDHE from 0.06 ± 0.02 to 0.34 ± 0.02 nmol/ml was observed in the ionophore-stimulated (2 μ*M*) rat platelets after a fish oil diet. In addition, 12-HETE production decreased from 4.49 ± 0.7 to 1.96 ± 0.2 nmol/ml. These changes in the formation of lipoxygenase products seemed to correlate with the changes in the fatty acyl composition of the platelets. The combined increases in HEPE and HDHE levels in the platelet may partially explain the decrease in arachidonic or thromboxane-induced aggregation observed in platelets obtained from humans on a diet rich in ω3 PUFAs.

Formation of Oxygenated Metabolites of ω3 Polyunsaturates

H.Y. Kim, J.W. Karanian, T. Shingu, N. Salem, Jr.
Laboratory of Clinical Studies, DICBR, NIAAA, Bethesda, Md., USA

Polyunsaturated fatty acids can be oxygenated by lipoxygenase to form hydroperoxy and hydroxy derivatives. Many of these metabolites are known to possess significant biological activity. In addition to arachidonic acid (20:4 ω6), ω3 fatty acid such as eicosapentaenoic acid (20:5 ω3) and docosahexaenoic acid (22:6 ω3), may also be metabolized in a similar fashion. Production of 14- and 11-hydroxy 22:6 ω3 (HDHE) and 12-hydroxy 20:5 ω3 (HEPE) by platelet lipoxygenase has been reported but the biological roles of these metabolites are not yet clear. However, we have found that 14- and 11-HDHE are capable of inducing smooth muscle contraction and antagonizing the contractile effects of the thromboxane-mimic, U46619. It has also been reported that 14- and 11-HDHE and 12-HEPE inhibit platelet aggregation induced by U46619. Therefore, any pharmacological intervention which can cause changes in the level of these metabolites may result in important biological consequences. In addition to lipoxygenation, other processes such as cytochrome P_{450}-dependent oxygenation or autooxidation may produce derivatives of similar structures, but they would be expected to be racemic mixtures of the possible stereoisomers.

In our studies, we first characterized hydroxy product formation in platelets and brain using reversed phase HPLC, thermospray LC/MS, chiral phase HPLC and GC/MS analysis. Subsequently the production of various hydroxy fatty acids, and especially the ω3 metabolites, was examined following ischemia, alcohol exposure and dietary manipulation. Experimentally induced ischemia resulted in increased brain production of hydroxy derivatives. These products were racemic mixtures suggesting that lipoxygenation was not the major route of their formation. Increased production of hydroxy derivatives was also observed when ω3 fatty acids were incubated with platelets from alcohol-treated rats. Following fish oil supplementation, an increase in 12-HEPE production with a concomitant decrease in 12-hydroxyeicosatetraenoic acid (12-HETE) was the most prominent change in rat platelets. These data demonstrated that the production of hydroxy fatty acids may be altered during alcohol exposure or ischemia. In addition, the profile of platelet ω3 and ω6 lipoxygenase products may be drastically altered by manipulation of dietary polyunsaturates.

Effect of a Fish Oil Diet on the Molecular Species Composition and the Topological Distribution of Aminophospholipids in Rat Erythrocyte Membranes

F. Hullin, M.J. Bossant, N. Salem, Jr.
Laboratory of Clinical Studies, DICBR, NIAAA, Bethesda, Md., USA

In order to more precisely define how fish oil supplementation leads to many biochemical and physiological modifications in mammalian systems, we studied changes in aminophospholipid, i.e. phosphatidylethanolamine (PE) and phosphatidylserine (PS), molecular species distributions in red blood cells (RBC) as well as the localization of these species in the plasma membranes. This was based on treatment of intact RBC with the membrane-impermeant reagent trinitrobenzenesulfonic acid (TNBS). A reverse-phase HPLC method was developed for separating the various molecular species of trinitrophenylated PE and PS with UV detection at 338 and 342 nm, respectively. The fatty acid composition of each species was confirmed by gas chromatographic analysis of each peak. The RBC membrane of rats fed with a chow diet containing 10% fat displayed a quite random distribution of the PE molecular species over the two membrane leaflets. However, the arachidonate species were slightly enriched in the inner leaflet (45% of all the PE species in the inner leaflet versus 40% in the outer). We also noted the preferential localization of ω3 polyunsaturated fatty acids (PUFA) in the alkenyl-PE subclass whereas ω6 PUFA was enriched in diacyl-PE. When rats were fed for 21 days with a fish oil-based diet, there was an increase in ω3 species and a decrease in ω6 species in both PE and PS in comparison to a corn oil-based diet. The increase in ω3 PUFA was primarily observed in the alkenyl-PE species as about 80% of the ω3 increase in PE was found in this subclass. A slight preference for this ω3 increase in the alkenyl-PE of the outer leaflet was observed whereas diacyl ω3 species were randomly distributed. The selective incorporation of ω3 PUFA into alkenyl-PE and its accentuation by an ω3-rich diet would be expected to influence the physical properties of cellular membranes as well as their availability as substrates for the formation of bioactive substances.

Retroconversion vs. Desaturation: Alternative Metabolic Fates for C22 ω6 and ω3 Fatty Acids in Human Cells

Miriam D. Rosenthal, Martha C. Garcia, Howard Sprecher
Biochemistry, Eastern Virginia Medical School, Norfolk, Va.;
Physiological Chemistry, Ohio State University, Columbus, Ohio, USA

The present study has used [3-^{14}C]docosatetraenoate (22:4 ω6 and [3-^{14}C]docosapentaenoate (22:5 ω3) to investigate the extent of retroconversion to C_{20} fatty acids and of delta-4 desaturation in isolated human cells.

Fetal skin fibroblasts (GM-10) and retinoblastoma cells (Y79) were cultured for 1–5 days in complete medium + 10% FBS with ^{14}C-fatty acid (2.5 μM). Methyl esters of ^{14}C-fatty acids derived from cellular glycerolipids were analyzed by gas-liquid chromatography.

Both cell lines actively retroconverted [^{14}C]22:4 ω6 to [^{14}C]20:4 ω6. Within 24 h, chain shortening was 29% for fibroblasts and 54% for retinoblastoma cells. In fibroblasts, retroconversion reached 46% in 96 h. By contrast, retroconversion of [^{14}C]22:5 ω3 was only 12% for fibroblasts and 6.5% for retinoblastoma after 24 h and did not increase after that time. Parallel experiments indicated that both fibroblasts and retinoblastoma cells elongated [^{14}C]eicosapentaenoate (20:5 ω3) more extensively (2–3 fold) than [^{14}C]arachidonate (20:4 ω6).

Retinoblastoma cells rapidly and extensively desaturated [^{14}C]22:5 ω3, producing 51% [^{14}C]22:6 ω3 within 6 h and $>$ 90% at 96 h. Desaturation of [^{14}C]22:5 ω3 was also observed in fibroblasts but much more slowly, producing $<$ 2% [^{14}C]22:6 ω3 at 48 h and 20% at 96 h. Similar results were observed with [^{14}C]22:5 ω3 elongated from [^{14}C]20:5 ω3. Retinoblastoma cells also desaturated [^{14}C]22:4 ω6 to [^{14}C]22:5 ω6 (28% in 24 h, 38% in 96 h). Essentially no desaturation of [^{14}C]22:4 ω6 was observed in fibroblasts.

These results suggest that the ratio of C_{22}/C_{20} fatty acids in human cells is regulated by the rates of elongation and retroconversion. Elongation to C_{22} is favored for ω3 fatty acids; retroconversion to C_{20} for ω6. Furthermore, the delta-4 desaturation of 22:5 ω3 may be regulated differently in fibroblasts than in retinoblastoma cells.

Metabolism of ω6 and ω3 Fatty Acids in Liver and Hepatocytes

Anne Voss, H. Sprecher
Department of Physiological Chemistry, Ohio State University, Columbus, Ohio

Molecular species analysis of choline- and ethanolamine-phosphoglycerides (PC and PE) from livers of rats fed corn oil (CO) versus a corn oil-fish oil (FO) diet showed that the decline in 20:4 ω6 was not confined to a specific molecular species. The 16:0/20:4 and 18:0/20:4 molar ratios in PC and PE from CO-fed rats were respectively 1.5 and 0.7, while in FO-fed rats they were 1.6 and 0.6. The 16:0/20:5 molar ratio in PC and PE in FO-fed rats was 1.9 and 0.7. PC and PE from CO-fed rats contained 1.8 and 4.2% 22:6 ω3. These values from FO-fed rats were 4.2 and 19.9%. Again molecular species analysis showed that 22:6 ω3 paired with 16:0 and 18:0 in about the same ratio in the two groups of rats. Although dietary FO alters fatty acid composition these studies show that the pairing of 20:4, 20:5 and 22:6 with 16:0 and 18:0, on a molar ratio basis, was independent of dietary fat. Hepatocytes from rats fed a chow or a fat-free diet were then incubated with either 3-^{14}C-labeled 22:4 ω6 or 22:5 ω3 in an attempt to further define why, in general, there is preferential synthesis and acylation of 22-carbon ω3 versus ω6 acids in liver phosphoglycerides. In hepatocytes from chow-fed rats both substrates were either acylated into lipids or retroconverted to 20:5 ω3 and 20:4 ω6 followed by esterification. When [3-^{14}C]22:4 ω6 was the substrate, 72% and 28% of the radioactivity in triglycerides (TG) was associated respectively with 22:4 ω6 and 20:4 ω6. When [3-^{14}C]22:5 ω3 was the substrate, 69% and 31% of the radioactivity in TG was respectively in 22:5 ω3 and 20:5 ω3. A different labeling pattern was observed in total phosphoglycerides. When [3-^{14}C]22:4 ω6 was the substrate, 21% and 79% of the radioactivity was associated respectively with 22:4 ω6 and 20:4 ω6. Conversely, when [3-^{14}C]22:5 ω3 was the substrate, only 43% of the radioactivity was retroconverted to 20:5 ω3 and esterified. Hepatocytes from rats fed an essential fatty acid-deficient diet also desaturated both substrates. These studies suggest that the 4-desaturase is regulated by diet and that substrate specificity for retroconversion may play a role in defining intracellular concentrations of 20- and 22-carbon ω3 and ω6 acids.

This study was supported by NIH grants DK18844 and DK20387.

Effects of EPA and DHA Ethyl Esters on Plasma Fatty Acids, and on Fatty Acids, Eicosanoid, Inositol Phosphate Formation, and Functional Parameters in Platelets, PMN and Monocytes in Healthy Volunteers

C. Galli, S. Colli, C. Mosconi, L. Medini, M. Gianfranceschi, M. Canepari, P. Maderna, C. Sirtori, E. Tremoli
Institute of Pharmacological Sciences, University of Milan;
E. Grossi Paoletti Center for the Study of Hyperlipidaemias, University of Milan, Italy

We have studied the effects of the administration of EPA (2.7 g/day) and DHA (1.1 g/day) ethyl esters, in 1 g capsules (Norsk Hydro, Porsgrunn, N), to 5 male volunteers (aged 26–51 years) with no clinical sign of disease and free from drug treatments, for 6 weeks, on plasma lipids and lipoproteins, and on platelet biochemistry and function.

The following parameters were analyzed: plasma triglyceride (TG) and cholesterol (C); VLDL-, LDL- and HDL-C; composition and levels of fatty acids in plasma lipids, lipoproteins and platelets; the aggregatory responses of platelet-rich plasma (PRP) to various agonists and the sensitivity of platelets to the antiaggregatory activity of exogenous prostacyclin; formation of TxB_2 by collagen- and thrombin-stimulated PRP and generation of inositol phosphates by ^{3}H-myoinositol-labelled washed platelets after stimulation with thrombin, generation of superoxide anion and expression of procoagulant activity (PCA) by isolated monocytes.

The effects on plasma lipids and lipoprotein levels were negligible. EPA and DHA were differentially accumulated in plasma lipid and lipoprotein classes, and their incorporation in platelets, up to about 1.8 and 1.2% of total fatty acids, respectively, on the average, was associated with a similar reduction of platelet arachidonate. Platelet TxB_2 formation was markedly inhibited (about 60%) and this was associated with reduced aggregability and enhanced sensitivity of platelets to prostacyclin. The generation of inositol phosphates (IP_3, IP_2 and IP) by non-stimulated and stimulated platelets was also significantly decreased by the administration of ω3 fatty acids. In addition, O_2 generation and expression of PCA (basal and LPS-stimulated) by adherent monocytes was significantly reduced. The data indicate that ω3 FA affect several functional parameters in the same type of cell, suggesting a central role of long-chain ω3 fatty acids in the modulation of responses to various stimuli. The relationships between the various parameters will be discussed.

Cytochrome P-450 Metabolites of Docosahexaenoic Acid Inhibit Platelet Aggregation without Affecting Thromboxane Production

M. VanRollins
Hypertension Research Division, Henry Ford Hospital, Detroit, Mich., USA

Diets enriched with the major ω3 long-chain fatty acid docosahexaenoic acid (DHA) inhibit platelet aggregation without affecting thromboxane A_2 synthesis. In this study, we examined whether cytochrome P-450 epoxygenase metabolites and their diol hydrolysis products can mediate the effects of dietary DHA. Cytochrome P-450 epoxygenase metabolites of both DHA and the major ω6 long-chain fatty acid arachidonic acid were chemically synthesized, and their effects on platelet aggregation and thromboxane production compared. Washed human platelets were isolated according to Radomski and Moncada [Thromb Res 1983;30:383] and incubated in the presence of human fibrinogen (0.5 mg/300 μl) and DHA epoxides or diols. Aggregation was induced by arachidonic acid at a concentration that would produce 85% of the maximal aggregation response (EC_{85} = 1.7 ± 0.1 μM (SEM), n = 27) as determined daily. The extent of aggregation was measured as the area under the turbidometric curve during the first 3 min following the addition of arachidonic acid. Platelets were suspended in 9 vol indomethacin (1 μg/ml water) and immediately frozen in liquid nitrogen. Thromboxane B_2 was measured by radioimmunoassay.

Overall the DHA epoxides and diols inhibited platelet aggregation 2 and 10 times, respectively, more potently than comparable arachidonic acid metabolites. The concentrations of the 5 DHA epoxides and 5 diols inhibiting platelet aggregation by 50% (IC_{50}) ranged from 0.7 to 1.5 and 3.4 to 11.7 μM, respectively. However, DHA epoxides were on the average only $^1/_{10}$ as potent as arachidonic acid epoxides in inhibiting thromboxane formation. The IC_{50} values for thromboxane formation were 6–78 μM for DHA epoxides and > 30 μM for DHA diols. Thus the epoxygenase metabolites of DHA blocked platelet aggregation long before they inhibited thromboxane A_2 synthesis. By making platelets non-responsive to normal thromboxane A_2 levels, DHA epoxides and diols may contribute to the early anti-thrombotic effects of dietary DHA. In addition, the DHA epoxides and diols may provide complementary anti-thrombotic actions during fish oil diets by reducing platelet responsiveness to what little thromboxane A_2 is produced.

Peroxisomal Retroconversion of Docosahexaenoic Acid (22:6 ω3) to Eicosapentaenoic Acid (20:5 ω3) in Human Fibroblasts and Rat Hepatocytes

T.-A. Hagve, M. Grønn, E. Christensen, B.O. Christophersen
Institute of Clinical Biochemistry, Rikshospitalet, University of Oslo, Norway

A high intake of polyunsaturated ω3 fatty acids, especially of eicosapentaenoic acid (EPA), probably causes the lower plasma triglyceride levels and lower incidence of atherosclerotic disease in Greenland Eskimos. Fatty fish and fish oils used in clinical trials contain both EPA and DHA (docosahexaenoic acid).

We have shown that [4,5-^{3}H]DHA is retroconverted to EPA in normal human fibroblast cultures by using radiogas-chromatography. This reaction was deficient in fibroblast cultures from patients with peroxisomal diseases (Zellweger syndrome and Neonatal adrenoleucodystrophy) which lack peroxisomes.

In isolated rat liver cells we found that 20–25% of added [4,5-^{3}H]DHA substrate was retroconverted to EPA by one cycle of β-oxidation probably with 4-enoyl-CoA reductase as auxiliary enzyme. The conversion was not stimulated by (–)carnitine which stimulates mitochondrial fatty acid oxidation, and was not inhibited by addition of (+) decanoylcarnitine which inhibits mitochondrial fatty acid oxidation in isolated liver cells.

[1-^{14}C]EPA on the other hand was metabolized to a large extent to docosapentaenoic acid in hepatocytes from fed animals. In hepatocytes from fasted animals little EPA was converted to docosapentaenoic acid.

It was concluded that DHA is retroconverted to EPA at a high capacity by peroxisomal β-oxidation in liver and by microperoxisomes present in fibroblasts.

Conceptional Aspects of Hypotriglyceridemic Action of ω3 Fatty Acids in Man

P. Singer[a], *J. Hueve*[b]

[a]Heckmannufer 6, Berlin; [b]Scientific Division, Omega Pharma, Berlin, Germany

It is commonly accepted that the hypotriglyceridemic action of ω3 fatty acids is based on a reduced hepatic triglyceride formation and decreased VLDL secretion by the liver (hepatic factor). Moreover, an increased clearance of triglyceride-rich particles is assumed (posthepatic factor). However, the role of free fatty acids (FFA) as the substrate for hepatic triglyceride synthesis (prehepatic factor) was widely ignored.

The prerequisite of the assessment of FFA within clinical studies are strictly standardized conditions. We, therefore, besides blood glucose and insulin estimated plasma FFA within oral glucose tolerance tests (75 g) before and after 2 weeks of supplements with sunflowerseed oil (60 ml/day providing 45 g/day of linoleic acid), linseed oil (60 ml/day providing 38 g/day of alpha-linolenic acid), mackerel diet (2 cans/day equivalent to 5.0 g/day of ω3 fatty acids) and herring diet (2 cans/day equivalent to 2.8 g/day of ω3 fatty acids) in patients with hyperlipoproteinemias.

Whereas blood glucose and insulin remained unchanged by the diets, the level of FFA was reduced especially after the mackerel diet providing the higher dose of ω3 fatty acids. In the same dietary group the decrease of serum triglycerides was most pronounced.

Although other factors must be considered, it can be speculated that the reduction of plasma FFA might be a hitherto ignored contribution to lower serum triglyceride levels by dietary ω3 fatty acids. A schedule summarizing hypotriglyceridemic factors is presented.

Blood Pressure-Lowering Effect of Fish Oil, Propranolol and the Combination of Both in Mildly Hypertensive Patients

P. Singer[a], *J. Hueve*[b]

[a]Heckmannufer 6, Berlin; [b]Scientific Division, Omega Pharma, Berlin, Germany

From the data available, mild hypertension might be a novel indication of fish oil supplement. So far, no comparison for blood pressure-lowering drug treatment and fish oil supplementation was made. Therefore we compared the effects on blood pressure serum lipids, plasma noradrenaline and thromboxane B_2 formation in a controlled double blind study.

Forty-seven male patients with mild essential hypertension were randomly allocated to 3 subgroups. After a run-in period of 4 weeks group P (n = 16) received propranolol (80 mg/day) for 36 weeks followed by a placebo period of 4 weeks. Group F (n = 15) after a run-in period of 4 weeks was supplemented with encapsulated fish oil (9 g/day) for 36 weeks with a subsequent period of 4 weeks during which fish oil placebo was given. Group P + F (n = 16) after a run-in period of 4 weeks received propranolol (80 mg/day) for 12 weeks, propranolol (80 mg/day) plus fish oil capsules (9 g/day, equivalent to 1.8 g/day of eicosapentaenoic acid and 1.1 g/day of docosahexaenoic acid for 12 weeks, propranolol + fish oil placebo (same doses) for 12 weeks with a subsequent period of 4 weeks with propranolol placebo alone.

In group P a significant decrease of systolic and diastolic blood pressure during propranolol treatment was found. This was associated with a significant decline of plasma noradrenaline (NA), thromboxane B_2 (TXB_2) formation and plasma renin activity (PRA). After propranolol placebo (4 weeks) the values returned nearly to the basal levels. Serum triglycerides were elevated after 24 and 36 weeks of treatment on a low level of significance.

In group F likewise a significant fall of systolic and diastolic blood pressure was seen. In addition, plasma NA and TXB_2 formation appeared lower, whereas PRA was elevated after fish oil supplementation. Serum triglycerides, total and LDL cholesterol were reduced, but HDL cholesterol was increased.

In group P + F propranolol treatment over 12 weeks resulted in a significant decrease of systolic and diastolic blood pressure, plasma NA, TXB_2 formation and PRA. The addition of fish oil (over 12 weeks) led to a further decline of blood pressure associated with a further reduction of plasma NA and TXB_2 formation as well as an increase of PRA. During concomitant fish oil supplementation serum triglycerides, total and LDL cholesterol were significantly decreased, whereas HDL cholesterol appeared elevated.

The data indicate a blood pressure-lowering effect of fish oil, which is comparable to that of propranolol. The concomitant intake of fish oil plus propranolol is more effective than propranolol or fish oil alone in patients with mild essential hypertension. The beneficial effects on serum lipids and lipoproteins, in addition to other dietary means, justify the consideration of fish oil as a nonpharmacological measure for the treatment of mild hypertension.

Eating the Fresh-Water Fish, Silver Back, Improves Blood Lipid Parameters in Hyperlipaemic Humans

T. Farkas[a], *I. Joo*[b], *I. Csengeri*[c],
[a]Institute of Biochemistry, Biological Research Center, Szeged;
[b]City Hospital, Kiskunfelegyhaza; [c]Fish Research Institute, Szarvas, Hungary

Silver back, *Hypophtalmitryx molitrix,* a phytoplankton-filtering freshwater fish, containing 2.7 g/kg EPA, 2.8 g/kg DHA and 11.8 g/kg total ω3 faty acids in its body, was supplied to healthy (n = 16) and hyperlipaemic (Fredericson IIA) patients (n = 76) in different doses (0.4, 0.8, and 1.2 kg) once a week. Total cholesterol (I), the atherogen index: total cholesterol – HDL cholesterol/HDL cholesterol (II), the antiatherogen index:

HDL_2 cholesterol/HDL_3 cholesterol (III), and the blood pressure (IV) was studied, the latter in hypertonic patients. I decreased significantly after the 3rd meal in patients consuming 1.2 kg fish/week while those consuming 0.8 kg fish/week responded only after the 15th week. II and III changed favourably after the 6th meal with patients eating 0.8 or 1.2 kg fish/week. Changes in III were more pronounced when the patients were on a cholesterol-rich (meals prepared with lard) than on a low-cholesterol (meals prepared with vegetable oil) diet. In the former case the group consuming 0.8 kg fish/week gave significant response already after the 3rd meal. Patients receiving 0.4 kg fish/week gave similar responses but only after a prolonged period. These effects persisted at least for 3 weeks following cessation of the diet. Each dose proved to be hypotensive. Results indicate that less than 1 g daily intake of EPA and DHA is sufficient to improve blood lipid parameters if consumed in the form of specific fish.

Acknowledgements: This work was supported by the Ministry of Agriculture, Budapest Hungary, and Repro-Gen International, Vaduz, Liechtenstein.

Fish Oil Lowers Blood Pressure in Normotensive Elderly Subjects on a Low-Sodium Diet

L. Cobiac, P.R.C. Howe, P.J. Nestel
CSIRO Division of Human Nutrition, Adelaide, Australia

Increasing consumption of fish oil and restricting sodium intake are both potentially useful dietary strategies to control mild hypertension. We have now examined the combined effect of these strategies in the elderly population, who have a higher risk of hypertension and are more sensitive to the hypertensive effects of sodium.

Fifty volunteers completed a dietary intervention comprising a run-in phase (2 weeks) followed by consumption of 8 capsules of Himega fish oil (4 g of ω3 fatty acids) or sunflower oil (5 g of ω6 fatty acids) per day for 12 weeks. During this period, sodium intakes were adjusted by adopting a low-sodium diet ($<$ 70 mEq per day) and taking either 'Slow Sodium' tablets (8×10 mEq sodium per day) or placebo tablets, administered in a double-blind protocol.

After the first 4 weeks of intervention, blood pressure fell significantly (systolic/diastolic: –8.9/–6.0 mm Hg) in subjects taking fish oil who reduced sodium intake (from 140 to 80 mEq/day). No consistent changes were seen in subjects taking fish oil with normal sodium or in those taking sunflower oil. At this stage, subjects changed to the alternative sodium tablets but remained on the same oil supplement. After a further 4 weeks, an effect of the crossover of sodium intake on blood pressure was seen only in the subjects taking fish oil; a 100 mEq/day change in sodium intake was associated with significant changes of 6.4 and 2.2 mm Hg in systolic and diastolic pressures. No significant effect of sodium on blood pressure was seen in subjects taking taking sunflower oil. Subjects remained on the same diets for the final 4 weeks of the study during which the effects on blood pressure were sustained.

In this short-term study, the combination of dietary sodium restriction with fish oil supplementation lowered blood pressure in normotensive elderly subjects, whereas neither strategy alone was effective. Further evaluation of the interactive effects of dietary sodium and ω3 fatty acids is warranted.

Fish Oil Lowers Blood Pressure and Reduces Vascular Resistance in Stroke-Prone Hypertensive Rats (SHRSP)

P.R.C. Howe[a], *P.F. Rogers*[a], *Y. Lungershausen*[a], *J.F. Gerkens*[b], *R.J. Head*[a], *R.M. Smith*[a], *P.J. Nestel*[a]
[a]CSIRO Division of Human Nutrition, Adelaide;
[b]Discipline of Clinical Pharmacology, University of Newcastle, Australia

We have examined the effects of fish oil extracts rich in ω3 fatty acids on blood pressure in an animal model of hypertension. When SHRSP were fed diets containing 5% MaxEPA or SanOmega from weaning, we observed a small but consistent suppression of blood pressure development. Mean arterial pressure (MAP) measured in conscious rats after 3 months averaged 165 ± 3 mm Hg (n = 35) compared with 174 ± 3 mm Hg (n = 36) for SHRSP fed olive oil. This difference persisted after ganglion blockade (111 ± 2 and 119 ± 3 mm Hg for fish and olive oil, respectively) but could be prevented by increasing dietary sodium intake.

Subsequent studies in adult SHRSP with established hypertension showed that administration of fish oil in the diet or by subcutaneous injection for 1–2 months caused similar reductions of blood pressure which were enhanced by lowering sodium intake. E.g. MAP of SHRSP injected with Himega fish oil for 6 weeks fell from 174 ± 6 to 154 ± 3 mm Hg when fed a low-sodium diet. Corresponding blood pressures for SHRSP injected with olive oil were 190 ± 5 and 176 ± 5 mm Hg (n = 7 in each case).

In preliminary experiments, basal vascular tone was examined in the blood-perfused mesentery in situ by measuring the perfusion pressure at different flow rates (0.5–3.0 ml/min). The pressure/flow gradient was 20% lower in young SHRSP fed fish oil for 3 months than in SHRSP fed diets containing olive oil. In adult rats fed low-sodium diets containing either fish oil or olive oil, an even greater difference in the pressure/flow gradient was seen.

These findings suggest that the antihypertensive effect of fish oil is influenced by the dietary intake of sodium and is accompanied by a fall in vascular resistance which may be due to a reduction of intrinsic vascular tone or altered levels of a blood-borne vasoactive substance. These possibilities are now being investigated.

Effect of Eicosapentaenoic and Docosahexaenoic Acids on Blood Pressure in Hypertension

K.B. Bønaa, K.S. Bjerve
Institute of Community Medicine, University of Tromsø, and Department of Clinical Chemistry, Regional Hospital, University of Trondheim, Norway

Marine polyunsaturated fatty acids may have hypotensive effects, but interpretation of the existing studies is constrained by variation in their design. We conducted a population-based, randomised, 10-week dietary supplementation trial in which 6 g/day of 85% eicosapentaenoic/docosahexaenoic acid was compared with 6 g/day of corn oil in 157 men and women with previously untreated stable mild hypertension. Blood pressure was

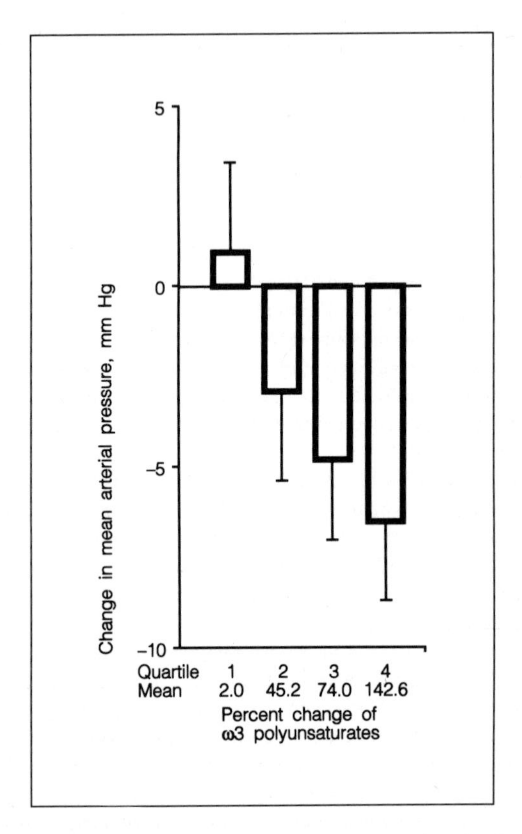

Fig. 1. The figure shows that the decrease in blood pressure in the marine oil group was related to increase in concentrations of plasma phospholipid ω3 fatty acids in a dose-dependent manner (linear trend $p = 0.027$).

measured with an automatic device. All subjects completed the study and compliance was demonstrated by significant changes in plasma phospholipid fatty acid concentrations.

The mean systolic blood pressure fell by 4.6 mm Hg ($p = 0.002$) and diastolic pressure by 3.0 mm Hg ($p = 0.0002$) in the marine oil group, compared to no change in the corn oil group. The difference between the groups remained significant for both systolic (−6.4 mm Hg; $p = 0.0025$) and diastolic (−2.8 mm Hg; $p = 0.029$) pressure after controlling for anthropometric, lifestyle and dietary variables (fig. 1).

Marine oil supplementation did not change mean blood pressure in those subjects who consumed three or more fish meals per week in their usual diet, or in those who had a baseline plasma phospholipid ω3 fatty acid concentration above 175.1 mg/l.

We conclude that eicosapentaenoic and docosahexaenoic acid reduce blood pressure in essential hypertension in a dose-dependent manner.

Effects of ω3, ω6, and ω9 Fatty Acids on Vascular Smooth Muscle Tone

M.B. Engler[a], *J.W. Karanian*[b], *N. Salem, Jr.*[b]
[a]University of California at San Francisco, San Francisco, Calif.;
[b]Laboratory of Clinical Studies, DICBR, NIAAA, Bethesda, Md., USA

It has been proposed that a diet rich in ω3 fatty acids derived from fish or fish oil may be beneficial in the prevention and treatment of cardiovascular disease. Eicosapentaenoic acid (20:5 ω3) and docosahexaenoic acid (22:6 ω3) are the ω3 fatty acids which are thought to be the active components of fish oils. Therefore, it was of interest to determine the effects of 20:5 ω3, 22:6 ω3 and other unsaturated fatty acids (ω3, ω6, ω9) on vascular smooth muscle tone. Isolated rat aortic ring preparations were immersed in Krebs-Ringer bicarbonate buffer solution (pH 7.4) and mounted between a fixed stainless steel hook and an isometric force-displacement transducer. The rings in smooth muscle baths were gassed continuously with 95% O_2, 5% CO_2 at 37 ± 0.5 °C. Equilibration of the vascular rings occurred at a preload tension of 1.5 g for 60–90 min. Isometric force was measured. Cumulative concentration-response curves in precontracted vessels were generated at 1–255 μ*M* for the ω3, ω6, and ω9 fatty acids. Administration of 20:5 ω3, 22:6 ω3, arachidonic acid (20:4 ω6), and cis-linoleic acid (18:2 ω6) to the bath resulted in significant relaxation. Relaxation induced by 20:5 ω3 was noted at 31 μ*M* (18 ± 1%) and above, e.g. 255 μ*M* (63 ± 4%). 22:6 ω3 induced significant relaxation at all concentrations (1–255 μ*M*). The relaxation induced by 22:6 ω3 was demonstrated at physiologic levels, including 1 μ*M* (7 ± 2%), 3 μ*M* (12 ± 2%), 7 μ*M* (17 ± 3%), 15 μ*M* (21 ± 3%) and above, e.g. 255 μ*M* (61 ± 4%). The relaxant effects of 20:5 ω3, 22:6 ω3, and 18:2 ω6 were not altered by indomethacin or NDGA. However, the relaxant effects of 20:4 ω6 were decreased in the presence of these inhibitors. Other polyunsaturated fatty acids, 20:3 ω6, 18:3 ω6, and 18:3 ω3 induced relaxation only at the highest concentration (255 μ*M*) tested. Fatty acids from the ω9 family, 20:1, 22:1, and 18:1 did not alter smooth muscle tone. The relaxant effects of ω3 fatty acids at concentrations which are physiologically obtainable in plasma may partially explain the possible benefits of fish oil on the cardiovascular system.

Influence of Dietary ω3, ω6 and ω9 Polyunsaturated Fatty Acids on Aortic and Platelet Fatty Acid Composition in the Rat

M.B. Engler[a], *J.W. Karanian*[b], *N, Salem, Jr.*[b]
[a]University of California at San Francisco, San Francisco, Calif.;
[b]Laboratory of Clinical Studies, DICBR, NIAAA, Bethesda, Md., USA

Low levels of ω3 and ω6 fatty acids in adipose tissue and platelets have been associated with increased risk of coronary heart disease. Since vascular tissue and platelets play an important role in thrombotic and atherosclerotic processes, the effects of dietary 18-carbon ω3, ω6 and ω9 fatty acids on the fatty acid composition of aorta and platelets were investigated. Sprague-Dawley rats were fed for 7 weeks purified diets containing 11% (weight) of either a blend of linseed and safflower oil, a source of 18:3 ω3, borage oil rich in 18:3 ω6, or sesame oil rich in 18:1 ω9. The oils were also enriched with similar amounts of 18:2 ω6. Total lipids were extracted from aortae and platelets and analyzed by gas chromatography. Feeding linseed/safflower oil led to a lowered 20:4 ω6 content and an increased level of ω3 fatty acids (18:3 ω3, 20:5 ω3, 22:6 ω3) in aortic tissue and platelets. Consumption of borage oil increased the levels of aortic and platelet 18:3 ω6, 20:3 ω6, and 20:4 ω6. The same sesame oil diet elevated the content of 18:1 ω9 in aortic and platelet total lipid extracts. These data suggest that the fatty acid composition of vascular tissue and platelets can be selectively modified by the administration of either 18-carbon ω3, ω6 or ω9 fatty acid enriched diets. Dietary treatment with 18-carbon fatty acids may have important implications for coronary heart disease.

Apolipoprotein B Metabolism in Miniature Pigs Treated with Dietary Fish Oil and a Cholesterol Synthesis Inhibitor

Murray W. Huff, P. Hugh Barrett, Dawn E. Telford
Department of Medicine, Robarts Research Institute, University of Western Ontario, London, Canada

VLDL triglyceride concentrations are consistently reduced by dietary fish oil, whereas their effect on LDL cholesterol is less consistent. In many studies, LDL concentrations are either unchanged or increased. Since plasma LDL is derived from VLDL catabolism, we investigated this paradox utilizing apolipoprotein B kinetic studies in miniature pigs. In previous studies [Huff and Telford, Arteriosclerosis 1989;9:58–66], diets containing fish oil (MaxEPA, 30 g/d, 18 days), when compared to corn oil (30 g/d, 18 days) reduced VLDL apo B concentrations, but the conversion of VLDL apo B to LDL increased 3-fold. In addition, MaxEPA reduced LDL apo B fractional catabolic rate (FCR) by 22%. Methyl-LDL turnover studies demonstrated that this effect is due to decreased hepatic B, E receptor activity. Simultaneous injection of ^{125}I-methyl-LDL and ^{131}I-native LDL demonstrated that MaxEPA reduced receptor-mediated LDL clearance by 50%. To determine if this effect could be reversed by HMG-CoA reductase inhibition, VLDL and LDL apo B kinetic studies were subsequently carried out in pigs fed MaxEPA (30 g/d, 18 days) and again following the addition of 30 mg/d lovastatin (L). Results were analyzed using SAAM multicompartmental analysis. MaxEPA + L further reduced the VLDL apo B pool size by 25% which was due to a decrease in synthesis. The conversion of VLDL apo B to LDL was reduced by 33% and VLDL apo B direct removal decreased by 5%. MaxEPA + L decreased LDL apo B concentrations by 15% due to a 36% reduction in LDL production. However, paradoxically, LDL apo B FCR was not affected. Therefore, MaxEPA + L reverses the decrease in B, E receptor activity caused by MaxEPA, such that receptor affinity for VLDL remnants is enhanced, but is unchanged for LDL. This combination may prove useful for the treatment of patients with combined elevations of VLDL and LDL.

Effect of ω3 Fatty Acids on Lipoprotein (a)

E.B. Schmidt, I.C. Klausen, S.D. Kristensen, H-H. Lervang, O. Færgeman, J. Dyerberg
Department of Clinical Chemistry, Aalborg Hospital and
Department of Medicine I, Aarhus Amtssygehus, Denmark

Marine ω3 polyunsaturated fatty acids (ω3 PUFA) may be of value in the prevention of coronary heart disease (CHD). Lipoprotein (a) [Lp(a)], described by Kåre Berg in 1963, has been shown to be an independent risk factor for development of CHD and acute myocardial infarction. Lp(a) level is considered to be unaltered by diet, but preliminary data have suggested that ω3 PUFA may lower Lp (a).

We have studied the effect of daily supplementation with 4 g of ω3 PUFA for 6 weeks on Lp(a) in healthy persons (n = 22) and in patients at high risk for CHD (patients with diabetes, n = 10 , hypertension, n = 10, hyperlipidaemia, n = 17 or angina n = 14). Finally, we examined the effect of 3 different doses of ω3 PUFA given for 6-week periods to 10 healthy males and the effect of supplementation with 4 g of ω3 PUFA for 9 months in 24 healthy volunteers on Lp(a) levels.

Lp(a) was measured using a radioimmunoassay (Pharmacia) and was unaltered in all groups after intake of ω3 PUFA. The effect of ω3 PUFA in patients with Lp(a) values above 20 mg/dl (n = 21) were analyzed separately, as such subjects are considered to be at increased risk for CHD. No consistent effect on Lp(a) was seen. However, in the subgroup of healthy volunteers with plasma Lp(a) above 20 mg/dl (n = 7) given ω3 PUFA for 9 months, a significant 15% decrease in Lp(a) was demonstrated.

We conclude, that short-term supplementation with ω3 PUFA does not alter Lp(a) levels. Long-term intake of ω3 PUFA may, however, reduce Lp(a) in subjects with high plasma levels of Lp(a).

Effects of Fish Oil Capsules or Emulsion on Chylomicronemia and Lp(a)

William S. Harris, Sheryl L. Windsor
Department of Medicine, University of Kansas Medical Center,
Kansas City, Kans., USA

Chylomicrons (chylo), their remnants and lipoprotein (Lp) (a) may all play a role in atherogenesis. Fish oils (FO) containing ω3 fatty acids (FA) have retarded the atherogenic process in both animal and human models. To study the effects of FO on these factors, 15 normal volunteers took 2.3 g ω3 FAs/day (provided from either 11 FO capsules or 2 tbsp of a FO emulsion; Dale Alexander Omega-3 Oil, Twin Labs, Inc.) for 4 weeks. Oral fat loads containing 1 g fat (predominantly cream)/kg were given in the fasting state once before and once after supplementation, and chylo triglycerides (TG) and ω3 FA levels were measured over 10 h postprandially. Supplementation with ω3 FAs lowered chylo TGs by 36% (from 589 ± 73 to 377 ± 70 mg/dl·10 h, $p < 0.01$). Fasting VLDL cholesterol and total TG levels were reduced by 25% ($p < 0.05$). HDL cholesterol levels rose from 56 ± 3 to 62 ± 5 mg/dl ($p < 0.01$) due entirely to an increase in the HDL-2 subfraction. Lp(a) levels were measured in 10 subjects and were unaffected by the FO supplement (11 ± 2 to 10 ± 1 mg/dl). There were no differences between the capsule and emulsion groups in plasma lipid, lipoprotein, or apoprotein responses, or changes in bleeding times or red blood cell deformability. ω3 FAs were absorbed at the same rate and to the same extent from capsules and emulsion.

Reductions in postprandial lipemia and increases in HDL-2 cholesterol levels resulting from an increase in ω3 FA intake may play a role in the reported ability of these FAs to retard atherogenesis. In addition, emulsified FO appears to be a reasonable alternative to FO capsules for subjects wishing to increase their intake of ω3 FAs.

Combined Treatment with Fish Oil and Aspirin Affects Bleeding Times and Platelet Function

William S. Harris, Scott Silveira and Carlos A. Dujovne
Department of Medicine, University of Kansas Medical Center,
Kansas City, Kans., USA

Aspirin (ASA) and fish oils (FO) both increase the bleeding time (BT). The effects of combined therapy have not been fully evaluated. We studied the impact of these agents, both singly and in combination, on platelet function in human volunteers. Eight healthy males took ASA (162 mg/d) for 3 days, then 9 g of FO (4.5 g ω3 fatty acids; SuperEPA, Pharmacaps, Inc.) for the next 14 days, and then ASA plus FO for the final 3 days. At baseline, after each therapy alone, and after combined therapy, BT and platelet aggregation (PLT-AGG) were measured. Compared to baseline, BT was increased by 34% ($p < 0.01$) after ASA, 9% (NS) after FO and 78% ($p < 0.01$) after the combination. ASA alone decreased platelet sensitivity to collagen and arachidonic acid but not platelet activating factor (PAF). FO alone had little effect on PLT-AGG when stimulated by any of the agonists. FO + ASA treatment did not make platelets less sensitive than did ASA treatment alone when exposed to low doses of dual agonists (collagen or PAF plus arachidonic acid). Compared to the effects of ASA alone, the combination of FO and ASA prolonged BTs but did not enhance PLT-AGG. Since PLT-AGG measures only platelet-related factors, and BTs assess the interaction of platelets and vessel wall factors, we conclude that ASA and FO may affect hemostasis by different mechanisms; ASA by a strong direct inhibition of platelet cyclooxygenase, and FO by a weak platelet effect combined with an effect on the vessel wall.

Fish Oil Supplement and PUFA Profile in Coronary Occlusion

Ralph T. Holman
Hormel Institute, University of Minnesota, Austin, Minn., USA

As part of a general study of effects of disease upon the fatty acid profile of plasma phospholipids, four cases of coronary occlusion were studied. Each had experienced coronary occlusion and coronary bypass surgery, and all were again experiencing some degree of stress or limitation of activity. All volunteered to take menhaden oil and to contribute periodic samples of blood for analysis of the fatty acids of plasma phospholipids. Prior to supplement, the composite profile of plasma PL showed a general paucity of polyunsaturated acids. Total ω6 and total ω3 PUFA were significantly low when compared to 100 normal healthy students and staff of this University. Among the saturated fatty acids, there was a skewing according to chain length: shorter-chain saturated acids were above normal, whereas longer-chain acids were suppressed despite significant elevation of saturated acids as a group. A similar skewing effect was found for the monoenoic acids. The mean chain length of the fatty acids was less and their melting point was elevated, both significantly. After one year of daily supplement with menhaden oil, the composite profile was greatly different. Total ω3 acids were elevated but total ω6 acids were suppressed, both significantly. The skewing of the saturated and monounsaturated acids persisted, and the mean chain length and mean melting point of the phospholipid fatty acids were not significantly changed.

Reduction in Microalbuminuria in Diabetics by Eicosapentaenoic Acid Ethyl Ester

T. Hamazaki[a], *E. Takazakura*[b], *K. Osawa*[b], *M. Urakaze*[a], *S. Yano*[a]
[a]First Department of Internal Medicine, Toyama Medical and Pharmaceutical University, Sugitani, Toyama; [b]Department of Internal Medicine, Kurobe City Hospital, Mikkaichi, Kurobe, Toyama, Japan

Diabetes mellitus is a major cause of chronic renal insufficiency, and the prevention of development of diabetic nephropathy is of vital importance. The appearance of microalbuminuria in diabetics is an early sign of diabetic nephropathy, and the reduction in microalbuminuria probably prevents or slows down the development of diabetic nephropathy. Here we report the effect of eicosapentaenoic acid (EPA) on albuminuria of diabetics.

EPA ethyl ester (1.8 g/d) was administered to 16 diabetic patients (5 insulin-dependent and 11 non-insulin-dependent diabetics) for 6 months. EPA mol% in total plasma fatty acids increased significantly from 4.0 ± 2.4% (mean ± SD) to 7.5 ± 3.1% ($p < 0.001$). Albumin excretion rate measured with spot urine was significantly reduced from 65 to 36 mg/g creatinine (geometric means, $p < 0.001$) during the study. Fasting blood sugar levels, glycohemoglobin, body weight and blood pressure did not change significantly during the study. There were also no significant changes in serum levels of creatinine, urea nitrogen, total cholesterol and triglycerides. Although no overt hemorrhage was observed in the patients, hematocrit was reduced significantly from 42.6 ± 2.8% to 41.0 ± 3.9% ($p < 0.02$). Ten other similar diabetic patients (4 insulin-dependent and 6 non-insulin-dependent diabetics) were followed as a reference group, not concomitantly, for 6 months with neither EPA ethyl ester nor placebo. The parameters mentioned above were not changed significantly during 6 months in this group.

EPA administration might retard the appearance of overt diabetic nephropathy.

Experimental Radioimmunotherapy of Breast Cancer is Modified by Polyunsaturated Fatty Acids[1,2]

Roberto L. Ceriani, Edward W. Blank, Chu Y. Lin
John Muir Cancer Research Institute, Walnut Creek, Calif., USA

BALB/c nude mice (nu/nu) were fed diets containing fat from either fish oil (MaxEPA), corn oil, or lard. Experimental animals carried established transplantable human breast tumors (MX-1) that were treated with either an unconjugated mixture (cocktail) or an ^{131}I-labeled cocktail of four (Mc1, Mc3, Mc5 and Mc8) anti-human milk fat globule monoclonal antibodies (MoAbs). Mean tumor volume of MaxEPA-fed mice was reduced to 36% of both corn oil and lard fed groups, respectively. However, injection of unconjugated MoAbs reduced tumor volumes only in corn oil (high 18:2 fatty acid)-fed nude mice when each group was compared to their respective controls. In contrast, injection of ^{131}I-MoAbs produced large tumor volume reductions in all dietary groups, showing a synergistic effect and the greatest tumor reduction with MaxEPA (low 18:2 fatty acid) and ^{131}I-labeled MoAbs. No differences in uptake or retention of the dose in tumor or other tissues was shown in biodistribution studies with the different diets. The different diets did not cause any significant change in total target antigen content per cell nor in its surface-cytoplasmic distribution, as determined by flow cytometry. Fatty acid analysis of serum, liver and tumor demonstrated the incorporation of characteristic fatty acids into each tissue and confirmed dietary manipulation. The low amount of tumor promoting 18:2 fatty acid and increased amount of 20:5 and 22:6 in MaxEPA that were incorporated into the tumors that were targets for the experimental radioimmunotherapy may be responsible for the increased effectiveness of ^{131}I-MoAbs.

Reversal of Cancer Cachexia by ω3 Fatty Acids

M.J. Tisdale and J.K. Dhesi
Cancer Research Campaign Experimental Chemotherapy Group,
Pharmaceutical Sciences Institute, Aston University, Birmingham, UK

Cachexia is common in malignancy and is a significant contributory factor for the severe morbidity and mortality in this disease. Patients with cachexia are characterized by severe weight loss and depletion of host reserves of both lipid and protein. We have utilized a transplantable murine adenocarcinoma (MAC16) as an experimental model of human cancer cachexia. This tumour produces profound weight loss in the host with extensive loss of both adipose tissue and lean body mass at tumour burdens less than 2% of the host body weight. Moreover, weight loss occurs without a reduction in either water or caloric intake, suggesting a metabolic effect of the tumour on the host.

We have investigated the effect of substitution of the carbohydrate component of the diet by calories derived from lipid in the form of fish oil on both host body weight loss and tumour growth rate in the MAC16 model. Diets containing fish oil significantly reduced host body weight loss in a dose-related fashion, with complete protection occurring when the fish oil comprised 50% of the calories. The effect occurred without untoward effects, and without an alteration in the total calorie consumption or nitrogen intake suggesting that the ω3 fatty acids in the fish oil were blocking the metabolic effects of the tumour on the host. The reduction of host body weight loss was associated with an increase in total body fat and muscle mass. In addition a significant reduction in tumour growth rate was observed, although the reduction in host weight loss was greater than might be expected from a smaller tumour burden. The effect appears specific to the type of fat since comparable results were not obtained with a gamma linolenic acid-enriched diet. When compared with cyclophosphamide and 5-fluorouracil, the fish oil diet exerted a similar antitumour effect at the maximum dose employed, but whereas the anitumour effect of the former agents was achieved with considerable host toxicity, the latter produced no toxicity, and almost completely abolished the cachectic effect of the tumour.

These results suggest that fish oil is a non-toxic highly effective, anticachectic agent with the added advantage of antitumour activity.

Trial of Fish Oil-Supplemented Diet in Ulcerative Colitis

W.F. Stenson, D. Cort, W. Beeken, J. Rodgers, R. Burakoff
Jewish Hospital of St. Louis, St. Louis, Mo; University of Vermont, Burlington, Vt.; Albany Medical College, Albany, N.Y.; Winthrop University Hospital, Mineola, N.Y., USA

Leukotriene B_4 (LTB_4) is an important mediator of inflammation in ulcerative colitis. LTB_4 is produced by activated neutrophils and is thought to recruit additional neutrophils from the bloodstream into the mucosa. Dietary supplementation with eicosapentaenoic acid, a fatty acid found in fish oil, decreases LTB_4 production by human neutrophils. We have conducted a study of the effects of fish oil supplementation on ulcerative colitis. Patients with active ulcerative colitis were entered in a double-blind crossover trial of fish oil supplementation (18 MaxEPA tablets per day) compared with placebo (18 vegetable oil capsules per day). Each diet period was 4 months long, the two diet periods were separated by a 1 month wash-out period. Previous medications were continued. Patients were evaluated for symptoms, sigmoidoscopy score, rectal histology, myeloperoxidase levels in rectal mucosa, and PGE_2 and LTB_4 levels in rectal dialysates. Patients were evaluated at baseline, after the first diet period, and after the second diet period. After the fish oil-supplemented diet period, sigmoidoscopy scores improved in 15 patients, were unchanged in 2 and worsened in 4 ($p < 0.05$ using sign test). After the placebo diet period, sigmoidoscopy scores improved in 8, were unchanged in 4 and worsened in 8. After the fish oil-supplemented diet period, global clinical assessment improved in 9, was unchanged in 9 and worsened in 1 ($p < 0.05$). After the placebo diet period, global clinical assessment improved in 3, was unchanged in12 and worsened in 3. Analysis of the biochemical parameters and histology is in progress.

Conclusion: Preliminary analysis shows statistically significant improvement in sigmoidoscopy score and global clinical assessment after 4 months of fish oil-supplemented diet compared to placebo diet in active ulcerative colitis.

Reduction of AZT-Induced Fatigue in AIDS by Dietary Supplementation with Combinations of ω3 and ω6 Fatty Acids

Barry Sears, Paul Kahl
BioSyn, Inc., Marblehead Mass.; Medic, Inc., Pittsburgh, Pa., USA

AZT has significant side effects, in particular the reduction of red blood cell count which gives rise to significant fatigue due to the lack of oxygen transfer to the peripheral tissue. We hypothesized that dietary intervention with essential fatty acids could increase cardiovascular efficiency in patients with AZT-induced anemia, and thereby provide a corresponding reduction in their fatigue.

Eight patients were recruited for a double-blind, placebo-controlled study. All patients had chronic fatigue and were using 1,200 mg AZT per day. Those in the active group received a total of 8 g combined marine and vegetable oils containing 960 mg eicosapentaenoic acid (EPA) and 192 mg gamma-linolenic acid (GLA) per day. Those in the placebo group received 8 g olive oil as a control.

CBC and Chem 20 tests were run prior to supplementation and at every 2 months until the end of the study at 6 months. Fatigue was self-assessed by patients every 2 months on the following scale: -2 (significantly more fatigue), -1 (more fatigue), 0 (no change), $+1$ (less fatigue), and $+2$ (significantly less fatigue). At the end of 6 months, the code was broken and the results were analyzed by Students t test. Those in the active group had an overall fatigue rating of $+1.0 \pm 0.4$, whereas those in the placebo group had a fatigue rating of -0.5 ± 0.3. The differences between the two groups had a statistical significance of $p \leq 0.025$. During this time period, the red cell counts of the patients in active and placebo groups did not change significantly nor did elecrolyte levels in the plasma.

These results indicate that dietary supplementation with combinations of ω3 and ω6 fatty acids can reduce AZT-induced fatigue in AIDS patients. The possible mechanisms will be discussed.

Rheological Characteristics of Red Blood Cells (RBC) from Humans Given Fish Oil and Vitamin E

A.R. Mangels, V.C. Morris, P.P. Nair, H.N. Bhagavan, J.T. Judd, R. Ballard-Barbash, O.A. Levander
USDA Human Nutrition Research Center, Beltsville, Md.;
Hoffman-LaRoche, Nutley, N.J.; NCI, NIH, Bethesda, Md., USA

Vitamin E (VE) deficiency in rats decreases RBC filterability (Levander et al, Ann NY Acad Sci 1980;355:277). Such impaired RBC deformability is directly related to lipid peroxidation and can be exacerbated by prooxidant stress. Here we report the effect of fish oil (FO) and VE supplementation of the filterability of human RBC under varied conditions of incubation and storage and after in vitro exposure to prooxidant stressors. Forty men were maintained for a total of 28 weeks on a controlled basal diet providing 40% energy from fat and an estimated 10 IU/day of dietary VE. During the first 10 weeks they were given placebo capsules (15×1 g/day containing a blend of fats simulating that in the diet and 1 IU/capsule of VE). This was followed by a second 10-week period during which they received 15 × 1 g capsules of fish oil/day (containing 50% ω3 fatty acids and 1 IU/capsule of VE). During the final 8 weeks of the study the subjects continued the 15 × 1 g capsules of fish oil/day and received an additional 200 IU/day of VE (FO + VE). RBC filterability of a subset of subjects was measured at the end of each period following 0 or 6 h incubation in buffered saline, after 4 days storage as heparinized blood, and after incubation in vitro with primaquine, hydrogen peroxide, or phenylhydrazine.

The time required to filter fresh RBC that were incubated for 6 h in buffered saline tended to decrease somewhat (i.e., filterability improved) during the FO + VE period (data presented below as M ± SE). Storage of RBC as heparinized blood markedly impaired filterability of cells from the placebo period, an effect that tended to be reversed during the FO period. Supplementation with FO + VE resulted in a marked improvement in filterability of stored RBC.

Storage treatment	Filtration time (s) after 6 h incubation		
	placebo	FO	FO + VE
Fresh	10±1	9±1	7±1
4 days at 8 °C	356±66	222±44	35±5

RBC filterability decreased after incubation with all prooxidants tested but was not affected by FO supplementation while additional vitamin E (FO + VE) resulted in some improvement (data not shown).

These results show that supplementation of human volunteers with fish oil did not alter the susceptibility of their RBC to a variety of oxidant and metabolic stresses. The enhanced stability of RBC under a variety of incubation and storage conditions from subjects receiving both the fish oil and vitamin E supplements (FO + VE) may have important implications for handling of human blood in vitro (blood preservation, culture of malarial parasites in RBC, etc.).

Antimalarial Activity of a Marine ω3 Free Fatty Acid Concentrate (FFAC) in Mice Fed Graded Dietary Levels of Vitamin E (VE)

O.A. Levander, A.L. Ager, V.C. Morris and R.G. May
USDA Human Nutrition Research Center, Beltsville Md.;
Center of Tropical Parasitic Diseases, U Miami, Miami, Fla., USA

Malaria is the most serious parasitic disease in the world today, causing an estimated 300 million cases and 1 million deaths annually. Particularly worrisome is the rapid global spread of parasite strains that are resistant to commonly used antimalarial drugs. Previously we have shown that menhaden oil (MO) has a strong antimalarial effect even against a chloroquine-resistant strain of the parasite when fed to mice in diets severely deficient in VE (Am J Clin Nutr 1989;50:1237). Here we report the antimalarial efficacy of FFAC fed in diets containing graded levels of VE. Weanling male mice were fed 20% casein diets containing either 4% FFAC or MO. The diets also contained 1% stripped corn oil, 20 mg/kg *p*-aminobenzoic acid and VE (*d*-α-tocopheryl acetate) added as indicated. Chow-fed mice were included as additional controls. After 4 weeks, all mice were inoculated i.p. with *Plasmodium yoelii* and parasitemia (% PARA) and survival were monitored:

Diet	Diet VE mg/kg	% PARA, days postinjection			Survival, days postinjection		
		6	13	20	20	27	45
MO	0	<1	6	13	10/10	10/10	10/10
MO	38.4	4	29	36	6/10	3/10	2/10
FFAC	0	<1	<1	<1	10/10	10/10	10/10
FFAC	4.8	1	31	21	10/10	9/10	9/10
FFAC	9.6	1	44	37	10/10	9/10	9/10
FFAC	19.2	1	36	21	9/10	7/10	7/10
FFAC	38.4	6	44	42	5/10	4/10	4/10
Chow	–	20	23	41	2/10	1/10	1/10

As reported previously, FFAC exerted a stronger suppressive effect (lower % PARA) than MO when fed in a diet severely deficient in VE although ultimate survival was similar with both diets. Feeding as little as 4.8 mg VE/kg in the FFAC diet (one eighth the current NRC recommended level) caused marked elevations in % PARA. Nonetheless, survival was enhanced in mice receiving as much as 19.2 mg VE/kg in the FFAC diet (one half the NRC level). The beneficial effect of FFAC seen in mice only marginally deficient in VE suggests that fish oil or fish oil concentrates may have a useful role in the prophylaxis and/or therapy of human malaria.

(Partial support: US Army, DAMD 17-85-C-5077).

Simopoulos AP, Kifer RR, Martin RE, Barlow SM (eds): Health Effects of ω3 Polyunsaturated Fatty Acids in Seafoods. World Rev Nutr Diet. Basel, Karger, 1991, vol 66, pp 537–572

Poster Session II

(Abstracts)

Modification of Fatty Acid Composition of Rat Colostrum and Mature Breast Milk by Dietary α-Linolenic Acid Supplementation

P. Guesnet, G. Pascal, G. Durand
Laboratoire de Nutrition et Sécurité Alimentaire, INRA, Jouy-en-Josas, France

ω3 Polyunsaturated fatty acid (ω3 PUFA) requirement was studied during the foetal and postnatal rat growth, a physiological status of high development of membrane structure (particularly nervous membranes). The aim of the present work was to determine α-linolenic acid (18:3 ω3) requirement during pregnancy and lactation by the measurement of PUFA contents from colostrum and mature breast milk lipids. Wistar female rats from a ω3 PUFA-deficient lineage (2nd generation) were used. The α-linolenic acid-deficient (or -low) diet supplied about 1,000 mg of linoleic acid/100 g diet but low amount (about 5 mg/100 g diet) of 18:3 ω3. Two weeks before mating and until postpartum day 14, they were divided into five equal groups in which dietary lipids respectively supplied 0, 100, 200, 400 and 800 mg of 18:3 ω3/100 g diet. To determine the importance of maternal diet before and after parturition on PUFA milk composition, two further groups were constituted in which dietary lipids, providing ω3 PUFA or not, were exchanged at birth. Fatty acid composition of milk lipids was determined by the PUFA lipid stomach content of pups at postpartum day 1 (colostrum), then at postpartum days 3, 7, and 14 (mature breast milk). Fatty acid composition of pup brain lipids was also investigated.

The results showed that: (1) In milk lipids, the arachidonic (20:4 ω6) and docosahexaenoic (22:6 ω3) levels, abundant in colostrum (respectively 2.5 and 1 %), were rapidly diminished until the lactation day 7. The proportion of 22:6 ω3 reached a plateau when the diet provided 400 mg of 18:3 ω3/100 g diet. Meanwhile, this proportion was below the plateau value when the diet only provided 200 mg, and that of 22:5 ω6 increased abnormally, an indication of subnormal α-linolenic supply. On the other hand, when the diet provided 800 mg/100 g diet, the level of eicosapentaenoic acid (20:5 ω3) increased, and that of 20:4 ω6 dropped, an indication of excess α-linolenic supply. Lastly, when the dietary lipids were exchanged at birth, PUFA composition of mature breast milk corresponded to that of switched dietary lipids as early as postpartum day 3. (2) In brain lipids of pups, the proportion of 22:6 ω3 reached a plateau when the diet supplied 400 mg/100 g diet. Below 200 mg, we observed an increase of 22:5 ω6, and above 400 mg, an increase of 22:5 ω3.

We conclude that in the female rat during the end pregnancy-beginning lactation period (1) the α-linolenic requirement is higher than 200 mg/100 g diet, but lower or equal to 400 mg; (2) the PUFA level in milk is rapidly modified by the lactating maternal diet whatever the maternal diet used during pregnancy; this sets the question about the importance of PUFA maternal stores in a nonruminant mammal like the rat.

Effects of Age and Dietary 18:3 ω6 and 18:4 ω3 on the Delta 6-Desaturations in Rat Liver Microsomes

J.P. Blond[a], *P. Guesnet*[b], *L. Ulmann*[a], *C. Maniongui*[a], *J. Bézard*[a], *J.P. Poisson*[a], *G. Durand*[b], *G. Pascal*[b]

[a]Université de Bourgogne, Dijon; [b]I.N.R.A., Jouy-en-Josas, France

The present study compares the effect of a partial replacement of dietary linoleic acid (18:2 ω6) or alpha-linolenic acid (18:3 ω3) by gamma-linolenic acid (18:3 ω6) or stearidonic acid (18:4 ω3), respectively, on the metabolism of 18:2 ω6 and 18:3 ω3 in adult rat liver at different ages. After weaning, rats were fed a control diet (containing ca. 980 mg of 18:2 ω6 and 188 mg of 18:3 ω3 in 100 g of diet), for 8, 19 and 29 weeks. At each age, two groups of 6 animals were given, for an age-dependent time, a diet including, for 100 g of diet, 170 mg of 18:3 ω6 (Borage group = Bg) or 207 mg of 18:3 ω6 + 38 mg of 18:4 ω3 (Blackcurrant group = Bc) while a third group (Control group = C) continued receiving the control diet. All the diets presented ω6 and ω3 contents similar to that of the control diet and a similar ratio ω6/ω3 near to 5.2. For each age (3, 6 and 9 months) and each group (C, Bc and Bg), delta 6-desaturation activities with [^{14}C]18:2 ω6 and [^{14}C]18:3 ω3 as substrates at a saturating concentration were measured in liver microsomes. Fatty acid composition of microsomal phospholipids was also determined.

The delta 6-desaturation reactions were not influenced in an identical manner by the two factors studied – age and diet. The desaturation rate of 18:3 ω3 was practically independent of age and diet, whereas that of 18:2 ω6 was lower at 3 months and increased at 6 months, without any significant effect of diet. On the contrary, at 9 months, this level was maintained with the diet containing 18:3 ω6 and unexpectedly was decreased with the control diet. In microsome phospholipids, the percentage of 20:4 ω6 slightly increased with age while the ratio 20:4/18:2 did not change. However, this ratio was more elevated in the diets containing 18:3 ω6, as was (slightly) the percentage of 20:4. On the contrary, age and diet hardly had any influence on the level of 22:6, the major ω3 fatty acid.

Our results show that substitution of a low amount of 18:4 ω3 for the same amount of 18:3 ω3 in the diet does not significantly change the content of polyunsaturated ω3 fatty acids of microsome phospholipids, whereas the same operation in the ω6 series induces a significant effect.

Influence of Dietary Fish on Eicosanoid Metabolism in Man

G. Hornstra[a], *A.C. van Houwelingen*[a], *G.A.A. Kivits*[b], *S. Fischer*[c], *W. Uedelhoven*[d], *M. Thorngren*[e]

[a] Department of Human Biology, Limburg University, Maastricht, The Netherlands;
[b] Department of Biosciences, Nutrition and Safety, Unilever Research, Vlaardingen, The Netherlands;
[c] The Medical University, Innenstadt Hospital, Munich, FRG;
[d] Institut für Prophylaxe der Kreislaufkrankheiten, Munich, FRG;
[e] Department of Geriatrics, University Hospital, Lund, Sweden

To investigate the effect of a reasonable amount of fish in the diet on the eicosanoid metabolism in man, two groups of 40 volunteers were given a dietary supplement consisting of 135 g of mackerel or meat (= control) paste per day for 6 weeks.

Compliance measured by the urinary recovery of lithium added to the supplements was about 80% in both groups, and the daily intake of timnodonic acid (20:5 ω3, TA) and cervonic acid (22:6 ω3, CA) from the mackerel supplement was about 1.3 and 2.3 g, respectively.

In collagen-activated platelet-rich plasma the production of HHT and HETE from arachidonic acid (AA) greatly reduced in the mackerel group, whereas the formation of products derived from TA (HHTE and HEPE) increased slightly, but significantly. Amounts of HHT and HHTE, measured by HPLC, correlated significantly with amounts of TxB_2 and TxB_3, respectively, measured by GC/MS.

Formation of $irTxB_2$ in clotting blood was significantly decreased in the mackerel group. In collagen-activated citrated blood, $irTxB_2$ formation tended to be reduced in the mackerel-supplemented volunteers, but the reduction did not attain statistical significance. Since the TxB_2 response was significantly correlated to the dietary adherence, a higher fish consumption would certainly have caused a significantly lower TxB_2 formation in collagen-activated blood. In blood oozing from a bleeding time skin incision, the $irTxB_2$ content was significantly diminished by the mackerel supplement.

The mackerel supplement had a variable effect on the urinary content of the major PGI_2 metabolite 2,3-dinor-6-keto-$PGF1_{\alpha}$. The amount of urinary 2,3-dinor-Δ_{17}-6-keto-$PGF1_{\alpha}$, the major metabolite of PGI_3 increased significantly.

No change was observed in the total daily turnover of prostaglandins E and F, as reflected by the amount of their tetranor metabolites in urine.

Fatty acid composition of the major classes of platelet phospholipids (PL) revealed that, in the mackerel group, TA and CA greatly increased in most PL classes at the expense of AA. However, these changes hardly affected the total degree of PL unsaturation, as reflected by the unsaturation index (UI). In the sphingomyelin fraction, on the other hand, the mackerel supplement caused an almost 20% increase in the UI, which was mainly due to significant reductions in most saturated fatty acids (20:0, 22:0 and 24:0 in particular), in association with a large increase in 24:1 ω9. The functional implications of these findings will be discussed.

Supported by a grant from the Netherlands Heart Foundation. The International Association of Fish Meal Manufacturers provided the mackerel, used in this study.

Fish Fatty Acids and Pregnancy-Induced Hypertension: Is There a Link?

G. Hornstra[a], *J.M. Gerrard*[b], *A.C. van Houwelingen*[a], *M.D.M.A. van der Schouw*[a], *Y.T. van der Schouw*[a]

[a] Department of Human Biology, Limburg University, Maastricht, The Netherlands;
[b] Department of Pediatrics, University of Manitoba, Winnipeg, Canada

In normal neonates born of Dutch mothers, the amount of osmond acid (22:5 ω6, OA) is higher in phospholipids of umbilical arterial walls as compared to umbilical venous walls, whereas all other ω6 fatty acids are lower. Since an increased OA content is a biochemical marker of a relative cervonic acid (22:6 ω3, CA) deficiency, this positive arterio-venous (A-V) difference for OA suggests that the CA status of peripheral fetal tissue may not be optimal.

In umbilical vessel walls of neonates born of Inuit mothers, whose diet is relatively enriched with marine ω3 polyenes as compared with that of Dutch mothers, the A-V difference for OA is negative, just as that for the other ω6 fatty acids. This strongly suggests that the neonatal CA status can be improved by increasing the maternal intake of marine polyenes.

In pregnancy-induced hypertension (PIH), the A-V difference for OA is much more positive than in normotensive pregnancy (NP). This indicates a higher fetal CA requirement under PIH conditions, which may form the rationale for a CA supplementation therapy in women at risk of PIH. This view is supported by the fact that, retrospectively, the blood pressure at the end but not at the beginning of pregnancy is significantly lower in Inuit mothers from 'fish eating communities' than from 'Caribou eating settlements'.

Recently we demonstrated that indications of PIH are less common in pregnant women from 'fish communities' as compared with 'meat communities'. This difference persisted after correction for other variables known to influence the risk of PIH. This finding supports a possible preventive role of fish polyenes in PIH.

In contrast, Carter et al. [South Med J 1987; 80: 692–697] found PIH to be extremely rare in a community of vegans, eating no animal-derived products and, consequently, no fish fatty acids. Since Sanders et al. [Am J Clin Nutr 1978; 31: 805–813] demonstrated that the CA status is very poor in vegan mothers and their babies, these results are not in line with a role of ω3 polyenes in PIH prevention. Moreover, we recently demonstrated that the maternal CA status is enhanced in PIH as compared with NP. This was associated with an improved fetal CA status, as reflected by a lower OA content in umbilical vein phospholipids, and the latter could also explain the more positive A-V difference for OA in umbilical vessel walls under PIH conditions.

These results indicate PIH to be associated with an enhanced CA status, which would argue against a CA supplementation therapy in patients at risk of PIH. Further studies are required to unravel the relationship between CA and PIH, if any, before fish oil supplementation to pregnant women can be considered.

Interaction between Dietary ω3 or ω6 Polyenes and Saturated Fatty Acids: Effect on Arterial Thrombosis in Rats in Vivo

G. Hornstra[a], *A.D.M. Kester*[b], *A.A.H.M. Hennissen*[a]
Departments of [a]Human Biology and [b]Medical Information and Statistics, Limburg University, Maastricht, The Netherlands

Although the importance of dietary lipids and arterial thrombosis in the development of cardiovascular disease is generally recognized, the lack of a suitable in-vivo technique prevents a systematic study of the influence of dietary fatty acids on arterial thrombosis tendency in man. We, therefore, used a rat model, based on measuring the time (obstruction time, OT, h) between insertion and complete thrombotic obstruction of a polythene, loop-shaped cannula, inserted into the abdominal aorta of the animals [Atherosclerosis 1975;22:499]. Extensive validation of this method demonstrated that the results reflect human arterial thrombosis tendency in a reliable way.

Diets (50% fat), containing various vegetable oils (n = 19) or oils and fats from marine (n = 8) or terrestrial (n = 2) animals, were given for 8–10 weeks. Fourteen different studies were performed, in each of which sunflowerseed was included as a reference oil. In total 60 groups of 16–20 animals were involved in the study; some experimental oils were tested repeatedly. Fatty acid compositions of the lipids, extracted from the complete diets, were measured by gas-liquid chromatography.

The rank correlation between the relative OT (% of the 'reference OT' = y) and the sum of all dietary saturated fatty acids, S (= x) was highly significant: $y = 53.2 - 0.74x$; $r_s = -0.74$; $p < 0.0001$). The same was true for the sum of all polyunsaturated fatty acids (P): $y = 8.8 + 0.71x$; $r_s = 0.71$; $p < 0.0001$. The ω6 and ω3 polyenes were equally antithrombotic; ω6: $y = 16.1 + 0.47x$; $r_s = 0.47$; $p < 0.001$; ω3: $y = 16.2 + 0.47x$; $r_s = 0.47$; $p < 0.001$. Linoleic acid, 18:2 ω6, appeared highly antithrombotic ($y = 8.8 + 0.46x$; $r_s = 0.46$; $p < 0.001$), but the fish fatty acids 20:5 ω3 and 22:6 ω3 were not significantly related to arterial thrombosis tendency. The same was true for the main monoene, oleic acid, 18:1 ω9.

A strong, linear relationship was observed between the P/S ratio of the dietary lipids (= z) and the relative OT: $y = 95 + 24.6 \log z$; $r = 0.71$; $p < 0.0001$. The slope of the relationship curve was significantly ($p < 0.01$, F-test) steeper for the marine lipids (y =

101 + 107.1 z; r = 0.73; p = 0.0387) than for the terrestrial polyenes (y = 91 + 21.1 z; r = 0.80; $p < 0.0001$). This implies that the antithrombotic effect of the marine polyenes may be more sensitive than the terrestrial polyenes to the prothrombotic effect of saturated fatty acids.

These studies demonstrate that all common S increase arterial thrombosis tendency. Linoleic acid reduces arterial thrombosis, whereas oleic acid does not have such an effect. Dietary marine oils are antithrombotic only if they combine a high polyene with a low saturate content.

Oral ω3 Fatty Acid Supplementation Suppresses Cytokine Production and Lymphocyte Proliferation: Comparison in Young and Older Women

S.N. Meydani, S. Endres, M.M. Woods, B.R. Goldin, C. Soo, A. Morrill-Labrode, Ch.A. Dinarello, S.L. Gorbach
USDA Human Nutrition Research Center on Aging at Tufts University, Boston, Mass., USA

Animal experiments and clinical trials have indicated a potentially beneficial effect of ω3 fatty acid (FA) supplementation on atherosclerosis and atherothrombotic disorders, autoimmune, and inflammatory diseases, the prevalence of all of which increases with age. Aging is associated with an altered regulation of the immune system especially of T cell-mediated functions. Proliferative response of lymphocytes to T cell mitogens and interleukin (IL)-2 production have been shown to decrease with age. Short-term supplementation with ω3 FA has been shown to decrease the inducible production of IL-1 and tumor necrosis factor (TNF). The effect of supplementation with low levels of ω3 FA for long periods of time on cytokine synthesis and lymphocyte proliferation of healthy subjects has not been studied. Furthermore, the question of age difference in ω3-induced changes has not been addressed. Therefore, the effect of ω3 fatty acid supplementation on cytokine production and lymphocyte proliferation was studied in 6 young (ages 23–33) and 6 older (ages 51–68) women. Each subject supplemented her typical American diet with 2.4 g/day ω3 FA for a period of 3 months. Blood was collected before and after 1, 2, and 3 months of oral supplementation. ω3 FA supplementation reduced total IL-1β synthesis 2-fold (p = 0.01) in young but 10-fold (p = 0.004) in older subjects; TNF was reduced by 58% in young and 70% in older women (p = 0.02). IL-6 production was reduced in young women by 30% (p = 0.004) but by 60% (p = 0.001) in older women. Older women produced less IL-2 (p = 0.01) and had lower mitogenic responses to PHA (p = 0.04) than young women. ω3 FA supplementation reduced IL-2 production in both groups; however, this reduction was only significant in older women (63%, p = 0.05). PHA-stimulated mitogenic response was significantly reduced by ω3 FA in older women only (36%, p = 0.04) but was not affected in young women. Thus, long-term ω3 supplementation reduces cytokine production and T cell mitogenesis in young and older women. The reduction is more dramatic in older women than young women. The reduction in cytokine synthesis associated with ω3 FA supplementation may contribute to their reported anti-inflammatory effect. However, since T cell-mediated function declines with age, ω3 FA-induced suppression of IL-2 production and lymphocyte proliferation may not be desirable in older adults.

Comparative Effects of Fish Oil and Vegetable Oils Rich in ω3 Fatty Acids on Platelet Aggregation, Red Blood Cell Deformability and Fatty Acid Composition in Elderly People

F. Driss[a], *Ph. Darcet*[a], *O. Henry*[b], *J.B. Leblond*[b]
[a] G.R.N.V., Hôpital Sainte Périne, Paris;
[b] Hôpital Georges Clémenceau, Champcueil, France

Vegetable oils such as a low-erucic acid rapseed oil (LEAR) or blackcurrant oil have been proposed as a source for long-chain ω3 fatty acids, although alpha-linolenic acid is known to be poorly converted into EPA and DHA in man. This study was carried out in elderly subjects to compare efficiency of low doses of EPA, provided either by fish oil or vegetable oils, on biological parameters involved in thrombotic disorders (platelet aggregation, red cell deformability, cholesterol, apoproteins).

Group A: 25 elderly subjects (mean age 78 ± 8 years) were asked to replace in their diet usual sunflower oil by equivalent amounts of LEAR oil rich in alpha linolenic acid. The ratio of linoleic/linolenic acid in the diet was then decreased from 33.5 to 6.5.

Group B: 20 subjects matched for sex and age were volunteers to supplement their initial comparable diet with 2 g/day of encapsulated fish oil (18% EPA + 12% DHA). Diet was followed up for 6 months. Results were as follows:

- Red cell deformability expressed as rigidity index (RI) was improved in group A (from 9.62 ± 1.75 to 7.44 ± 1.05; $p < 0.001$) and group B (from 9.21 ± 1.60 to 6.44 ± 1.37; $p < 0.001$).
- Platelet aggregation slightly increased in group A (from 25.5 ± 8.7 to 27.1 ± 8.20) and was significantly decreased in group B (from 28.9 ± 9.1 to 14.8 ± 7.5; $p < 0.01$). Absolute number of lymphocytes were significantly enhanced only in group B (1,100/mm^3 to 2,000/mm^3; $p < 0.01$).
- Fatty acid pattern in plasma and red cell membrane was modified by the diet: in group A: DHA and AA increased respectively from 4.19 ± 1.69 to 6.67 ± 1.94 ($p < 0.001$) and from 12.10 ± 2.76 to 14.4 ± 1.68 ($p < 0.01$) at the expense of palmitic acid which decreased from 31.7 ± 2.07 to 26.9 ± 2.27 ($p < 0.001$). EPA level remained constant; in group B: Linoleic acid content significantly decreased from 14.47 ± 0.51 to 10.17 ± 0.61 ($p < 0.01$) while EPA was enhanced (0.62 ± 0.02 to 0.86 ± 0.03; $p < 0.05$).

These results suggest that under our experimental conditions alpha-linolenic acid is not converted into EPA but into DHA which play an important role in red cell deformability. Weak amounts of EPA provided by fish oil are more potent to reduce platelet aggregation than alpha-linolenic acid.

High Doses of Fish Oil May Be Less Clinically Effective than Expected because They Lower Dihomo-Gamma-Linolenic Acid Concentrations

Y.-S. Huang, D.F. Horrobin
Efamol Research Institute, Kentville, Nova Scotia, Canada

ω3 Essential fatty acids (EFA) in the forms of fish oil and concentrates are being widely investigated, particularly with regard to therapeutic use in cardiovascular and inflammatory disorders. The rationale for this is, in part, to reduce the levels of arachidonic acid (AA) by inhibiting its formation from linoleic acid and displacing it from phospholipids. AA has potentially pro-inflammatory, pro-thrombotic and vasoconstrictor actions. However, high doses of ω3 EFAs also reduce the concentrations of dihomo-gamma-linolenic acid (DGLA) and its immediate precursor, gamma-linolenic acid (GLA). Since DGLA is a potent anti-inflammatory, anti-thrombotic and vasodilator agent, this reduction of DGLA may limit the favorable effects of high-dose fish oils. It is often forgotten that Eskimos have been reported to have high DGLA levels, possibly because of impaired delta-5-desaturation: this is in sharp contrast to what happens when other people take high doses of marine oils. This effect in Eskimos can be mimicked in non-Eskimos by providing GLA or DGLA in addition to the marine oil. GLA by-passes the rate-limiting delta-6-desaturation step which is inhibited by ω3 EFAs and is rapidly elongated to DGLA. DGLA is converted to AA only slowly, if at all, partly because delta-5-desaturation is inherently slow, and partly because the ω3 EFAs inhibit it. Administering GLA/DGLA together with a high dose of ω3 EFAs successfully imitates the situation in Eskimos: it prevents the depletion of DGLA in non-Eskimos.

Metabolism of Arachidonic (AA) and Eicosapentaenoic (EPA) Acids by Fish Brain Cells

D.R. Tocher, J.G. Bell, J.R. Sargent
NERC Unit of Aquatic Biochemistry, School of Natural Sciences,
University of Stirling, Stirling, UK

In addition to very high levels of 22:6 ω3, and unlike mammals, fish neural tissues contain relatively large amounts of EPA and low levels of AA. EPA and, especially, AA are concentrated in the phosphatidylinositol (PI) fraction of the membrane phosphoglycerides. These facts suggest that EPA and PI may play significant roles in eicosanoid metabolism in fish neural tissues. We have studied the incorporation and metabolism of AA and EPA in isolated brain cell suspensions from rainbow trout (*Salmo gairdneri*). Cell suspensions were prepared from chopped brains by mechanical sieving through nylon gauzes. Cells were incubated with an equimolar (2 μ*M*) mixture of ^{3}H-AA and ^{14}C-EPA (^{3}H:^{14}C = 11.7) at 15 °C and incorporation into individual phosphoglycerides determined by lipid extraction, HPTLC separation, scraping and scintillation counting. Results were corrected for quench, overlap and efficiency.

The incorporation of the acids into total cell lipid paralleled each other and plateaued by 4–6 h incubation. The ^3H:^{14}C ratio in total lipid was 14.3 ± 1.3 (mean ± SD, n = 4) after 15 min incubation but steadily declined reaching 7.8 ± 0.2 at 6 h. Incorporation into phosphoglycerides showed the same time course as for total lipid with the mass of both acids being incorporated primarily into phosphatidylethanolamine (54.9 ± 3.2% and 57.2 ± 1.9% for AA and EPA at 6 h, respectively). Incorporation of AA after 6 h was greater into PI (19.7 ± 3.6% vs. 9.5 ± 0.7%, $p < 0.001$) and less into phosphatidylcholine (PC) (18.4 ± 1.2% vs. 27.1 ± 1.4%; $p < 0.002$) compared to EPA. Expressing the incorporation per μg of lipid class showed that both acids were preferentially incorporated into PI. However, the specificity was greater for AA as the ^3H:^{14}C ratio in PI averaged 24.0 ± 3.5 compared to 11.7 ± 0.7 in the supplemented acids ($p < 0.001$). In contrast, the ^3H:^{14}C ratio in PC averaged 8.4 ± 1.0 ($p < 0.002$ compared to the supplemented ratio) with the ratio in other phosphoglycerides not significantly different to the supplemented ratio.

The ^3H:^{14}C ratio of total eicosanoids (including free acid) extracted from supernatants after stimulation of dual-labelled cells with 20 μ*M* calcium ionophore A23187 for 30 min at 15 °C was 8.1 ± 1.1. Similarly, the ratio in lipoxygenase products (after HPLC) was 8.3 ± 1.8. This suggested that PI was an unlikely source for the released eicosanoid precursors and that there was no specificity in the conversion of the released acids via lipoxygenase to hydroxylated derivatives. The 12-hydroxy derivatives of AA and EPA were the major lipoxygenase products accounting for 38.1 ± 6.5% and 45.7 ± 4.7% of the totals, respectively.

Alterations in Cerebral Microvessel Fatty Acid Composition and Arachidonate Cascade Induced by High Dietary Fish Oil in Rats

J. Kalman[a], *J. Gecse*[b], *T. Farkas*[c]
[a] Department of Neurology and Psychiatry, [b] Department of Pathophysiology, Albert Szent-Györgyi Medical University, Szeged, [c] Institute of Biochemistry, Biological Research Center, Szeged, Hungary

Brain cortical microvessels are rich in the enzymes of arachidonate cascade. The cyclooxygenase and lipoxygenase metabolites of arachidonic acid may mediate some cerebrovascular dysfunctions, such as stroke, brain trauma and seizures.

Male weanling Wistar rats were maintained on one of two semisynthetic diets, differing only in the type of oil used: 10% by weight fish oil (FO group), containing 20% EPA and 17% DHA, or 10% by weight sunflower oil (SFO group). Control group was kept on the standard diet for 4 weeks. Blood-free microvessels were isolated from 300 mg brain cortex by a rapid micromethod, and their fatty acid composition was determined by gas chromatography. It was found that the proportion of ω3 fatty acids (including EPA and DHA) increased significantly in the microvessels of the FO group, accompanied by a decrease of the ω6 fatty acid series. The changes in fatty acid composition of endothelial cells were not significant in the case of SFO group as compared to the control.

The amounts of lipoxygenase and cyclooxygenase metabolites were determined after preincubating the microvessels in TC medium 199 at 37 °C for 10 min. The enzyme reaction was started by addition of 1-^{14}C-arachidonic acid (3.7 kBq, spec. act. 2.04 GBq/mmol) as tracer to the incubation medium. Radioactive arachidonate metabolites were separated by OPTLC and quantitated by liquid scintillation counting. Dietary fish oil decreased the percentage of total products of arachidonate by 50%, while the SFO diet had no effect on it. The amount of lipoxygenase products in the FO group decreased significantly from 16,930 ± 3,130 dpm to 6,399 ± 357 dpm/300 mg wet weight of brain. Significantly less $PGF_1 1\alpha$, $PGF_2\alpha$ and 12-hydroxyheptadecatrenoic acids were found in the capillaries of FO-treated animals, compared to the SFO group. The ratios of vasoconstrictor and vasodilator metabolites of arachidonate cascade were not modified by the diets.

Our results suggest that fish oil diet reduces the arachidonate cascade in cerebral microvessels, which might be an explanation for the effectiveness of ω3 fatty acids in vascular diseases.

Inhibition of Phagocytic Function by Eicosapentaenoic and Docosahexaenoic Acid of Human Neutrophil Granulocytes in vivo and in vitro

S. Sipka[a], *J. Csongor*[b], *G. Szegedi*[a], *T. Farkas*[c]
[a]Department of Internal Medicine, and [b]Central Research Institute of University Medical School, Debrecen; [c]Institute of Biochemistry, Biological Research Center, Szeged, Hungary

Previous works have demonstrated that both EPA and DHA inhibit the activity of phagocytes. In this study, we have measured the effects of EPA and DHA on the chemoluminescence of human neutrophil granulocytes (HNG) stimulated by opsonized Mannozym (zymosam) either after in vivo administration by feeding the fresh water fish, silver back, containing 2.7 g/kg EPA and 2.8 g/kg DHA, to healthy humans (1 kg fish/week) or by incubating HNG in the presence of various forms (free, complexed to BSA, incorporated into soybean phosphatidylcholine vesicles) of these fatty acids. Feeding the fish resulted in a reduction of phagocytic activity of the isolated HNG. Both EPA and DHA depressed this phenomenon also under in vitro conditions as compared to the slight inhibition caused by oleic acid (as control) if they were complexed to BSA. When EPA and DHA were incorporated into phospholipid vesicles, the inhibitory effect was much less. Similar results were obtained when phosphatidylcholine, prepared from the fish, was supplied to the cells in the form of vesicles. It is suggested that in these cases the phospholipid vesicles were phagocytosed (but probably not liberated), and this rendered the cells unable for further phagocytosis (of zymosam), or the incorporated EPA and DHA were not accessible for the metabolism, whereas if complexed to albumin they could modulate the lipid composition of cell membranes immediately, resulting in the observed decrease of phagocytic activity of HNG.

Acknowledgments: This work was supported by Repro-Gen International, Vaduz.

Longevity in BHE Rats as Affected by Feeding 2 or 10% Fat as Corn Oil, Menhaden Oil or Beef Tallow

Carolyn D. Berdanier
University of Georgia, Athens, Ga., USA

A longevity study using 360 male genetically diabetic/lipemic BHE rats was conducted. Weanling rats were fed a sucrose-based diet containing 2 or 10% fat. The fat component consisted of 1% corn oil and 99% as corn oil, beef tallow or menhaden oil. Glucose tolerance, serum lipids, and histopathology of liver, spleen, kidneys, pancreas, heart, aorta and lungs were evaluated at 300, 500, 600 and 700 days of age. Mean arterial pressure (MAP) and heart rate were determined using direct methods via an indwelling catheter. Body composition was determined at 300 and 600 days. Spontaneous deaths were assigned a cause by the pathologist. Renal disease was the most common cause of death. Longevity was affected by fat type and level. At 700 days of age only 6.6% of the rats fed 10% menhaden oil diet were alive; 22.2% of those fed 10% corn oil and 31.1% of those fed beef tallow were still alive. In contrast, in rats fed the 2% fat diets, 38.5% of those fed menhaden oil, 16.7% of those fed corn oil, and 31.3% of those fed beef tallow were still alive. MAP rose as the animals aged. Those fed the beef tallow diets generally had lower MAP than those fed other diets. Glomerulonephrosis was more severe in the menhaden oil-fed rats. Lesion in the kidneys appeared earlier and at death were more severe. There was a good correlation between kidney weight, pathology and blood pressure. Glucose intolerance and blood lipids rose with age. Fat level and type affected these parameters.

Supported by Sea Grant NA88AA-D-5G098.

Egg Yolk as a Dietary Source of ω3 Polyunsaturated Fatty Acids for Infants

R.A. Gibson, D.J. Farrell
Department of Paediatrics, Flinders Medical Centre, Adelaide, Australia

Human breast milk contains only small amounts of ω3 and ω6 long-chain (20 and 22 carbon) polyunsaturated fatty acids (LC PUFA). However, the quantities present are adequate for the total theoretical LC PUFA requirement during perinatal growth. Infant formulas, on the other hand, are devoid of these LC PUFA and it has been demonstrated that infants raised on formulas have lower levels of ω3 LC PUFA in the membranes of their erythrocytes than infants fed human milk.

After 3 months of age, infants are often given supplementary foods, some of which contain LC PUFA. We have determined that egg yolk, a common weaning food, contains both ω6 and ω3 LC PUFA but at levels which would need excessive consumption (up to 100 g egg yolk/day for a 7-kg infant) to match the intake of these same LC PUFA from breast milk.

To improve the nutritive value of eggs, we have fed laying hens diets supplemented with vegetable (linseed) or marine (MaxEPA) oils (10%) w/w). Supplementation with linolenate (ALA) or eicosapentaenoate (EPA) plus docosahexaenoate (DHA) increased the ω3 LC PUFA level 6- to 8-fold. ALA-containing oil (linseed) was as effective as marine oil in increasing DHA levels, but only marine oil effectively increased the level of EPA. The highest level of ω3 LC PUFA in egg yolk occurred in hens whose diets had the lowest ω6:ω3 ratio: the ω6 LC PUFA levels were only marginally affected by diet.

Diet	Polyunsaturated fatty acids, % total fatty acids (mean ± SD)				
	ω3			ω6	
	ALA	EPA	DHA	LA	AA
Reference[1]	0.2 ± 0.1	0	0.3 ± 0.05	12 ± 1.2	1.5 ± 0.2
Linseed	3.3 ± 0.6	0.1 ± 0.05	1.8 ± 0.8	13 ± 1.6	1.6 ± 0.4
Fish Oil	0.2 ± 0.05	1.8 ± 0.7	2.4 ± 0.6	8 ± 1.0	1.1 ± 0.4

[1] Commercial layer mash.
AA = Arachidonic acid.

Thus egg yolks from hens fed diets with an optimal ω6:ω3 ratio can provide an important source of ω3 LC PUFA for the formula-fed infant. A single 20 g egg yolk can provide up to 80% of the ω3 LC PUFA provided by breast milk.

Cholesterol and Fatty Acid Content of Australian Chicken and Duck Eggs

A.J. Sinclair, G.W. Hopkins, K. O'Dea
Department of Applied Biology, RMIT, Melbourne, Victoria, and Department of Human Nutrition, Deakin University, Geelong, Victoria, Australia

The present investigation was initiated following the observation of a considerable variation in the cholesterol content of eggs analyzed for a dietary experiment. Eggs were sampled from the major producers in five sites from four states (Vic., NSW, Qld, WA); the eggs were derived from 15 different layer strains. The same strains from the same sites were sampled on two occasions, 6 months apart. At each sampling, 6 eggs of about 50–55 g were requested from each layer strain as well as information on the major feed ingredients of the layers. The age of the layers ranged from 30 to 77 weeks. A total of 291 eggs were examined from chickens and in a preliminary study 24 duck eggs were also analyzed.

Information on egg weight, yolk weight and total edible weight was recorded for all samples and the lipids were extracted from the yolk by a standard chloroform-methanol procedure. An aliquot of the extract was saponified together with 5a-cholestane and the cholesterol content was determined by capillary gas-liquid chromatography (GLC). The recovery of cholesterol through the procedure was routinely determined. The fatty acid composition was measured on a representative number of samples of chicken and duck eggs by capillary GLC. The mean results for the chicken and duck eggs, respectively, are detailed as follows: total egg weight 58.4 vs. 75.9 g; yolk weight 18.2 vs. 27.4 g; yolk/egg 31.1 vs. 36.1 %; lipid/egg 10.5 vs. 12.6 %; cholesterol/egg 196 vs. 370 mg, cholesterol/100 g edible weight 380 vs. 552 mg.

The distribution of cholesterol values in the chicken eggs was skewed towards the lower values with 75 % of results below 400 mg/100 g and 95 % below 450 mg/100 g. The cholesterol content of eggs in the Australian Tables of Food Composition is 450 mg/100 g. Six strains (110 observations) consistently gave cholesterol values below 415 mg/100 g, whereas for the other 9 strains (181 observations) approximately one third of the values were above 415 mg/100 g. The increased lipid content of the duck eggs was largely due to a higher concentration of 18:1 in these eggs, together with higher concentrations of 16:0, 20:3, 20:4 and 22:5; in addition these eggs contained lower levels of 18:2 and 22:6.

Examination of the data for all chicken eggs revealed the following relationships: (a) As the eggs became heavier the weight of the yolk increased, and the percentage of the yolk increased. (b) The lipid content and the yolk weight were highly correlated ($r = 0.859$), however in addition the percentage of lipid increased as the yolk weight increased. (c) Yolk and egg weights were significantly correlated with cholesterol/egg. (d) The stepwise regression of the cholesterol content per egg with the other variables established the following equation: cholesterol/egg = 75.5 ± 34.1 lipid (g/egg) – 8.5 egg lipid %.

Fatty Acid Profiles of Rat Tissue Phospholipids following a Change in the Dietary Level of ω3 Polyunsaturated Fatty Acids

Carl-Erik Høy, Gunhild Hølmer
Department of Biochemistry and Nutrition, Technical University of Denmark, Lyngby, Denmark

Polyunsaturated fatty acids (PUFA) of the ω3 family compete with ω6 PUFA for incorporation into the 2-position of membrane phospholipids. Dietary ω3 PUFA are available from marine sources containing 20:5 ω3, 22:5 ω3, and 22:6 ω3 or from vegetable sources containing 18:3 ω3.

In this experiment we have examined the effects of changes in the dietary levels of ω3 fatty acids of marine or vegetable origin on the fatty acid profiles of liver, heart, and kidney during a period following a sudden shift of dietary fats.

Four groups of rats were fed initial diets containing 20 % (wt) fat for 6 weeks and thereafter shifted to final diets:

Group	Initial fat, %			Final fat, %		
	18:2 ω6	18:3 ω3	EPA+DHA	18:2 ω6	18:3 ω3	EPA+DHA
1	10	0.3	0	10	4	0
2	10	0.3	0	10	0.3	4.5
3	10	4.0	0	10	0.3	0
4	10	0.3	4.5	10	0.3	0

The fatty acid profiles of PC, PE, PS, PI, and CL as well as the phospholipid species were analyzed at days 0, 1, 2, 4, 9, and 17 following these changes of diets.

It was demonstrated that the incorporation or depletion of ω3 PUFA occurred selectively with regard to the chain length of the dietary ω3 PUFA as well as the rat organ, the phospholipid class, and the phospholipid species considered. In liver major increases in the levels of DHA, from 17% to 35%, were observed in PC, PE, and PS immediately following the shift to fish oil diets, whereas in kidney primarily EPA was incorporated. No changes in the fatty acids in the 1-position of the phospholipids were observed indicating that not compensation for possible changes in membrane fluidity following incorporation of ω3 PUFA was operative.

Fortification of Foodstuffs with Marine ω3 Fatty Acids

Margit Nøhr, Jan Vælds
Danochemo, Ballerup, Denmark

Overwhelming evidence points to the fact that an increased intake of fish will be of benefit to the population in the Western world. Fish oil is particularly important as it has been shown to have a lowering effect on the risk of heart disease.

Although massive campaigns have been initiated by health authorities and heart associations in the attempt to change peoples' eating habits, these changes only happen very slowly if at all. Many people have an aversion to eating fish due to bones, smell and/or bad childhood memories of cod liver oil. In view of this, the fortification of foodstuffs not normally associated with fish seems an obvious way of offering these people the benefit of fish oil without the negative aspects which they dislike.

Fortification of a specific foodstuff with fish oil raises three crucial questions: percent coverage of recommended daily dose, organoleptic quality, and stability. Recommended daily dose will be discussed in detail with a special view to the fortification of foodstuffs.

Analytical data will be presented, proving the practical possibility of enriching bread with microencapsulated fish oil in powder form. It will be shown that enriched bread does not differ from ordinary bread with regard to organoleptic quality and stability.

Production of EPA and DHA Oils from Microalgae

David J. Kyle, Ray Gladue, Sue Reeb, Kim Boswell
Martek Corporation, Columbia, Md., USA

Oceanic phytoplankton are a subset of a diverse group of microorganisms known as microalgae which represent the primary food source for all sea life. This includes the fish whose oils contain abundant amounts of eicosapentaenoic (EPA) and docosahexaenoic

(DHA) acids, which have been associated with reduced incidence of coronary vascular disease. Fish oils, however, contain additional polyunsaturated fatty acids in significant quantities which have biological effects different from EPA or DHA and, in some cases, may even be antagonistic to these effects. Even EPA and DHA have different biological effects in man with EPA principally affecting the cardiovascular system through prostaglandin metabolism, while DHA is an important structural lipid in highly active nervous tissues. Indeed, there may be specific reasons for dietary supplements of either of these fatty acids, but not necessarily both simultaneously.

Microalgae can produce as much as 40–50% of their biomass in extractable triglyceride, and many of these species produce ω3 polyunsaturated fatty acids. In fact, the marine microalgae actually represent the largest pool of EPA and DHA in the biosphere. Our objective has been to develop a single cell 'designer' oil process using microalgae wherein the product triglyceride contains significant quantities of only a single bioactive fatty acid, either EPA or DHA. MK8805 represents a marine microalgal strain which has been optimized for growth on glucose in conventional fermentors. This strain produces an oil enriched in myristic, palmitic and oleic acids as well as about 30% DHA. A unique feature about this oil is that it contains no unsaturated fatty acids other than oleic and DHA in quantities over 2% in the final oil. This process has now been scaled up and development quantities of the oil are available for testing.

Atlantic Salmon as an ω3 Source: What Can Be Done through the Feed?

Ø. Lie, R. Waagbø, K. Sandnes, G. Lambertsen
Institute of Nutrition, Directorate of Fisheries, Bergen, Norway

The natural source of ω3 PUFA is the intake of fatty fish fillets, and salmon is the preferred food for this purpose. The contents and FA composition of lipids in farmed salmon depend on the feed, i.e. on its oil component. Various fish oils are used in fish feeds in different countries, and even vegetable oils may be used according to economic considerations. Farmed Norwegian Atlantic salmon are given feeds containing North-atlantic fish oils. The salmon fillets normally have 6% 20:5 ω3 (EPA) and 9% 22:6 ω3 (DHA) in the lipids and give 2–3 g of ω3 PUFA per 100 g.

Results from a feeding experiment with Atlantic salmon are shown, based on an 18-month growth period with a weight increase from 30 to 1,700 g. The feed oils were a normal and a high-PUFA fish oil and a vegetable oil, and the 3 feeds had 1.0, 2.5 and 5.0 g ω3 FA/g. While no weight gain differences were observed, we found widely different fatty acid compositions in the fillets, reflecting the dietary lipids.

The specific requirement of salmon for PUFA is met through the lipid content of the fish meal in the feed. The fatty acid composition of individual phospholipid classes in salmon fillets are shown to demonstrate the specific incorporation of PUFAs of the ω3 as well as the ω6 family. Lastly, the effect of the sea water temperature on the fatty acid composition of the salmon fillet lipids is shown.

Comparative Effects of Dietary Fish Oil (FO), Safflower Oil (SO) and Palm Oil (PO) on Selected Enzymes of Hepatic Lipid Metabolism

M.A. Halminski, J.B. Marsh, E.H. Harrison
Medical College of Pennsylvania, Philadelphia, Pa., USA

Studies were conducted to further elucidate the mechanism(s) by which dietary FO decreases hepatic very-low-density lipoprotein (VLDL) secretion. Thirty rats (10/group) were fed purified diets containing 10% fat as either FO, SO or PO for 10 days. Plasma triglyceride (TG), an indicator of hepatic VLDL secretion, was lowest in the FO group followed by the SO and PO groups. The liver capacity to oxidize fatty acids was assessed by separate, in vitro assays of the mitochondrial (MOX) and peroxisomal (POX) betaoxidation pathways in whole liver homogenates. Additionally, the key regulatory enzymes of phosphatidylcholine (PC) biosynthesis (phosphocholine cytidylyltransferase, CT) and TG synthesis (phosphatidate phosphohydrolase, PPH) were assessed in microsomal and soluble fractions of the homogenates. MOX was unaffected by type of dietary fat while FO produced a 45% increase in the much smaller POX component of total oxidation compared to SO and PO. FO and SO significantly lowered soluble PPH activity by 48 and 25%, respectively, compared to PO. Similarly, microsomal CT activity was reduced by 52 and 34% by FO and SO, respectively. Both PPH and CT activities significantly ($p < 0.01$) and directly ($r = 0.78$ and 0.84) correlated with plasma TG levels. These results suggest that dietary FO suppresses VLDL secretion by decreasing PC and TG synthesis rather than by increasing fatty acid catabolism.

Effect of Chain Length of Dietary ω3 Fatty Acids on the Fatty Acid Distribution in Monkey Liver Phospholipids

G. Hølmer[a], S.G. Kaasgaard[a], C.-E. Høy[a], J. Beare-Rogers[b]
[a]Department of Biochemistry and Nutrition, Technical University of Denmark, Lyngby, Denmark; [b]Nutritional Research Division, Food Directorate, Ottawa, Canada

The necessity of feeding fish oil instead of α-linolenic acid-containing plant oils to ensure a balanced production of prostanoids known to influence the thrombotic tendency in humans has been discussed. Furthermore it is of interest to know how fish oil feeding affects lipid composition in various tissues of primates.

We have therefore made an experiment in which groups of monkeys were fed diets rich in linoleic, linoleic + α-linolenic, and linoleic + ω3 fatty acids from fish oil, respectively.

Previously, the influence on fatty acid distribution in total liver phospholipids was reported [Proc. 15th Scand Symp Lipids Rebild, Denmark, 1989, pp 53–60]. However, various phospholipid classes do have specific functions in the cell, and a special fatty acid composition seems to be linked to the various phospholipid classes. In this context we have examined the effect on some of the minor phospholipids, phosphatidyl serines, phosphatidylinositols and cardiolipins, all known to have functional purposes in membranes.

As a main result it was found that a dietary supplement of fish oil increased the levels of eicosapentaenoic acid and docosahexaenoic acid especially in phosphatidylserine and to some extent in phosphatidylinositol, whereas only a minor effect of the corresponding amount of α-linolenic acid was observed.

For cardiolipins, in which fatty acids with longer chain than C18 are normally present in minor amounts, only about 0,5% C20:5 and 0,5% C22:6 were incorportated after fish oil feeding, whereas dietary α-linolenic resulted in a deposition of 4–5% C18:3 but no C20:5 and C22:6. It remains to be established whether these different responses to ω3 fatty acids of varying chain length are crucial for membrane function.

Biological Effects of Fish Oil on Rats: Changes in Lipid Levels and Composition, in Eicosanoids and in Peroxisomal Function

C. Skorin[a], *A. Miguel*[a], *G. Vera*[b], *F. Solis de Ovando*[a], *C. Barriga*[a], *V. Guasch*[a], *M.E. Kawada*[a], *N. Velasco*[a], *M. Bronfman*[a], *A. Arteaga*[a], *A. Valenzuela*[b], *C. Necochea*[a], *F. Leighton*[a]

[a]Universidad Católica de Chile; [b]Universidad de Chile, Santiago, Chile

Fish oil, presumably through its high content in ω3 very-long-chain polyunsaturated fatty acids, elicits various biological responses, among them an apparent decrease in the risk of death in humans from coronary heart disease which has prompted the search for the biochemical basis of its effects. As part of a research program that also involves studies in cultured cells and in humans, we have explored biochemical responses in rats. Male Sprague-Dawley rats were fed 3 weeks with semisynthetic diets containing either 10% or 18% fat by weight. The fat contained variable ratios of ω3/ω6, as follows: 0; 1/5; 1/1; and 5/1. The content of ω3 fatty acids in these diets was 0, 1.6, 3.6, and 4.9 g/100 g diet, respectively, for the 18% fat diet. The results, at increasing ω3/ω6 ratios, showed: A dose-dependent increase in the content of 20:5 ω3 and 22:6 ω3 in phospholipids from serum lipids and from platelet and red blod cell membranes. The serum levels of triglycerides and cholesterol did not change. Serum thromboxane B_2 levels (RIA) decreased. Strikingly, the polymorphonuclear leukotriene B_4 release (Ca^{2+} ionophore-induced) did not decrease, it showed a several-fold increase at the higher ω3/ω6 ratios; these results were obtained both by radioimmunoassay (RIA) and by HPLC measurements.

In order to evaluate peroxisomal function, the activity of the peroxisomal fatty acyl-CoA oxidase in liver was measured: it increased 2-fold at the highest ω3/ω6 ratios. In addition, the fatty acid oxidation capacity of both, the mitochondrial and the peroxisomal systems, was evaluated in isolated hepatocytes prepared from the rats fed the different diets. The results confirm the enzymatic measurements since a significant increase in peroxisomal oxidation was also detected in whole cells (substrate-dependent H_2O_2 and ketone body generation employing 12:0, a peroxisomal and mitochondrial substrate;

dicarboxylic 12:0, a peroxisomal substrate; and 16:0, a mainly mitochondrial substrate).

These data confirm that ω3 fatty acids are incorporated into membrane phospholipids in a dose-dependent fashion; they also show that ω3 fatty acids modify eicosanoid metabolism, but in our observations, a striking increase in LTB_4 raises the possibility of a proinflammatory response in rats with our experimental design (in humans, with the same oil preparation, we do find a decrease in LTB_4); and finally, they demonstrate that ω3 fatty acids supplied in the diet are peroxisomal inducers which enhance, in whole cells, the contribution of peroxisomes versus that of mitochondria in fatty acid oxidation.

Supported by projects UNDP CHI/88/017 and FONDECYT 250/88.

Effect of ω3 Fatty Acids on Membrane Fragility of Rat Erythrocytes

Tor-Arne Hagve, Bjørn O. Christophersen
Institute of Clinical Biochemistry, University of Oslo, Rikshospitalet, Oslo, Norway

Increased intake of dietary ω3 fatty acids has been shown to have an inhibitory effect on the development of cardiovascular disease and certain forms of cancer. The mechanisms of these beneficial effects are, however, still uncertain. A growing body of evidence indicates that long-chain polyunsaturated fatty acids in phospholipids may play a major role in the regulation of membrane-bound enzymes and receptors, probably by modulating membrane fluidity. A connection between increased level of ω3 fatty acids in membranes, membrane-related cell-functions, and disease has thus been postulated. In order to study the effect of ω3 fatty acids on physical properties of the membrane, we have fed rats a ω3-rich diet and correlated the membrane fragility of erythrocytes to the phospholipid composition of the cells.

Male rats (150–180 g) were fed a diet supplemented with 1.5 ml ω3 concentrate (35% 20:5 ω3, 30% 22:6 ω3 for 21 days. Osmotic resistance of the erythrocytes was tested as a measure of membrane fragility. The cells were washed and analyzed for phospholipid content and fatty acid composition.

With salt concentrations ranging from 0.37% to 0.41% the fragility of erythrocytes from rats fed the ω3 diet was increased by 20–40% compared to control-fed animals. The eicosapentaenoic acid (20:5 ω3) content in phospholipids was increased from 1.8 ± 0.3 nmol/10^{12} cells in rats fed the control diet to 5.8 ± 0.4 nmol/10^{12} cells in ω3-supplemented rats with a concomitant decrease of 18:2 ω6. More or less the same changes were observed in serum. No differences were found in docosahexaenoic acid (22:6 ω3) level in phospholipids of either erythrocytes or serum.

A small but still significant increase in total phospholipid concentration was measured in the ω3-enriched erythrocytes compared to controls (129.0 ± 2.1 and 115.9 ± 1.4 nmol/10^{12} cells, respectively), while no difference was found in the relative level of cholin-, ethanolamin-, serine- and inositol-phosphoglyceride fractions.

It is thus concluded that the increased membrane fragility of erythrocytes from rats fed a ω3-rich diet mainly is due to increased 20:5 ω3 levels in membrane phospholipids.

Changes in Fatty Acid Composition in Blood and Organs after Infusion of Docosahexaenoic Acid Ethyl Ester into Rats

K. Yamazaki, T. Hamazaki, S. Yano
First Department of Internal Medicine, Toyama Medical and Pharmaceutical University, Sugitani, Toyama, Japan

Docosahexaenoic acid (DHA, 22:6 ω3), a major polyunsaturated fatty acid of fish oils, is found in brain, retina, sperm and heart of land animals as a major fatty acid component in phospholipids. DHA is probably essential for the functional development of the nervous system including retina. DHA has some other biomedical effects such as depression of platelet aggregation, augmentation of efficacy of anticancer treatment, stabilization of coronary arteries and probably prevention of arrhythmia due to ischemia. We have developed an infusible DHA emulsion which could increase DHA concentration in tissues very rapidly. Here we report to what extent DHA is incorporated in various organs after the injection of DHA emulsion into rats.

An emulsion of DHA ethyl ester (DHA-EE) was prepared. One hundred ml of the emulsion contained 10 g DHA-EE (90% pure), 1.2 g egg yolk phospholipids as an emulsifier and 2.5 g glycerol. 3 ml of the emulsion was infused into tail veins of 22 Wistar rats weighing about 300 g. They were sacrificed 1, 6 and 24 h, and 3 and 7 days after the infusion (4 or 5 rats each time), and fatty acid composition of various organs and plasma was analyzed. Another group of 4 similar rats, which had not been infused with DHA-EE, served as baseline data.

The infusion of DHA-EE did not cause any macroscopic abnormality of organs including lipidosis when checked during the sacrifice. DHA levels reached their peaks within 24 h after DHA infusion in plasma fractions of phospholipids, cholesteryl esters, free fatty acids and triglycerides (including EE), and in the phospholipid fraction of liver, kidney, spleen, aorta and lung. DHA did not increase at all in cardiac phospholipid fraction. However, DHA levels increased markedly (from 0.7% to 11%) in the free fatty acid fraction of heart 1 h after the infusion. In liver eicosapentaenoic acid levels were increased rapidly reaching their peak 1–6 h after DHA infusion, which indicated a rapid retroconversion of DHA.

This method of administering DHA might be useful for patients in whom a rapid increment in DHA in tissues seems beneficial.

Effect of Dietary Supplementation of 18:3 ω3 and 18:4 ω3 on Plasma Fatty Acids in Rats

K. Yamazaki, M. Fujikawa, N. Nakamura, T. Hamazaki, S. Yano
First Department of Internal Medicine, Toyama Medical and Pharmaceutical University, Sugitani, Toyama, Japan

There are many studies which investigate the effects of dietary supplementation of 18:3 ω3, 20:5 ω3 and 22:6 ω3. However, there is no report about the supplementation of 18:4 ω3. In the present study we compare the effect of dietary supplementation of 18:3 ω3 and of 18:4 ω3 on plasma fatty acid composition.

Eighteen male Wistar rats (4 weeks old) were divided into 3 groups (A, B and C) and fed the following diets for 3 weeks. Diet A: lipid-free diet (90%) + lard (no ω3 fatty acids contained, 10%); diet B: lipid-free diet (90%) + lard (9%) + pure 18:3 ω3 ethyl ester (1%); diet C: lipid-free diet (90%) + lard (9%) + pure 18:4 ω3 ethyl ester (1%). Arterial blood samples with EDTA were obtained from the abdominal aorta after 3 weeks on diets. Lipids extracted from plasma were separated by thin-layer chromatography and triglyceride (TG), free fatty acid (FFA) and phospholipids (PL) fractions were obtained. Fatty acids of each fraction were transmethylated and analyzed by gas chromatography.

In PL fraction of groups B and C, 20:4 ω6 (AA) was decreased (group A: 18.6%; group B: 10.2%; group C: 9.5%), 20:5 ω3 (EPA) was increased (A: 0.02%; B: 0.7%; C: 1.3%) compared to group A. In TG fraction of groups B and C, AA was decreased (A: 1.7%, B: 0.7%; C: 0.7%), and EPA was increased (A: 0.1%; B: 0.9%; C: 1.9%) compared to group A. 18:3 ω3 was increased in group B (A: 0.2%; B: 1.6%; C: 0.2%), and 18:4 ω4 was increased in group C (A: 0%; B: 0.07%; C: 0.3%). In FFA fraction of group B, 18:3 ω3 was increased (A: 0.3%; B: 2.2%; C: 0.3%), and AA was decreased (A: 2.1%; B: 0.6%; C: 1.5%). In FFA fraction of group C, 18:4 ω3 (A: 0%; B: 0%; C: 1.1%) and EPA (A: 0%; B: 0.01%; C: 0.7% were increased. 20:4 ω3 was detected only in group C (PL: 0.1%; TG: 0.2%; FFA: 0.2%).

Conversion to EPA was easier with 18:4 ω3 than with 18:3 ω3. 20:4 ω3 was not detected in groups A and B. Retroconversion of 18:4 ω3 to 18:3 ω3 was not observed.

Blood Pressure-Lowering Effect of Fish Oil in Mildly Hypertensive Patients

U. Schmidt[a], *N. Schenk*[a], *P. Singer*[b]
[a]Arbeits- und Forschungsgemeinschaft für Arzneimittel-Sicherheit e.V., Köln, FRG; [b]Heckmannufer, Berlin, Germany

There is increasing evidence that fish oil exerts a blood pressure-lowering effect not only in normotensive volunteers, but also in (mildly) hypertensive subjects. Sixty-five patients with mild essential hypertension in a multicenter, double-blind study were allocated to a group receiving 6 g fish oil/day and to a group receiving placebo (6 g olive oil/day). 4 patients were eliminated because of non-compliance (2) or adverse effects (fishy taste, exanthema-1; pyrosis-1).

Sixty-one patients (36 in the fish oil group, 25 in the placebo group) completed the study after a period of 12 weeks followed by a placebo period of 1 week. Systolic and diastolic blood pressure appeared significantly lower already after 2 weeks of fish oil supplementation and decreased further to the 12th week. In the placebo group diastolic blood pressure was likewise significantly decreased after 2 weeks and revealed a further decline till the end of the trial. This was, however, less pronounced when compared with the fish oil group. Likewise, serum triglycerides and cholesterol were significantly reduced, which appeared more pronounced in the fish oil group.

The data indicate that not only blood pressure, but also serum lipids are favorably influenced by fish oil in patients with mild hypertension. This might be of great practical importance, since hypertensive subjects frequently have lipid disorders which might be at least as atherogenic as elevated blood pressure.

Capillary Blood Flow and Serum Lipoprotein Changes after Dietary Fiber, Fish or Safflower Oil Intervention

G. Bruckner, K. Gannoe, A. Atakkaan, P. Webb, P. Oeltgen, D. Richardson
University of Kentucky, Lexington, Ky., USA

Fish oil supplementation has been shown to lower primarily serum triglycerides while dietary soluble fiber may preferentially decrease serum cholesterol. Furthermore, fish oil supplementation, in normolipidemic humans, has been shown to increase nailfold capillary blood cell velocities (CBV). Therefore the objective of the proposed study was to evaluate the independent and the combined effects of fish oil (MaxEPA-FO), safflower oil (SO), soluble fiber (Oat Bran-SF) and insoluble fiber (Wheat Bran-IF) on serum lipoproteins and CBV in hyperlipidemic human subjects in a crossover design balanced for residual effects. Thirty-six hyperlipidemic male subjects, 30–65 years of age, were recruited for the study (initial fasting cholesterols > 220 and triglycerides > 240 mg/dl). They were randomly assigned to the following initial treatment groups: (a) FO and SF, (b) FO and IF, (c) SO and SF or (d) SO and IF. The oils were supplemented at 1.5 g/10 kg body weight and the bran at 10 g total fiber/day for 2 months. This supplementation period was followed by a 1-month washout and crossover to the next 2-month treatment protocol. All lipid and CBV measurements were taken prior to and after every 2-month supplementation. Capillary blood cell velocities were significantly increased over baseline values by both FO and SO dietary supplementation, and these values remained elevated throughout the subsequent treatment periods in spite of the 1-month washout periods. FO significantly and consistently lowered serum triglycerides (TG) during all treatment periods and in particular decreased very-low density lipoproteins. SO also significantly lowered TG in the third treatment period but these effects were not consistent throughout the study. SF and IF did not significantly alter CBV or serum lipoprotein. Our findings indicate that changes in CBV which result from dietary oil supplementation may be evident even after 1 month following discontinuation of these oils.

Effect of Menhaden Oil and Other Reference Oils upon Serum Lipids and Platelet Aggregation in Rats

T.R. Watkins, P. Lenz, M.L. Bierenbaum
Kenneth L. Jordan Research Group, Montclair, N.J., and Fairleigh Dickinson University, Madison, N.J., USA

Attempting to seek terrestrial sources of ω3 fatty acids that work as efficiently as fish oil in decreasing serum lipids and platelet aggressiveness, an 8-week study in a rat model fed a 2% supplement of cholesterol was done comparing Canola oil, menhaden oil, safflower oil and partially hydrogenated soya oil to standard chow. The menhaden oil group was the only one to show a significant reduction in serum cholesterol (from 49.8 $\pm$1.7 to 38.7 $\pm$ 1.6 mg/dl, $p < 0.005$) and triglycerides (59.3 $\pm$ 7.9 to 33.8 $\pm$ 26 mg/dl, $p < 0.005$) when compared to chow and at the same time yielded the most uniform decrease in platelet aggregability. However, this same group was also the only one to show a decrease in serum tocopherol levels (0.36 $\pm$ 0.01 to 0.20 $\pm$ 0.01 mg/dl, $p < 0.01$), suggesting that improving an atherogenic risk profile may require antioxidant supplementation. In this model, despite its considerable content of linolenic acid, Canola oil did not yield the same results expected from using menhaden oil.

Inhibition of Platelet Aggregation by Combinations of ω3 and ω6 Fatty Acids

Barry Sears
BioSyn, Inc. Marblehead, Mass., USA

Prostaglandins derived from arachidonic acid (AA) have a significant role in the process of platelet aggregation. Dietary supplementation with high levels of ω3 fatty acids such as eicosapentaenoic acid (EPA) can reduce platelet aggregation. However, the amount of these dietary EPA levels required for reduction of platelet aggregation are unrealistic for long-term compliance. Prostaglandins (primarily PGE_1) derived from dihomo-gamma-linolenic acid (DGLA), an ω6 fatty acid, also demonstrate a significant reduction in platelet aggregation. DGLA levels can be increased by dietary supplementation with gamma-linolenic acid (GLA), but any increase in DGLA can lead to a corresponding increase in AA. EPA is the feedback inhibitor of the enzyme that catalyzes the formation of AA from DGLA. Therefore, appropriate combinations of GLA and EPA should be able to increase DGLA levels in platelets, simultaneously decreasing AA levels. The end result of dietary supplementation with combinations of EPA and GLA should be a reduction in platelet aggregation at a much lower total edible oil intake than observed with fish oils containing EPA.

10 volunteers were placed on daily dietary supplementation program of 80 mg GLA and 640 mg EPA for 2 weeks. Blood samples were obtained immediately before the subjects began the supplementation and after the 2-week period. The results are shown in the table.

Aggregation agent	Pre-test	Post-test	Significance
ADP (1 μ*M*)	19±18	17±12	n.s.
U46619 (500 μ*M*)	117±18	88±27	$p < 0.01$
Collagen lag (0.5 μg/ml)	66±23	79±36	$p < 0.05$

Although the responsiveness of the platelets to ADP-induced aggregation was reduced, it was not statistically significant. On the other hand, platelet responsiveness to

U46619 (a thromboxane A_2 agonist) was significantly reduced, as was the lag time to the onset of aggregation from the addition of collagen.

These results indicate that dietary supplementation with relatively low doses of combinations of ω3 and ω6 fatty acids can have a statistically significant effect on the reduction of platelet aggregation.

Lipoprotein Cholesterol and Hepatic LDL Receptor Activities of Endogenous Hypercholesterolemic Rabbits Supplemented with ω3 and ω6 Fatty Acid-Containing Dietary Oils

John E. Bauer, Katalin Vienne
Department of Physiological Sciences, J. Hillis Miller Health Sciences Center, University of Florida, Gainesville, Fla., USA

Casein feeding of rabbits results in hypercholesterolemia and the suppression of high-affinity hepatic low-density lipoprotein (LDL) receptor activities. We have previously shown that dietary safflower oil protects against the endogenous hypercholesterolemic response but that dietary menhaden oil does not. To further understand reasons for this observation we have now investigated the effect of dietary ω3 and ω6 fatty acids on serum lipoproteins and hepatic lipoprotein receptor activities in this casein-fed rabbit model.

Twelve male New Zealand white rabbits were fed a casein-corn starch semipurified diet (BHC) to induce hypercholesterolemia. After 40 days of feeding, 4 of the hypercholesterolemic rabbits were orally supplemented with safflower oil and 4 were given menhaden oil (1 ml/kg). Cholesterol had been added to the safflower oil in an amount equal to that present in the menhaden oil. The remaining rabbits were maintained on the BHC diet as a source of LDL for subsequent studies. After 3 weeks of supplementation, a 50% reduction in mean serum cholesterol concentration was noted in the safflower group; no change was seen in rabbits given the menhaden oil supplement. Lipoprotein cholesterol distributions determined after single-spin density gradient ultracentrifugal fractionation of serum indicated marked elevations of LDLs and moderate decreases of high density fractions with menhaden oil supplementation. Controlled studies in humans have recently found similar dietary ω3 fatty acid effects.

Hepatic apo-B, -E receptor activities of liver membranes from the oil-supplemented rabbits were determined using radioiodinated LDL from the BHC-fed animals. Results confirmed that the EDTA-sensitive binding sites of casein-fed rabbits are greatly reduced under this condition. More importantly, however, it was found that safflower oil supplementation caused a specific increase in high-affinity receptor binding. While menhaden oil supplementation increased receptor activity to some extent, a 2-fold difference in absolute binding between safflower and menhaden oil groups was observed.

These data demonstrate that the hypocholesterolemic effect of dietary safflower oil in endogenously hypercholesterolemic rabbits is due, in part, to the stimulation of high-affinity LDL receptor activity. Menhaden oil supplementation did not restore this activity to the same extent. This latter finding is consistent with only a slight decrease in serum cholesterol concentration in the menhaden oil group at the end of the supplementation period. Whether these differential effects can be further explained by differences in the polyunsaturated-to-saturated fatty acid ratio or a specific dietary fatty acid remains to be investigated.

Differential Response to ω3 and ω6 Essential Fatty Acid Depletion and Repletion in Normotensive and Hypertensive Rats

Y.-S. Huang, D.E. Mills, D.F. Horrobin
Efamol Research Institute, Kentville, Nova Scotia, and Department of Health Studies, University of Waterloo, Waterloo, Ontario, Canada

Spontaneously hypertensive rats (SHR) when compared with the normotensive Wistar-Kyoto rats (WKY) exhibit abnormal fatty acid patterns. Evidence has also shown that essential fatty acid (EFA) deficiency exacerbates, whereas EFA supplementation attenuates the development of hypertension in SHR. To examine whether there is a different lipid metabolism between these two strains, the effect of depletion and repletion and in vivo oxidation of ω3 and ω6 EFAs on tissue phospholipid (PL) fatty acid composition in SHR and WKY were compared. During a 12-week EFA depletion, we observed an earlier onset and a greater severity of biological EFA deficiency symptoms, but a lower biochemical EFA deficiency indicator, 20:3 ω9/20:4 ω6 ratio, in SHR than in WKY. This result indicated that in SHR, the biological symptoms did not correspond to the biochemical indicators. Following a 7-week EFA depletion, SHR and WKY rats were supplemented with either an ω6 fatty acid-rich safflower oil, an 18:3 ω3-enriched linseed oil, or a 20:5 ω3- and 22:6 ω3-enriched fish oil diet for 7 days. SHR, like the WKY, effectively incorporated ω6 fatty acids, but unlike the WKY, less effectively incorporated ω3 fatty acids into tissue PL. Studies with radio-labelled 18:2 ω6 and 18:3 ω3 demonstrated that the 24-hour oxidation rates of ω3 acids were faster than ω6 acids and were greater in SHR than in WKY. In other words, ω3 fatty acids in SHR were preferentially oxidized and therefore less were being incorporated into the tissue PL. It is thus concluded that the metabolism of ω3 and ω6 fatty acids differs significantly between these two strains. Whether this difference plays a significant role in the etiology of hypertension in SHR remains to be elucidated.

Mackerel Oil and Regression of Atherosclerosis in Pigs

J.M.J. Lamers[a], *L.M.A. Sassen*[b], *J.M. Hartog*[b], *P.D. Verdouw*[b]
[a]Department of Biochemistry I and [b]Laboratory for Experimental Cardiology, Thoraxcenter, Erasmus University Rotterdam, The Netherlands

The purpose of this study was to assess the effect of regular intake of mackerel oil on the regression of atherosclerosis. In 35 pigs atherosclerosis was induced by balloon abrasion of parts of the aorta and coronary arteries and a diet containing 2% (w/w) cholesterol and 7% (w/w) lardfat. After 4 months of induction 9 animals were killed for analysis of the extent of atherosclerosis, while the diet of the other 26 pigs was changed to a low-cholesterol diet containing either 9% (w/w) lardfat (L), 9% mackerel oil (M) or a mixture (ML) of 4.5% (w/w) lardfat and 4.5% (w/w) mackerel oil. This diet was continued for 3 months to induce regression of atheroclerosis. The hyperlipidemic diet increased plasma total cholesterol from 2.0 to 4.3 m*M* whereas the triglyceride level remained unaffected (0.6 m*M*). Return to normolipidemic diet led to a rapid decrease of plasma total cholesterol to normal level in all three groups, and in the mackerel oil-fed pigs (M and ML) triglyceride initially decreased as well (to about 0.4 m*M*). No regression of atherosclerosis was observed in the previously denudated aortic wall because the increase of lipid content of aortic lesions were similar in all pigs (induction group and L, ML and M). Evidence for regression of atherosclerosis was obtained by light microscopic analysis of the coronary arteries. In the previously denudated coronary arteries of the induction group the mean luminal encroachment was 11 ± 2%. This was similar in L (13 ± 4%) but significantly lower in ML (6 ± 2%) and M (3 ± 1%). In the non-abraded coronary arteries of the induction group the luminal encroachment was 1.3 ± 0.3%. For M and ML similar values were found, but in L there was an increase to 11 ± 3% after 3 months on a normolipidemic diet which indicates that progression of atherosclerosis continues in L despite the normal diet. ADP-induced platelet aggregation, measured in whole blood, was lower in ML and M than in L. Thromboxane A_2 production under these conditions was reduced in M but not in L, while the production of the weak agonist thromboxane A_3 was larger in M than in ML and no production was present in L. It is concluded that mackerel oil retards the progression of and causes regression of coronary atherosclerosis.

This study was supported by grant 86-086 from the Netherlands Heart Foundation.

ω3 Fatty Acids Increase LDL Cholesterol without Affecting the Extent of Aortic Lesion in the Casein-Fed Rabbit Model of Atherosclerosis

R. Adelstein, K.A. Rogers
Department of Anatomy, The University of Western Ontario, London, Ont., Canada

The reported effects of ω3 fatty acids on plasma lipid levels and atherogenesis in experimental models of hypercholesterolemia have been inconsistent across and within species. The disparate findings may partially reflect the variety of protocols employed in testing for ω3 fatty acid effects, particularly with reference to the designation of appropriate control groups and the nature and/or severity of the induced hypercholesterolemias. In rabbits the use of casein diets reliably produces moderate hypercholesterolemia in which the excess cholesterol is carried largely in the LDL fraction. We have examined the effects of a cholesterol-free fish oil supplement on plasma lipid profiles and atherosclerotic lesion formation in male New Zealand White rabbits (1.5 kg) maintained on a low-fat, cholesterol-free casein diet for a period of 11 months. Animals were rank ordered using baseline plasma cholesterol values and subsequently block randomly assigned to either an experimental or a control group. The rabbits were treated with one of two oil supplements administered daily by intrabuccal gavage. The experimental group (n = 14) received a cholesterol-free fish oil preparation (MaxEPA) at a dosage of 0.5 ml/kg body weight. The control group (n = 15) received the same dosage of a control oil mixture of corn, palm, and safflower oils (1:1:1 by weight) calculated to have a P/S ratio equivalent to that of the MaxEPA (P/S = 2.1, manufacturers information). Blood collection and plasma lipid analyses were conducted monthly. Bimonthly, plasma samples were separated by sequential ultracentrifugation to yield the lipoprotein fractions VLDL, IDL, LDL, and HDL. Cholesterol determinations were performed spectrophotometrically using commercially available enzymatic kits. At 11 months, animals were sacrificed and perfusion-fixed thoracic aortas were removed, stained with oil red 0, and photographed en face. Grossly visible lesions were traced from photographs using an imaging analysis tablet and the percentage lesion of each intimal surface was automatically computed.

The casein diet was well tolerated and over weekly assessments both groups demonstrated similar weight gain patterns. An analysis of variance on plasma cholesterol using a repeated measures design revealed no differences attributable to the oil supplement; the means ($\pm$ SEM) for the experimental and control groups were 169 $\pm$ 21 mg/dl and 155 $\pm$ 15 mg/dl, respectively. A similar analysis on LDL cholesterol, however, revealed that the experimental group had a mean LDL concentration 40% higher than that of the control group (92 $\pm$ 14 mg/dl vs. 65 $\pm$ 8 mg/dl respectively, $p = 0.05$). Comparisons of mean percentage lesion in the aortic arch and descending thoracic aorta of experimental vs. control animals revealed no significant group differences. These findings suggest that dietary ω3 fatty acids may exacerbate a moderate LDL hypercholesterolemia without affecting the atherogenic process.

Supported by the Heart and Stroke Foundation of Ontario.

Fish Oil, a Potent Inhibitor of Platelet Adhesiveness

X. Li, M. Steiner

Division of Hematology/Oncology, Memorial Hospital of Rhode Island, and Brown University, Providence, R.I., USA

The effect of fish oil administration on platelet function was studied in 8 normal individuals, 4 men and 4 women, who received fish oil equivalent to 6 g eicosapentaenoic acid (EPA) per day for a period of 24 days. Platelet aggregation, platelet adhesion, phospholipid and fatty acid distribution were measured at periodic intervals before, during and after the period of fish oil administration. Platelet aggregation induced by arachidonic acid, epinephrine, ADP and collagen was not affected by the administration of fish oil. Platelet adhesion to fibrinogen and collagen I, which were studied at low shear rates (25 sec^{-1}) in a laminar-flow chamber, on the other hand, showed a 60–65% decrease ($p < 0.0005$) in response to the dietary supplementation. The change in adhesiveness could be correlated with the number of pseudopodia formed in response to agonistic stimulation. Scanning electron microscopic examination of the adherent platelets showed an overall reduction of pseudopodia which appeared short and stubby upon fish oil administration. The profile of the fatty acids extracted from plasma confirmed compliance of the volunteers with their dietary supplements. Analysis of phospholipids showed changes in sphingomyelin, lysophosphatidylcholine and phosphatidylcholine between pseudopodia and platelet cell bodies. Fish oil administration did not affect their overall distribution, except for a moderate decrease in phosphatidylethanolamine in platelet pseudopodia. Changes were also recognized in the total fatty acids extracted from platelets, affecting primarily arachidonic acid, EPA and docosahexaenoic acid (DHA). We conclude from these studies that fish oil, at least when administered over a limited period of time, is an effective inhibitor of platelet adhesiveness.

Biological Effects of Fish Oil in Normal Subjects and Dyslipidemic Coronary Heart Disease Patients: A Dose-Effect Study

A. Arteaga[a], *C.L. Villanueva*[a], *C. Necochea*[a], *N. Velasco*[a], *A. Maiz*[a], *A.M. Acosta*[a], *A. Miguel*[a], *F. Solis de Ovando*[a], *C. Barriga*[a], *V. Guasch*[a], *A. Valenzuela*[b], *M.N. Morales*[a], *M. Bronfman*[a], *C. Skorin*[a], *F. Leighton*[a]
[a]Universidad Católica de Chile; [b]Universidad de Chile, Santiago, Chile

To study biological effects, dose response, and possible mechanisms of action of ω3 fatty acids in humans, observations were made on control subjects and on dyslipidemic patients (IIA, IV, IIB) with coronary heart disease. Fish oil containing 30% ω3, 3.5% ω6 and 33% saturated fatty acids was refined by molecular distillation, stabilized with 1 mg/g *DL*-α-tocopherol, and encapsulated in 1 g capsules. The subjects, after a 30-day basal period, received either 2, 4, or 6 g daily for 60 days. A normal placebo group received 5 g daily of grape seed oil. Clinical and laboratory evaluation was made at days 0, 30 and 60 of oil administration. Groups with 6–8 subjects were studied. Diet composition was determined by 24 h recall procedure at the end of the basal and oil supplementation period.

Changes in serum lipid levels were observed, in general in both main groups, only with 4 and 6 g of oil. The larger changes, in agreement with previous reports, are on triglycerides, with up to 50% reduction in total or VLDL triglycerides (TG). At the higher dose, LDL cholesterol increases, and VLDL cholesterol decreases. The fatty acid composition of serum, platelet and red blood cell phospholipids was evaluated; EPA and DHA are incorporated in a dose-dependent manner at all doses; arachidonic acid tends to decrease at the higher doses of ω3, while linoleic decreases in the three sources of phospholipids.

A decrease in serum Thromboxane B_2 was observed at the higher 6 g dose, while Leukotriene B_4 release from PMN (Ca^{2+}) is decreased after 4 and 6 g ω3 dietary supplementation.

To explore for evidences of oxidative stress induced by ω3 supplementation, serum vitamin E levels (which others have found decreased) and total plasma antioxidant capacity were measured.

As a preliminary approach to determine the relative role of mitochondria and peroxisomes in fatty acid oxidation, plasma levels of ketone bodies, long-chain acyl carnitine, acetyl carnitine, free carnitine and free acetate were measured in some groups. The measurements do not support the possibility of an increase in mitochondrial activity.

These data confirm that phospholipid fatty acid composition is the most sensitive parameter to follow ω3 administration and that plasma total and VLDL TG levels are reduced at daily doses of 4 g ω3 or more, while LDL cholesterol increases and VLDL decreases at the high (6 g) dose. Attempts to detect evidences of oxidative stress gave negative results, and metabolic preliminary evaluations do not support a mitochondrial oxidative role in the decreased TG levels. For both, oxidative stress and metabolic evaluations, additional parameters should be evaluated to reach definitive conclusions.

Supported by projects UNDP CHI/88/017 and FONDECYT 250/88.

Short-Term Effects of Fish Oil on Plasma Triglyceride, Eicosapentaenoic Acid and Low-Density Lipoprotein Cholesterol Levels

William S. Harris, Mario Sztern
Department of Medicine, University of Kansas Medical Center,
Kansas City, Kans., USA

Fish oils (FO) are useful in the treatment of hypertriglyceridemia, but low-density lipoprotein cholesterol (LDL-C) levels often increase. To determine how quickly FO begins to lower triglyceride (TG) and raise eicosapentaenoic acid (EPA) levels in the plasma, we recruited 9 normal subjects who were given a single dose of FO (0.25 g/kg providing 50 mg/kg ω3 fatty acids; Dale Alexander omega-3 Capsules, Twin Labs, Inc.) and 5 other subjects who were given placebo. Plasma TG and phospholipid EPA levels were measured at baseline and at 24, 48 and 72 h post-dosing.

In a group of 5 hypertriglyceridemic patients, 12 capsules of FO/day were given for 14 days. Plasma TG, EPA and LDL-C levels were measured at baseline and every 2–3 days thereafter including a 10-day washout period.

In the single dose study, TG levels decreased by 15% at 24 h and 16% at 48 h ($p < 0.01$ both), and began to normalize at 72 h. EPA levels rose as TG levels fell and vice versa. TG levels in the placebo group did not change significantly. In the multiple-dose study, FO began lowering TG levels by day 1 and continued doing so through day 14 (from 709 ± 121 to 407 ± 63, $p < 0.01$). LDL-C levels began to increase immediately, plateauing after day 4 (from 97 ± 12 to 141 ± 10, $p < 0.01$). EPA levels rose 5-fold by day 14 and had nearly returned to baseline after 10 days of no treatment. Although both TG and LDL-C levels began to normalize in the washout period, baseline levels were not reachieved within 10 days. We conclude that FO treatment altered TG, EPA, and LDL-C levels very rapidly, but TG and LDL-C levels did not return to baseline as quickly as did EPA levels. The latter observation suggests that tissue EPA levels (and not serum levels) influence TG and LDL-C levels.

Effects of Encapsulated Corn and Fish Oils in Japanese Quail: Vessel Histology

J.G. Chamberlain, H. Olson
Anatomic Sciences, University of the Pacific School of Dentistry,
San Francisco, Calif., USA

The majority of ω3 PUFA animal models have utilized bulk fish oil added to the diet. Quail respond to ω3 fatty acid treatment and are useful in atherosclerosis research [Chamberlain and Belton, Atheroscler 1987;68:95–103; Fann et al, Cardiol Res 1989;23: 631–638]. We report here feeding inbred Japanese 'SEA' quail (Artery, Inc.) soft-gelatin capsules for 6–7 months while on a turkey starter diet (29% protein, 5.6% fat, 15% calories from fat, 0.10 kcal/mg, 30 mg cholesterol/100 g). 10 birds each were randomly assigned to 3 groups and fed weekly either a single corn oil or one of two types of fish oil (F_1 or F_2) soft capsules[1] (1.0 g). After an overnight fast, birds were anesthetized and killed

(103 ± 2 g at autopsy). Tissues were removed for HPLC/MS (in progress) or profused, fixed (formalin), step-sectioned (5 μm) and stained (HE; Masson's Trichrome) for light microscopy (× 40–400). Hearts and great vessels (i.e. coronaries, bracheocephalics, aorta, carotids) were subsequently examined (710 sections).

All birds fed F_1 capsules showed numerous fatty streaks and intima fatty infiltration (0 normal). There was both a quantitative and qualitative difference between groups ($p = 0.05$). 50% of the F_2 animals showed only occasional small streaks (5 normal) while 30% of the corn oil-fed animals showed only small fatty streaks or points (spots) (7 normal). These results indicate the varying effects of different fish oil preparations on animal tissues, especially PUFA varying in cholesterol, saturated fat and ω3 fatty acid content.

[1] Corn oil (<0.05 mg cholesterol, 10 mg ω3, 13% saturated fat) and F_1 fish oil (Menhaden; 4.1 mg cholesterol, 250 mg ω3, 26% saturated fat) capsules kindley supplied by NIH/NOAA/NMFS (B.T.M. program) (V87196V; A87196A); F_2 capsules (Sardine; <1.0 mg cholesterol, 624 mg ω3, 5% saturated fat) supplied by Pharmacaps, Inc., Elizabeth, New Jersey (Supra EPA 600).

Intravenous Infusion of Trieicosapentaenoyl-Glycerol into Rabbits and Leukotriene $B_{4/5}$ Production by Leukocytes

S. Sawazaki, T. Hamazaki, N. Nakamura, M. Urakaze, K. Yamazaki, S. Yano
First Department of Internal Medicine, Toyama Medical and Pharmaceutical University, Sugitani, Toyama, Japan

Oral administration of fish oil can reduce the production of leukotriene (LT) B_4 by polymorphonuclear leukocytes (PMNL). Eicosapentaenoic acid (EPA), the major fatty acid of many fish oils, is thought to be the major effective component. Although the reduction in LTB_4 production seems to be beneficial to patients suffering from acute myocardial infarction or rejection of transplanted organs, oral administration of EPA may not help them at all because it usually takes at least 1–2 weeks for EPA to take effect. Thus, we developed infusible EPA emulsion. In the present study we show how fast EPA emulsion takes effect with regard to the inhibition of LTB_4 production by PMNLs.

An emulsion of trieicosapentaenoyl-glycerol (EPA-TG, > 90% pure) was prepared. 100 ml of the emulsion contained 10 g EPA-TG, 1.2 g egg yolk phospholipids as an emulsifier and 2.5 g glycerol. 30 ml of the emulsion was infused into ear veins of 6 Japanese white rabbits weighing about 3 kg. Heparinized blood samples were obtained from central ear arteries before, and 6, 24 and 168 h after the infusion. PMNLs were prepared from these samples and stimulated by calcium ionophore to produce $LTB_{4/5}$, which was measured by reversed-phase HPLC.

Averaged LTB_4 production was significantly reduced to 60% of the baseline value 6 h after EPA infusion. The production of LTB_4 24 h and 168 h after the infusion was 72% and 86% of the baseline value, respectively. Averaged LTB_5 production by PMNL's, which was below the detection limits before the infusion, was increased markedly up to the similar values of LTB_4 6 h after the infusion, then reduced gradually during the experimental period. In control rabbits, LTB_4 production did not differ significantly between before and after the infusion of the soybean oil emulsion (30 ml); LTB_5 production was not detected throughout the experiment.

Consequently, EPA emulsion can reduce LTB_4 production in 6 h by 40% and may be beneficial to patients who are suffering from pending myocardial infarction or rejection.

Inhibition of Mixed Lymphocyte Reaction and Proliferative Response of T_H Clone by Dietary ω3 Fatty Acid Supplementation

M Fujikawa, N. Yamashita, A. Yokoyama, T. Hamazaki, S. Yano
First Department of Internal Medicine, Toyama Medical and Pharmaceutical University, School of Medicine, Sugitani, Toyama, Japan

Recently, many investigators have examined the effect of eicosapentaenoic acid (EPA)-rich fish oil in the alleviation of glomerular nephritis, rheumatoid arthritis, lupus erythematosus and in modulating the immune function and immune disease. We have examined the effect of EPA on mixed lymphocyte reaction and antigen presenting cell function.

A group of C3H/He mice (EPA Group) was fed EPA-rich diet (lipid-free diet, 90% w/w, plus lard, 8%, and EPA ethyl ester, 2%). Another group (LA Group) was fed in the same way except that EPA ethyl ester was replaced by safflower oil. C3H-responding spleen cells were cultured together with BALB/c-stimulating spleen cells which had been irradiated. Allo-mixed lymphocyte reaction of murine spleen cells was inhibited by more than 50% in EPA group compared to LA group.

Murine helper T cell (Tw/H) clones were used to dissect the effect of EPA on antigen-presenting cell population. The proliferative responses of the T_H clones were determined with cultures containing irradiated spleen cells and appropriate antigen for 3 days. The clones' proliferative responses were also inhibited by more than 50% in EPA group compared to LA group.

These data demonstrate that dietary ω3 fatty acid supplementation can inhibit mixed lymphocyte reaction and antigen-presenting cell function.

Effects of Fatty Acids on Inflammation and Immune Responses

R.B. Zurier, D. DeMarco, D. Santoli
Rheumatology Section, University of Pennsylvania School of Medicine, Philadelphia, Pa., USA

Dietary gamma-linolenic acid (GLA) is converted readily to dihomo-gamma linolenic acid (DGLA), precursor to the monoenoic prostaglandins. Administration of GLA in plant seed oils suppresses inflammation and tissue injury in several experimental models, and results in reduced production of PGE_2 and LTB_4 and increased production of PGE_1. In the rat subcutaneous air pouch model, GLA suppressed significantly the cellular phase of acute inflammation (polymorphonuclear leucocyte accumulation, and lysosomal enzyme activity), but it had little effect on the fluid phase (exudate volume and protein concentration). In contrast, an eicosapentaenoic acid (EPA)-enriched diet suppressed the fluid phase but not the cellular phase of inflammation. The findings indicate that the fluid and cellular phases of acute inflammation can be controlled independently. A combined diet of fish oil and plant seed oil (EPA-enriched and GLA-enriched) reduced both the cellular and fluid phases of inflammation.

It is likely that many biological effects of essential fatty acids are independent of their conversion to eicosanoids. We examined the effects of fatty acids on activation and function of human peripheral blood leucocytes and of T lymphocytes from synovial fluid and synovial tissue. DGLA and arachidonic acid (AA) suppress interleukin-2 (IL-2) production and IL-2-dependent T cell proliferation by a mechanism which is direct and independent of their conversion to PGE. EPA inhibited IL-2 production by peripheral blood mononuclear cells from only some donors and was not an effective inhibitor of IL-2-driven T cell proliferation. Incubation of cells with oleic, stearic and palmitic acids did not influence IL-2 production or T cell proliferation. DGLA (but not AA) also reduced expression of activation antigens (IL-2 receptor, HLA Class II molecules) on T cells.

Because T cell activation and proliferation in vivo depend on an interaction with accessory cells, we investigated the effects of DGLA and EPA on peripheral blood monocyte (PBM) function. When incubated with DGLA or EPA, LPS-stimulated PBM exhibited reduced cell-associated and -released tumor necrosis factor α, compared to untreated control cells.

These studies indicate that small changes in cellular fatty acids may alter profoundly function of cells important to immune responses.

Interaction between Vegetable and Fish Oils in Relation to Leukocyte Eicosapentaenoic Acid (EPA) Content and Leukotriene B Production

L.G. Cleland[a], *R.A. Gibson*[b], *M.J. James*[a], *J.S. Hawkes*[a], *M. Neumann*[b]
[a]Rheumatology Unit, Royal Adelaide Hospital, and
[b]Dept. of Paediatrics, Flinders Medical Centre, Adelaide, Australia

Dietary fish oil supplements containing EPA have been associated with reduced symptoms and reduced production of LTB_4, a potent neutrophil chemoattractant, when given to rheumatoid subjects as an adjunct to conventional therapy [Cleland LG, et al, J Rheumatol 1988;15:1471]. However, incorporation of dietary EPA into human neutrophil membranes is relatively inefficient. We have examined the effects of other dietary fats on EPA incorporation and neutrophil LTB production.

Weanling rats were fed diets containing fish oil 5% w/w alone, or with added sunflower oil 5%, olive oil 5%, linseed oil 5% or further fish oil up to a total of 10%. The 5 and 10% fish oil diets were associated with similar incorporation of EPA and similar LTB_4 production by leukocytes stimulated in vitro. The olive oil- and linseed oil-supplemented diets yielded substantial incorporation of EPA (more than 70% of levels with the fish oil only diets). The sunflower oil-supplemented diet achieved the least EPA incorporation (less than 30% of the fish oil only diets). In line with our previous observations, there was a close association between the ratio of EPA and arachidonic acid in cell membranes and the ratios of production of leukotriene B_5 (a product of EPA which is relatively inert) and LTB_4 (a potent neutrophil chemotaxin).

The findings suggest that dietary vegetable oils can greatly modify the effect of dietary fish oil on leukocyte EPA content and LTB production.

Effect of ω3 Fatty Acid Diet Therapy on Autoimmune Disease with and without Calorie Restriction

G. Fernandes, J.T. Venkatraman, A. Fernandes, V. Keshavalu, V. Tomar
Department of Medicine, The University of Texas Health Science Center at San Antonio, San Antonio, Tex., USA

Autoimmune-prone (NZB × NZW)F_1 mice are known to develop systemic lupus erythematosus (SLE)-like disease which is hormonally influenced and characterized by disordered immune regulation that regularly leads to the development of very severe renal disease and short lifespan. However, these mice are known to live significantly longer either by restricting their calorie intake from weaning or, alternatively, by feeding a diet containing high levels of ω3 fatty acids. Both kinds of diet therapy appear to act by inhibiting the early occurrence of autoimmune disease by lowering prostaglandin E_2 production and changes in several immunological functions, particularly preventing the rise in B-cells and Ly1/B-cells with age. The present study was undertaken in B/W female mice to observe the longevity by feeding isocaloric diets containing high (20%) and low (5%) dietary lard (ω9), corn oil (ω6), and menhaden fish oil (ω3). Both ω6 and ω3 oil contained equal levels of antioxidants (alpha- and gamma-tocopherols and TBHQ) to prevent lipid peroxidation. The lifelong ad libitum feeding revealed the following survival rate: lard (5%) – 247 ± 6 days; lard (20%) – 249 ± 8 days; corn oil (5%) – 267 ± 10 days; corn oil (20%) – 256 ± 9 days; fish oil (5%) – 381 ± 33 days; and fish oil (20%) – 434 ± 29 days.

The results clearly indicate that fish oil at both 5% and 20% levels exerts protection better than lard and corn oil diets. Low levels of lard and corn oil diets did not show any protection against autoimmune disease. In order to study the effect of calorie restriction, diets containing both 5% and 20% corn oil or fish oil were fed daily at approximately 40% less than ad libitum to observe the lifespan. Ongoing studies indicate that both 5% and 20% fish oil/restricted calories appear to be highly beneficial as compared to both levels of corn oil/restricted diets. For instance, at 450 days, 20% corn oil/restricted diet showed 100% deaths as opposed to 80% deaths in 5% corn oil/restricted diets, whereas both 5% and 20% fish oil/restricted groups have caused less than 20% and 40% deaths, respectively, indicating that ω3 lipid diet in combination with calorie restriction appears to be much more effective in inhibiting autoimmune disease in B/W mice. Also, survival rate of all of the above dietary groups were closely correlated with their serum cholesterol levels and anti-DNA autoantibody levels. Both ad libitum and calorie restriction of fish oil 5% and 20% diets lowered cholesterol levels significantly (100–150 mg/dl in all fish oil groups versus 200–350 mg/dl) both in lard and corn oil diets.

In summary, ω3 lipids, with adequate levels of antioxidants, appear to exert better protection against autoimmune disease much more efficiently with combination of calorie restriction. Furthermore, ω3 lipid diet was also found to significantly inhibit the rise of both serum cholesterol levels and anti-DNA autoantibody levels in autoimmune-prone B/W female mice.

Supported by NIH grant AG03417.

Effect of Dietary Oils on Passive Cutaneous Anaphylaxis

G. Crozier, M. Fleith, M. Bonzon, R. Fritsche
Nestlé Research Centre, Nestec Ltd., Vers-chez-les-Blanc, Lausanne, Switzerland

Dietary fat is now known to affect many different aspects of immune function. In this study we examined the effects of different oils on passive cutaneous anaphylaxis (PCA), a Type I immediate hypersensitivity reaction. Guinea pigs were fed one of three semi-purified experimental diets containing 10% black currant seed oil (BCO), 9% lard + 1% grapeseed oil (LARD), or 2% fish oil + 7% lard + 1% walnut oil (FISH) for 6–8 weeks. PCA reactions were then elicited in these animals with homologous reaginic antibodies (anti-β-lactoglobulin) and allergen (β-lactoglobulin). Intensity of the reaction was evaluated by measure of the diameters of the spots marked by Evan's blue dye. Sensitivity of skin was evaluated by dose response to histamine. Skin samples were analysed for fatty acids by gas liquid chromatography.

Results showed that the FISH-fed animals had greater cutaneous anaphylactic response (12.6 ± 0.6 at antibody dilution 1/400) than did the BCO (10.6 ± 0.5) or LARD (9.3 ± 0.3) animals. These differences were statistically significant ($p < 0.05$). Other antibody dilutions gave similar results. BCO feeding resulted in a moderate response which was always greater than the LARD group and less than the FISH group although not always significantly. Walnut oil controls did not show elevated reactions. The same relative responses were obtained when ovalbumin/anti-ovalbumin were used. Diet did not alter skin sensitivity to histamine.

Skin fatty acids which are precursors of the prostaglandins and/or leukotrienes were significantly influenced by the dietary oil. Eicosapentaenoic acid (20:5 ω3) was found only in the FISH group. There was more dihomo-gamma-linolenic acid (20:3 ω6) in the BCO group than in the LARD or FISH. Arachidonic acid (20:4 ω6) was the same in all groups.

In conclusion, these studies demonstrate that dietary fat modulates cutaneous immediate hypersensitivity reactions of the guinea pig. Changes in available bioactive lipid metabolites resulting from changes in available fatty acids may be involved in this modulation.

Improvement of Psoriatic Symptoms by Fish Oil

U. Linker[a], *M. Staender*[b], *K. Oette*[c], *P. Singer*[d]
[a]Hauptstrasse, Köln; [b]Clinic for Psoriasis, Bad Bentheim;
[c]Institute of Clinical Chemistry, Köln; [d]Heckmannufer, Berlin, Germany

In a randomised double-blind study under the conditions of a dermatological practice 60 patients with stable chronic psoriasis were supplemented with either 9 g/day of fish oil or 9 g/day of placebo (olive oil) in addition to their standard therapy over 12 weeks. The patient's compliance was confirmed by gas-chromatographic analyses of total serum lipids. The most relevant symptoms were evaluated by a score including 3 degrees of severity. Moreover, the extent of psoriatic lesions (in cm) was assessed.

The score values of erythema, scaling, infiltration and itching as well as the extent of individual lesions and the sum of all psoriatic lesions were reduced, which appeared more pronounced after adjuvant fish oil supplementation than after placebo. The frequency of severe lesions was likewise decreased at the end of the fish oil period.

The results suggest a beneficial effect of adjuvant fish oil supplement in stable chronic psoriasis when standard therapy is continued.

Effects of ω3 and ω6 Fatty Acid-Enriched Diets in MCA-Induced Sarcoma Bearing Rats

P.D. de Rooij, R.N. Younes, N.A. Vydelingum, F. Scognamiglio, L.F. Andrade, M.F. Brennan
Surgical Metabolism Laboratory, Department of Surgery,
Memorial Sloan-Kettering Cancer Center, New York, N.Y., USA

Diets rich in linoleic acid (C18:2 ω6) induce a tumor-promoting effect in most experimental models, while diets rich in ω3 fatty acids (C20:5 and C22:6) seem to have a dose-dependent, inhibitory effect on tumor development and tumor growth. These inhibitory effects of ω3 fatty acids, however, have been primarily described in epithelial tumor models and the reported data are not consistent.

The present study was designed to assess the effect of both types of fatty acids on the development and growth of a non-epithelial tumor model. Two million cells derived from a methylcholanthrene sarcoma were subcutaneously injected into male F344 rats fed two different semi-purified isocaloric diets. Diet A contained 14% w/w MaxEPA oil (30–40% ω3 and 2–4% ω6) and 8% corn oil (50–60% ω6), while diet B contained only 8% corn oil. Animals in group I were fed diet A for 4 weeks prior to tumor injection, in group II for 2 weeks and animals in group III had their standard laboratory Purina Chow diet replaced with diet A on the day of tumor injection. Animals in groups IV and V were fed diet B for 4 and 2 weeks, respectively, before tumor injection. After tumor injection the animals were maintained on their respective study diets (groups I, II, III: diet A, and groups IV, V: diet B) and sacrificed 4 weeks later. Animal weight and tumor volume were recorded daily.

Table 1. Tumor weight (TW) and tumor burden (TB) expressed as mean ± SEM

Group	I	II	III	IV	V
n	9	11	11	11	11
TW, g	64.4 ± 3.6	27.5 ± 1.9	27.5 ± 3.1	60.4 ± 3.2	25.1 ± 3.1
TB % of total BW	18.8 ± 1.2*	8.8 ± 0.5	9.6 ± 1	18.2 ± 0.8**	9.1 ± 1

* $p < 0.0001$, I vs. II, III and V; ** $p < 0.0001$, IV vs. II, III and V; t-test

Results: Food intake and animal growth curves were similar in the five groups. There was also no difference in time of appearance (take) of tumor in the different groups.

There were significant effects on tumor growth and thus tumor burden (table 1).

Differences in TB observed in groups I and IV are likely to be related to the high content of ω6 fatty acids in the diets. However, these effects occur only after a pre-feeding period of 4 weeks.

The expected inhibitory effects of 14% w/w MaxEPA oil were thus completely overwhelmed by these ω6 promotor effects. It is also possible that the MCA-induced sarcoma is insensitive to ω3 fatty acids at the concentration used in this study.

This study was supported by NCI Grant 38858.

Effect of Fish Oil on Incidence of Mammary Tumors in DMBA-Treated BALB/c Virgin Female Mice

M.C. Craig-Schmidt, M.T. White, P. Teer, H.W. Lane

Departments of Nutrition and Foods and Pathobiology, Auburn University, Ala., and Biomedical Operations and Research Branch, NASA/JSC, Houston, Tex., USA

To test the hypothesis that consumption of diets high in ω3 fatty acids (ω3 FA) can inhibit mammary tumorigenesis, 7,12-dimethylbenz(a)-anthracene (DMBA)-treated BALB/c virgin female mice were fed one of three diets for 9 months, beginning at 6 weeks of age. Six doses of DMBA (1 mg in 0.2 ml corn oil) were administered intragastrically for 6 consecutive weeks beginning at 8 weeks of age. Mice were randomized into each of the three treatment groups and fed a semi-purified, casein-based diet varying only in the type of fat: (1) 20% corn oil; (2) 18% fish oil + 2% corn oil; and (3) 18% coconut oil + 2% corn oil. The fish oil used in treatment 2 was menhaden oil containing 13.8% 20:5 ω3 and 8.3% 22:6 ω3. Linoleic acid (18:2 ω6) was 7% of the total fatty acids in the mixtures of oils used in treatments 2 and 3 and 58% in treatment 1. All diets were balanced with respect to antioxidants and cholesterol.

There were fewer ($p < 0.025$) mammary tumors (histologically confirmed) in the mice fed the coconut oil or fish oil diets than in the corn oil-fed animals. Tumor incidence was 77% (20/26) in the corn oil dietary treatment group, compared to 45% (13/29) in the coconut oil-fed group and 36% (10/28) in the fish oil-fed group. Because the amount of linoleic acid (18:2 ω6) was approximately the same in the coconut oil and fish oil dietary treatments, the inhibition of tumorigenesis observed may be due to the lower amount of linoleic acid in the fish oil and coconut oil diets compared to the corn oil diet, rather than the presence of ω3 fatty acids in the fish oil.

Supported by Auburn University Grant-in-Aid Program and AICR.

Inhibition of Mammary Tumorigenesis by Dietary Combinations of Difluoromethylornithine and ω3 Fatty Acids

O.R. Bunce, S.H. Abou-El-Ela
Department of Pharmacology and Toxicology, University of Georgia, College of Pharmacy, Athens, Ga., USA

In a previous study, we have shown that the consumption of dietary ω3 fatty acids reduced mammary tumor incidence, ornithine decarboxylase (ODC) activity and eicosanoid synthesis during mammary tumor promotion in 7,12,dimethylbenz(a) anthracene (DMBA)-induced tumors in female Sprague-Dawley rats. When 0.5% *D,L*-2-difluoromethylornithine (DFMO) an inhibitor of ODC, was fed with the 20% fat diet [15% menhaden oil (MO) + 5% corn oil (CO)], tumor promotion, ODC activity and eicosanoid synthesis were additively inhibited. However, feeding of 0.5% DFMO with the 20% fat diet showed significant reduction in body weight. The objective of this study was to establish the minimal and non-toxic level of DFMO which can give an additive or synergistic inhibitory effect when fed along with dietary ω3 fatty acids to female Sprague-Dawley rats in DMBA-induced mammary tumors. Four dietary levels of DFMO (0, 0.125, 0.25, and 0.500%) were fed in diets containing 20% fat as either CO or a combination of MO and CO. Dose-response effects on mammary tumorigenesis (tumor incidence, number of tumors and tumor burden per tumor-bearing rat) as well as body weight gain were observed. As noted previously, a failure to gain weight was observed when both 20% CO diet and 15% MO + 5% CO diet were fed in combination with 0.5% DFMO. There were no significant differences in food consumption between rats fed no DFMO and rats fed 0.5% DFMO when the 20% CO diet was fed. However, when 15% MO + 5% CO was fed in combination with 0.5% DFMO, there was a significant reduction in food consumption. At lower levels of DFMO food consumption and weight gain were not significantly different. Tumor burden was not significantly altered among the eight diets; however, the number of tumors/rat was significantly altered by increasing levels of DFMO. Supporting data on mammary tumor eicosanoid synthesis and ODC activity will be presented. The oils used for diet preparation were kindly provided by Zapata Haynie Corporation, Reedville, Va. and Traco Labs, Champaign, Ill. Difluoromethylornithine was provided by Merrell Dow Pharmaceuticals Inc., Cincinnati, Ohio.

Author Index

(A) = Abstract

Subject Index